Cosmology, Physics, and Philosophy

Cosmology, Physics, and Philosophy

Benjamin Gal-Or

Springer—Verlag New York Inc.

First Printing 1981
Second Printing 1983

Benjamin Gal-Or
Technion — Israel Institute of Technology, Haifa, Israel

Sole distributors in the USA and Canada:
Springer-Verlag New York Inc.
175 Fifth Avenue
New York, NY 10010
U.S.A.

Library of Congress Cataloging in Publication Data

Gal-Or, Benjamin.
  Cosmology, physics, and philosophy.

  (Recent advances as a core curriculum course; v. 1)
  Includes index.
  1. Physics — Philosophy.   2. Cosmology.   I. Title.
II. Series.
QC6.2.G34   530'.01   81-5257   AACR2

With 61 illustrations.
Printed in Israel
ISBN 0-387-90581-2 Springer—Verlag New York

To Leah, Amir, Gillad, David, Yetti and Tzvi.

TABLE OF CONTENTS

PART II

From Physics To Philosophical Crossroads and Back

PART III

From Physics to Cosmological Crossroads and Back

Lecture VI

Lecture VII

Lecture VIII

PART IV

Beyond Present Knowledge

Lecture IX

VOLUME II

Critique of Western Thought

FOREWORD I
by
Sir Karl Popper

This is a great book, and an exciting book. I say so even though I happen to disagree with the author in many minor points and one or two major points. Some of the minor points are merely terminological, and therefore *very* minor. I dislike the term 'dialectic', because of its use since Hegel and Marx; and I dislike the term 'gravitism', perhaps without a good reason. Thus I dislike the name which Professor Gal-Or has given to his theory. But the theory seems to me a great and a very beautiful theory, so far as I can judge. Other minor points of disagreement are connected with Gal-Or's original and remarkable views of the great philosophers, including Spinoza and Kant. A major point of disagreement is that Gal-Or, following Einstein, is a scientific determinist, while I cannot but regard determinism as a modern superstition. Of course, he may be right and I may be completely mistaken.

I mention these critical points rather in order to emphasize how strongly I am impressed by Professor Gal-Or's great book. Even in the very unlikely case that, wherever we disagree, he should be in the wrong and I right, even if that should be the case (which is improbable in the extreme), it would remain a great book: readable, worth reading and enlightening; with a most fascinating cosmological story of time, expansion, and gravitation.

FOREWORD II
by
Sir Alan Cottrell

For several years Professor Gal-Or has devoted himself to the physics of time and especially to the question of time's arrow. This has taken him into many far-flung branches of science, such as cosmology, thermodynamics, elementary particle physics, physiology and psychology.

He has now brought together in this fascinating and highly original book his considered views on these and related topics. His net is widely cast—it even reaches as far as science policy and political philosophy—but it is centred mainly round the Einsteinian view of space, time and matter, and the explanation of the directionality of time in terms of cosmological processes. The book thus provides a most stimulating survey of a large area of modern physical science and the reader can study in it many subjects, such as astrophysics, fluid mechanics and general relativity, at the same time as he is beguiled into taking a universal view of things and provoked into thinking afresh about the foundations of physics.

Science used to be called *natural philosophy,* a term still used in Scottish universities. Yet, science and philosophy today stand rather far apart. To the laboratory scientist, the philosopher seems to be merely playing with words. Equally, to the philosopher, such a scientist seems to be merely playing with things. Yet, in the end, science and philosophy cannot do without each other. Without the scientific fountain of empirical facts, philosophy indeed becomes mere word-spinning. And equally, science without philosophy can descend into mere factgrubbing and technicality.

Of course, some people have tried to hold the two together, have tried to sketch a grand system of the world in terms of modern science. Professor Gal-Or is one of them. This has necessarily led him to some of the hardest philosophical problems facing modern scientific theory—such as the role of the observer in quantum mechanics, the nature of perception, above all the nature of time itself—but the chapters are stratified so as to be accessible, in part or whole, to a wide range of readers. Gal-Or's general outlook is grounded securely in skeptical science and his bias is towards the more classical and objective forms of scientific theories. The reader can thus learn a lot of orthodox science at the same time as his vision of the world is expanded.

PREFACE

There are a number of useful sequences by which this book may be read. This is, of course, left to the reader's interests, discretion and background. However, readers whose main interests are in the *'humanities'*, especially those interested in socio-political studies and in the potentials of education, philosophy and ideology in a time of crisis, may prefer to begin with Volume II. As to *condensed courses,* one alternative may be designed in the following order:

Volume I	Volume II
Introduction	Introduction
Lecture I	Lecture X
Lecture IV	Lecture XII
Lecture VI	Lecture XIII
Lecture IX	Lecture XIV

Volume I contains a core of knowledge that, according to the latest trends in higher education, should be mastered by all advanced students in the natural sciences, or in philosophy. It departs from traditional texts in its emphasis on the interconnectedness of up-dated information in astronomy, astrophysics, cosmology, physics, and the theory of knowledge, as well as in its critique of present academic methodologies. Hence, teachers and students in the natural sciences, or in philosophy, may find it useful as a textbook for a basic *'Core Curriculum Course'*.

Part of the text is an expanded version of three university courses entitled, *'Introduction to Astrophysics', 'Thermodynamics',* and *'Philosophy of Science',* given at the Technion-Israel Institute of Technology, The Johns Hopkins University, the University of Pittsburgh, and the State University of New York, over a period of years.

However, today the college student is not content with *isolated* courses on astrophysics, classical physics, relativity, quantum mechanics, thermodynamics, evolution, and philosophy; he wants to know not only the physical laws that lie behind the observational data but also how to use these laws as tools to further discoveries, or as guides to a better understanding of observational data in other disciplines; that is to say, to gain a better understanding of nature at large; an all-embracing view which so often results from a close cooperation between the various sciences. This, somewhat loosely, defines the aim of this book.

For the sake of unification and simplicity the physical laws are first introduced in the main text in as simple a manner as possible. For the more advanced students I supply *Accessory Sections* that demand a greater degree of competence in astrophysics, philosophy of science, general relativity and classical physics. Consequently, two tracks through the subjects are supplied.

The first track focuses on 'interdisciplinary' material of general interest. As such, it is suitable for a *one-semester course* at the junior or senior level or (combined with "track two" material) *in graduate school*. Much of "track one" material should prove fairly easy reading, though a few mathematical formulations of some difficulty had to be included in their turn.

The second track is intended to provide a greater degree of competence in a few selected fields of study. Readers and teachers—especially those in need of enrichment material on physico-philosophical foundations, physico-mathematical formalisms, or some observational data that have recently gained special importance—are invited to select these sections of 'track two' that interest them the most. With a few exceptions, "track two" material can be understood by readers who have studied only "Freshman" physics and mathematics, and, in addition, have read the earlier "track one" material. Both tracks combined may suffice—depending on the students' background—*for a basic two-semester course*.

All "track two" material is marked with ●. It includes the more specific mathematical formulations and various technical details.

It is believed that such an ordering will help those readers who are either unfamiliar with, or uninterested in detailed formulations and empirical data, and would rather consider the general arguments involved without interruption. Hence, a substantial part of this book is also suitable for the general educated reader.

*   *   *

The subject matter of this book has never before been investigated or presented as a unified field of study, though some fragments of it may be found scattered throughout the literature on specific problems. *As a result, different authors have approached the problems discussed from widely varying viewpoints, employing disjointed concepts to what should be a unified system.*

A few years ago, during a course of lectures (Weizmann Institute of Science, Yale, Princeton and Harvard) I found that the three combined subjects—cosmology, physics and philosophy—command the attention not only of astrophysicists, astronomers, physicists and philosophers of science, but of other disciplines as well. This trend has been reflected in recent years by the growing number of papers and books which refer to, or reflect on new theories which deal with these three subjects as well as with their wide implications in other fields of modern human thought.

Because of this growing interest and because of the somewhat surprising rate at which a unified approach to these three disciplines has recently been gaining wide acceptance, a very definite need has been felt for *an up-to-date book that gathers together separate though related strands of modern thought and research, combine them into an interdisciplinary field of study, with a uniform terminology and a careful attention to the elucidation of updated scientific concepts, and, then, offer critical and/or new viewpoints on the issues involved.*

VOLUME I

COSMOLOGY, PHYSICS, AND PHILOSOPHY

> The telescope at one end of his beat,
> And at the other end the microscope,
> Two instruments of nearly equal hope, . . .
>
> **Robert Frost**

> Nothing puzzles me more than time and space;
> and yet nothing troubles me less,
> as I never think about them.
>
> **Charles Lamb**

> If you can look into the seeds of time,
> And say which grain will grow and which will not,
> Speak then to me . . .
>
> **William Shakespeare** (Macbeth)

INTRODUCTION

1.1 THE REVIVAL OF RELATIVISTIC COSMOLOGY VS. MODIFIED CONCEPTS IN PHYSICS AND PHILOSOPHY.

Recent explorations of the external world, as conducted by astronomers, astrophysicists and cosmologists, have generated a revolution in our understanding of the physical world; a scientific revolution without parallel in the whole recorded history of mankind; a fresh look at nature, which opened up new horizons and generated *new needs to re-examine all previous philosophical convictions.*

At the very core of this revolution we find *Einstein's general theory of relativity; the cornerstone of modern cosmology and astrophysics; the indispensible theory for understanding most of the new discoveries.* Indeed, it is Einstein's physico-philosophical ideas, combined with the new observations, that will guide us in these lectures.

It is an unfortunate fact, however, that, although some of the greatest scientists (from Newton to Einstein) studied cosmology and made substantial contributions to it—until recently it had *a rejecting speculative image.* Indeed, most scientists used to reject its importance, using such arguments as: *Why should we rely on any theory or philosophy based on information that had originated in remote, personally inaccessible regions of space and time?—Should not we complete our local physics before we use strange astronomical data to re-examine the very foundations of science and philosophy?*

To begin with it should be stressed that such *geocentric*, or rather *anthropocentric* conceptions are as common today in science and philosophy as they had been in the times of *Copernicus* (who dethroned the *Earth*), *Shapley* (who dethroned the *Sun*), and *Baada* (who dethroned our *Milky Way*). And they make the science of modern cosmology a difficult intellectual subject even to mature scientists and philosophers.

Indeed, astronomy, astrophysics and cosmology—unlike local physics, chemistry, and biology, are *observational* rather than *experimental* sciences, since they deal with objects at such great distances as to be beyond the reach of direct, man-made experimentation. Moreover, until now these observations have been too scarce,

2

and the ratio of speculation to fact too high, for cosmology and astrophysics to qualify as *"hard sciences."* Recent discoveries however, have, *almost overnight*, transformed the situation by yielding solid new data and drastically circumscribing both speculations and cosmological models. This transformation is due largely to new *infrared, radio-* and *X-ray techniques*, including the newest generation of unmanned, spacecraft-based telescopes. *The new discoveries of violently active galaxies, binary X-ray sources, cosmic (microwave) radio background radiation, neutron stars, pulsars, quasi-radio sources, etc.—have stimulated new interest in astrophysics, general relativity and relativistic cosmology.*

Indeed, of late, we have witnessed a new revival of astronomy, general relativistic cosmology and astrophysics, a renaissance sustained by almost daily inflow of verified empirical information on the dynamic universe around us. To deny today the central role of astronomy, cosmology and astrophysics in modern science is to deny the very methodology of science and to reject a large portion of its empirical evidence. There is nothing too insignificant for careful study, nothing unworthy of scientific investigation. Modern science recognizes no established scale or size as a limit to its activity. Every item of carefully recorded information is a proper object for analysis. A drive towards novelty and discovery impels scientific inquiry to explore all corners of the universe. In fact, the whole structure of fundamental physics is not dissociated from the detailed inquiries of astronomy, astrophysics and modern cosmology (see below).

Closely related to the unrestricted content of modern science is its unrestricted *questioning* of all earlier convictions. *No belief is too sacred or too firmly established to prevent radical evaluation of its soundness on any scale of observation.* A serious scientific attitude premises willingness to admit new hypotheses, even those that seem to contradict intuitive, geocentric and anthropocentric theories. *There is something provisional about all scientific theories.* They have to be confirmed by fresh information from any subfield (including astronomy), hence are subject to constant *revision* and even replacement; in this, each succeeding generation takes a measurable step beyond the position of *its predecessors*.

1.1.1. The Problem of Ordering

Thanks to the *subdivision* of knowledge into *fragmented* 'disciplines', we often fail to perceive the *interconnectedness* between 'self-centered' sciences, to judge their *collective* importance and to estimate their *inherent structure and ordering*. In trying to overcome this problem, we may begin with the problem of ordering these very lectures. Here I am simultaneously faced with subjectivistic, empirical, and cosmological discourses; subjectivistic, because all thought is, to some extent, ordered by personal bias; empirical, because all rational-empirical science singles out regularity and order; cosmological, because it is in line with our main Assertions (see below).

More specifically, I am faced here with the immediate questions: Should this course begin with the evolution of the vast universe around us, and gradually concentrate on the subjective perceptions of the individual "here-now", or vice versa?

Should it be directed from "innate", or *"a priori ideas"* of subjective human knowledge, to "external" objective concepts; or in reverse order? Should it be from historical speculations to vindications by modern empirical science, or vice versa?

It is in the Assertions below that I postulate the idea that *large (cosmological) systems dominate the evolution and development of smaller systems, not vice versa.* It is in Lecture VI where this postulate is first vindicated by employing recent empirical data that span a broad range of modern science. And it is in subsequent lectures that this postulate is confirmed further and employed to arrive at some entirely new concepts of science and philosophy. Therefore, I shall try to employ it here *as a general guiding idea in ordering the main discussions of this course.*

This means that the road from objective science to a better understanding of the origins of subjective knowledge must always be from external to internal worlds, i.e., *from a strict scientific analysis of the most remote regions (in both space and time) of the external "outside" to the most innate spheres of subjective memory, perception, and brain-mind of the individual.* Here the concepts "external" and "outside" need philosophical clarifications. This will be done later.

Consequently the following discussion is so ordered that we shall proceed from the origin of the universe to the origin of life and from there to the origin of knowledge. It is only by passing these initial stages that we shall be prepared for a rational examination of old-new disputes about such philosophical questions as the following one;

Is perceptual organization and ordering physiologically inborn? In other words, *is it inherent in innate-isolated aspects of brain functioning rather than depending on a synthesizing process of learning through repeated experience with external objects?*

Therefore, in adopting this universal order of occurances we utterly reject the tendency of most Western philosophers to leave the study of the vast external world to the care of the scientists and to search for the foundations of the universe in their own minds.

> *Faust:* 'Tis writ: 'In the beginning was the Word!'
> I pause, to wonder what is here inferred?
> The Word I cannot set supremely high,
> A new translation I will try.
> I read, if by the spirit I am taught,
> This sense: 'In the beginning was the Thought'.
> This opening I need to weigh again,
> Or sense may suffer from a hasty pen.
> Does Thought create, and work, and rule the hour?
> 'Twere best: 'In the beginning was the Power!'
> Yet, while the pen is urged with willing fingers,
> A sense of doubt and hesitancy lingers.
> The spirit come to guide me in my need,
> I write, 'In the beginning was the Deed!'   **Johann Wolfgang von Goethe***

* *Faust I.* Transl. *Philip Wayne* (Penguin Classics; London).

1.1.2 How Did it all Start?

The Bible answers this question in three sentences:

> *In the beginning God created the heavens and the earth. The earth was without form and void, and darkness was upon the face of the deep; and the spirit of God was moving over the face of the waters. And God said, "Let there be light", and there was light.*
>
> **--Genesis 1**

Most astrophysicists, cosmologists and astronomers agree that the biblical account of the beginning of cosmic evolution, in stressing "a beginning" and the initial roles of *"void"*, *"light"* and a *structureless* state, may be uncannily close to the verified evidence with which modern science has already supplied us.

In the beginning, according to astronomy and empirical relativistic cosmology, the highly curved space-time geometry expanded as the time-reversal of gravitational collapse into a black hole (Lecture VIII). In other words, the early stages of the universe are, in many senses, analogous to a *"white hole"* in which radiation-dominated (structureless) fluid expanded away from extremely high density and temperature to end up in the present matter-dominated era in which the structureless fluid has gradually developed into the present *hierarchy of structures* (see below).

It will be mentioned in later lectures that in the universe accessible to our observations a black hole *cannot*, in principle, turn into a white hole, but a white hole *can* gradually develop (contained) black holes or even turn into one eventually. This fact alone appears to ensure the unidirectional evolution of cosmic time and time anisotropies (Lecture IV) which dominate all subsequent evolutions. Here *gravitation* plays an apparently two-fold role.

Firstly, it brings about *non-static* evolution, i.e., expansion of the universe as a whole. This result emerges from the fact that Einstein's field equations, as well as Newtonian gravitation, accept no plausible *static* (cosmological) solutions. It is a remarkable confirmation of present-day physics that these predictions agree with all astronomical observations (§2.4 below and Lectures III and VI).

Secondly, the growth of density perturbations by *unidirectional* gravitational *attraction* is responsible for the present existence of a *hierarchy of structures*, namely, from the clusters of galaxies, galaxies, star clusters, stars, planets, and temperature-composition gradients in planetary atmospheres, down to the evolving structures of biological systems (§3 below and Lecture IX).

But this *separation* is erroneous, for, as will be shown below, *both* phenomena are caused and controlled by the *same* mechanism; i.e., the cosmic expansion, in itself, is caused and controlled by the gravitational field. The details of this mechanism are to be stressed below.

However, in response to these claims, one is tempted to pose the following questions:

1) How certain is this science?
2) What evidence supports it?

3) How is this cosmic history being linked to all presently observed processes on earth?

Perhaps the most extraordinary development of all modern sciences, during the last decade or so, has been the emergence of astrophysics, astronomy and modern cosmology as genuine *empirical* sciences. There are several reasons for this renaissance. Partly it stems from the recent dramatic advances in infrared, radio, X-ray, gamma-ray and neutrino astronomy (Lecture I). Partly it stems from recent vindications of Einstein's general theory of relativity (Lecture III). But it has also been stimulated by the utilization of modern computers and new mathematical techniques. *As a result there is now a general consensus that the universe has been expanding from an "initial state" of extremely high densities and temperatures to its present state, and that any reliable theory of cosmic history must include the effects of general relativity.* In other words, the so-called "steady-state theory"—which postulates the "creation" of matter *by the cosmic expansion*—has been totally rejected on grounds of incompatibility with major sources of empirical evidence (e.g., the cosmic background radiation, radio-source counts, etc.—see below).

However, at this point a word of caution must be added. According to the known equations of state and the equations of general relativity, as one examines earlier and earlier cosmic epochs, the density and temperature of the cosmological fluid increases without limit. At present few scientists are confident of extrapolating the known laws of physics to the first 10^{-23} seconds or even to the first 0.0001 seconds of cosmic evolution (stages 1 to 3 below).

Moreover, we do not know if there was "at the beginning", any single singularity involved or perhaps a few ones. Nevertheless, it will be seen that the lack of knowledge about the first 0.0001 second is not too limiting for understanding the rest of cosmic evolution.

One last remark about the validity of general relativity. It was by employing Einstein's general relativity, that astrophysicists calculated the *chemical compositions* of the material emerging from the initial stages of the universal expansion. And, it is widely regarded as *one of the more important triumphs of the theory* that it enables scientists to predict that this material is about 27% helium and 70% hydrogen — in a remarkable agreement with all available astronomical observations (Lectures VI, I and III). This result is not only a remarkable confirmation of the "big bang" theory, but it also solves a long-standing problem in theories of nucleogenesis of the chemical elements: because, whereas the origins of heavier elements can be adequately explained by nucleosynthesis associated with the thermodynamics of stellar evolutions, such as super-nova explosions (Lecture I), it had been a problem for a long time to explain the high abundance and relative uniformity of *helium*. Once the 2.7°K microwave background radiation was discovered in 1965 (§2.4 below), the entire thermal history of the universe became very clear.

Using this information—and in agreement with all available evidence concerning the validity of Einstein's general relativity—the outline of the history of the universe is currently calculated to be as follows.

1.1.3 The First Seven Stages

In the following outline we mark the "age of the universe" by cosmic-temperature arrow of time, or by earth seconds (years).

In these calculations general relativity and statistical physics predict that the temperatures, densities and geometrical curvatures should become arbitrarily high as we approach close to the very initial moment.

Stage 1: Up to an "age" of 10^{-43} seconds the curvatures of geometry and the temperatures were so high that perhaps even Einstein's general relativity may not be an adequate description (for one alternative proposal, which, however, may not be valid, see the description of "super-space" in § IV.13).

Stage 2: Up to an "age" of 10^{-23} seconds the curvature of space-time suggests that various kinds of particle pairs can be "created" out of the gravitational field in an *analogous manner* to the known and verified creation of electron-positron pairs in strong electromagnetic fields. However, adequate description of this early stage is not yet possible.

Stage 3: Up to an "age" of O.0001 seconds of cosmic history (when the temperature was still above $10^{12°}$K), modern relativistic cosmology still encounters great difficulties. These are mainly associated with the difficulty of finding proper equations of state to describe the so-called "thermal equilibrium" of strongly interacting elementary particles—including photons, neutrinos, leptons, mesons, and nucleons and their antiparticles*.

Stage 4: This stage is marked by a drop in temperature down to $10^{11°}$K. During this cooling stage (about 0.01 seconds) the universe contained photons, neutrinos, antineutrinos, muons, antimuons, electrons and positrons. As the temperature dropped to about $10^{11°}$K the muons and antimuons began to annihilate and at $10^{11°}$K the neutrinos and antineutrinos *decoupled* from the other particles. Consequently, this stage may be called "*the stage of neutrino decoupling*".

Stage 5: As the "age" of the universe increased from 0.01 seconds (about $10^{11°}$K) to around 4 seconds ($5 \times 10^{9°}K$) the electron-positron pairs began to annihilate and *the cooling of the neutrinos and* antineutrinos froze the neutron-proton ratio at about

---

* Nevertheless there is a promising mechanism which explains the thermodynamics involved here, and, in particular, the *dissipation of energy*. This is provided by the extremely interesting phenomenon of *bulk viscosity*. We shall see in Lecture VI that *shear viscosity* can play no role in a pure radial expansion of matter and/or radiation.

If the present dissipation of energy in the universe (as calculated from the observed background radiation, etc.) was *not* due to bulk viscosity alone, then perhaps it was produced by dissipation effects of shear viscosity or heat conduction in an initially anisotropic or inhomogeneous expansion. Here, Misner, Zeldovitz and others (Lecture VI) have shown that neutrino viscosity, acting before the cosmic temperature dropped to about $10^{10°}$K, would have reduced the anisotropies of the background radiation, produced in an initially homogeneous but *anisotropic* expansion, to less than the value now observed. Therefore, this stage may be called "*the stage of isotropisation.*"

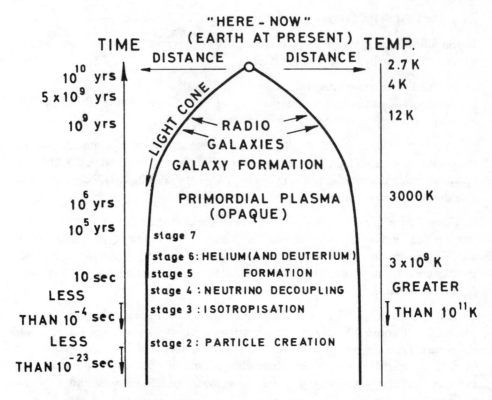

The Cosmological Evolution of the Physical World
(as a *Past-Light Cone* in space-time coordinates, cf. FIG. IV.4.) Cosmological time can be measured in terms of earth years, cosmic-temperature arrow of time (in terms of degrees), or the universal time R (in terms of centimeters) (Lecture VII). Note the cosmic times and temperatures from particle creation (stage 2) to galaxy formation. For a discussion of these stages see §1.1.3.

1 : 5. This age may, therefore, be called "*the age of neutron-proton freezing*". The mathematical expression of the cosmological arrow of time is given by eq. (VI.2) as

$$\frac{T(t_p)}{T(t)} = \frac{R(t)}{R(t_p)} \tag{VI.2}$$

where $R(t)$ is the linear cosmological scale factor at time t, $R(t_p)$—its value at the *present* time (which may be thought of as representing the average distance between clusters of galaxies), $T(t)$ and $T(t_p)$—the corresponding temperatures of the black-body radiation which decrease with the expansion according to eq. (VI.2). Now, eq. (VI.2) means that

$$T(t) \propto \frac{1}{R(t)} \tag{1}$$

From elementary physics we know that, for $t < 1$ seconds, the temperature is high enough for the photons to create electron-positron pairs. At "equilibrium" (at a given constant T and density), electrons and positrons are created and annihilated at the same rate. But it is due to the expansion that they disappear giving back their energy. Not so for neutrinos. For $t < 1$ seconds, the so-called "statistical equilibrium" between neutrons, n, and protons, p, depends on the *weak interaction* with electronic neutrinos, v_e, and antineutrinos $\bar{v}_e$, according to (see also Assertion 2, §3.1)

$$n + e^+ \rightarrow p + \bar{v}_e \tag{2}$$

$$p + e^- \rightarrow n + v_e \tag{3}$$

Thus, these require the presence of electron-positron pairs. Since at $t = \sim 1$ second these pairs disappear abruptly, the n/p ratio is frozen at a value of 1 to 5 (about 15% neutrons) for a few hundred seconds, after which it is *reduced* as a result of *neutron decay.* But, before this happens the neutrons start to fuse with the protons to form deuterium and gamma radiation according to

$$n + p \rightarrow D + \gamma \tag{4}$$

However, this reaction does not become fast enough before the temperature has fallen below $10^{9\circ}K$ (at about 100 to 180 seconds). It is only at these lower temperatures that the photons stop the rapid disintegration of the newly formed deuterons. But this cosmic temperature marks the next stage in cosmic history.

Stage 6: As the "age" of the universe increased from 4 seconds (about $5 \times 10^{9\circ}K$) to 100–180 seconds (about $10^{9\circ}K$) the neutrons fused rapidly with protons to form deuterium according to reaction (4).

As soon as this reaction is complete, further (and rapid) nucleosynthesis forms the nuclei of about 25 to 27% helium by weight and traces of He^3 and other heavier elements.* The presently *observed* abundance of helium in the universe is between 25% and 30% and this agreement is considered a *remarkable confirmation* of this sequence of events! Consequently, the frozen-in ratio of n/p that took place in Stage 4 at around $10^{10\circ}K$ has remained as a *"relic"* of this early epoch till the present time. Hence, this stage may be called the *"stage of helium formation."*

Massive empirical evidence that has been accumulated by various other observations involving elementary particles supports these calculations and makes them highly reliable.

Stage 7: As the free expansion of photons, neutrinos, anti-neutrinos and nuclei continued, the ("plasma") temperature dropped from $10^{9\circ}K$ (about 100 to 180

---

*

$$D \xrightarrow{+P} He^3 \xrightarrow{+He^3} He^4 + 2p$$
$$\xrightarrow{+He^4} Be^7$$

$$B^8 \longrightarrow Be^8 + e^+ + v \longrightarrow He^4 + He^4$$

$$Li^7 \xrightarrow{+P} Be^8 \longrightarrow He^4 + He^4$$

seconds) to around 3000–4000°K. *This cooling lasted not less than about 100,000 years!* Then, at the end of this stage something quite dramatic happened. To understand it let us keep in mind that so far the universe has been entirely *structureless* and *opaque*; for *radiation and matter* have been *intimately coupled* to each other; *electrons scattered photons* and everything was *uniform in space. Not so in time*; for the cosmic expansion continued to cool this opaque cosmological fluid. No self-gravitating systems, not even protogalaxies or protostars existed yet. The only *non-uniformity*, and in fact, the *only "clock", the only time asymmetry* that has existed throughout the first 100,000 years of the world's history is the *unidirectional expansion away from a dense-hot initial state, toward lower temperatures and densities.*

Now, what happened around 3000–4000°K? At these temperatures the free electrons *were captured* by the cooling nuclei of the cosmic plasma and were (irreversibly) *confined to atomic orbits.* In these circumstances they no longer scattered photons, and matter and radiation were finally decoupled from each other and have remained so. Then, almost suddenly, and for the first time in cosmic history, the universe became *transparent* to photons.

From this moment in cosmic history till the present epoch, the evolution has proceeded along two complementary tracks, which, eventually, also brought the conditions for the emergence of life.

From §III.2.7.3 and §III.3 one can see that, according to general relativity, a radiation-dominated model gives

$$R(t) \propto t^{1/2} \tag{5}$$

Combining with Eq. (1), one obtains a simple relation between the temperature of radiation and the cosmic time in seconds, i.e.,

$$T_{\text{radiation}} = \frac{10^{10}°\text{K}}{t^{1/2}} \tag{6}$$

When radiation ceases to dominate—and matter is decoupled from it—the latter cools much faster according to the relation

$$T_{\text{matter}} \propto \frac{1}{R(t)^2} \tag{7}$$

Here matter (i.e., atomic hydrogen and helium below 3000°K) is assumed to obey the simple perfect gas equation [Eq. (7) should be compared with Eq. (1)].

Down from a temperature of 3000°K, the universe proceeded to expand in a course in which the very expansion becomes the *prime cause for gradual building up of hierarchy of structures*, namely: clusters of protogalaxies, clusters of galaxies, galaxies, clusters of stars, planets, planetary atmospheres and eventually biological structures. As the radiation-dominated era comes to a close, the (energy) density of radiation (photons and neutrinos) drops below the (rest-mass) density of matter (mainly hydrogen and helium) and new phenomena come into play.

1.1.4   The Present Matter-Dominated Era

A singular common feature of the present matter-dominated era is the *growth of gravitational instabilities* which gradually lead to the formation of self-gravitating aggregations of matter that emit radiation into an expanding, and *therefore*, (proof in Lecture VI) *unsaturable* intercluster space.

Unfortunately, in spite of the fact that we now have fairly good information about stellar evolution, stellar cores, stellar atmospheres, stellar dynamics, and about sizes, distances, shapes, temperatures and nuclear reactions in stars, galaxies, etc., we still do not have an accepted theory of *formation* of galaxies anywhere near so complete and plausible as our present theories of the origin of the cosmic abundance of helium or the cosmic background radiation (see below).

The first serious theories of growing instabilities of an *expanding* universe (the first was *Jean's theory* which is restricted to a *static* universe), were reported by Lifshitz* in 1946 and by Bonner** in 1956. *According to these theories, and to more recent ones,*** the expansion of the universe emerges as the very cause for the formation of protogalactic fluctuations.* Coupled with the effects of the cosmic black-body radiation and cooling due to expansion, the amplitudes of the fluctuations grow under the influence of its own gravitation. Here the effects of dissipation of energy may also become important (Lecture VI).

Once the fluctuations grow into self-gravitating systems, the evolution follows the various processes to be described in Lecture I. In the meanwhile let us turn to another subject: methodology.

2.1   THE EINSTEINIAN METHODOLOGY: A PRELIMINARY REMARK

It is Einstein's general theory of relativity, his attempts to develop an all-embracing unified field theory, as well as his physico-philosophical reflections that have influenced many of the concepts presented below. For this reason I proceed with a preliminary note related to the *Einsteinian methodology*.

Perhaps no one is better equipped to describe Einstein's methodology than Professor Philipp Frank (1884–1966). Himself a leading authority in physics and

---

*   E. Lifshitz, *J. Phys.* (U.S.S.R.) *10*, 116 (1946)

**   W. B. Bonner, *Z. Astrophys. 39*, 143 (1956)

***   S. Weinberg, *Gravitation and Cosmology: Principles and Applications of the General Theory of Relativity,* Wiley, N. Y. 1972 pp. 578–588. See also G. B. Field *"The Formation and Early Dynamical History of Galaxies"* and K.C. Freeman *"Stellar Dynamics and the Structure of Galaxies"* in *Galaxies and the Universe,* (Vol. IX in *Stars and Stellar System*), A. Sandage, M. Sandage, and J. Kristian, eds., The University of Chicago Press, Chicago, 1975.

philosophy, he knew Einstein intimately and long. His understanding of Einstein's methodology may be best demonstrated by the following:

> "Since Einstein was chiefly interested in the general laws of physics or, more precisely, in deriving logically the immeasurable field of our experiences from a few principles, he soon came into contact with a set of problems that are usually dealt with in philosophical works. Unlike the average specialist, he did not stop to inquire whether a problem belonged to his field or whether its solutions could be left to the philosophers" [Philip Frank, *Einstein: His Life and Times*, Alfred Knopf, 1953, 1972]

Today this methodology is almost totally forgotten. Consequently, we now live in a post Einsteinian era that is characterized by "scientism", specialism, declining academic standards and acute crisis (Appendices III and VI). "Back to Einstein" is one possible remedy to the present crisis. But, as I emphasize in the text, we must also take steps beyond Einstein, and, at the same time, return to some classical methodologies that preceded him. This is not a simple task, for most of us end our university studies with an ingrained aversion to philosophical inquiry. Whether we are persuaded that the nature of the physical world may be expressed through mathematics (following, for example, Jean's maxim: "*The Great Architect of the Universe now begins to appear as a pure mathematician*"), or whether we are persuaded that it is abstruse to human understanding, we are inclined to dismiss philosophical inquiry as "impractical" and time-wasting. Is this a symptom of the failure of Western methodology and the declining role of its philosophy?

The justification of the "point of departure" of a given scientific theory is a matter entailing far more ambiguity than the technicalities of the derivation and application of the theory. But such justifications are bound up with *philosophical* considerations, and the latter, according to most contemporary Western scientists, are not *bona fide* topics for the exact sciences. *In the recent past, these philosophical considerations have shaped the frame of research of the greatest scientific thinkers; today the university student in the West is almost specifically trained by his professors not to ask fundamental questions, as a result of which inconsistent or outright incorrect premises are given a better chance of perpetuating themselves.*

As things stand at present, a student in the natural sciences who commences his studies in the firm belief that such fundamental concepts as time, space, irreversibility, causality, entropy, complementarity, probability, gravitation and retarded potentials are clearly defined and comprehended by his teachers — or at least by the hallowed authorities quoted by them — is extremely perplexed, not to say embarrassed, to discover, sooner or later, that these experts and authorities have never been really certain about the origin, or even the proper meaning of the concepts underlying their own theories. *Those who seek a more consistent explanation of these concepts eventually realize that it entails innumerable links between the philosophy of science and the more mundane problems of thermodynamics, quantum statistical mechanics, elementary particles, electromagnetism, relativity, and cosmology! Many give up at this stage, often blaming themselves unnecessarily for their inability to grasp what others seem to regard as trivial; a few dare to continue the interminable search, which calls for questioning, reexamination, screening, and at times radical rejection of "accepted", or "established" theories which fail to justify themselves through their own logic, consistency, universality and testability.* However, a small, but steadily growing number of Western scientists, have recently come to believe that this *anti-philosophical attitude,* which is *a barrier* to the refinement of the core of most of our theories, should be reexamined and revised.

2.2 THE WITHDRAWAL OF PHILOSOPHY FROM PHYSICS (AND OF PHYSICS FROM PHILOSOPHY)

Can we identify the causes of this aversion? Until around the mid-Thirties of this century, philosophical courses (in Europe!) were made as "scientific" as possible, and physics propped up its foundations by turning to philosophy. That attitude was in part motivated by the desire to appropriate for physics the great historical *"pres-tige"* of philosophy, and for philosophy the new *"status"* of science. European professors vied with one another in presenting science as a unified system founded on established philosophical grounds, and philosophy as a *"science among sciences",* or even as the *"sum of the other sciences".*

Western philosophy has since withdrawn from the game. "Science and technology" have meanwhile gained the highest prestige in our society; no longer do they need borrow it from philosophy. *Today, especially in the United States (Appendix III), professors in the natural sciences vie with one another in presenting technical lectures devoid of any philosophical content, for they can no longer hope to achieve popularity by injecting philosophical inquiry into science. This turn of events has resulted in overvaluation of technical science, empty scientism, absolutation of specific interpreta-tions in physics, and the common inclination to reject any philosophy of science from physics.* A pity. For these trends can only push physics into stagnation (cf. Appendices III and VI).

Nevertheless, these trends may be reversed, if and when we regenerate our metho-dology and thinking. On one hand regeneration calls for continuous philosophical revolutions and a fresh outlook on nature; on the other hand, we must bear in mind that new philosophies are not necessarily more advanced than old ones (see below).

The origin of this problem is intimately linked to the *separate* roles played in West-ern academe by scientists and philosophers. Nowadays most Western philosophers are ignorant of modern advances in the exact sciences, of their mathematical formula-tions, and technical and empirical methods. They are likely to resort to *a priori,* or superficial answers to complex problems whose detailed scientific implications are beyond them. Consequently, most of them withdraw to the *"study of knowledge",* the *"human mind"* and *"the structure and logic of language".* On the other hand, the contemporary scientist, a stranger to philosophy, can only deal with problems by applying those scientific methods which are familiar to him through the non-philo-

The aversion of the Western scientist to philosophical inquiry.

Who attends the school of wisdom now? . . . Who has regard for philosophy or any liberal pursuit,
except when a rainy day comes round to interrupt the games, and it may be wasted without loss?
And so the many sects of philosophers are all dying out for lack of successors.
The Academy, both old and new, has left no disciple.

Lucius Annaeus Seneca (c. 4 B.C.–65 A.D.) (in **Physical Science in the Times of Nero,** by J. Clarke,
Mcmillan, London, 1910)

sophical (and sometimes even anti-intellectual) methods of education in our universities. The resulting vicious circle opens unbridgeable gaps between philosophers
and scientists, and, gradually, *isolates* each discipline. It is this *separatism* which
causes declining standards, deterioration and empty academicism (Appendices III
and VI).

*In contrast, the history of physics testifies that the greatest advances were made
when philosophy and physics maintained close dialogue with one another; when scientific
-philosophical thinking was a two-way street. Until the mid-Thirties many eminent
European physicists were engaged in proposing new concepts for advancing the
(philosophical nature of the) foundations of physics; very few physicists have dared
to do so since World War II.* Perhaps this "modern" tendency is a major cause of
the partial *"stagflation"* that has overshadowed physics (excluding astrophysics
and biophysics) since the War. Of course, physics is technically richer and more
diversified today than in the first third of this century, but this explains only part of
the subtleties involved.

In the not-too-distant past this difference in ideology and policy was remarkable.
Unlike the elite European universities, which were generally working on subject-
oriented, research problems, American physics stressed *"the production of research-*

ers". The Cavendish Laboratory at Cambridge, for example, was notorious for its lack of concern with *"institutional arrangements"*, *degrees, matriculation, and the like*; the American universities, on the other hand, were in the business of producing PhD's. Thus, in spite of intellectual infusion U.S. academe has received from Hitler's émigrés in the 1930's, it has mainly relied on 'Big Science', with its big machines, large 'PhD teams', and voracious appetite for money. Somewhere along the road American academe has forgotten the real meaning of a PhD (cf. Appendices III and VI).

Thus, the shift of academic influence, from Western Europe to North America, has caused serious philosophical study of physics to fall out of fashion. Indeed, the dichotomy in American pragmatism (Appendix III) brought a murderous thrust at traditional modes of physico-philosophical progress, and, contrary to intent, a damaging blow to original physico-philosophical thought. Hidden taboos have helped generate an academic atmosphere in which a respectable Western physicist can almost never further his reputation by writing on the philosophy of of science — in fact, the usual result is to damage it.

Such taboos have produced another trend — an "inner-directed" trend towards the study of the *"individual"*, a trend which affirms idealism, pragmatism, subject-ivism, transcendentalism and the role of "man's mind and consciousness", a trend which seeks subjectivistic meaning and purpose in the world and often associates them with a dogmatic *"philosophy of language"* (Appendix III). In addition, this trend led many Western thinkers in the postwar era to embrace the individual as represented in the philosophies of existentialism, nihilism, neo-anarchism, syndical-ism and other pessimistic doctrines. Even those deeply interested in logical positivism (Appendix III), usually paid little, or no attention, to competent modern studies in the philosophy of the exact sciences. Thus, leaving the study of the vast external world to the care of the scientists, Western philosophers decided to concentrate on the internal worlds of the mind, toying with the analysis of language, symbols, sensationalism, operationism, logomachy, etc. They did not want to be called "literary physicists." Instead, they became "amateur psychologists." Indeed, the decline in the use of the Einsteinian methodology began when our philosophers started to search for the foundations of the universe in their own minds. Their efforts to escape the influence of physics caused separation between scientsts and philosophers (Appendices III and VI), the decline of Western thought, and the rise of mediocrity and monotonous accounts of private perceptions and sensations.

What can be gained by changing the present situation?

The constant search for sounder physico-philosophical theories, resting on founda-tions more unified and more universal than those of their present-day counterparts, is not merely an *academic, aesthetic* and *logical* necessity. In fact, it stems from the *remarkable failure* of most theories to resolve and eliminate serious paradoxes, in-consistencies, and incoherencies which exist today at the fundamental level.[4]

2.3   THE GREATEST AMBITION OF PHYSICS

2.3.1 Unification of Initial-Boundary Conditions First? Unification of Fields Second?

Present physics recognizes four basic forces in nature. In order of increasing strength they are the *gravitational*, the *'weak'*, the *electromagnetic*, and the *'strong'* (Table I).

The greatest ambition of physics is to unify *the general theory of relativity*, which describes the gravitational field, with *quantum field theories*, which provide a context for dealing with the other three known forces (Assertion 2 in §3 below). If this will be achieved, all four basic fields of force would be described by the common concepts of a dynamic geometry of space-time (Lecture III). However, so far all attempts to unify all four fields have mainly made their incompatibility more apparent. Nevertheless, a number of recent developments (including *'supergravity-supersymmetry'* — Assrt. 2 below), demonstrate that *Einstein's grand aim of achieving such a unification is not an impossibility*. (Einstein devoted much of the last part of his life to an unsuccessful search for a unified field theory of gravitation and elctromagnet-

TABLE I

Present physics recognizes four basic forces-fields in nature. In order of increasing strength they are the *gravitational,* the *'weak'*, the *electromagnetic,* and the *'strong'*. Note that although the gravitational force-field is exceedingly feeble, it 'acts on' (see, however, below) *all* particles observed in nature and is also *the only* one of the four forces-fields that is both *long-range* and *attractive* between *all* pairs of particles. Moreover, it is *reducible*, within the scope of present physics, *to 'pure' geometry*, i.e., *to the dynamic properties of curved space-time* (Lecture III, Table IV, and Assertion 2 below).

	The gravitational force	The 'weak' force	The Electromagnetic force	The 'strong' force
Range (see, however, §III.5 and IV.12)	long	short	long	short
Particles-objects 'acted on' (Assert. 2 and §V.6)	all	mainly neutrinos electrons quarks	only on electrically charged	only on hadrons and quarks
Particles-objects exchanged (cf. §III.5 and §V.6)	gravitinos? gravitons?	intermediate vector bosons?	photons	gluons?
Spin of exchanged particles	2	1	1	1
Relative strength	1	10^{34}	10^{37}	10^{39}
Principles of symmetry (cf. §V.6, §III.5 and Lecture IV)	local; time reversal	possible time reversal violation?	local; time reversal	time reversal; isotopic symmetry

ism). This possibility is, to some extent, grounded on the fact that *no fundamental field in physics can avoid running into the common concepts of space, time, symmetry-asymmetry and initial-boundary conditions* (see below and Assertion 2, §3).

Yet, as the reader will see (perhaps with surprise), after centuries of trying, occupying the attention of the greatest scholars, it has become clear that *these concepts are now back in the melting pot!* This is partly the result of the impact of new information about the universe around us, as supplied by modern observations in *astronomy* and *high-energy physics*; partly, it is a product of bizarre new evidence concerning violation of the time-reversal symmetry involving *subatomic* particles known as *kaons* (Lecture IV); partly it is caused by an erroneous view of the nature of the theories and problems to be analyzed; and partly, as we shall see, it reflects the recent progress associated with unification and with a new approach that might be called *gravitism* (see below).

Critical inspection of the scientific literature which deals with the aforementioned concepts reveals a blurred picture of confusing ideas, which, like a mirage in the desert, misguides wandering scientists in search of the origin and meaning of these fundamental concepts of physics and philosophy of science. In fact, it shows that physics is not a complete logical system. Rather, at any epoch, it spans a great variety of ideas, some of them survivors, like folk epics, from the heroic periods of the past.

Much of the prevailing confusion stems from lack of precision as to which phenomenon is actually being analyzed, and in what manner explanations of the observed phenomena are unified under the auspices of a reasonable consistent theory. To begin with we, therefore, base much of our methodology on the *Einsteinian approach to nature*; i.e., we *assume* that *there is an objective need to construct a broad physico-philosophical framework which does not a priori deny the possibility of a unified field theory* that would describe all four basic fields of force in terms of a common space-time terminology. Gradually, this methodology will lead us to unification of some other physico-philosophical concepts in terms of four concepts only, namely;

SPACE-TIME *and* SYMMETRY-ASYMMETRY

Thus, such concepts as irreversibility, initial and boundary conditions, as well as gravitation and other fields will be 'reduced' to these four concepts *wherever possible! We shall also try to demonstrate that the ideas of symmetry and asymmetry are the most powerful ones in science and philosophy; that their proper use can revolutionize many of our traditional concepts; that they can be the best tools for the unification of apparently different disciplines.*

2.3.2 Should Unification Begin With Differential Equations?

All attempts to unify the four fields have, so far, concentrated on the unification of the mathematics of the *differential equations* which describe the spread of these

TABLE II

The global structure of theoretical physics in terms of symmetric 'laws' (differential equations), asymmetric initial-boundary conditions, and 'solutions' that are comparable with observations (groups I, II, and III respectively). Note that a comparison with observations *cannot* be made with differential equations (even if we consider an ultimate, fully-unified field equation), unless we combine them with *asymmetric* initial-boundary conditions (to obtain 'solutions', integro-differential equations, etc.). In light of the difficulties encountered in the unification of the differential equations, *by themselves*, we first pose the questions concerning *the methodology of unification*, namely;

Should unification begin with differential equations? Should not we unify first the apparently different asymmetric initial-boundary conditions? How can the asymmetric conditions be unified and then combined with symmetric differential equations?

(For details see Lectures V, VI, VII, and IX and also Assertion 2 below.)

GROUP I	GROUP II	GROUP III
Time-Symmetric 'Laws' (differential equations)	Asymmetric initial and boundary conditions	'Solutions' comparable with observations
General relativistic field equations; Maxwell's equations of electromagnetism; the various conservation and balance equations of momentum, mass, energy, etc.; any unified field equation, e.g., QED, QFD, QCD, or SUPER-SYMMETRY (Assert. 2, §3 below)	Dissipation of energy; viscosity; irreversibility; arrows of time (time asymmetries); causation; 'symmetry-breaking'; classical, statistical, and quantum statistical thermodynamics; entropic growth; information transfer; retarded potentials; time-reversal violations in kaonic systems; cosmological instabilities; biophysical structures; recording; memory; etc.	Mathematical combinations of symmetric GROUP I with asymmetric GROUP II; integro-differential equations; energy equation with embedded asymmetric dissipation; any solution of a field eq., etc.

fields in terms of space and time, space-time, or their equivalents (§III.5 and §IX.2.20). This methodology requires a careful examination.

First and foremost we note that physics is composed of *three major groups* (Table II), namely;

■ *GROUP I*

This includes the so-called *'laws of physics'* (e.g., Einstein's field equations [Lecture III], Maxwell's equations of electromagnetism [Lecture VI, §10], the various conservation and balance equations of momentum, mass, energy, etc. [Lecture II], Schrödinger equation [Lecture V], *QED, QFD, QCD* and *SUPERSYMMETRY field theories* [Assertion 2, §3 below]).

■ *GROUP II*

This group may be called the *asymmetric* group of initial-boundary conditions. It contains such concepts as (Table II) dissipation of energy, viscosity, irreversible thermodynamics, arrows of time (time-asymmetries), 'symmetry breaking' (§III.5), quantum statistical irreversibility (Lecture V), entropic growth, information transfer, retarded potentials in electrodynamics (§VI.10), time-reversal violations in the sub-atomic particles known as kaons (§IV.II), cosmological instabilities (Lectures I, III, IV, VII and VIII), biophysical structures (§3 below and Lecture IX), recording, memory, perception of space and time, mode of thinking and perhaps also the concept of pre-geometry (Lecture IX). Most important, this group includes *the physics of time* — a most difficult subject in human understanding of nature (Lectures IV, VI and VII).

■ *GROUP III*

Unless we combine the *symmetric* differential equations with the *asymmetric* initial-boundary conditions, i.e., unless we combine *GROUP I* with *GROUP II*, we *cannot* make a comparision with physical *observations*! Thus, in theoretical physics, as well as in any mathematically-based science, we construct *GROUP III* (Table II), which, *inter alia*, contains *integro-differential equations* and the *'solutions'* of the equations.

In light of the difficulties encountered in the unification of the differential equations, *by themselves*, we first pose the questions concerning *the methodology* of unification, namely:

Should unification begin with differential equations? Should not we unify first the apparently different initial-boundary conditions? How can the asymmetric conditions be unified and then combined with symmetric differential equations?

Indeed, if unification of initial-boundary conditions is required first, what are the problems which prevent, if at all, such a unification? *Are these problems associated with identification of a master (unified) initial-boundary condition?* In trying to answer these questions one may first proceed through the following prerequisites:

■ PREREQUISITE 1 : Identification of the most universal master asymmetry.

■ PREREQUISITE 2 : Transformation of all the 'independent' asymmetries observed in nature into the language of this asymmetry.

■ PREREQUISITE 3 : Resolving the apparent contradiction between present symmetric 'laws of physics' and present asymmetric initial-boundary conditions.

We shall return to these questions later. Meanwhile let us turn our attention to recent advances in astronomy and astrophysics.

2.4 THE GREAT PHYSICO-PHILOSOPHICAL GAINS FROM THE DISCOVERY OF THE COSMIC BACKGROUND RADIATION

How can the very foundations of physics, or the principles of modern philosophy of science, be modified in response to the new discoveries in astronomy and cosmology?

This subject is taken up in Lectures VI to IX. Here we only stress a single example and outline the main principles involved.

Perhaps the most important discovery in astronomy during the last few decades is the discovery, by *Penzias* and *Wilson*, of the cosmic background radiation. Grand and simple discoveries are not frequent in science. Penzias's and Wilson's discovery was one. And, for the first time, it made cosmology a respectable scientific field based on precise observations. (To stress the great significance of their contribution to physics, Arno A. Penzias and Robert W. Wilson of Bell Laboratories, were awarded the 1978 Nobel Prize in Physics.)

But what gives this discovery its great significance?

Einstein published his theory of gravitation—the general theory of relativity, in 1916. In 1917 he added a cosmological constant to his original field equations for, as he wrote, it led 'to a natural solution of the cosmological problem'.* In 1922, a Russian mathematician, *A Friedmann*,** proved that (with spatial isotropy) *Einstein's original field equations of general relativity do not admit static solutions; the slightest fluctuation would cause any world model based on them to expand or contract!* Einstein was horrified. How can one live in an unstable universe?

Thus, to stop the unwanted expansion-contraction predicted by his 1916 theory, Einstein further justified the addition of the so-called cosmological constant (for it also stops the unwanted expansion-contraction). He was then relaxed, for it was a fashionable idea among scientists and non-scientists alike, that they live in a *static* ('stable') world. That peace of mind lasted only for a few years. During the years 1924–1929, the American astronomer *E. P. Hubble*, using the methods of distance measurements employed in modern astronomy (Lecture I), showed that the redshift in the radiation arriving from distant galaxies is proportional to distance (cf. Lectures I and VI). Since the redshift is also proportional to the relative velocity between the radiation source and the observer, the result means that the more distant the galaxy, the more rapidly it is moving away from the earth, i.e., *that the distances between galaxies, or rather clusters of galaxies* (Lecture I), *are increasing with time. This is the first cosmological arrow of time* (to be discussed later).

Hubble's discovery led to the conclusion that *our universe is in a state of continual expansion*.*** Einstein was horrified again. It led him to regret that he had introduced into his original theory of general relativity the cosmological constant. Indeed, to reject the predictions of his original theory was a mistake, or as he put it, '*THE worst blunder*' of his life. Henceforth he rejected the cosmological constant from his general theory of relativity.

---

* A. Einstein, *The Meaning of Relativity*, pp. 127 and 111–112 (1956), Princeton University Press, Princeton, N. J.

** A. Friedmann, 'Über die Krümmung des Raumes', *Zeitscher. f. Physik, 10*, 377–386 (1922).

*** Note that this does *not* mean that there exists anything like a center of gravity of the total amount of expanding matter and radiation (cf. Lectures III and IV), nor that our galaxy or the solar system expand.

TABLE III

Our Address and Motion in Space *(Including the 1977 "Aether Drift" Experiment, which shows Uniformity Departures in the Cosmic Background Radiation, Revealing the Earth's Motion with Respect to the Universe as a Whole).* See also Fig. VI.1.

features \ structure	OUR CLUSTER OF GALAXIES	OUR GALAXY (THE MILKY WAY)	The SOLAR SYSTEM
Absolute motion with respect to the background radiation — a non-Hubble motion.[63]	~ 600 km/sec	~ 600 km/sec	~ 400 km/sec
Dimensions, distances, and some other relative motions, (cf. Lecture I).	contains ~ 20 galaxies. 1.6–25 Milky Way's diameters away from each other (Table I.5)	~ 10^{11} stars. ~ 100,000 light-years in diameter. ~ 1000 l–ys thick near solar system. — 80 km/sec towards the galaxy Andromeda.	~ 30,000 light-years away from the galactic center. Sun's velocity around galaxy 250–300 km/sec. — 270 km/sec to Andromeda. + 1150 km/sec to the Virgo cluster (recession veloc.)
Recession velocities of some other clusters (+ indicates recession velocity — and is mainly due to Hubble motion — the expanding universe). Other remarks.	possibly rotating. Intercluster space filled with background radiation, neutrinos, etc. (Lecture VI)	250×10^6 years to complete a single circuit. Interstellar space filled with radiation, gas and dust.	+ 5400 km/sec (Perseus) + 19500 km/sec (Leo) + 41000 km/sec (Ursa Major II) + 60600 km/sec (Hydra) [Table I.5]
Distance to the neighboring cluster Virgo; remarks.	~ 55×10^6 light-yrs, increasing with time	Only intercluster space expands (Lecture VI)	No expansion on solar, galactic, and single-cluster scales.

Hubble's discovery has changed the physicist's concept of the universe, *but did not settle its precise evolutionary stages!* Indeed, speculation concerning the origin and evolution of the universe was not quite respectable without an undisputed observational basis on which to decide among various cosmological theories (all of which took the observed expansion into account.).

In 1965 Penzias and Wilson provided such a basis by finding an isotropic microwave radiation that comes not from the sun, not from our galaxy, not from any identifiable individual source, *but from the depths of space!* Their observational results have confirmed that this radiation has a spectral distribution corresponding to that of an electromagnetic radiation in equilibrium with a blackbody at a temperature around 2.7 degrees Kelvin.

A few preliminary remarks will now be made on the physical meaning of the 2.7° cosmic background radiation.

2.5   THE EXPANDING UNIVERSE

A very important feature of the 2.7° cosmic background radiation is its *isotropy*, i.e., the uniformity of the radiation arriving from different directions in space. But even more instructive is the *slight departure from isotropy* that has been discovered in 1977, for it indicates that our galaxy is moving through space with the surprisingly high velocity of 600 km/sec (see below).[63]

Combined with Hubble's discovery, the properties of the cosmic background radiation give the strongest evidence for the expansion of the universe from an initial hot and dense state (which occurred some 20 billion years ago), till the present epoch. *Not only is it the most ancient signal ever observed in astronomy; it is also the most distant, coming from the most remote spaces known.* Consequently, it allows us to calculate *the earliest stages of the evolution of the world*, starting from the first 10^{-23} seconds, or at least from the first 0.0001 seconds, after the beginning of the expansion —which is also *the beginning of physical time*.

But what should be stressed again at this early stage of the course is the fact that Einstein's field equations predict that the *"amount" of space between "objects" (in an isotropic world) would never be fixed.* And indeed, *astronomical observations show that it is expanding*, that the expansion involves only the space between *clusters* of galaxies *(intercluster space)* and that *no cosmological expansion exists on terrestrial, solar, galactic or a single-cluster scale* (Table III). (The reader may consult Fig. VI.1 for a simplified picture of this state of affairs).

These facts lead us immediately to two intriguing questions:

■ *Can local physics be isolated from the processes that take place (Lecture VI) in expanding intercluster space?*

■ *Is the unification of the four basic fields of nature — and, in particular, unification of asymmetries — linked, directly, or indirectly, with the radiative cosmological asymmetries produced by the very phenomenon of expanding intercluster space?* What is the evidence which supports or disproves such a contention?

These questions are central to our understanding of nature and, therefore, will be given special attention in the rest of this course.

The reader, however, should be warned against falling into traditional traps associated with some confusing (popular) concepts, such as *'finite'* or *'infinite'* spaces, *'creation'* or *'end'* of the world, etc. Fortunately, the formalism and the conclusions regarding the foundations of physics *do not depend on questions like 'finiteness' or 'infiniteness' of the universe*. All that is required to keep in mind is the fact that *distances* between neighboring *clusters* of galaxies *are increasing with time*, and that this cosmological expansion of the universe does *not* at all mean that there is *'outer edge'* to the distribution of matter, or that the universe expands into *an hitherto unoccupied space* (§IV.8). Indeed, the *rate* of this expansion slows down with time according to the presence of matter and radiation in the expanding space, and, consequently, if the average mass-energy density in galactic and intercluster spaces is *less than a critical value* (about 10^{-29} gr/cubic centimeter), the expansion would go on forever, and if it is *more than the critical value*, the expansion will, gradually, slow to a stop and then turn into implosion (Figs. III.4 and IV.6). Yet, when we deal with problems of static vs. non-static worlds; of evolving vs. unevolving space-times; of symmetric vs. asymmetric phenomena; of reversible vs. irreversible processes in intercluster space, *we do not need to know precisely the value* of the *rate* of the expansion. All we need establish without doubt is its very existence. *And it is on this basis alone that we make many of our conclusions.* (Incidentally, astronomical observations have, *so far*, showed an average mass-energy density *less* than the critical mass, thereby giving a *tentative* support to the conclusion that we live in an *ever-expanding* world.)

2.6 THE 1977 "AETHER DRIFT" DISCOVERY

Closely related to the aforementioned subject is a highly instructive discovery, made in 1977 aboard a U-2 aircraft equipped with sensitive horn antennas that can detect slight departures from isotropy of the background radiation.

■ The cosmic background radiation reaching the earth today is *a relic* from the period it 'decoupled' from matter some 100,000 to 500,000 years after the beginning of the expansion (Stage 7 in §1.1.3). Previously this radiation *was cooled by the expansion of space* while it was scattered and rescattered by the free electrons of the expanding hot plasma. (Hence the black-body characteristics).

■ Since then it expanded and cooled without scattering (for the previously opaque plasma became 'clear' after its temperature dropped to around 4000° Kelvin, thereby causing electrons to be bound, forming atoms and subsequently galaxies and stars). The high velocity of the expanding shell of this radiation, or more properly the high rate at which space between us and the shell has been increasing, may then be viewed as *the cause* for its cooling down to 2.7°K (or, more properly, for its red-shifting by a factor of 1,500) (Lecture VI).

■ All of space is therefore fully saturated (or 'filled') with this 2.7°K uniform radiations, i.e., *it bathes the earth as it moves through space.* Hence, by measuring slight *differences* in the temperature of this radiation, as the antennas are oriented in different directions into outer space, one may find the direction as well as the absolute motion of the earth with respect to this uniform radiation that fills the whole of expanding and non-expanding spaces outside planets, stars, galaxies and clusters of galaxies.

■ The 1977 U-2 results show that the earth's net motion in space is about 400 km/sec. The direction of the earth's net motion lies in the same plane as its orbit around the sun and at angle of $61°$ tilted upward (northward) from the plane of the milky way (our galaxy). As the sun's gravitational captive, the earth is being swept around the center of the milky way at around 300 km/sec, while it travels around the sun at a speed of about 30 km/sec.

Our galaxy completes one revolution in about 250 million years. Being located about 30,000 to 33,000 light-years away from the galactic center, the earth therefore participates in this rotation (which also involves a large portion of the 10^{11} stars that make up the galaxy—see Lecture I for more details).

Taking all these motions into account (including all inclination angles), the 1977-U-2 discovery shows that the *galactic center* moves at the surprisingly high velocity of about 600 km/sec *with respect to the cosmic background radiation*[63].

■ A word of caution may be due here to avoid confusing this velocity with the *recession velocities of clusters from each other* (the expanding universe). Table I (and Table I.5 in Lecture I) list a sample of these (Hubble) recession velocities. Taking into account all these volocities, the 1977-U-2 experiment shows that our galaxy, as well as our local group of some odd 20 galaxies (which includes Andromeda), *is deviating from the cosmological uniform expansion motion by about 600 km/sec*. In itself a small deviation compared to the large cosmological (Hubble) recession velocities of distant clusters, it shows, nevertheless, *that our local group is traversing and/or rotating with respect to the whole universe at about 600 km/sec*.

What is even more interesting is the observational evidence (Lectures I, VI, VII and VIII), that *the cosmic background radiation is not the only medium through which the earth is hurtling. The whole of outer space is also filled* with *varying* densities of *neutrino radiation* (see *'neutrino telescopes'* in Lecture I), *light, gas, dust, and, possibly, gravitational radiation* (Fig VI.1).

■ The discovery of the background radiation provides also a *most reliable evidence for rejecting some of the more speculative models of the universe* (such as the *'steady-state universe'*, or the *'spinning universes'*). For example, if the universe is spinning, it would show up clearly *as a particular departure from isotropy* in the cosmic background radiation.

2.7 VERIFICATION OF PHYSICAL LAWS BY ASTRONOMY AND ASTROPHYSICS

What physical laws can be tested and verified by astronomical-astrophysical methods? Does outer space provide us with physical laboratories that we can never hope to reproduce on the earth? Indeed, what are the observational, theoretical, and historical arguments which support or disprove such possibilities?

Any elementary textbook on physics, or the history of science, demonstrates that the magnificent development of classical mechanics, as well as its elegant presentations in the eighteenth and nineteenth centuries, had *resulted* from the investigations of celestial objects; and certainly our present comprehension of gravity (through the Newtonian or the Einsteinian theories), owes much to astronomical research. Similarly, during the present century, the discoveries of astronomers, astrophysicists and cosmologists, have resulted in substantial contributions to *the physics of space-time*, to the *laws of radiation*, and even to a deeper penetration into *atomic processes* at extremely hot and dense states.

Indeed, especially in the last decade, the research of stellar interiors has considerably supplemented the work of nuclear physicists and has helped us in our understanding of the *synthesis* of *all* the chemical elements that we find in nature (Lectures I and III). *In fact, the stars and galaxies—whose radiation we analyse—are nothing less than unsurmountable physical laboratories that we can never hope to reproduce here on earth, particularly as far as enormously high temperatures and highly condensed states of matter are concerned*; densities of matter that reach billions and hundreds of billions of metric tons per cubic centimeter; temperatures that may reach 10^{10} degrees Kelvin. Thus, the only laboratories available to us *"to test and verify"* our physical theories under such extreme conditions are exemplified by such astronomical phenomena as *neutron stars, pulsars, supernova, quasars, clusters of galaxies, galactic interiors, intergalactic gas and dust, the background cosmological radiation*, cosmic rays, etc. (Lectures I and III).

All this means that the radiation arriving to earth, constitutes, in itself, a link between cosmology and 'local physics'. Understanding the physical nature of this link, or rather a complexity of interconnected links, is, therefore, of great importance in any attempt to re-examine the foundations of physics in the light of the new discoveries.

One of the main objectives of these lectures is, therefore, to demonstrate the new need to *re-examine* the 'laws of local physics' in light of the recent advances in astronomy, astrophysics and cosmology. It is the need to assimilate new physical observations in an up-dated philosophy of science; a philosophy that takes into account some hitherto undetected physical links amongst the parts and the whole (Lecture IX); a philosophy that does not *ignore*, nor *isolate* important sections of empirical knowledge from each other; a system of thought that is based on the *Einsteinian methodology* as well as on the *Einsteinian physico-philosophical convictions*.

Hence, first and foremost, we must think afresh about the *unified* physical meanings of time, time asymmetries, cosmological arrows of time, acceleration-gravitation-structure, etc. In light of the recent discoveries, these concepts must be assimilated anew. *However, the very concepts of time and gravitation are the most difficult of all human concepts*. Why? The answer to this question is given below and in Lectures IV, VI, VII, and IX. (For these reasons we have also introduced brief reviews of the theories of general relativity, thermodynamics, quantum theory and the physics of time asymmetries as integral parts of the text).

2.8   EINSTEIN'S CLASSIFICATION OF PHYSICAL THEORIES

When we predict the behavior of a specific set of natural phenomena (as in *micro-physics*), we usually mean that we have found a *"constructive theory"* covering this set. When we find that other sets of phenomena are incompatible with that theory, we tend either to generalize or to modify it, or, failing that, to seek an alternative theory. To this *"constructive"* category Einstein opposes the so-called *"theories of principles"* (exemplified by *thermodynamics* and by the theory of *relativity*), whose point of departure and foundation are not *hypothetical constituents*, but *empirically observed* general properties from which mathematical formulae are *deduced* so as to apply to *every case or scale of observation which presents itself!* According to Einstein, the merit of the constructive theories lies in their comprehensiveness, adaptability, and clarity; that of the "theories of principles"—in their logical perfection and universality and in the grand observational scale of their foundations.

The failure of statistical mechanics (both classical and quantum) to deduce and explain the origin of irreversibility, entropy growth, and time asymmetries—as well as its philosophical and applicative limitations and its lack of large-scale universality — are now well known (Lecture V).

In the absence of a quantum-statistical solution, we can find both qualitative and quantitative solutions to many fundamental problems in science and philosophy within the framework of a new school which unifies the foundations of (apparently) separate doctrines in physics and philosophy with those of relativistic cosmology and the theories of time and gravitation. Being based on new evidence in geophysics, astrophysics, astronomy, and microphysics, it deduces empirical laws from new causal-mathematical links which it systematically relates to gravitation, and to gravitation alone.

An abstract covering some of the main assertions of this new school of thought is given below.

3.1   SOME TENTATIVE ASSERTIONS

Before we proceed with the formal lectures, it may be useful to group together, in the form of a non-technical *'preview'*, a few of the assertions to be proved later. Reference to these Assertions will help us maintain a kind of broad perspective as we pass through the lengthy stages of detailed formulation and analysis. Later, in particular in Parts III and IV, these Assertions will become useful in fusing the entire spectrum of diverse subjects into a single consistent theory. It should be stressed, however, that the exposition of the reasonings and grounds of these Assertions must be postponed to Part III (for reasons explained in Part II).

Thus, to avoid premature conclusions, the assertions given below should be considered as *tentative* and *inseparable* from the main text.

Assertion 1: Pluralism and Economy of Physical Laws

■ *Scientists who claim that they operate without reliance on philosophical principles are self-deluded. Scientific theories always advance, stagnate or decline under the domination of a philosophy, whether declared or undeclared. To undertake a scientific approach to the whole of nature one must consciously evolve a unified and universal philosophy of science and not pretend to avoid it.*

■ *Intellect is given us to speculate, to observe, to compare, and to speculate again; this is scientific-philosophical progress; almost all the rest is technocracy or chimera.* In this process we are guided by two principles: one, empirical, according to which the conclusions drawn from a theory must be confirmed *by experience*, the other, semi-logical and semi-aesthetic, according to which the fundamental laws should be *as few and as unified* as possible and *compatible with logic.*

■ *Bold and adventuresome philosophical speculation is at best self-indulgence, a passing culture occuring when philosophers speak a language that is not accessible to nor intended for empirical verification, mathematical exactitude and the unity of science and philosophy.* For synthetic, empirical and speculative methodologies must seek to comprehend nature, by putting it all together in the service of repeatable verification, exactitude and the unity of human thought (see below). In seeking to comprehend nature we must bear in mind, on the one hand, that the greatest snare for creative thought is *uncritical* acceptance of traditional assumptions, and, on the other, that *new philosophies are not necessarily more advanced than old ones.*

■ It is meaningless to ask, What is the world made of? Materialism, dialectical materialism, idealism, positivism and the like *are isolated oversimplifications in a pluralistic—but unified—science.* Actual scientific work requires the *simultaneous* use of several concepts and methods. This methodological pluralism explains why a given scientific theory can be considered "valid" and yet *relative* rather than absolute. Unified science is relative and may vary from time to time; in other words, verifiable, universal-unified knowledge grows with historical time, as is illustrated by our widespread success in applying it in practice. This evolution occurs through the accumulation of unified science, which teaches us *the connections* between *different* descriptions of *one and the same reality.*

T-INVARIANCE

"SPONTANEOUS SYMMETRY BREAKING"; Physicists attack the unwieldy problem of time invariance in elementary particles physics using oldfashioned symmetry-asymmetry concepts. (drawing adapted from Mal's untitled caricature, The New York Times, 1975)

Assertion 2: The Search For Higher Symmetries and Asymmetries

The underlying axiom of all *conservation laws* in physics is that space is stricly *symmetric* under translations, rotations, etc., i,e., if one part of space is the same as every other, why should one place be the origin of a sudden creation of a particle? Or why should a particle start spinning without *external* propulsion? Hence, no object can start linear motion, or rotation, or spinning 'spontaneously'.

This symmetry principle is most powerful, for it leads us to *physical conservation laws*, on one hand (cf., conservation of mass, energy, linear and angular momentum [Lecture II], special and general theories of relativity [Lecture III] and quantum physics [§V.6.2]), and to ever larger symmetric spaces-structures — *up to cosmology*, on the other hand (§V.6.2.1).

But symmetry is not only the basis of conservation laws and 'outer-space' symmetry or asymmetry; it is the most powerful *unifying principle* in physics and philosophy (§V.6.2 and below). Yet, not all symmetries are *exact* nor *universal*. [For instance, the weak interaction 'destroys' time-reversal, parity, and isotopic symmetries (§V.6.2.2 and §IV.10–11).] Hence the fundamental meaning of each '*symmetry breaking*', *or asymmetry*, should always be uncovered; for it not only raises other fundamental problems, like *asymmetries* and certain *couplings among different fields*, but it also shows the way for the discovery of new, *higher* symmetry principles.

The search for higher symmetry levels of reality is marked by the collapse of lower symmetry concepts. It is an unabated war against 'established' physico-philosophical traditions, and habits, that prevent higher levels of unification of the laws of physics.

At stake is not only unification but the very basis of all physical concepts: from the highest levels of conservation laws, to a better understanding of the subatomic world; from a more universal view of the macroworld to everyday concepts of *structure, order and perception*.

But there is more to it. The search for higher symmetry principles is strongly linked to another search; the one that seeks *higher, or more universal asymmetries!* This paralleled search is also marked by the collapse of lower, or more restricted, asymmetry concepts, and by continual battles against obsolete physico-philosophical views ranging from subatomic to macroscopic structures, and from initial-boundary conditions in *mathematics, thermodynamics, and cosmology* to asymmetric concepts of time, space, gravity, structure, order and perception associated with the search for a *MASTER* asymmetry.

Indeed, the concept *SUPERSYMMETRY* (see below) is not dissociated from the search for *MASTER ASYMMETRY*, or *MASTER ARROW OF TIME*. Hence we view supersymmetry concepts as tentative, or at least as strongly coupled to the super*asymmetry* concepts to be discussed in assertion 3.

Assertion 2a: Symmetry-Asymmetry and the Unification of the Laws of Physics

We have already noted that the concepts symmetry and asymmetry are the most powerful tools in physics and the philosophy of science. In contrast, most scientists think that these concepts can be easily defined and understood. Hence, they do not pay much attention to comprehensive studies of symmetry and asymmetry. Well, at least those who search for a greater degree of unification in human understanding of nature should reject this habit. Why?

Part of the answer is based on the fact that these concepts are, next to the concept of time, the most difficult of all human concepts. It is not easy to explain the difficulties involved for, on one hand, the whole of present physics is concerned with them (see below), and, on the other hand, most scientists have been influenced by distorted or traditional concepts concerning symmetry and asymmetry (see Appendix I for a brief historical review).[62, 64]

To begin with we list below a number of physical problems associated with symmetry and asymmetry:

■ Symmetry and the *general principle of covariance* (which is an alternative version of the *principle of equivalence* in the general theory of relativity—as will be explained in Lecture III).

■ Dynamical symmetry, form-invariance conditions, *'gauge symmetry breaking'*, *'spontaneous symmetry breaking', global and local symmetries in quantum theories of gravitation and in supergravity-supersymmetry unified theories* (see below).

■ *Maximally-symmetric spaces*, isotropy, homogeneity, Killing vectors, privileged observers, and the cosmological principle in relativistic cosmology (see below).

■ *CPT invariance* (Charge-Parity-Time symmetry theorem in particle physics, §IV.II), time-reversal invariance, reversibility-irreversibility theorems in thermodynamics, quantum statistical mechanics, etc. (Lectures IV and V).

■ Dynamical symmetry and *conservation laws in physics* (Lecture II), as well as chiral symmetry[61] (which governs the interactions of the pi-meson field).

Even a single glance on this list reveals that the whole of present-day physics is required to explain and define symmetry and asymmetry. But, conversely (as we shall also demonstrate in Lecture IV in respect to the concept of time), symmetry-asymmetry concepts are proper physical guides for unification and for a critical examination of phenomena ranging from the smallest scale of atomic physics to the largest distances of the universe accessible to our observations. It is through symmetry and asymmetry that physics, cosmology, and philosophy become one and the same subject for fundamental studies.

Hence, we must proceed carefully in any preliminary attempt to define these concepts. Thus, the following statements should be viewed as tentative and preliminary.

Our first problem is to understand the mathematical meanings of the concepts *'invariance'* and *'transformations'*. Any reliable textbook, such as Steven Weinberg's book[61], gives detailed explanations of these concepts. What we shall do below is to summarize some of the more general conclusions by a non-technical terminology. (For additional details see Lecture III).

To construct physical theories that are invariant under general coordinate transformation, we must first know *how* the quantities described by the theories behave under these transformations *in space and time*. For that purpose we employ vector and tensor algebra, and, in addition, introduce such concepts as 'invariance'. *Now, any equation remains invariant under general coordinate transformations if it states equality of two tensors (with the same upper and lower indicies).* (However, not everything with indicies is a tensor—see, for instance, the affine connection in Lecture III. A statement that a given tensorial expression *vanishes* under general coordinate transformations is also an invariance statement, *since zero may be viewed as any kind of tensor that we want.*) A zero invariance statement becomes important in *conservation laws* which are expressible in terms of *partial derivatives in space-time*, viz;

$$\frac{\partial \mathbf{T}^{\mu v}}{\partial x^v} = 0 \qquad\qquad\qquad\qquad\qquad\qquad (8)$$

This (symmetric) zero invaniance sums up the conservation laws developed in Lecture II. We can rewrite it another way;

$$\mathbf{\nabla} \cdot \mathbf{T} = 0 \qquad\qquad\qquad\qquad\qquad\qquad (9)$$

This is equation (39) from Lecture III. Such compact equations are simply (symmetric) differential formulations of the *laws of conservation* of mass, energy, and momentum, as generalized from *classical physics* (Lecture II) into the domain of the *theory of relativity* (Lecture III). (Here $T^{\mu v}$, or simply $\mathbf{T}$, is the so-called *energy-momentum tensor* and ∂x^v, or $\mathbf{\nabla} \cdot$, represents the derivatives in space-time — the so called *divergence operator*. It is this energy-momentum tensor which appears on the right-hand side of Einstein's field equations—Lecture III). `

Now, any physical principle which takes the form of an invariance principle under certain mathematical transformations, but whose content is actually limited to a restriction on the interactions of one particular field, is called a dynamic symmetry.[62]

There are a number of dynamic symmetries of importance in physics, notably the *local gauge invariance* (which governs the interactions of the electro-magnetic field and is also important in unified field theories).[55-60]

To proceed, we now need to introduce the concept of *'local symmetry'* as contrasted with *'global symmetry'*. The simplest way to view the difference between them is to visualize two equal spheres whose surfaces are painted with an arbitrarily-spaced set of coordinates.

■ *Global symmetry* is exhibited if the first sphere is rotated about some axis. This rotation is a 'symmetry operation' because the form of the sphere remains invariant (unchanged); it is a global symmetry for the positions of all the points on the surface *are changed by the same angular displacement!*

■ In contrast, *local symmetry* is exhibited if the second sphere is not rotated as a whole, but instead *stretched* while keeping its spherical shape; i.e., in local symmetry the points on the surface of the sphere move *independently* (thereby introducing *'stress forces'* between points).

■ Now, the *special theory of relativity* exhibits *global symmetry*, for it is limited to observers (or coordinate systems) that are moving with *constant* relative velocity with respect to each other (i.e., the possibility of acceleration and gravitation is excluded). Global symmetry in this case is exhibited by the statement that the laws of physics, for these particular observers, take *the same form* (Poincaré invariance*), even if the observers are shifted and rotated with respect to one another (with the same uniform velocity). Thus, in special relativity space-time is *uncurved-unstretched* (i.e., it is *'flat'*), *and the transformation is the same for all points in space-time* (Lecture III).

■ In contrast, *the general theory of relativity* performs a transformation from *global* to *local* symmetry whenever *acceleration or gravitation are present*, i.e., *the very transition from global to local symmetry describes the origin of the gravitational field!*

■ In fact, *both* Einstein's general theory of relativity and Maxwell's theory of electromagnetism are based on *local* symmetries. Hence, the transition from global to local symmetry may also be associated with *the origin of the electromagnetic forces and the properties of photons* (§V.6).

Now back again to terminology.

■ Theories with *local symmetries* may also be called *gauge theories*. Thus, Einstein's theory of gravitation and Maxwell's theory of electromagnetism may also be expressed as gauge theories.

Assertion 2b; Supersymmetry-Supergravity and Unified Gauge Theories

The recent attempts to unify the 'weak' and the electromagnetic forces [7,8,55] are also based on *local symmetries* and may, therefore, be called gauge theories. Their great success has given rise to a new assertion, namely; *any theory unifying all the four forces in nature should also have a local symmetry as a basic requirement. This means that each of the four basic forces in nature (Table I) may arise from the*

---

* Poincaré invariance states that all laws of physics take the same form in any two coordinate systems, even if they are shifted and rotated and moving with respect to one another, provided they have a **constant relative velocity** (Hence Poincaré invariance is the **global** space-time symmetry underlying the special theory of relativity. In fact it preserves the simple quadratic form of the Minikowski metric [§IV.9] and comprises the *Lorentz transformation*).

requirement that the differential equations remain invariant under a local symmetry transformation. A great number of such gauge theories have been proposed in recent years. But they all show that *additional* requirements must be *added* to achieve complete unification (§V.6).

In one methodology *general relativity* is expressed in the language of *quantum field theory*, in another it is formulated in such terms as forbidden interaction symmetries, resummation, etc.[55,64].

An interesting set of theories (which express the general theory of relativity in the language of quantum field theory), derives the gravitational force from the exchange of massless spin-2 particles (called *gravitons*) and incorporates also the concept of *supersymmetry-supergravity.*

■ *Supersymmetry* relates the two broad classes of elementary particles, namely the *fermions* (such as the electron, the proton and the neutron) and the *bosons* (such as the photon—Table I). Any fermion and boson with adjacent spin can then be regarded as alternative manifestations of a single 'superparticle' with an asymmetry in an auxiliary space. In general, the introduction of the requirement of invariance under local symmetry can only be met by introducing '*new fields*' which give rise to '*new forces*'. The 'new fields' may then be called '*guage fields*' that are associated with '*new particles*' whose exchange gives rise to the corresponding 'forces' (§V.6).

■ *Local supersymmetry* becomes possible if two 'new fields' are introduced; they are the field of the spin-2 graviton and a new spin-3/2 field (in which spin-3/2 *gravitinos* are exchanged). Since gravitation appears naturally in the local supersymmetry theory it is also called '*supergravity*' (§V.6).

■ In some supergravity theories the gravitino is massless (and is coupled to other particles only by the feeble force of *microscopic gravitation*); in other theories it may acquire a mass through a special mechanism called '*spontaneous symmetry breaking*' (see below). Here the strong forces arise between the gravitino and other particles, and any gravitino can be transformed into any other gravitino by an internal symmetry operation (see below). Thus, by combining supersymmetry and internal symmetry the entire group of elementary particles may be unified.

■ *Internal symmetry* is another concept employed in the unification process. In the case of electromagnetism it is the local symmetry; in nuclear physics it relates particles that have the same spin *(isotopic symmetry)* (§V.6.2.2).

■ The most important fact, from our point of view, is that *the predictions of the general theory of relativity for long-range interactions in nature remain unchanged in these theories; new effects are predicted only at microscopic scales! Thus, the newly proposed unified theories add no correction to the general theory of relativity at the macroscopic and cosmological levels!* But according to these theories the gravitational force arises entirely from the exchange of *gravitons* (in long-range interactions) and *gravitinos* (in short-range interactions at the quantum level). *It should be stressed, however, that neither the graviton nor the gravitino has been yet observed experimental-*

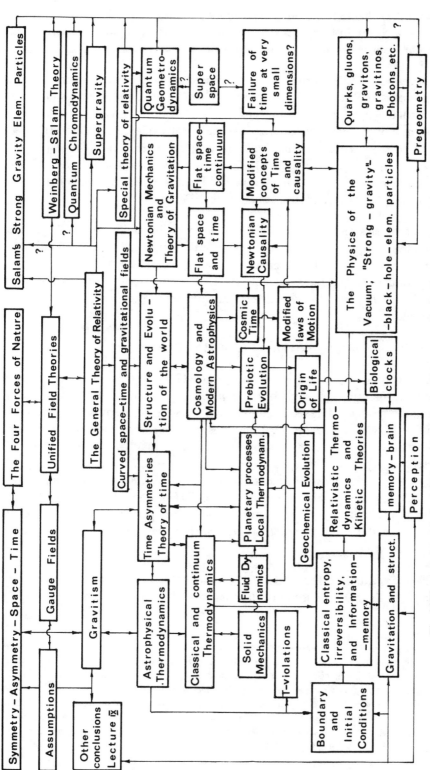

Fig. 1: **The main interrelationships in gravitism.** It should be noted that gravitationally-induced thermo-dynamics (or the so-called astrophysical school of thermodynamics)[1-5], is part and parcel of gravitism. In turn, gravitism is part of a broader view that may be called *Havayism* (Lecture IX).

ly! The same applies to many other elementary particles whose existence is predicted by the new theories. In turn, these theories may raise the question *"Which particles in nature are elementary, if at all?"* (q.v. Lectures IV and IX).

■ But a serious problem arises from the imposition of the controversial *"cosmological constant"* on the equations while transforming from a global to a local symmetry. What is worse, the value of the "cosmological constant" predicted by the theory exceeds the upper limit derived from astronomical evidence (Lecture III). *Even more conspiciuous is the intriguing link between the imposition of an asymmetric "spontaneous symmetry breaking" and the value of this cosmological constant. Does it demonstrate a valid or a false link between the unidirectional dynamics of the large-scale world (Lecture VI), and the "local" asymmetries in physics?*

These problems must ultimately be resolved by further unification of the foundations of physics and by comparison of large-scale with small-scale phenomena. In trying to clarify the present confusion surrounding these concepts we shall examine available evidence concerning possible causal links between *symmetry-asymmetry and gravitation, in particular those associated with cosmological ond other large-scale asymmetries.*

Assertion 2c: Is 'Symmetry Breaking' Legitimate?

■ Can we assume that which we set out to prove? Can we get asymmetry from a symmetric formalism? Are some of the *difficulties* encountered in supersymmetry-supergravity theories, and in old quantum physics (lecture V), derived from an *a priori* imposition of *'symmetry breaking'*; from a postulate of *asymmetry* which is, *in itself,* equivalent to the very results that the mathematical analysis is aimed *'to prove'*? *Is the very concept asymmetry related to fields, to physical laws, to supergravity? Or is it an entirely different problem that must be treated as an initial-condition problem?* And if, indeed, it is a problem linked to *GROUP II* of Table II, should unification begin with local symmetry, with supersymmetry, or, perhaps, *with unification of asymmetries*; with the identification of *'the' global asymmetry, 'the' local asymmetry, or 'the' superasymmetry?*

■ Unification entails other important problems; from 'spontaneous symmetry breaking'*, to the old questions of determinism-indeterminism; from the *testability*

---

* The guage particles of the Yang-Mills field theory acquire mass by this "mechanism" (called also the "Higgs mechanism"). Initially the quanta of the fields are massless. But "spontaneous symmetry breaking" provides a "mechanism" by which some quanta, such as the weak field quanta, can aquire masses. As a consequence new particles called Higgs bosons, are assumed to exist.

The mass of the Higgs boson responsible for "symmetry breaking" in these unified field theories depends on a number of factors, such as the masses of the quarks (q.v. Table IV), and cannot be specified with present knowledge. For a discussion of symmetry-asymmetry and symmetry breaking see Lecture II, §IV.10, §IV.11, §IV.12, §IV.2.4.1, §IX.2.8.1, Lectures VI to VII and Appendix I.

and *validity* of various *quantum-mechanical interpretations***, to some general physico-philosophical problems associated with asymmetries, time and gravitation. These problems are treated in the following lectures and, to some extend, in the following assertions.

Assertion 2d: Elementary Particles As 'Strong-Gravity' Black Holes?

Abdus Salam (who won the Nobel prize for physics in 1979, with *Steven Weinberg*, for his contribution to the advance of unified field theories) has recently proposed (at the 1977 International Symposium on Frontiers of Theoretical Physics) a new theory which is of particular interest to the physico-philosophical foundations of gravitism. In it Salam and his collaborators consider unification of the strong interactions with the rest of physics. These interactions play the leading role in binding together the proton, neutron and a host of other subnuclear particles (see below). The strong force (between the nuclear particles) is *attractive* in nature, *like gravity*, but very much stronger at *short* ranges (excluding massive stars where the gravitational force may become stronger — cf. Lecture I)· Hence, Salam has used the concept '*strong gravity*' (or the *f-gravity*) to describe this force. Using this analogy with gravity, Salam maintains that *the strong force also operates through Einstein-like equations, but with a very much stronger gravitational constant*, i.e.,

$$G_f \sim 10^{40} G, \tag{11}$$

where G_f is the '*strong gravity*' constant and G the gravitational constant of the *Newtonian* or the *Einsteinian* theory of gravitation (Lecture III).

But, most important, this analogy allows Salam to treat 'elementary particles' as 'strong-gravity' *black holes*. (The physical nature of black holes is to be described in Lectures I and VIII). At this point we only stress the fact that their size is given by the so-called *Schwarzschild radius* (Lecture VIII), which, in the Salam theory, *becomes similar to the characteristic sizes associated with observed subnuclear particles!* However, as such, the Salam's strong gravity theory has *not* yet been confirmed *experimentally!* Moreover, other theories, in particular *quantum chromodynamics*, are now emerging as interesting candidates for unification.

Assertion 2e: Quantum Chromodynamics, Gauge Theories and Gravitation

In assertions 2a and 2b we defined theories with *local symmetries* as *gauge theories* and the introduction of the requirement of 'invariance under local symmetry' as the introduction of '*gauge fields*' that are associated with 'particles' (whose 'exchange'

** Quantum theory cannot explain the origin of time asymmetries, irreversibility, and boundary conditions in nature [1-2], nor the dynamics of large cosmological or astrophysical systems[4]. On the other hand the gravitational field may be endowed with properties of the quantum particles (e.g., an intrinsic spin angular momentum). This makes the theory of general relativity, in principle, the best candidate for accommodating all other interactions at any scale. In fact, past attempts to formulate cosmological Hamiltonians have only demonstrated the impossibility of old quantum mechanics becoming the best framework for unification.

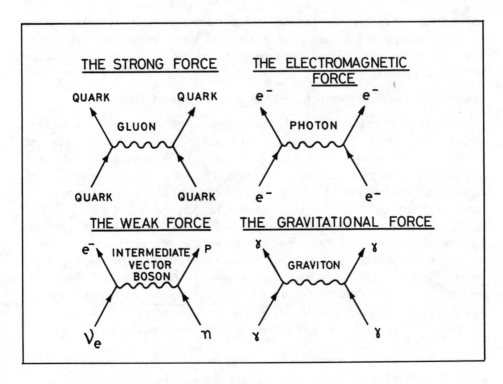

Figure 1a **PICTORIAL DESCRIPTION OF THE FOUR FORCES OF NATURE:**

Supergravity is a *gauge theory* developed in the past few years to describe gravitation in terms of a *unified theory* for the four fundamental forces of nature. Unlike the old quantum theory it uses the concepts of quantum *field* theory, which explains the forces acting on a particle in terms of other, *exchanged particles,* that can be emitted or absorbed (photons, gravitons, gravitinos, gluons and intermediate vector bosons). Thus, the exchanged particles, or quanta, of the electromagnetic force are *photons;* those of the strong interactions between *quarks* (the supposed constituents of subnuclear particles- see below and in Table IV) are mediated by quanta called *gluons.* The weak force (which is responsible, for instance, for the beta decay of radioactive elements—see below) is transmitted by *intermediate vector bosons* (that are somewhat similar to photons but have a large mass). The quantum of long-range gravitation is the *graviton,* a massless particle. The mass of the exchanged particle determines the *range* of the force; only forces transmitted by *massless* particles can have a *long* range. Hence, the electromagnetic and gravitational forces are long-range in nature. Therefore, they dominate our physical surrounding and are the most familiar.

On the other hand, both strong and weak forces are extremely *short* range (they dominate physical reality only over distances $\lesssim 10^{-13}$cm), so their relevance is primarily in *nuclear and high-energy physics.*

The weak force is responsible for phenomena such as beta decay of neutrons into protons and neutrino interactions with matter (compare the diagram shown for the weak force with equation 3—*Stage 5* in the evolution of the universe. Note also that the reverse of this process is possible; it is called the beta decay process. For the nature of the neutrino see Lectures I and II).

As for the *strong force* it should be stressed that while photons do not self-interact (in electromagnetism), the exchanged particles of *quark-quark* strong interactions do, i.e., *gluon-gluon* strong interaction via the exchange of a virtual gluon is possible (see §V.6.2.4 and V.6.2.5).

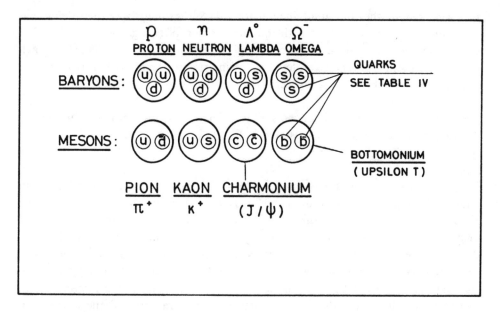

Figure 1b **PICTORIAL DESCRIPTION OF A FEW HADRONS; THE QUARK MODEL IN QUANTUM CHROMODYNAMICS (QCD).**

The term *hadron* refers to any strongly interacting particle. There are hundreds of known hadrons (an infinite number if excitation levels are included). They are divided into two classes; *mesons*, which have an integer spin (0, 1, 2, . . .), such as the pion and the kaon, and *baryons*, which have half-integer spin (1/2, 3/2, . . .). such as the proton, neutron, lambda, and omega. The mesons and baryons can be further grouped into multiplets called the *eightfold way* (which provides hadrons with organisation somewhat similar to that of the Periodic Table for the chemical elements). This regularity led *Gell-Mann* and *Zweig* to the idea that all hadrons are bound states of a few fundamental spin-1/2 constituents called *quarks*.

All quarks carry *fractional* electric charge (as shown in Table IV with other quark properties). So far five distinct species of quarks, or *flavors*, have been required to explain the properties of known hadrons (only 3 were needed for all hadrons discovered before 1974). However, a sixth quark, the Top, may be required (and perhaps even more ones) to explain the properties of some newer hadrons. Hence, *if exist,* the quarks may *not* be 'elementary particles' for they may still have internal constituents (perhaps like those of *Salam's 'strong-gravity' black-hole elementary particles*— Assertion 2c).

Note also that inside hadrons quarks act like free *point-like* 'particles' and that the bar in the pictorial description indicates *antiquarks*. Thus, in the quark model *mesons* are bound states of a quark (q) and antiquark ($\bar{q}$), while *baryons* are constructed from 3 quarks qqq (antibaryons are then constructed from 3 antiquarks $\bar{q}\bar{q}\bar{q}$). Here an antiparticle has the same mass as a particle but exactly opposite quantum numbers. But most important is the fact that experiments have *not* yet revealed an *isolated free quark*. Yet, there is a growing conviction that the strong interactions between quarks may be described by the new quantum field theory called *quantum chromodynamics* (see below, in the main text and, in particular §V.6.2.5).

QCD explains the dynamics of strong interactions which the quark model failed to account with (even when appended with an additional quantum number called *color*—see below). The key new idea of *QCD* is that the additional quantum number called *color symmetry* must be *a local gauge symmetry*. This, in turn, requires the introduction of *8 gauge fields* called *colored gluons*. These gluons are massless spin-1 quanta which mediate the strong interaction between quarks. When the celebrated *Weinberg-Salam unified field theory* (§V.6) is appended to *QCD* the road is paved for the ultimate unification with *supergravity gauge theories* (§V.6). Moreover, quarkless hadrons (glueballs), may be formed out of gluons[67].

may give rise to the four fundamental forces). *Supergravity* was then defined as *local supersymmetry* associated with the introduction of two gauge fields: the field of the spin-2 *graviton* and the field of the spin-3/2 *gravitino*. The exchange of gravitons and gravitinos is then considered as 'giving rise' to the force of gravity in long and short-range interactions, respectively (see also §V.6).

Using similar gauge-field concepts all four forces of nature may be visualized by drawing such diagrams as those shown in Fig. 1a (see also §V.6).

However, the simple quark models shown in the drawing are inadequate. In addition to flavor, quarks must carry *a quantum number called color* (§V.6.2.4). Each quark flavor comes in three distinct colors. Hence, there are three times as many quarks as shown in Table IV. It is perhaps interesting to note that the color degree of freedom was originally required to reconcile problems of *symmetry and asymmetry*. As quarks have spin $\frac{1}{2}$, they should obey Fermi statistics, that is all quantum states should be anti-symmetric (change sign) under the interchange of two identical quarks (cf. the Omega particle in the drawing and the footnote in §V.6.2.4).

The quark model (appended with the color symmetry), has provided a reasonable description of the observed spectrum of hadrons. *However, it failed to account for the dynamics of the strong interaction, and, in addition, has left unanswered some important questions* (e.g., why, contrary to the behavior of all other known fundamental forces, the strong interaction diminishes as one probes nearer the quark? — but see §V.6.2.4).

It was the new theory of QCD which resolved these problems. Based on the same general idea of local guage symmetry as previously described, QCD fits beautifully with the quark model (and with the need for three hidden quantum numbers called color). Thus, *the key new idea of QCD was that the color symmetry must be a local gauge symmetry! This, in turn, requires the introduction of eight gauge fields called colored gluons! These gluons are massless spin-1 quanta which mediate the strong interaction between colored quarks* (§V.6.2.5).

The great success of QCD is that it incorporates all the earlier successes of the quark model, while allowing also further unification with the other forces of nature. It includes all the observed strong interaction symmetries, parity and time reversal invariance (see §IV.10 and 11), and no additional unobserved symmetries. *Moreover, when the celebrated Weinberg-Salam unified gauge field theory is appended to QCD, it does not disturb the strong interaction rules, except by very small calculable values, and, vice versa, the strong interactions do not disturb weak and electromagnetic phenomenology. These recent advances have paved the way for the ultimate unification with supergravity gauge theories* (§V.6).

Assertion 2f: Solitons and Instantons

Another important sub-assertion is based on the success gained recently by a new class of gauge field configurations called *instantons and solitons* (see §III.5). They may provide a better understanding of the nature of 'elementary particles,' and of the complicated *vacuum states* surrounding the hadrons (§IX.2.21). Indeed, the geometro-mathematical considerations of instantons may give the best explanation of such phenomena as *tunnelling effects* (which take place between topologically distinct vacuum states), *violation of chiral symmetry*[55-61], as well as some subtle *coupling effects* between the strong and weak interactions. In fact, the new discovery of gauge field instantons not only elucidates new properties in unified field theories, but also suggests, most persistently, that further surprises are in store.

Assertion 2g: Reduction of Everything to Space-Time-Symmetry-Asymmetry?

Philosophical Conclusions: A First Look

■ Any universal-unified doctrine (in physics or philosophy of science) must somehow include the concepts of symmetry-asymmetry and space-time or their equivalents. But given the limitations mentioned above, we must unequivocally reject traditional physico-philosophical concepts that are in clear contradiction to experimental evidence.

■ Such a doctrine must, therefore, be guided by two additional principles: one is semi-philosophical and semi-empirical, according to which *we shall never have any unified experience or science which we cannot interpret in terms of space-time, symmetry-asymmetry, or their equivalents.*

■ The other is empirical, according to which *all space-time is permeated with the gravitational field* (and with neutrinos, see Lectures I and II). Therefore, no physical system can, *in principle*, be analyzed as *"entirely isolated"* (Lecture III). Consequently, we cannot have any reliable science (or philosophy of science), if we forget that

TABLE IV

Quark species ("flavors") and their properties (Note that the weak interactions can change the flavor of quarks—see also §V.6).

Quark "flavor"	Symbol	Mass ratios*	Charge	Baryon no.	Spin
Up	u	1	$+2/3$	$1/3$	$1/2$
Down	d	2.5	$-1/3$	$1/3$	$1/2$
Strange	s	50	$-1/3$	$1/3$	$1/2$
Charm	c	375	$+2/3$	$1/3$	$1/2$
Bottom	b	1125	$-1/3$	$1/3$	$1/2$
Top (?)	t	3375 (?)	$+2/3$	$1/3$	$1/2$

* With respect to m_U. For antiquarks, mesons and baryons see the explanations under the pictorial description of a few well-known hadrons, and in §V.6.

space-time-matter-energy and the gravitational field are part and parcel of a *single whole* that *cannot* be subdivided into *isolated* systems (§III.5). In other words, *we cannot arrive at complete knowlege of a given object, in itself, for it cannot be isolated from the gravitational field and the rest of nature;* and by "the rest of nature" we also mean historical origins and cosmic times. *Therefore, complete knowledge of a given object is unattainable since it would involve the whole space-time of the world. Nevertheless, the study of the whole of nature should be given the first priority*

To illustrate the meaning of this assertion I shall briefly review some of its well-known consequences (cf. Appendix I).

■ From Anaxagoras, Plato, Aristotle and Descartes we inherited the idea of the separation between "mind" and "matter"; and from *Newton* the false ideas that matter consists in static bits which exist and move *in* absolute space and absolute time; that energy is somehow something which acts upon objects, not something composing them; that gravitation is a "single-potential force" acting in perfectly "flat" *Euclidean* space; that space is the "stage" *upon which* physical processes are displayed to us during the *"flow of time"* and that "absolute", "uniform" time and a *"universal now"* exist.

■ *Einstein* brought about the collapse of these archaic ideas. His theory of relativity has since been tested and verified in endless situations in microphysics and macrophysics, in solar-system experiments, in astrophysics and astronomy and even in cosmology. To date Einstein's general theory of relativity is the most accurate theory; it has been verified in almost every scale, size, distance, mass, density and accuracy (see Lecture III for specific tests and their results). Both the special and general theories of relativity have emerged from each of these tests unscathed—a remarkable tribute to Einstein's intellect. (For specific results concerning modern *quantum mechanics* see Lecture V and also below).

■ Thus, *no longer need we think in terms of an absolute separation between geometry, space-time, matter, energy, momentum and gravitation. In Einstein's theory we have a semi-unified, pluralistic system in which we are even entitled to perceive and analyze the world as composed of matter without matter, of elementary particles without elementary particles, of clocks without clocks, or of gravitation without gravitation* (Lecture III and below). *Indeed, Einstein's geometric theory of gravity treats regions of space and time as governed by the curvature of space-time geometry and the dynamic changes of this curved geometry is what many mean today by that old concept "gravitation".*

■ Much verified information has been accumulated by physicists, astrophysicists and astronomers enabling them to claim, at first with hesitation but now with confirmation (Lectures I and III), that many fundamental concepts, involving classical physics, mathematics, geometry, microphysics and cosmology, *are all coupled together!* Indeed, in the domain of the special and general theories of relativity, *there is no a priori requirement to distinguish, fundamentally, between changes in geometrical properties of space-time and changes in all other physical properties in nature. The whole world, or any part of it, can now be treated as a unified system composed of no-*

thing but dynamically curved space-time or "superspace" (§IV.13) or as nothing but gauge fields, or Salam's 'elementary particles'.

■ The emergence of special relativity, verified *geometrodynamics* (Einstein's general theory of relativity), and relativistic cosmology, modified other anthropomorphic concepts about the structure of the world, our place in the universe, Newtonian causality,* "accelaration," "force,"** "change," "the age of the universe,"*** "becoming," "geocentric thermodynamics,"**** "geocentric philosophy," etc. They became part and parcel of a new structure: *curved space-time geometry.* Embedded in this web we live; in it we "spend our lifetime"; in it the anthropomorphic concept of "the same moment" in two different places (or in two different minds) is without absolute meaning (Lectures III, IV and V). Since nothing can exist or be conceived of as *apart* from *space* and *time* we must treat them as *the ports of departure to our doctrine.*

Assertion 3: Cosmology→Local Physics Principle of Equivalence

Traditionally scientists distinguish among a number of time asymmetries, notably among the thermodynamic, electromagnetic, cosmological, biological, and (more recently) the microscopic arrows of time (Fig. 2). This separation not only breaks up the foundations of physics into disjointed channels, but it has created a serious crisis at the basis of most of our theories. This crisis, as well as the problems described in this volume, manifest themselves most clearly by new attempts to provide answers to certain fundamental questions, a sampling of which follows:

Are the origins of thermodynamic, electromagnetic, and microscopic time asymmetries known? Can they be physically linked with gravitation and cosmology, and deduced from general relativity?

Is the separation among the causes of the aforementioned time asymmetries justified? Or are they no more than different manifestations of one and the same effect? I.e., in what sense might these arrows be physically correlated with each other and with a single "master asymmetry"? What are the observational and theoretical arguments which support or disprove such possibilities?

Can irreversibility (or the 'established' statistical law of increase of entropy, "*H*-theorems," "mixing," "Markov processes," etc.) be deduced from the mathematics of statistical mechanics, or was it fitted into the theory as a fact-like postulate without any rigorous proof? Indeed, how is irreversibility smuggled into the various theories of statistical mechanics? Is it an accident that no one has found a proof of irreversibility by using *H*-theorems, etc., after almost a century of trying? If it turns out (as we show in Lecture V) that statistical mechanics, with all its powerful ap-

---

* See §IV.6 and §IV.7
** See §VII.3
*** See §III.3
**** See §VII.7

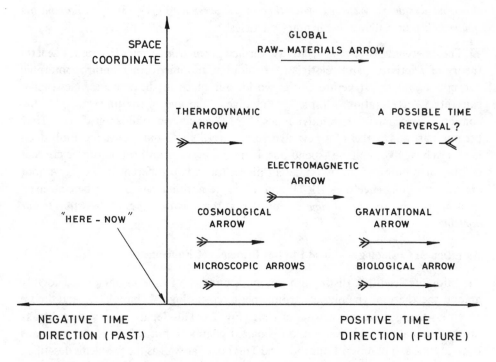

Fig. 2: **Gravitation generates and controls time** *(Lectures IV, VI and VII). Traditionally, scientists distinguish between many (apparently "independent") time asymmetries* (known also as *"arrows of time"*). These include the thermodynamic, electromagnetic, geological, cosmological, gravitational and biological time asymmetries. Many other macroscopic and microscopic "arrows of time" are frequently associated with "disciplinary" ("isolated") studies of nature [e.g., the growth rate of *scientific publications* (Appendix IV); *the socio-historical evolution* (Appendix II); the *"microscopic arrow of time"* (§IV.11 and §VII.8); a (man-made) *"statistical-quantal arrow of time"* (§V.5); *the "geological, geophysical and geochemical arrows of time"* (§I.1 and §IX.2); the *"global raw-materials arrow of time"* (Appendix II) and the *"linguistic"* and *"thought-related"* arrows of time (§IX.2.8.1). *Why are all time asymmetries in evidence mutually consistent in pointing in the so-called positive direction? Is this our own choice, a mere coincidence, or does one of these time asymmetries cause the rest to point in the same direction? If so, which of these is the "master arrow"?* These questions are taken up in Lectures IV, VI, VII and IX.

plications, has totally failed to deduce irreversibility and time asymmetries in nature, what about its claim to fundamentality?

In trying to answer these questions, I have found it useful to demonstrate first the indirect links between cosmology and thermodynamics. These links are to be illustrated by our laboratory-universe principle of equivalence, which is described in Lecture VI.

It is based on two sets of observations, both of which have been thoroughly confirmed by independent methods (Lecture VI). These are:

(i) The universe accessible to our observations contains a very large number of self-generating galactic energy sources, which, on sufficiently large scale, are homogeneously (though randomly) and isotropically distributed (the cosmological principle).

(ii) The universe accessible to our observations is evolving (i.e., the observed isotropic recession in the systems of galaxies and supergalaxies). These observations verify the theoretical predictions of Newtonian theory of gravitation and of Einstein's general relativity, namely, that the universe expands (Lectures VI and III).

These conclusions, and others that follow, largely rely on recent discoveries that include the discovery of cosmic microwave background radiation (which is isotropic to within less than 0.15% and is a relic of earlier, hotter epochs), on the extended Hubble relation, on radio source counts, and on verified transformations of energy in stellar cores by nucleosynthesis (Lectures VI, VII and I).

Next I illustrate our principle by employing a simplified analog within the domain of Newtonian theory. It is, however, only an analog. Imagine a large number of uniformly distributed masses of about equal size, each with its own self-generating energy source, located inside a rigid *(nonexpanding)* laboratory room with reflecting (isolated) walls. It is evident from our observations that after a finite time (depending on the size of the laboratory and on details of the masses), a thermal equilibrium will be established between the sources and the surroundings (air or vacuum), at which stage all temperature gradients will vanish.

By contrast, a laboratory room with flexible walls, continuously maintained in a state of sufficiently fast *expansion,* cannot achieve equilibrium (since the density of radiation near the emitters decreases with time), the result being one-directional, unrelaxed, net energy fluxes from emitters to the unsaturable surroundings. This cause-and-effect link is *indirect* and involves no action at a distance, so that if expansion ceases, both the fluxes and the gradients persist, albeit only for a finite relaxation time during which the system approaches equilibrium (and which increases with room size and with decreasing number of sources). Since the distribution of the sources is *isotropic* (though random), a *static* room of astronomical size would reach equilibrium at all its parts at about the same finite time, provided all sources have the same maximum temperature.

Since energy always flows in opposite directions along a line of sight connecting any two neighboring sources, there exists a point on this line at which the net flux vanishes and each source is enclosed in an imaginary *adiabatic envelope* consisting of an infinity of such points (Fig. VI.1). Each such adiabatic *cell* can now be considered as an expanding room, *equivalent* to the original room but of smaller size, and its evolution away from or toward equilibrium would be also *equivalent* to that of its original counterpart. Extrapolating, accordingly, to any astronomical scale, each source assumes the dimensions of a typical supergalaxy, and a typical abiabatic cell represents the statistically averaged *cell volume per supergalaxy* (Fig. VI.1).

In fact, it is now immaterial whether the original "large laboratory" is bounded (i.e., closed) or infinite (i.e., open), since for both cases the average evolution observed inside a typical cell is equivalent to that of the "larger laboratory" and would suffice for reaching *the same* general conclusion regarding its dynamics namely, that the observed astronomical laboratory has *not* been *static* a long time*. Changing the shapes of the sources, resolving them into billions of smaller units, rotating a part of or the entire galaxy, introducing clouds, galactic dust, local explosions, black holes,** cosmic rays, electromagnetic waves, electromagnetic dissipation, etc.—all these, in themselves, cannot change the aforementioned conclusion, which is also supported by additional and independent observations (Lectures I, VI, VII). For instance, this effect is confirmed by observations inside and near our own galaxy. Furthermore, the rotational motion of newly created neutron stars (Lecture I) is found to be subject to electromagnetic "braking," with mechanical energy dissipated into radiant energy in the corona of the star and eventually lost in space (Fig.I.6).

Observed electromagnetic energy emissions by supernova remnants (Lecture I) also suggest that similar dissipation effects are due to fluid and radiation stresses, magnetic fields, and turbulent viscosities. Even recourse to curved space and distribution of the galaxies according to size and type leave the general result unchanged insofar as the "laboratory" accessible to our observations expands and remains isotropic.

Recent investigations of a possible initial cosmic anisotropic evolution have shown that it must rapidly vanish by dissipation (Lecture VI). Some of these complexities may cause local distortions or even a net flow of energy from a too small cell to another, thereby forcing us to resort to a larger cell until isotropy has been restored.***

Summarizing the deductions from our principle, we arrive at the following additional conclusions:

■ The observed expansion causes space to act as an *unsaturable* thermodynamic sink.

■ The *origin* of irreversibility can be traced all the way back in time and out into *expanding space* and gravitation (see below).

---

* Whether the process consists in "reality" of expansion or contraction is essentially a matter for the definition of time (see below). What is important here is the negative conclusion.

** Black holes, if verified, provide only additional sinks to the already unsaturable sink of expanding interstellar space. Recent observations strongly indicate, however, that black holes may play a role in generating irreversibility by supplying a sink which remains unsaturated even if the expansion reverses into a contraction (see also Lecture VII on the origin of time for a discussion of the effect of contraction on our theory)

*** Gold's star in a (static) box model has only a weak connection with the cosmological arrow and may still lead to symmetric boundary conditions. This restriction is obviated in the present laboratory-cell-universe equivalence, which leads to a supergalaxy in an expanding adiabatic envelope (q.v. Lecture VI).

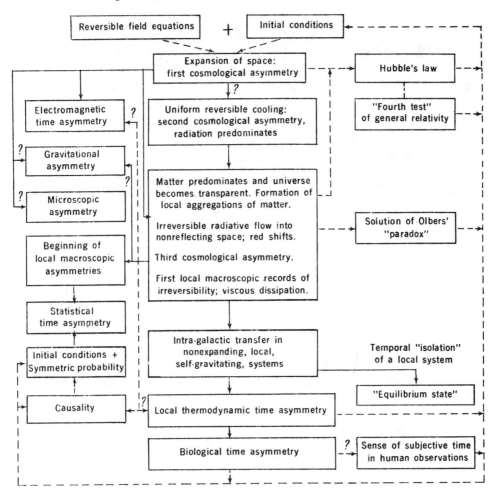

Fig. 3: **Some preliminary links between concepts, showing both theoretical and observational foundations of gravitism** (Lectures IV, VI, VII and IX). [Reproduced, with permission, from the author's article in *Science, 176*, 11 (1972)].

■ The *first cosmological time asymmetry* (Fig. 3) (according to our school, the *master asymmetry*) is that associated with the expansion of the universe proper.

If, for instance, we photograph a group of supergalaxies at instant $t_1 > t_2$, their different mutual separation R_1 and R_2 would tell us whether $t_1 > t_2$, or $t_2 > t_1$. This arrow of time is independent of any cosmological model (including the steady-state model, which also takes it into account). For time-dependent models, we define the decreasing density of radiation and matter as the *second*, and the one-directional energy flow into unsaturable space as the *third cosmological time asymmetry* (Fig. 3).

■ The second and third time asymmetries are (independently) created and dominated by the master asymmetry. To the best of our knowledge, reverse links are physically impossible, i.e., no physical theory based on observations and present information can be devised in which the observed expansion would be caused (or dominated) by the decreasing density of radiation and matter or by the one-directional flow of energy from hot galactic sources to cold space.

■ The very existence of a universal sink (which remains unsaturated as long as space keeps expanding) is the *indirect* cause of the observed, unlimited, one-directional energy losses from the surfaces of all galactic stars and nebulae. Gravitation and unsaturable space become the indirect causes (see below) of energy-density and temperature gradients originating at the cores of galactic emitters (where, provided the temperature is sufficiently high, nuclear reactions generate the energy to be transported to outer regions by radiation, convection, conduction, etc.), and extending through the outer layers of the various planets, stars, and nebulae and throughout the galaxy itself (cf. Fig. VI.1 and Lecture I).

These causal links are independently supported by both macrocosmic observations and predictions of general relativity, as well as by the theories of nuclear reactions, electrodynamics, stellar structures, and hydrodynamic stability. Therefore, if relativity is unacceptable as the universal framework of thermodynamics, one's argument can still be well based on purely nonrelativistic observations and Newtonian gravitation (cf. §VI.5).

Assertion 4: The Dialectical Method and Gravitism

The *"dialectical"* method of debate by *"question-and answer"* may be conceived as the *"science of first principles"*, which differs from other sciences by its universal and cosmological implications (cf., e.g., Platonism). The term may also express the *"dynamic interconnectedness"* of nature, the universality of all *temporal changes* and the *unified* approach to all disciplines of human thought and practice (see also Havayism in §IX.2). Moreover, if the grand aim of all science is to cover the greatest possible number of empirical facts by logical inductions-deductions from the smallest possible number of hypotheses or axioms, then the dialectical method should play a central role in *selecting* the proper inductions and deductions, in the *re-examination* of the axioms and their role in the general scheme of intellectual thought, and in the *study* of the nature of science, especially its methods and its concepts and presuppositions.

The immediate task is two-fold. On the one hand, it involves the re-examination and critical study of all the basic notions of human thought and practice—including notions which are commonly held by professional scientists but require constant re-examination in the light of the most recent advances in scientific and other intellectual "disciplines". On the other hand, it involves similar critical study of certain presuppositions, such as the popular belief in the central role that *"mind"*, *"consciousness"*, and *"uncertainty"* play in our analysis of nature. Presumably such a study should include both the *empirical* and *rational* sciences and also attempt to ascertain the *limits* of *disciplinary* sciences, to disclose their interrelations, and to examine their practical implications in a more unified framework.

Following *Aristotle* one may also distinguish here between *dialectical reasoning*, which proceeds syllogistically from general accepted ideas, and demonstrative reasoning, which begins with primary premises. In addition, Aristotle holds that dialectical reasoning is a *"process of criticism wherein lies the path to the principle of all inquires."* (cf. Appendix VI)*

But most important, the dialectical method, as used here, means that all quantities must be investigated in terms of their origin, causes, evolution and histories; the important examination is not of the present state, but of the "time rate," origin, "direction" and outcome of universal changes. Thus, the necessity of understanding the concepts of time, time rates, time anisotropies, history, evolution, and "prediction" in every field is ontologically grounded in our approach, and is based on, and coupled to empirical observations in every field of science. It rejects a priorism, any theory without practice, and any practice without theory.

The roles of time, time rates, time anisotropy, symmetries, asymmetries, structures, memory, gravity, relativity, causality, causation, irreversibility, determinism, probability, information, "uncertainties," quantum mechanics, "free will", "observation", and "prediction" will, therefore, be stressed in all these lectures.

Assertion 5: The Failure of Quantum Physics to Explain or Unify Asymmetries.

The greatest ambition of physics is to unite the two overarching theories of present science—the general theory of relativity and the quantum theory. Hence, there is an objective need to construct a broad physico-philosophical framework which does not deny this ambition (and may even serve to guide further developments of such a unified approach to nature). Consequently, it is imperative to re-examine, with a critical eye, many traditional, or "well-established" interpretations of the quantum theory, especially those that might prevent, or contradict, such a unification!!

The first group of physico-philosophical concepts which requires re-examination includes the relativistic causality and determinism versus the probabilistic, statistical, indeterministic and "uncertainty" views of "established" quantum — mechanical interpretations. Other physico-philosophical concepts that require re-evaluation (and perhaps modification), include the quantum-mechanical concepts of time asymmetries, irreversibility, "information", "memory", initial conditions, "pre-

---

* But one must be careful and selective in applying *Hegel's dialectic*, which is primarily the distinguishing characteristic of speculative thought; thought which exhibits the structure of its subject-matter *("the universal"*, "system") through the construction of synthetic "categories" *("synthesis")* which deal with the "opposition" between other "conflicting categories" ("theses and antitheses"). The same selectiveness applies to *Kant's dialectic* which deals critically with special difficulties ("antinomies," "paralogisms" and "ideas") arising out of the futile attempt ("transcendental illusion") to apply the "categories of understanding" beyond the realm to which they can apply, namely, the realm of objects "in" space and time ("phenomena") (see Lecture IX).

For many newer problems the more advanced methods of modern logic may be needed. Consequently, we don't *a priori* limit ourselves to any particular definition of the term dialectic. Also, we stress the need to re-examine this concept within the context of *Havayism* (Lecture IX).

diction", "retrodiction", "observation" and "observables", as well as the problematic quantum-mechanical views of "measurement" (Lecture V).

Take for instance the "origin of time asymmetries and irreversibility in nature",
as presented by almost all current quantum-statistical textbooks. *Here most physicists (and non-physicists alike) possess an unshakeable belief in the eventual possibility to explain and deduce this "origin" from the quantum-mechanical formalism.*
This belief stems, perhaps, from the fact that, today, most scientists are so thoroughly
conditioned to the artificial *imposition* of the quantum-statistical postulates, that
they hardly pause to consider their fathom consequences in a broader unified approach to nature. With the present aversion to physico-philosophical inquiry, an
attempt to displace this sacrosanct myth calls for more than a proof; it calls for
the full impact of a body of new empirical data in a wide variety of scientific fields.
*Indeed, "well-established" quantum-mechanical "conclusions" have gradually been
undermined since Einstein first attacked them; others have been exposed as myths or
wishful beliefs for which no experimental support was ever found.* These objections
are discussed in Lecture V. Here we briefly mention only one of them. It is based on
what might be called an axiom. It reads (Lecture V):

**All Fundamental Statistics and Probabilities Are Precisely Symmetric Under
Time Reversal.**

From this axiom (and from a series of propositions and scholia), I arrive at the
following results (Lecture V):

■ Statistical "prediction" is precisely symmetric with statistical "retrodiction".

■ *No* distinction as to the direction of time can be derived from statistics and
probabilities in themselves.

■ Recourse to statistics (or to any "averaging process over detailed molecular
motions", etc.), *cannot,* by itself, result in time asymmetry, "time's arrow", "evolution", irreversible behavior, "memory", "asymmetric structure", "information",
or in unification of asymmetries.

■ The origin of asymmetries, irreversibility, "memory", "information", structure,
and evolution in nature *cannot* be found in any formalism of quantum or classical
statistical mechanics (no matter how "sophisticated" the latter).

■ A rigorous statistical-mathematical derivation of *irreversible* macroscopic equations from *reversible* microscopic dynamics *is impossible.*

■ There is *no justification* for imposing an exclusive "observer" on physico-philosophical interpretations of "initial conditions" in quantum theory, and, at the same
time, to deliberately prevent this "procedure" in the interpretations of other (reversible) equations of the quantum theory.

■ Classification of the laws of nature as *initial (or boundary) conditions, memory,
time asymmetries, "information", and dissipation,* on the one hand, and as *differential*

equations on the other, is *mathematically equivalent* to *integration* (where the two groups are combined—q.v. Table II).

This integration harbors deep-rooted physico-philosophical reasons (§ IX.2.4.1 and § IX.2.8.1 and Table II).

Unlike any other theory the general theory of relativity tells us about *both* symmetric causality and time asymmetries or about *both* symmetric differential equations and irreversibility (q.v. Table 1 in Lecture V and, in particular, Lecture VI).

The total failure of classical and quantum-statistical mechanics to deduce the observed time and structural asymmetries, "memory", "direction", and irreversibilities in nature demonstrates its *lack of universality* as well as its *inability* to explain possible relationships among the thermodynamic, electromagnetic, cosmological, biological, "information", and microscopic irreversibilities (Lectures V, VI and VII). Einstein was right; Bohr, Heisenberg, Von Neuman and others were misled by a philosophy that would stand neither the test of time nor that of the theory of time and gravitation (Lectures IV to IX).

Statistics and probability theories, of themselves, are time symmetric. Hence they cannot yield any time-asymmetric results. Therefore, irreversibility, retarded causality, thermodynamics and evolutionary theories cannot, in principle, arise from any probabilistic consideration. Consequently there is absolutely no meaning in such concepts as "statistical evolution", "statistical irreversibility", "statistical arrow of time", "statistical increase of entropy", etc. (Lecture V).

Assertion 6: Observation, Asymmetries and Gravitation

All laws of nature are abstractions coupled to observations. These laws are divided into *time-symmetric "laws"* and time-asymmetric "boundary conditions" (IX.2.4). Therefore, time-asymmetries cannot, alone, be attributed to the very act of "measurement". They are, nevertheless, linked to the very process of human thinking, which is irreversible (§IX.2.8.1). It is only *by combining* symmetry with asymmetry that we can describe, in agreement with observational evidence, a constantly *evolving* world in which *there are no truly static entities*. It is *gravitation* which lies behind the *"non-static"* nature of the world at large (see below and also Lectures IV to VII and Lecture III on *relativistic* cosmology.)

The world comprises an *"interconnected whole"* in which *no entity can be completely isolated, even in principle.** The *gravitational field* comprises "an interconnecting agent" of the whole. Subject to the relativism and the pluralism of science, and in agreement with all observational evidence to date, including the cosmic background radiation, the extended Hubble's empirical law (§I.7), radio-source counts, transformation of energy in stellar cores by nucleosynthesis (§I.2 and Fig. VII.2), etc., both the Newtonian and the Einsteinian theories of gravitation

---

* See the roles played by *neutrinos* and gravitation as described in Lectures I, II, III and IX.

admit *only* time-evolving (non-static) cosmological solutions (Lectures VI and III). Therefore, the evolution of the world has its origin explained by the Einsteinian or Newtonian theories of gravitation (§2.4–5).

Assertion 7: Gravitation As The Origin of All Asymmetries in Nature

Large (cosmological) systems dominate the evolution and motion of smaller systems, not vice versa (§VI.3 and §VII.2). The dynamics of *"unsaturable"* (intercluster) space** affect all processes on the earth, in the solar system and in all galactic media. *The origin of all observed irreversibilities in nature—of time* (§VII.2), *of all time anisotropies, of energy dissipation, of T-violations in "elementary particles",* (§IV.11), *of retarded potentials in electrodynamics* (§IV.10), *of biological clocks, of biological arrows of time* (§IX-2.3.3) *and of life itself—is one; it is cosmological and non-local, i.e., the expanding universe (for proof and details see Part III)*.

But above all, it is *gravitation* which generates "boundary conditions" on planets, stars and nebulae; which causes proto stars to condense, contract, warm up and initiate *outward*[+] flow of energy, whereby *gravitational* and nuclear energies are converted into radiant energy and lost in "unsaturable" (expanding) intercluster space,[++] creating additional sets of "boundary and initial conditions" in our atmosphere.

A few additional results emerge from this analysis:

A new primary standard of time, defined in terms of "the master asymmetry" (eqs. 14, 16, and 43 in Lecture VI, and eqs 5 and 7 in Lecture VII). This *"master time asymmetry" is independent of any cosmological model-decelerating or accelerating, expanding or contracting, "steady-state" or evolutionary* (§VI.3 and §VII.2, 3).

The "master time asymmetry" may be converted into a "spatial asymmetry." This resolves a number of standing paradoxes in the physics of time, and allows transformation of any time asymmetry into a spatial asymmetry, spatial irreversibility, or "asymmetric structures".

The master time-space asymmetry allows the construction of a few theorems in various domains of science and philosophy (see "Other Assertions" below and Assertion 9).

The new asymmetry allows the development of *"entropy-free"* thermodynamic and evolution theories (§VI.8—see also below).

It follows that any irreversible phenomenon is an asymmetric entity, which is intimately connected with gravitation and, indirectly, with the expansion of the universe. Thus, every change in the distribution of energy and matter creates an effect which is irreversible as far as the universe is non-static. Therefore, gravitation

---

** These concepts are defined in §VI.3 and are illustrated in Fig. VI.1.

[+] But see the discussion of black holes in Lectures VII and I.

[++] q.v. §VI.3 and Fig. VI.1.

and the expansion of the universe are indirectly connected with all irreversible processes down to the smallest scale. Furthermore, *one should also expect any fundamental theory to incorporate gravitation and the (observed) expansion of the universe into its fundamental principles. One is then led to the general conclusion that microphysics, macrophysics, thermodynamics, cosmology and the general theory of relativity are intimately related to each other.* But the *origin of all irreversibility in nature depends ultimately on gravitation alone.*

It also follows that all macroscopic changes in the world are asymmetric in time (i.e., irreversible), and occur in accordance with certain over-all regularities, or laws, that, in turn, are governed by a single universal master asymmetry associated with gravitation and deduced from the *"Irreversibility-Gravitation Principle of Equivalence"* (Assrt.3). Microscopic irreversibility in elementary-particle physics, if valid, is also affected by this master asymmetry (§V.II, §V.6.2.2 and §VII.8).

Assertion 8: Gravitation, Structure and Instabilities

The gravitational field on earth is the *weakest* of the four known fields-inter-actions-forces.* *Hence, many physicists consider it unimportant.* However, in the interiors of many stars and galactic cores the gravitational force becomes stronger than the other known forces in nature; i.e., it may get stronger than the strong force which binds subatomic particles together (causing the entire structure of atomic nuclei to 'break down' and collapse on itself—cf. gravitational collapse in Lecture I).

Physicists that *a priori* reject the importance of gravitation (and of the general theory of relativity) forget a few additional facts:

■ The central role of gravitation in building all terrestrial structures as described in Lecture I and in Assertion 9 below.

■ The incorporation of Newtonian gravitation into the general theory of relativity (Lecture III).

■ The possibilities of constructing unified field theory based on the Einsteinian ambition (Assertion2).

---

* *Gravitation* influences all particles and its range is unlimited. At the subatomic scale the *"weak"* force is many orders of magnitude stronger than gravitation and it also affects all kinds of matter-energy. The *electromagnetic* force, on the other hand, acts only on particles that have an electric charge (e.g., electrons, protons, muons and quarks). But electromagnetic forces bind atoms together and are responsible for almost all the gross properties of matter, including chemical properties. At these scales the *"stronger"* force is more than 100 times stronger than the electromagnetic (Table I).

The "strong" force does not act on *leptons* (these include only four known particles: the *electron*, the *muon* and two kinds of *neutrino* q.v. §II.12 and §I.8. There are also four *antileptons*, e.g., the antielectron, or positron). Only quarks and *hadrons* are affected by the "strong" force (hadrons, such as *protons* and *neutrons*, are comlex objects *assumed* to be made up of *quarks*. More than 200 kinds of hadrons have been identified so far). Quarks can interact with leptons through the "weak" and electromagnetic forces, but with each other they interact almost exclusively through the "strong" force.

■ The central role of gravitation in other fields of study, like thermodynamics, the physics of time, the theory of evolution, etc.

■ The potential of these to revolutionize many of our concepts in science and philosophy as demonstrated by a specific example below:

Contrary to current textbooks on thermodynamics, statistical physics, biophysics, philosophy, etc., the observed world does not proceed from lower to higher "degrees of disorder", since when all graviatationally-induced phenomena are taken into account the emerging result indicates a net decrease in the "degrees of disorder", a greater "degree of structuring", and more aggregation, form, directionality, recording, memory and asymmetry (Assertion 9 and Lecture IX).

It is only for a *subjectivistic mind,* and only from a narrow-minded, *anthropomorphic viewpoint,* that the so-called entropic processes appear to *"increase disorder"* and *"destroy the structure and order of the world".* But once the various effects, and "after effects", of gravitation are taken into consideration, the entire world picture changes and becomes *objective;* then, and only then, one realizes that, as a matter of observational fact, *the net result of the dissipation of all energies in unsaturable (Lecture VI) space is increasing aggregation, structure, differentiation, stratification, asymmetry, boundary conditions, and evolution* (q.v. erosion vs. the various gravitationally-induced galactic, stellar, planetary, geochemical, prebiotic and biotic evolutions, as well as nucleosynthesis, biogeochemical synthesis, sedimentary and volcanic structuring, stratification cooling, etc. mentioned below).

One of the questions that we therefore raise is *the identification of some hitherto unexamined conditions (and causes) for the building of (physical, geological, biological and social) structures through gravitationally-induced asymmetries.* This issue carries some new implications for almost any field of human thought (Lecture IX).

Assertion 8a: The Thermodynamics of Structure and Life

Classical equilibrium thermodynamics is essentially a theory of the *"destruction of structure".* Physicists often consider the entropy production (§IV.5) as a measure of the 'rate' of this destruction. But, in some way, such a theory has to be completed by a theory of *'creation of gravitationally-induced structures'.* All evidence indicates that gravitation is indeed the source of order, both in time and space. Hence, such considerations may ultimately contribute to narrow the gap which still exists today between biology and theoretical physics. *Do biological structures originate in some gravitationally-induced processes? This is quite a challenging problem which still requires a considerable amount of thought and study.* At least, what seems henceforth certain, is that the basic biological structures have evolved under the influence of the gravity vector (see Assertion 9 below), while all the chemical elements had actually been constructed by gravitationally-induced processes (Lecture I).

A new structure is always the result of an instability! The main point is therefore: *'What is the origin of instabilities in nature?'*

In trying to answer this fundamental question we may first examine the formation of new organizations and new structures through interaction with the outside world. This interaction may be expressed by such parameters as the *Rayleigh Number*, which includes the quantity *g—the acceleration due to gravity*. Three examples may be cited: the *Zhabotinsky reaction*, which produces *horizontal* space structures*; the *Leisegang ring patterns***; and the *Benárd problem* (involving thermal instability in *horizontal* layers of a fluid heated from *'below'*; for some critical value of the Rayleigh number, the state of the fluid at rest becomes unstable and *cellular* convection sets in).

These examples are of particular interest for two main reasons:

■ Biological systems show a high degree of structure, organization and order. Structures, oscillations and rhythms somewhat similar to the ones described above characterize living systems at the molecular, cellular and supracellular levels of organization.

■ The structures described above have been formed under the influence of the gravity vector, which, as we have stressed above, plays the central role in the generation and the orientation of all structures (Assert. 9).

One of our main objects is, therefore, to relate instability, structure and asymmetry to gravitation in order to obtain a better understanding of biotic evolution (Lecture IX). These are fascinating problems, and we are only at the very beginning.

Assertion 9: Gravitation, Gravimorphism, and Evolution

We now give a few well-known examples that show the cardinal role gravitation plays in establishing *structures, asymmetry, evolution, 'order', stratification and aggregation.*

Gravitation causes the universe to evolve into myriads of subsystems that acquire diverse structures and evolutionary patterns. It causes every proto-star, or star to act as a self-powered thermodynamic system in the sense that its entire evolutionary process is controlled by the gravitational energy that heats up the star *till nucleosynthesis starts.* Nucleosynthesis under gravitational attraction *generates all atomic elements heavier than helium* (Lectures I and VII).
Therefore the very generation of all the "building blocks" of the universe is due to

---

* During this reaction dissipative space structures, including spiral or flat wave patterns develop, starting from an initially homogeneous solution in which bromination of malonic acid occurs.

** In 1898 R. E. Leisegang analysed a *periodic precipitation* which occurs when a concentrated salt solution, such as lead nitrate, diffuses into an aqueous medium (often with a lyophilic gel such as agar-agar) containing, for example, potassium iodine. The resulting precipitate, in this case lead iodide, forms discontinuously in bands parallel to the diffusion front, which are now known as Leisegang rings.

gravitation. (Even the formation of helium is now fully explained within the framework of the theory of gravitation-Lecture IX).

The gravitational force *structures* all the newly formed elements in *concentric layers* (Fig.I.6a)-the heaviest in the star's core and the lightest (hydrogen) in its periphery. It also determines the evolution of any astrophysical system, including *supernova explosions* and the formation of *planets* and their *atmospheres* (Lecture I).

The heat balance of the earth, and all *atmospheric, hydrologic and geologic phenomena, are dominated by gravitation;-* without it seas, land, lakes, rivers, updrafts, wind, volcanoes, erosion, sedimentation, stratification and temperature gradients in planetary atmospheres are impossible (Lecture I).

Assertion 9a: Natural vs. Gravitational Selection

Many experimental data obtained from laboratory investigations have contributed to put on a firm scientific basis the assumption that the *genetic code and life itself* origanted from *nonliving* systems: in fact they also represent the strongest arguments for a *general theory of evolution based on gravitationally-induced asymmetries (Lecture IX).* At this stage we only stress the fact that the *origin* of biotic evolutionary processes can be traced back to the earliest creation of the presently observed *neutron-proton ratio*, to the stage of helium formation (§1.1.3), to the onset of the "*matter-dominated era*" (§1.1.4), to *nucleosynthesis*, to *chemical evolution*, and, then to the complicated stages of terrestrial *prebiotic evolution* (§IX.2.3).

Most evolutionary stages took place inside *non-expanding* galactic bodies such as advanced massive stars and supernova, and later, on our planet. It was always *gravitation* that played the central role in the *"selection"* of the structures to be produced: First in forcing the evolution through a succession of specific nuclear effects *in which evolution means the development of complex elements from simple ones,* i.e., of oxygen, carbon, silicon, etc. out of hydrogen and helium. Here gravitation was the *only suitable force* that could build all the elements in the periodic table by successively *adding* small increments of mass and electric charge *in selected combinations that are controlled by the force of gravity* (the mass of the star determines its rate of evolution, i.e., the attainment of specific temperatures of nucleosynthesis at given evolutionary times (Lectures I and VII).

Another criterion that may be brought into the theory of evolution and asymmetry is *"gravitationally-induced chemical change"*.

Following gravitationally-induced cosmological expansion and supernova explosions the complex nuclei cool and combine to form complicated molecules. The first local "aggregates" of matter in the solar planets already contained atoms and complicated molecules, including some based on carbon. It is gravitation which stratifies these compounds in horizontal layers of homogeneous materials according to their *specific gravity!* The geological strata of layers of rocks composed of one material, e.g., shale or limestone, lying between rock beds of other materials were thus obtained by the *"selective force of gravity* (Lecture I). But, most important, the *gravitationally-induced life envelope* of earth, comprising a very thin layer (surface and lower atmosphere) was created by the "selective force of gravity". Here we may *replace* the term "*natural selection*" by the prebiotic concept "*gravitational selection*" (Lecture I and IX).

Indeed, it is "gravitational selection" that puts carbon, nitrogen, oxygen, water vapor, phosphor and lightnings at one and the same layer for the beginning of biological evolution. And it is "gravitational selection" that protected the early products of this evolution from damaging radiation by supplying the upper protective layer of ozone, etc.

Assertion 9b: Gravitation and Life

All plants and animals evolved under the asymmetric influence of gravity. Their *form* as well as their *structural development* have been strongly influenced by this asymmetric vector. In turn, they have *"learned"* to exploit it and even to cope with it-learned in the evolutionary as well as in the ontogenetic sense of the development of the individual organism.

The sensing devices which plants and animals use for *"gravity perception"* (*"gravity receptors"*, *"g-perception"*, *"bio-accelerometers"*, *"gravity-induced biological clocks"*, etc.) are not yet well understood, even though a voluminous literature has been published on this subject*. But what we already know justifies the central role we expect gravity to play in all life processes. For instance, we know that if a growing higher plant is *displaced* with respect to the *"upright"* position, some tens of minutes later it will alter its growth in such a way as *to restore its original orientation in coincidence with the gravitational vector.* (If it is displaced only briefly and then restored to its original orientation well before the growth response can set in, it still responds to that displacement.)

Indirect, gravitationally-induced orientation of an organism occur when, for example, an organism orients itself by a gradient of *density differences* or *hydrostatic pressure*. Nevertheless, very small organisms (including all bacteria) may have *no means of sensing gravity! But if very small organisms have no detectable gravity receptors it does not mean that they are not affected by gravity!* Throughout all evolution there have been many functional interactions with gravity—some of which may be detected or recognized only by indirect and subtle reasoning.

The connection between gravitation and the *origin of time asymmetries and life*, on the one hand, and between gravitation and *form, structure, growth rate, growth direction, behavior, navigation,* and *space perception*, on the other, must therefore be studied. Indeed, a considerable empirical evidence has been accumulated so far on these subjects by biophysicists, biologists, plant physiologists, botanists, zoologists, neurophysiologists, etc*

---

* For a comprehensive review see *Gravity and the Organism* (Ref. No. 9). Further references to this book will be denoted by *G&O*. *Note Added in Proof*: The predictions made here & in Lectures I, VII and IX, have now been partially verified by Prof. H. Eyal–Gilladi of the Hebrew Univ. ("Ha-Aretz", Nov. 27, 1980.

This evidence now covers the following main topics:

■ Growth responses of the organism to gravity: *"geomorphism"*, *"gravimorphism"*, *"gravimorphogenesis"*, *"geomorphotic phenomena"*, *"geoepinasty"*, *"geotrophism"*, etc.[+]

■ *Weightlessness studies* (on earth and in biosatellites).

■ *"Space" orientation and gravity* (*"graviperception"*, etc.)

■ *Central nervous system responses to gravity and acceleration.*

■ *Structure and operation of gravity receptors, g-perception and bioaccelerometers.*[5]

■ *Gravity-induced biological phenomena* (including various "biological clocks").

■ The *evolution* of gravity receptors and gravity sensing mechanisms.

■ *Evolution of balancing, flying, and navigating capabilities.*

■ The effects of gravitationally-induced processes on *human perception* of space, time, asymmetries, structures and aggregations.

Thus, gravitationally-induced *asymmetry* (e.g., higher concentration of statolith-like particles in the lower halves of the cells or their morphological distortion) *has been correlated with asymmetric growth trends in higher plants. In other words, gravitational asymmetry has been correlated with "asymmetric gravity perception" in the organism and with asymmetric changes in its ultrastructure.*

Assertion 9c: Gravitation and Orientation

Above ground plant organs orient themselves primarily by their sensitivity toward gravity. (Phototropism, at least at high light levels, plays a less important role.[++]) *This includes the position of branches and leaves as well as their orientation in space. Flower (or wheat) morphology is also influenced by the direction of gravity; apical dominance depends on the position of a shoot in relation to gravity. Thus the form (habit) of a plant depends on the orientation of its parts in relation to gravity.*[+++]

[+] The evidence that special dense bodies falling inside living cells (including starch grains) serve as gravity receptors is overwhelming[**]. In general the physical action of gravity on higher plants consists in an interaction between the mass of the earth and the masses of some constituents of the plant—a plant cell or particles of the cell. That this is so has been confirmed by various experiments, including the substitution of centrifugal forces for gravity in producing "geotropic" responses.[***] The response of the organism may be initiated by changes in the distribution of pressure on sensitive loci, exerted either by the entire cell content or by particles (statoliths) heavier or lighter than the surrounding cytoplasm.

The response may also be caused by movements or reorientation of such particles. Acceleration (by gravity or centrifuge) causes stresses to be set up in the cell membranes or in the organ as a whole, which may lead to its physical distortion. In principle, *any of the following effects can act as the primary gravity detector:* stress on plasma membranes, displacement of cell parts or of portions of multicellular structures, and intercellular stratification.

[++] Jankiewics, L. S., p. 324, *G&O*.

[+++] According to Leopold, (p. 331, *G&O*), gravitationally-induced growth changes might be related to the production of ethylene by physical stress imposed on the tissue.

Structural changes result from prolonged stress, pressure or acceleration. There-fore, the form and structure of the organism are conditioned by gravity and gravity-induced processes. [+++]

Biological detectors of *sound modulation* have much to do with *sensors for earth gravity; both are special kinds of bioaccelerometers.* [++++] In other words, *gravitation, bioaccelerometers and* the various ways organisms are affected by *vibration* are interrelated![*]

Gravity-related biological phenomena must therefore be studied on several levels of organization-from those of the molecule, the organelle, the cell, and the individual organism to those of the population, the ecosystem, the biosphere, the solar system, etc.

Indeed gravity structures ecosystems and global phenomena ranging from mountain crests, tectonic folds, valleys, oceans and lagoons to springs, wells, swamps, glaciers and rivers, or the ecological systems connected with them. Eco-logical systems associated with human life are also affected, indirectly, by gravity, e.g., transportation systems (railroads, bridges, harbors, airports, etc.), agrarian communities (farm fields, etc.), irrigation systems, etc.

Life is therefore strongly tied up with gravitation. Organelles and nuclei of the living cells are "heavier" than the rest of the cell. Such asymmetry allows plants to grow "vertically upward". Thus, gravitational asymmetry controls "the direction of growth" in roots, leaves, branches, etc. It also affects the movement of all animals (in the sense of giving a reference for orientation)[**]

Indeed, recent experimental evidence collected in spacecraft proves that there is a *"loss of orientability"* in growing *seeds* under *weightlessness!* Animals are no exception. All human beings are *conscious* of their "up-down" surroundings and about their "weight", speed and acceleration. When conscious, we are aware of the direction of this force and, accordingly "orient" ourselves: "up", "down" and "hori-zontally"; "rise", "recline", "lean", "leap", "sit", "load", "lift", "walk", "float", "bal-

---

[+++] Brown, A. H., p. 6 in *G&O.*

[++++] *Ibid.,* p. 2.

[*] c.f., for instance, inertial changes of the fluid in the vertebrate inner ear.

[**] For instance, in a crystallizing suspension of organic spheres in water the earth's gravitational field produces an *elastic deformation that is readily observed through its asymmetric effect on the evolving crystal structure. These elastic forces play an important part in crystallized virus systems. Moreover, the gravitational force provides a useful biological function by excluding foreign particles such as antibodies from a crystal: hence gravity is a most powerful factor in any biological evolution.* [See for instance R. S. Crandall and R. Williams, *Science, 198,* 293 (1977); S. Hachisu, A. Kose, Y. Kobayashi, K. Takano, *J. Colloid Interface Sci. 55,* 1199 (1976); A. Klug, R.E. Franklin, S.P.F. Humphreys-Owen, *Biochem Biophys, Acta* ɪv, 203 (1959). For the connection between gravity and the crystal structure see L. D. Landau and E.M. Lifshitz *Theory of Elasticity* (pergamon, New York, 1959), Chap. 1].

ance", "overlay", etc. All these are normal experiences *directly* connected with gravity, motion and acceleration in the gravitational field***

But there is more to it. The very evolution of our bones, legs, etc. has been "in response" to the gravity vector. Nevertheless the cardinal role of gravitational control of time, perception and evolution has been largely overlooked by physicists, biologists and philosophers, partly because the role of the graviational field as the universal generator of asymmetries, form, time, motion, structures, recording, memory, clocks, complexity and organization has not yet been well understood.

Assertion 9d: Gravitation and Sociobiology

As another example we refer to the emerging conclusions of modern sociobiology. Here many aspects of animals social organization and evolution can be predicted on the basis of an understanding of a limited set of (gravitationally-induced) environmental variables. Since biogeochemical evolution causes key natural resources to be distributed non-uniformly in the spatial and temporal coordinates of the biosphere (§IX.2.3.1),resource monopolization develops* in all levels of sociobiological systems.**

*** "Motion sickness" and other phenomena associated with *loss of orientability vis-a-vis gravity* may also be cited here (see for instance M. Treisman *Science 197*, 493 (1977) and §IX.2.5). *Note also that acceleration and gravity are fundamentally equivalent* (§III.1.1a).

* For instance, monogamy, polygyny, polygamy, polyandry, etc. See J. Verner *Evolution, 18,* 252 (1964); G. A. Bartholomew *Ibid. 24,* 546 (1970) and R. D. Alexander in *Insects, Science and Society,* D. Pimental, Ed. Academic Press, N. Y. 1975 p. 35.

** Students find it instructive to name *things, motives, processes* and *concepts induced by gravitation;* for instance, in trying *to continue the following series:* Sky, high, low, heavy, light, heavenliness, heavenward, lightness, buoyancy, floating, swimming, flying, sinking, landing, falling, climbing, jumping, throwing, skiing, sitting, walking, resting, etc.; or bed, table, chair, stairs, roof, walls, floor, dome, arch, shelves, tides, rivers, waterfalls, birds, seeds, roots, trees, bridges, ships, mountains, lakes, roads, canals, etc.; or paleontology, sediments, biogeochemistry, clocks, fractional crystallization, geophysical periodicities, year, seasons, day, time, climate, fossil fuels, solar energy, solar system, etc.; or curved spacetime, geodesics, photon trajectories, principle of equivalence, gravitational collapse, neutron stars, X-ray sources, supernova, black holes, active galaxies, stability, aggregation, asymmetry, irreversibility, boundary conditions, memory, unified field theories; or evolution, origin of life, socio-environmental interactions, point of view, etc.

Gravitational and "anti-gravitational" leitmotifs (see below) may even influence *aesthetics* (cf., e.g., certain trends in *landscape gardening* and *landscape architecture* vs. *city and village structure,* or certain trends in *dadaism* and *surrealism* vs. *realism* and *naturalism*). Many *theological* leitmotifs are also associated with gravitation (e.g., heaven, heavenly, heavenliness, upper regions vs. lowest grounds, high altar, High Church, High Churchman, High Mass, highness, high priest, fallen angels, etc.).

The same applies to various *socio-economical concepts* (e.g., highborn, high commissioner, high-hat, high-minded, high-powered, highrise, high table, lowborn, lowbred, lowbrow, Low Churchman, lowdown, lower class, lower deck, lower house, low life, lowly, lowliness, far down, to fall to the ground, decline, rise, standup, break down, upright, withstand, tear down, downgrade, precipitate, settle, settle down, settle up, settle upon, showdown, uphold, bring down, upbringing, upgrade, upheaval, ascend, upkeep, uplift, upper class, upper court, uppity, uproot, uprush, upstanding, upstart, upswing, uptight, sink, go to the bottom, bottomless, atop, deep, earthy, earthly, earthiness, spiritually, lofty, elevated, towering, sublime, etc.

Thus *territoriality* develops in a given spatial region or in a given assembly of individuals *when the key resources are sufficiently abundant and stable through long periods of evolutionary or historical times.* For instance, one species (or one sex) may control a larger quantity of resources (say, scarce water ponds) than the other individuals until a small percentage of the population monopolizes all the key resources. Thus the limited space must be divided into *"lebensraums"* and *"well-defined territories"*, a constrain which may induce *"natural selection"* in some sociobiological systems; from animals' mating systems to animals' monopolies on water resources; from natural food administration to the control of key strategic ridges; from controlling fossil fuel resources to international disputes over territories. Indeed much of known sociobiological evolution and perhaps even much of human history may be explained along these lines. (For other 'external influences' see §IX.2.8.1 as well as Spencer's philosophy—Appendix II).

Natural Selection is therefore the result of continuous interaction with (gravitationally-induced) *environmental-ecological systems* which, in turn, means that all living beings are what thermodynamicists call *open systems*[+] (i.e. they persist only by a constant influx-outflux of energy, matter and "information"). Thus the fluctuations in the spatial distribution of important resources are key determinants of sociobiological selection and evolution.

Possible links between *human evolution* and *gravitationally-induced changes in the enviornment* (§VII.4) may also be illuminated by the empirical results that are emerging from recent studies in *geology* and *anthropology*. Many anthropologists are now searching for *fossil traces* of the human line among hominoids that lived *as long as 14 million years ago.* The new clues indicate that *changes in environment* and, as a result, in *feeding behavior* are most likely to give rise to the evolution of new species. [For instance as a (gravitionally-induced) *volcano* erupts a boundary in a *uniform* forest emerges, and a savanna develops at the base of the mountain on top of the lava. *As forest gives way to mixed environment, changes in feeding behavior develop, and with them changes in teeth, jaws, etc. Natural selection is then enhanced by the fact that maximum opportunities for survival are provided to species that live on the boundary between two environments and exploit both*].

There are other, indirect effects of gravitationally-induced processes. For example the origin of certain trends in the evolution of mankind can be traced back, at least in part, to *gravitationally-induced plate tectonics* (the movement of the outer crusted plates that float on the earth's semimolten mantle—cf. §VII.4).

Some 45 million to 50 million years ago, the plate that carries the Indian subcontinent was pushed up, gradually forming the massive mountain range of the *Himalayas*. This new barrier to global wind circulation helped change weather

---

[+] All natural systems (except some man-made pseudo-isolated systems) are open systems which do not approach equilibrium (Lectures VI and VII).

patterns, *altering average temperatures around the world. By about 14 million years ago, climates that had been tropical had turned largely temperate, jungles had thinned out, and fruits and nuts normally available year around began to appear only seasonally.*

The changing conditions for feeding, outside the forest in the resulting savanna, or grasslands, attracted forest-dwelling apes in search of food such as roots, seeds and finally the meat of other animals. Anthropologists theorize that once out of the forest these apes began to evolve rapidly. The new prevailing conditions of the grasslands favored the survival of those apes who could *stand up;* for an *erect position* enabled them *to see over* the tall grass to spot and hunt their prey—and to see and escape the animals that preyed on them. Thus they were able to survive longer and produce more offspring who shared their characteristics. After many generations they had evolved into the *upright-standing Homo erectus.* Intelligence was henceforth aided by other interactions with the environment and, especially, when the fore-limbs were freed from the tasks of walking. Thus, gradually, experience with gravitationally-induced factors grew in these creatures into asymmetric thinking—§IX.2.8.

Assertion 10: Methodology, Macro-Studies and Valid Philosophies

Many modern thinkers are losing their faith in scientific or material "progress", seeing the fate of mankind as a nightmare of confusing ideas, anarchy, wars, pollution, overpopulation, or the tyrany of uncertainty. Growing cynicism and the decline of old norms cast serious doubts on whatever has remained of Western philosophy. Some thinkers have, therefore, cultivated their despair to the degree of seeing all human affairs, including technology and science, as absolutely out of control, as patternless, incoherent, and without any orientation.

Nevertheless, most contemporary thinkers are more optimistic; they do *not* believe in 'inevitable decay', or the prospect of 'a West beyond repair'. Even those who speak of decline, or possible disasters, do so *to provoke* Western intelligentsia into taking steps *to prevent* the impending disaster. Most of the public shares this optimism. Why? One reason is:

Even if there is a master equation of history, individual minds constitute an integral part of it. Vary fundamental concepts, change minds, and you change the outcome.

In this connection our thesis is as follows:

In physics the Master Arrow of change extends from large to small individual systems. In intellectual thinking the process is frequently reversed; the Master Arrow of (human) change is directed from the individual to the large, macro-system. Nevertheless, as we shall demonstrate in Lecture IX, intellectual thinking *cannot* be isolated from *external* influences, nor from the physical master arrow of change.

Here one may distinguish between a *micro-substratum* and a *macro-superstructure.*

The former is related to such problems as *"observation," "observer," "measurement"* and causality in micro-physics (as well as in microbiological and micro-social systems). The *macro-superstructure* (Lectures I and IX) is entirely different. In the case of history it may be represented by global historical perspectives as those of *Vico, St. Augustine, Spencer, Hegel, Marx, Spengler* and *Toynbee* (and many other less known macro historians; see Appendix II, §B). Yet, most historians and philosophers are skeptical of such efforts (and part of their criticism has some validity). However, these critics offer no alternative solutions to the global issues raised by macro-historians and macro-evolutionists. Some critics simply say that a useful discussion on the global scale is impossible! *It is interesting to note that similar criticism has been directed against physicists conducting research on the large scale physical world around us, i.e., cosmologists, general relativists and astrophysicists.*

Proponents of macro-history (or of cosmology) argue that critics who dismiss the very possibility of such a history (or of our capabilities to understand the physical evolution of the world at large), are like scientists who use a *microscope* to examine a *large river* running down to the sea and then claim that *"it is impossible to determine the direction of the flow!"* Obviously, a scientist must learn about the *whole* river (and about *gravitation*) in order to arrive at correct conclusions regarding the questions of direction, sources and sink, trends, structures, streams, flows, movements and macrodynamics.

While I cannot accept any particular theory of macro-history, or of cosmology, as being proven, I recognize the objective *need* for such branches of modern thought. Hence I do not reject them on an *a priori* basis.

Time (including historical or cosmic) and *structure* (including social, gravitationally-induced, or biological) can be assigned with no *a priori* limit for the purpose of modern scientific and philosophical research. *We must not place a priori restraints on research into any length of time, scale, complexity, or space; and into any aspect of the (historical, cosmological, social, biological, etc.) worlds accessible to our observations.* The demonstrated *pluralism* of modern scientific, historical and philosophical methods explains why a theory can be valid and yet relative rather than absolute. Moreover, a given proposition may be universally correct only within the prospective of a unified approach to science, history and philosophy.

Hence, *macro-evolution, techno-economical change and macro-history, as well as modern relativistic cosmology, should have a status equal to that of other, more traditional fields, like classical history or classical physics* (Lecture IX).

Assertion 10a: Unification and the Task of Philosophy

Should there be any guiding idea in the ordering of such studies? This question was raised before. Here I stress a few additional points to those discussed in §1.1.1.

Macro-studies, especially within the context of havayism (Lecture IX), should be ordered, as far as possible, *along a common time coordinate; from early to later evolutionary stages, and from the whole to the parts.* The theories and philosophies of evolution *cannot,* therefore, be separated into *isolable* subtheories about 'inde-

pendent' evolutions along *separable* time coordinates. *Evolution is one in the same sense that time is one uninterrupted space-cosmos asymmetry directed from early to later; from universe to parts; from non-gravitated to gravitated; from unstructured to structured interconnected systems.* Hence, in the same sense that we must give up the idea of unrelated origins of time, irreversibility and structure, we must also give up the idea of isolable, separable, unrelated *"sources of scientific knowledge".*

Cosmological evolution is therefore the Master Arrow of change, and gravitation the Master Field affecting all evolution; all times; all processes; all studies. The theory of gravitation and the unified field theory emerging from it, should, therefore, be granted the status of Master Science, i.e., any valid philosophy cannot dissociate itself from havayism (although some philosophical studies may be distributed among various 'disciplines' there is an objective need to develop a tradition that cannot be associated with any specialism-separatism, because the ideas with which it deals are common to all studies and to *gravitism-havayism*) (Lecture IX).

Fragmented studies are, therefore, unreliable guides to the whole province of modern knowledge; instead, interconnected modes of thinking must be given the highest status in science and philosophy; *for interconnected thinking must always attempt to gather together (apparently) different strands of evidence that are related to each other through a central idea; and out of this ensemble to weave the infrastructure of one unified thought; with common terminology-logic, and careful attention to the fewest possible unified concepts.*

Unification presents difficult problems, since it maintains that concepts and things may not what they appear; for they are not the separate, independent, unrelated objects which constitute the foundations of *disciplinary* theories. Moreover, it requires exposition of common causes, origins, structures, substratums, organization and interconnectedness; and it requires the closing of the artificial gaps between philosophy and science; between science and culture; between the natural sciences and the humanities; between sociology, anthropology, history and the arts; between mind and body and between "internal" and "external" worlds; between society and academe and between academe and "non-scientific" values (Appendix VI).

Some of these ideas can be traced back to Spinozism, which, *inter alia,* supports the contention (§IX.2.18) that we are aware of *external* things only *"in relation"* to one another. Hence, we cannot know them *"as-they-are-in-themselves".* In Spinozism this leads to the assertion that all "sense experience" (the so-called mental correlate of the action of external things *on* our bodies) and all deductions based on sense experience alone are inadequate.

This last point leads us into an immense number of problems, some of which are dealt with in the main text. But nothing is less wanted here than to issue a clear warning against the dangers inherent in a number of popular schools of philosophy which isolate themselves from the problems that arise *outside* "professional philosophy"—a profession which nowadays stresses the *internal* roles of mind and language; schools which may justly be criticized as engaging in meaningless verbiage.

Perhaps no one has expressed this danger better and more clearly than Sir Karl Popper: "... every philosophy, and especially every philosophical 'school', is liable to degenerate in such a way that its problems become practically indistinguishable from pseudo-problems, and its cant, accordingly, practically indistinguishable from meaningless babble. This ... is a consequence of philosophical inbreeding. The degeneration of philosophical schools in its turn is the consequence of the mistaken belief that one can philosophize without having been compelled to philosophize *by problems which arise outside philosophy*—in mathematics, for example, or in cosmology, or in politics, or in religion, or in social life". And Popper adds:

"*Genuine philosophical problems are always rooted in urgent problems outside philosophy, and they die if these roots decay*". "In philosophy", according to Popper, "methods are unimportant; *any* method is legitimate if it leads to results capable of being rationally discussed. What matters is not methods or techniques but a sensitivity to problems, and a consuming passion for them; or as the Greeks said the gift of wonder."*

Today many Western philosophers are advocates of *logical positivism*—a doctrine which does not agree with this task of philosophy. 'What is then the task of philosophy?' In his famous *Tractatus*, Wittgenstein found a simple answer: *THE SOLE REMAINING TASK OF PHILOSOPHY IS THE CRITIQUE OF LANGUAGE.*

Whatever our opinion of the *Tractatus,* the fact remains that Western philosophy can never return to what it was before its publication. Many Western philosophers considered the *Tractatus* their "bible", and much of American and British *"activity in philosophy"* in our times (in particular in logical positivism) has been affected, in one way or another, by Wittgenstein's philosophy of language (Appendix III).

Assertion 10b: Feed-Back Causality vs. Inductivism—Deductivism

As for logical positivism based on Wittgenstein's philosophy of language Popper writes that "*it excludes from science practically everything that is, in fact, characteristic of it (while failing in effect to exclude astrology). No scientific theory can ever be deduced from observation statements, or be described as a truth-function of observation statements.*"** And he adds*** "The problem which comes first, the hypothesis (H) or the observation (O),' is soluble; as is the problem, 'Which comes first, the hen (H) or the egg (O)'. The reply to the latter is, 'An earlier kind of egg', to the former, 'An earlier kind of hypothesis'. It is quite true that any particular hypothesis we choose will have been preceded by observations-the observations,

---

* *Conjectures and Refutations,* Routledge and Kegan Paul, London 1963, 1972 p. 71.
** *Ibid.,* p. 40.
*** *Ibid.,* p. 47. For further criticism of logical positivism see also there in pp. 258–284.

for example, which it is designed to explain. But these observations, in their turn, presupposed the adoption of a frame of reference; a frame of expectations; a frame of theories." "There is no danger here" he adds "of an infinite regress. Going back to more and more primitive theories and myths we shall in the end find unconscious, *inborn* expectations. [+]

Popper is therefore an opponent of the widely popular *dogma of inductivism—of the view that science always starts from observation (O) and proceeds, by induction, to generalizations, and ultimately to theories* [++]. Thus, as an *anti-inductivist*, he asserts that, *initially*, the hypothesis cannot be directly supported by observations; *all we can do is try to falsify it later* (see Assertion 11).

The great debates on these issues [+++] may, however, become unnecessary once we re-examine them carefully within the context of the theory of causality and time asymmetry.

Firstly, I stress that *Popper's hypothetico-deductive method and the dogma of inductivism are both extreme views which, inter alia, infer that the evolution of knowledge is, in a sense, a one-dimensional continuum that leads from (H) to (O), or from (O) to (H) infinite chains on an (isolated) time scale of scientific evolution,* i.e., isolated from the common, large-scale time coordinate of evolution at large (Lecture IX).

Secondly, I refer to Assertion 1 where I stress that intellect is given us to speculate, to observe, to compare, and to speculate again as a continuous scientific-philosphical process.

Thirdly, I assert that a deeper cognition of causality and of the time scales involved in the growth of scientific knowledge leads to the concept of *mutual, 'feed-*

[+] This brings Popper to a discussion of the theory of inborn *ideas* (cf. Kantianism, etc., in Lecture IX), which, to his mind, is absurd, for he thinks that "every organism has inborn *reactions* or *responses*; and among them, responses adopted to impending events".

"Thus we are born with expectations; with 'knowledge' which, although not *valid a priori*, is *psychologically or genetically a priori*, i.e. prior to all observational experience." (cf. also §IX.2.7, §IX.2.7.2, §IX.2.10 and §IX.2.22). Popper stresses also the fact that "one of the most important of these expectations is the expectation *of finding regularity.*"

"This 'instinctive' expectation of finding regularities," he writes, "corresponds closely to the 'law of causality.'"

At this point our epistemological view starts to diverge from Popper's. While Popper tends somewhat to fall back here on Kantianism, our views are based on general-relativistic causality, strict determinism and modern interpretations of Spinozism (Lectures IV, VI, VII and IX). Moreover, when it comes down to the correct interpretation of probability, *'ensemble'*, and the alleged 'uncertainty' in quantum mechanics, our views coincide, but only to diverge again on the role of determinism (Lecture V).

[++] "Perfect" induction only sums up the observations and does not proceed beyond the facts observed (see below and Assertion 11). Modern inductivists believe that the hypothesis, however acquired, may be, initially, supported directly by evidence (not merely indirectly, by its surviving attempts to falsify it, see Assertion 11).

[+++] See, for example, *Ibid*, pp. 153–165 and Kirk, G.S., *Mind 69*, 318 (1960).

back causality' between elements which comprise the integral of human knowledge (at a given culture and historical time), *and that such a concept does not preclude the possible existence of (time-dependent) feed-back connections in which the "results" change the "causes", and in which (O) supports or leads to (H), or vice versa. A maze of mutual causal relationships results and, with cause and effect, (H), (O), and 'tradition' indistinguishable in many instances, no clear cut definition of the growth of knowledge can be made; all we can do is to study the "interconnectedness" between these processes and between them and environment, tradition, economics, etc.* (see also Assertion 11, and, in particular, our views of *Spinozism* in §IX.2.9).

Immediately following the rejection of the simple reductionist models of the *"mechanics"* of the growth of knowledge, the road is opened for a broader perception of the *"interconnectedness" between science and philosophy* (including a better understanding of the roles 'method', 'procedure', 'discovery', 'empiricism', 'rationalism' and 'intuition' play in scientific-philosophical evolution — see also Lecture IX)*.

* Induction is conceived by the British Philosopher, *J.S. Mill* (1806–1873), as essentially *evidencing process*. His view fits, therefore, with the ideas of empiricism (Append. III). Moreover, he thinks that inductive inference goes from particulars to particulars, i.e., to further instances, *not to generalization!* But another renowned British philosopher, *W. Whewell* (1794–1866), conceives it as essentially *discovery*, viz., discovery of some conception, not extracted from the set of particular facts observed, but nevertheless capable of "colligating" them, i.e. of expressing them all at once (or, better stated, of making it possible to deduce them). E.g., Kepler's statement that the orbit of Mars is an ellipse represented the discovery by him that the conception of the ellipse "colligated" all the observed positions of Mars. Although Whewell's work on the theory of induction was overshadowed by that of J. S. Mill, his lasting reputation rests on his *History of the Inductive Science*, 3 vol. (1837), and his *Philosophy of the Inductive Sciences* (1840) [the latter being eventually expanded into three books, *The History of Scientific Ideas*, 2 vols. (1858), *Novum Organon Renovatum* (1858) and *Philosophy of Discovery* (1860). On the other hand, *Charles Peirce* (see Append. III), viewing induction as generalization, contrasts it not only with *inference from antecedent to consequent* ("deduction") *but also with inference from consequent to antecedent*, called by him *"hypothesis"* (also called by him "abduction" but better termed "diagnosis"). [Peirce's view is closer to the mutual, feed-back interconnectedness stated above but is not identical with it].

Ordinarily, however, "induction" is used to mean *ampliative inference* (as distinguished from *explicative*), i.e., it is the sort of inference which attempts *to reach conclusion concerning all the members of a class from observation of only some of them.* Therefore, conclusions inductive in this sense are only probable, in greater or lesser degree, depending on the precautions taken in selecting the evidence for them (see Assertion 11 below).

Deduction, on the other hand, is usually considered as analytical inference, or as the mental drawing of conclusions from given postulates. In logic it is inference in which a conclusion follows necessarily from one or more given premises. [Definitions given have usually required that the conclusion be of less generality than one of the premises, and have sometimes explicitly excluded immediate inference. However, these restrictions do not agree well with the ordinary actual use of the word]. *"Empirical deduction"* is the factual explanation of *how concepts arise in experience and reflection.*

Hypotheses about objects *not directly observable* (such as *electrons*, magnetic, or gravitational fields) are sometimes called *transcendent hypotheses*. They *cannot* be reached or directly confirmed by simple induction. The very scientific procedure or method by which they are reached or confirmed is sometimes called *secondary induction*. But here I stress the fact that sometimes some grand overall premises are sought which can turn inductive arguments into deductive ones, making *both* indistinguishable.

Many have written about induction. To mention a few by name: Bertrand Russell (see also Appendix

This means that science cannot be isolated from philosophy, nor can it be sub-divided into uncoupled disciplines. Science is therefore the basis and backbone of philosophy, and vice versa. (Indeed, historical evidence demonstrates that science has not been what many philosophers have claimed it to be.) Thus, philosophy and science are continuous with each other, as the great Greek philosophers thought, but as only a few Western thinkers hold today. There is no well-defined borderline (or 'demarcation') between the two, and any attempt to impose one cannot stand the test of time. It is the absence of dialogue which builds walls between scientists and philosophers. But such walls should not be confused with a universally valid borderline.

Assertion 11: Popper's Epistemology

This last assertion is mostly based on Popper's epistemological ideas which are among the most instructive in modern philosophy. In his own words some of them read:**

"The way in which knowledge progresses, and especially our scientific know-ledge, is by unjustified (and unjustifiable) anticipations, by guesses, by tentative solutions to our problems, by *conjectures*. The conjectures are controlled by criticism; that is, by attempted *refutations*, which include severely critical tests. They may survive these tests; but they can never be positively justified: they can neither be established as certainly true nor even as 'probable' (in the sense of the probability of calculus)" (cf. also Assertion 4).***

"The criterion of the scientific status of a theory is its falsifiability, or refutability, or testability." "A theory which is not refutable, by any conceivable event, is non-scientific. Irrefutability is not a virtue of a theory (as people often think) but a vice". "Every genuine *test* of a theory is an attempt to falsify it, or to refute it. Testability is falsifiability; but there are degrees of testability; some theories are more testable, more exposed to refutation, than others; they take, as it were, greater risks".

"Some genuinely testable theories, when found to be false are still upheld by their admirers—for example by introducing *ad hoc* some auxiliary assumption, or by re-interpreting the theory *ad hoc* in such a way that it escapes refutation. Such a procedure is always possible, but it rescues the theory from refutation only at the price of destroying, or at least lowering, its scientific status".

I), Karl Popper, Paul Edwards, Wesley Salmon, Stephen Barker, Henry Kyburg, John Lenz, Richard Swinburne, Max Black, Peter Achinstein, Keith Campbell, Nicholas Maxwell, and R. B. Braithaite. For a review see *The Justification of Induction*, R. Swinburne, ed., Oxford University Press, 1974.

   ** *Ibid;* p. 36, 27, 28, 30.

   *** See also T.S. Kuhn's *The structure of Scientific Revolutions;* 1962; M. Masterman's *'The Nature of a Paradigm'*, in I. Lakatos and A. Musgrave (eds) *Criticism and the Growth of Knowledge*, 1970, and Lecture III. In Kuhn's later works *paradigms become "standard" forms of solutions to problems* (e.g., they become equations, formulae, etc.). *These solutions are then used for solving further problems, and so govern the forms these further solutions take. He is largely concerned with how shifts of paradigm occur as science develops.* In earlier work Paradigms were, basically, ways of looking at the world, shared as-sumptions which govern the outlook of a scientific epoch and its approach to solving scientific problems.

"There are no ultimate sources of knowledge. Every source, every suggestion, is welcome; and every source, every suggestion, is open to critical examination". "... all kinds of arguments may be relevant. A typical procedure is to examine whether our theories are consistent with our observations. But we may also examine, for example, whether our historical sources are mutually and internally consistent."

"There can be no point in trying to be more precise than our problem demands. Linguistic precision is a phantom, and problems connected with the meaning or definition of words are unimportant" (see also our Appendix III).

"If we ... admit that there is no authority beyond the reach of criticism to be found within the whole province of knowledge, however far it may have penetrated into the unknown, then we can retain, without danger, the idea that truth is beyond human authority. And we must retain it. For without this idea there can be no objective standards of inquiry; no criticism of our conjectures; no groping for the unknown; no quest for knowledge."

"Quantitatively and qualitatively by far the most important source of our knowledge—apart from inborn knowledge—is tradition. Most things we know we have learned by example, by being told, by reading books, by learning how to criticize, how to take and to accept criticism, ..."

"The fact that most of the sources of our knowledge are traditional condemns anti-traditionalism as futile. But this fact must not be held to support a traditionalist attitude: every bit of our traditional knowledge (and even our inborn knowledge) is open to critical examination and may be overthrown. Nevertheless, without tradition, knowledge would be impossible".

One last remark on the way in which scientific knowledge progresses. In general, when dealing with an imperfectly understood physical, biological or social phenomenon, we often try to remedy the situation with the aid of additional variables, introduced directly or indirectly; when, however, an attempt is made to reduce the number of variables and to unify apparently disparate phenomena, these additional variables generate even greater conceptual difficulties than those originally encountered. This striking paradox in the evolution of scientific knowledge may best be illustrated by the evolution of the theory of time as presented in Lecture VII. Other redundant concepts and variables include entropy (Lectures IV and VI), geocentric thermodynamics (VII.7), and endless 'independent' arrows of time.

Other Assertions

The remainder of our system is set forth in the text by way of elucidations and derivations of the assertions mentioned above. Additional assertions result from the various technical analyses and from detailed syntheses. To mention a few:

The physical links between *black holes,* the *cosmological* arrows of time and *thermodynamics* (Lecture VIII, §VII.2, §VI.5).

A theory of knowledge (§IX.2.4 to §IX.2.8.1), including the problems of innate knowledge, perception of space and time, asymmetry in thinking, innate patterns and simple orientation movement in the field of gravity, the failure of current *information theory*, asymmetry and *memory*, as well as fundamental problems associated with aggregated forms and structures in *language, mathematics, information, and verbal thinking.*

The fundamental effects of *neutrinos* (§I.8, §II.12, §VII.8), *expansion* (bulk), *viscosity* (§VI.7, §IX. §2.22), *T-violations* (§IV.11, §VII.8), *pregeometry*, and the physics *of the vacuum* (§IX.2.7; 2.7.1; 2.7.2; 2.8; 2.9; 2.10; 2.15; 2.16; 2.18; 2.19; 2.21; 2.11– 2.22).

The impossibility principle in biological evolution (IX.2.3).

The structure of physics in terms of differential equations, initial and boundary conditions, asymmetries, observables, gravitation and integrodifferential equations (Table I, Lecture V and §IX.2.4.1).

Also included in our system are a number of ideas that have been selectively extracted from other philosophers: from Plato to Maimonides; from Crescas to Albo (Appendix I), from Spinoza to Kant (Lecture IX), from Spencer (Appendix II) to Compte (Appendix III), from Popper to Jaspers (pp. 433–35), and from Mach to Einstein (Lecture IX).

Lecture IX contains, among other things, a few concepts which may, at first, appear similar to some concepts in dialectical materialism, but they do not originate in it. In fact, their origin is in classical Western thought, especially in Spinozism. It was mainly through Hegel's borrowing from Spinozism, that these concepts appear also in dialectical materialism. These include:

a) In the theory of knowledge: The assertion that consciousness is the reflection of objective reality. But this approach is essentially based on Spinozism which implies that *the world and its various phenomena exist objectively, independently of our senses and that any separation between mind, brain, and body or between fact, virtue, and value is impossible. All mechanistic models and any reductionism are therefore substituted by Spinoza's subtle interconnectedness of all reality* (§IX.2.9).

b) *The tendency to transcend philosophy through socio-historical and socio-political facts and theories, and to integrate it with the biophysical, socio-historical, and political sciences.*

c) *The attempt to transcend science by dealing with those theories and facts which apply to reality as a whole and to all fields in common.*

d) The physico-philosophical issues of determinism.

3.2. THE SKEPTIC'S OUTLOOK

A well-known attitude among some scientists is that both old and new philosophical assertions have no practical applications. Some ungentle reader may even inform us here that philosophy is as useless as chess and as obscure as religion. This attitude is thoroughly misconceived.

Obviously most scientists wish to resolve the whole into parts and to narrow their analysis into isolated disciplines. But physics without a workable philosophy, isolated facts without an all-embracing perspective, cannot save us from empty scientism. Science without philosophy; society without an all-embracing perspective generate a decaying culture; and such devastating processes in Western academe are evident today more than ever before. Thus, all I can offer to a skeptic is to examine carefully the present state of physics, philosophy and academe at large before berating me with such criticism (Volume II).

A second difficulty is associated with the fact that our theory has inherited from general relativity, astrophysics and philosophy *a number of unresolved questions.* For instance, we are still far from having a clear idea about the formation of self-gravitating matter in an expanding *uniform* medium. Is it through turbulence, bulk viscosity, weak interactions, or perhaps through a basic instability somehow due to the master time asymmetry?

We are also far from understanding the initial phases of the observed expansion of the universe and the theoretical difficulties associated with the principle of equivalence as well as with modern unified field theories (including the quantized theories of general relativity). Are the newly conceived *"pregeometry", "superspace", gauge theories*, etc., the arena to be considered if physical observations, in the unforeseeable future, penetrate into extremely small scales comparable to the so-called Planck length (10^{-33}cm)? However, the answers to these questions are irrelevant to our theory so long as the *"master asymmetry"* is observed while the Planck length is not. Even by a most conservative estimate, this state of affairs would continue for a much longer time than is normally presumed for any theory to remain valid.

A third difficulty is associated with a series of unanswered questions, a sample of which follows: What can our theory tell us about the origins of *knowledge itself* and, in particular, about the origin of *structures, information and human memory?* Moreover, what are the origins of subjective information? of *perception? of mind?* Is there any *"a priori knowledge"* about space and time, as for instance is claimed in Kantianism? But, most important, can gravitism be incorporated in a more general outlook which includes the whole of our present knowledge and perhaps goes even beyond it? These questions are taken up again in Lecture IX.

Let me make it clear from the outset that I do not claim that definite answers can be given here to all these questions; nor that the methodology that will be used in these lectures is entirely divested of temperamental bias. All that I do say is that in seeking answers to these fundamental questions we can try, as in all branches of science, to make successive approximations toward a better understanding of nature, in which each new stage results from an improvement, not rejection, of previous doctrines. Hence, in trying to seek answers to these questions we must, in due course, give credit to certain past doctrines in science and the philosophy of science.

The general and physico-philosophical views put forward here can only claim attention insofar as they offer new conclusions based on a consistent and "up-to-date" study of modern thought and research. Therefore, historical and other general ideas are discussed in these subject-oriented lectures only to the extent that they become essential.

Another reason behind the unorthodox structure of this volume is the following: It is usually claimed that either one must write to be understood by the educated general reader and the book may appear superficial and be criticized by the specialists; or one must write to please the specialists, but then it may be incomprehensible to others.

However, the facts cannot be eluded. Insofar as evidence about contemporary academe is concerned, we know that most contemporary specialists are fairly narrow in their professional occupation; they are quite *uncritical* and *provincial* in their overall philosophical and historical outlooks (Appendix VI). Outside the domain of their discipline they tend to respond like laymen. Since the subject matter of this volume covers so many disciplines in science and philosophy, *no reader or critic can be a "specialist" in all of them.* For these and other reasons that are explained in the text, I have been faced by a two-fold problem, to write exclusively for the student of philosophy and the generally educated reader, or to try to stress the need to develop a *"critical science of interconnectedness."* I chose the second alternative.

But, this requires the development of a new methodology — a sample of which is presented in this volume, especially in those sections dealing with gravitism and havayism. The interested specialist can, nevertheless, find here new ideas and mathematical formalism that form the basis of a new physico-philosophical theory.

References

1. Gal-Or, B., *Found. Phys., 6,* 407 (1976).
2. Ibid. *Found. Phys., 6,* 623 (1976);
3. Ibid. *Science, 176,* (1972); *Nature, 236,* 12 (1971); *234,* 217 (1971).
4. Ibid., ed. *Modern Developments in Thermodynamics* (MDT), Wiley, N. Y., 1974.
5. Ibid. *Space Science Reviews, 22,* 119 (1978).
6. Cline, D. B. and Mills, F. E., eds., *Unification of Elementary Forces and Gauge Theories,* Harwood Academic Publishers, London, 1978.
7. Weinberg, S., *Phys. Rev. Lett., 19,* 1264 (1967).
8. Salem, A. *Proc. of the 8th Nobel Symp.,* Almquist & Wiksell Stockholm, 1968.
9. Gordon, S. A. and Cohen, M. J., eds., *Gravity and the Organism.* The Univ. of Chicago Press, Chicago, 1971.
10. Gal-Or, B. *The new Philosophy of Thermodynamics,* in *Entropy and Information in Science and Philosophy,* Zeman, J., ed., Czechoslovak Academy of Sciences, Elsevier, Amsterdam, 1974.

11. Ibid., *Annals, N. Y. Acad. Sci. 196* (A6), 305 (1972); (N.Y.A.S. Award Paper).
12. Popper, K. *Conjectures and Refutations,* Routledge and Kegan, London, 1963, 1972.
13. Gold, T. in Ref. 4, p. 63.
14. Narlikar, J. V., in Ref. 4, p. 53.
15. Neeman, Y., in Ref. 4, p. 91.
16. Curran, P. F., in Ref. 4, p. xix.
17. Callen, H., in Ref. r, p. 201.
18. Noll, W., in Ref. r, p. 117.
19. Eringen, A. c., in Ref. 4, p. 121.
20. Ramsey, N. F., in Ref. 4, p. 207.
21. De Groot, S. R., W. A. van Leeuwen and Ch. G. van Weert, in Ref. 4, p. 221.
22. De Beauregard, O. C., in Ref. 4, p. 73.
23. Schulman, L. S., in Ref. 4, p. 81.
24. Grunbaum, A., in Ref. 4, p. 413.
25. Ikeda, K., in Ref. 4, p. 311.
26. Penrose, O. and A. Lawrence, in Ref. 4, p. 303.
27. Tauber, G. E., in Ref. 4, p. 247.
28. Kranys, '., in Ref. 4, p. 259.
29. Aharony, A., in ref. 4, p. 95.
30. Schmid, L. A., in Ref. 4, p. 143.
31. Katz, J. and G. Horwitz, in Ref. 4, p. 237.
32. Toda, M., in Ref. 4, p. 79.
33. Thorn, R. J., in Ref. 4, p. 343.
34. Frey, J. and J. Salmon, in Ref. 4, p. 391.
35. Gal-Or, B., in Ref. 4, p. 3, 429.
36. Ibid (ed., with E. B. Stuart and A. J. Brainard) *A Critical Review of Thermodynamics (Relativistic and Classical),* Mono Book Corp., Baltimore, 1970.
37. Prigogine, I., in Ref. 36, p. 1.
38. Guggenheim, E. A., in Ref. 36, p. 211.
39. Landsberg, P. T., in Ref. 36, p. 253.
40. Kestin, J. and J. R. Rice, in Ref. 36, p. 275.
41. Chu, B-T., in Ref. 36, p. 299.
42. Melehy, M. A., in Ref. 36, p. 345.
43. Fong, P., in Ref. 36, p. 407.
44. Keenan, J. H. and G. N. Hatsopoluos, in Ref. 36, p. 417.
45. Tribus, M. and R. B. Evans, in Ref. 36, p. 429.
46. Redlich, O. in Ref. 36, p. 439.
47. Quay, P. M., in Ref. 36, p. 83.
48. Griffiths, R. B., in Ref. 36, p. 101.
49. Tisza, L., in Ref. 36, p. 107.
50. Pathria, R. K., in Ref. 36, p. 119.
51. Hornix, W. J., in Ref. 36, p. 235.
52. Arzelies, H., in Ref. 36, p. 49.
53. Meixner, J., in Ref. 36, p. 37.
54. Balescu, R. C., in Ref. 36, p. 511.
55. Weinberg, S., *Gauge Symmetry Breaking,* in *Procd. Conf. on Gauge Theories in Modern Field Theory,* M.I.T. Press, Cambridge, 1976.
56. Ibid., *Phys. Rev. Lett. 36,* 294 (1976); *31,* 494 (1973).
57. Ibid., *Phys. Rev., 177,* 2604 (1969); *D7,* 1068 (1973); *D7,* 2887 (1973); *D8,* 605 (1973); *D8,* 3497 (1973); *D8,* 4482 (1973); *D9,* 3357 (1974); *D13,* 974 (1976); *D13,* 3333 (1976); *D17,* 275 (1978).
58. Ibid., *Phys. Lett. 62B,* 111 (1976); *Physics Today 30,* 42 April 1977.
59. Ibid., *Phys. Rev. Lett. 29,* 388 (1972); *29,* 1968 (1972).
60. Ibid., *Phys. Rev. Lett. 18,* No. 5, 188 (1967); *20,* No. 5, 224 (1968).
61. Ibid., *Gravitation and Cosmology: Principles and Applications of the General Theory of Relativity,* John Wiley, New York, 1972.
62. Wigner, E. P., *Symmetries and Reflections,* Indiana Univ. Press, Ind., 1967.
63. Smoot, G. F., M. V. Gorenstein and R. A. 'uller, *Phys. Rev. Lett., 39,* 898 (1977).
64. Hawking, S. W. and W. Israel, eds., *General Relativity,* Cambr. 1979.

65. Wilkinson, D., ed., *Progress in Particle and Nuclear Physics,* Vol. 7, Pergammon Press, 1981.
66. Peebles, P.J.E., *The Large-Scale Structure of the Universe,* Princeton University Press, 1980.
67. Perlmutter, A., ed., *Gauge Theories, Massive Neutrinos, and Proton Decay,* Orbis Scientiae 1981, Studies in Natural Science, Plenum 1981.
68. Nicolson, I. *Gravity, Black Holes, and the Universe,* Halsted Press, 1981.
69. Fayet, P., and S. Ferrara, *Supersymmetry, Phys. Rep.,* 32C : 249, 1977.
70. Weyl, H., *Symmetry*, Princeton University Press, 1952.

Part I

PRELIMINARY CONCEPTS

With the help of physical theories we try to find our way through the maze of observed facts, to order and understand the world of our sense impressions. We want the observed facts to follow logically from our concept of reality. Without the belief that it is possible to grasp the reality with our theoretical constructions, without the belief in the inner harmony of our world, there would be no science. This belief is and always will remain the fundamental motive for all scientific creation. Throughout all our efforts, in every dramatic struggle between old and new views, we recognize the eternal longing for understanding, the ever-firm belief in the harmony of our world, continually strengthened by the increasing obstacles to comprehension.

Albert Einstein and Leopold Infeld

FROM TERRESTRIAL GRAVITATIONAL STRUCTURES

TO BLACK HOLES AND NEUTRINOS IN ASTROPHYSICS

Philosophy is written in that great book which ever lies before our gaze—I mean the universe—but we cannot understand if we do not first learn the language and grasp the symbols in which it is written.

GALILEO GALILEI

Science is not just a collection of laws, a catalogue of unrelated facts. It is a creation of the human mind, with its freely invented ideas and concepts. Physical theories try to form a picture of reality and to establish its connection with the wide world of sense impressions. Thus the only justification for our mental structures is whether and in what way our theories form such a link.

ALBERT EINSTEIN AND LEOPOLD INFELD

Terminology and a Word of Caution

Cosmology, in the broadest sense of the word, is that branch of learning which treats the universe as an ordered system. The name is derived from the Greek κοσμος ("order", "harmony", "the world"), plus λόγος ("word", "discourse").This term is usually linked to a description of the salient features of the observed universe, in terms of such categories as space, time, matter, structure and gravitation. In these lectures we use the term cosmology in connection with problems concerning the origin and structure of the universe: namely in connection with *cosmogony* and *astrophysics*.

Cosmogony is a branch of science studying evolutionary behavior of the universe and inquiring into the origin of its various characteristic features.

The *cornerstone* of modern cosmogonical theories is the *astronomical observation* that the system of clusters of galaxies scattered through the vast space of the universe is in the state of *progressive expansion* (or dispersal). This fact, based on the observation of red shift of spectral lines in the light coming from distant galaxies (see below), on the discovery of the black-body radiation (Lecture VI), and on Einstein's theory of gravitation (Lectures III and VI), led to the basic hypothesis of modern cosmology—namely, that the present high degree of aggregation of matter in the universe accessible to our observations, and the complexity of structures and forms of various astronomical objects, must have resulted from a violent "explosion" and *subsequent* dispersal of the originally highly compressed and intensely hot homogeneous matter-energy.

Astrophysics, in comparison, is concerned with the detailed physical properties of celestial objects. It is, however, part and parcel of both physics and cosmology because it provides new information in three domains:

1) The behavior of matter and radiation under extreme conditions of temperature, density, magnetic and gravitational forces that cannot be realized in a terrestrial laboratory.

2) Verification of the physical laws discovered in our laboratories, when such laws operate over very long periods of cosmic time (up to about 20,000,000,000 years), and over enormous distances (over 10,000,000,000 light-years —q.v. §I.7 below).

3) Verification of new laws of nature, which, *otherwise*, might remain undetected under terrestrial conditions, and/or the *refutation* of theories, which, *otherwise*, might lead scientists and philosophers astray.

Astrophysics is also part of *modern astronomy* because it derives much valuable observational evidence from measurements of the positions, motions, constitution and evolution of clusters of galaxies, galaxies, stars, planets, satellites, etc. It also depends on *chemistry* for the study of the behavior of various types of molecules (in the atmospheres of the planets and the cooler stars), the formation of simple molecules and dust particles in interstellar space, the internal constitution of the planets and the chemical composition of meteorites. It is also related to *geology* and *geophysics*, not only in the study of the surface features of planets with craters, extinct volcanoes, gravitationally-induced mountains and other structures, but with the interpretation of terrestrial vs. astronomical structures, stratification, evolution, etc.

There is also a connection with *biophysics*, especially with regard to the *origin* and *evolution* of living organisms on earth and (probably) on other planets belonging to the families of other stars. Many *modern mathematical theories* (including advanced *computational methods*), have found applications in astrophysics which, in turn, promoted new research in *Einstein's field equations, thermodynamics, electromagnetism, quantum physics*, and *statistical mechanics*. The field of astrophysics is therefore highly diversified. In fact it is frequently subdivided into several groups, according to criteria such as methodology, application, theory, or observation.

Astronomy, in the broadest sense of the word, deals with observations of objects of the universe and, in particular, with their *positions, motions, constitution* and *evolution*. The analysis of the radiation emerging from the various celestial objects, constitues those parts of the science of astronomy known as astrophysics and cosmogony. *Hence astronomy, astrophysics, cosmogony and cosmology are intimately interconnected. So must be their proper discussions in modern science and philosophy.*

In fact, astronomy has an important part to play in the general advance of *physics*. For the galaxies, the stars and other galactic objects constitute physical laboratories where "experiments" performed by the force of gravitation can be observed under conditions that can never be obtained in any terrestrial laboratory. Indeed, *it is through astronomy that Einstein's general theory of relativity has revolutionized many of our conceptions in physics and philosophy. But, as we shall see in subsequent lectures, this revolution has not ended yet, for new astronomical evidence keeps coming in and further changing many of our views on the structure of the physical world around us. For this reason we shall stress the central role the theory of relativity plays in cos-*

mology, cosmogony, astrophysics, astronomy, physics, thermodynamics, quantum physics, philosophy and the modern theory of evolution.

A few *preliminary concepts* related to the *methods* employed by observational astronomy may be useful at this early stage of the course.

All astronomical observations depend on *electromagnetic radiation* which reaches the earth (light, radio waves, γ- and X-rays, etc.). The astrophysicst usually categorizes these electromagnetic data according to a number of major parameters:

1. *Direction* (including the newest methods of Very-Long-Baseline Interference (VLBI) radio telescopes).
2. *Frequency* (of emission or absorption).
3. *Intensity* (or luminosity).
4. *Polarization, interference, phase, amplitude, etc.*
5. *Variation over the period of measurement.*
6. *Independent methods of measurement from the same source.*
7. *Independent evaluations of the medium extending between source and the terrestrial absorber.*
8. *Distribution over space-time, form, structure, etc.*

From these parameters he may first be able to characterize the *physico-chemical state* of the radiating source (or the medium in between source and observer) in terms of *composition, temperature* and *density* as well as *position, velocity, size, mass, inclination* and *intensities of the electromagnetic and the gravitational fields.* Interactions between *multiobject* systems (e.g., binary systems —q.v. §I.4 below), as well as *evolutionary patterns* may then be deduced in connection with a *theory* and *other* observational data (e.g., by comparing the analyses from visible, radio and X-ray radiations emitted by the same or by neighboring sources). Intensity, frequency and variation (e.g., eclipse, or periodic variations) over the period of the measurements can then help him in deducing such quantities as distance (see §I.7 below). Generally speaking, the observed (photographed) *spectral lines* can yield a great deal of information about the emitting or absorbing medium, whether it is a laboratory sample, or a cosmic object (§I.4).

But at this early stage of the course I prefer to divert attention from the methodology of observational astronomy to the more general conclusions that are based on its most recent findings. Indeed, these conclusions carry some very interesting subtleties. These, as we shall see, are not dissociated from our *"local physics."* The first group of these subtleties may be best exemplified by the following questions: *Are the limits of physical knowledge linked to structural hierarchy? What is our status in the hierarchy of gravitationally-bounded structures in the universe?* Do we "belong" to a typical group of galaxies? Or do we occupy a special position in the cosmological hierarchy of structures? *Is there any fundamental link between cosmology, gravitation and the origin of asymmetries and structures in terrestrial phenomena?*

It is mainly the last problem which we plan to attack here. But before we plunge into details of astrophysics and terrestrial physics, a *preliminary* attempt will be made to give an answer to the first questions.

The Hierarchy of Structures

It is now generally believed that there is a fundamental tendency in nature for all things of a given class "to clamp together", thereby forming units of a new class of "greater order". In the nonliving world *"elementary particles"* clamp together to form atoms, atoms to form molecules, atoms and molecules to form planets and stars, and so on up to galaxies and clusters of galaxies. *At either end of this structural hierarchy we encounter with the limits of present physical knowledge!* On the one hand it is possible that the so-called *"elementary particles"* are *not* elementary at all, *for they have an internal structure,* i.e., they consist of association of *smaller* entities (quarks or even smaller entities—see Lecture IV). On the large-scale end there is a slight possibility that clusters of galaxies are bound into still *larger* associations known as *"superclusters"*.

Let us first focus attention on clusters of galaxies for they are the last firmly established levels at the top of the hierarchy of structures. Indeed, these structures provide us with huge laboratories, ranging several million light-years across, and densities that may *surpass* that of the atomic nucleus (i.e., greater than 10^{14} grams per cubic centimeter—see below). By studying the interaction between these structures (as well as their evolution on a grand scale), we may arrive at some general conclusions regarding the origin of asymmetry and direction in nature.

The Main Galactic Structures

My first task is, therefore, to stress the following *observational facts* about the galactic structures:

1) At least half of the galaxies in the universe are members of a *group* or a *cluster* of some size or other, ranging from "galaxy-poor" groups, such as our own, to "rich" clusters, consisting of thousands of galaxies. (The average number of galaxies per cluster is approximately 100.)

2) The 20-odd galaxies in our own "local group" lie within a sphere that has a diameter of roughly two million light-years. In comparison, the diameter of our galaxy—the Milky Way—is 10^5 Light years (see Fig. VI.1).

3) The nearest "rich cluster" is the one in *Virgo*, about 55 million light-years away. It consists of about 2500 galaxies embedded in cosmic gas and radiation. Inside clusters the average distance between neighboring galaxies is approximately 10^7 light-years.

4) Some 10% of all galaxies belong to "rich clusters". However, small clusters, like our "local group", are the commonest type of galactic association.

5) There is a vast "empty" space between the galaxies in each group. But the "empty space" is not empty at all, for it contains varying *densities of gas and radiation*. These densities are *highest* near the galaxies, and near the individual stars that

make up the galaxies, and *lowest* in the vast space *between* the clusters. (The lowest energy density of radiation in intercluster space is about 0.4 eV per cubic centimeter, which is approximately the energy density of the cosmic background radiation (see Fig. VI.1)).

6) At opposite ends of our "local group" are our galaxy (the *Milky Way*) and the "Great Nebula" in *Andromeda*, the galaxy M31. A typical galaxy consists of 10^{11} stars, a figure which also applies to our own galaxy. The average energy density of starlight inside our galaxy is approximately 0.5 eV cm^{-3}, while in the depths of intergalactic space it is only about 0.008 eV cm^{-3} (see also Fig. VI.1 for details). Total luminosity of the Milky Way is about 10^{10} solar luminosities. (Luminosity is defined in I.2.1 below.) Typical energy density of magnetic fields in our interstellar space is approximately 2 eV cm^{-3}. Ours is a spiral galaxy (see Fig. I.10) about 100,000 light-years in diameter and only about 1000 light-years thick in vicinity of the solar system (which is located about 30,000 light-years away from the galactic center - see §I.6.1).

7) The mass required to keep the clusters gravitationally bound is about 10 times greater than that observed in the main body of the galaxies. This generates the notorious problem of the "missing mass" (see below).

8) Recent X-ray studies show however, that clusters hold a considerable amount of gas. These new findings may help resolve the problem of the "missing mass" — together with some other subtleties involved in the computations (cf. Lect. III).

9) Many rich clusters contain *a centrally located supergiant galaxy* surrounded by an extended halo of faint stars. In turn the centers of many galaxies harbor ultra-compact objects which may be massive black holes embedded in dense, swirling mass of stars, gas and dust (cf. §I.5.2).

10) Finally we stress the observational fact that distances between clusters of galaxies are increasing with time, i.e., the clusters are moving away from each other. This is the so-called *"expansion of the universe"*, which is discussed in Lectures III, VI and VII. *How do we measure the rate of this expansion?* This is explained in detail in §I.7. For the moment we stress only that it is less than 25 km per second per million light-years, and that it should be described in terms of the mutual recession of clusters from each other rather than of individual galaxies.

At this point I add a word of caution, namely, that new astronomical information keeps coming in even as this lecture is being printed. Hence, its contents will not remain up-to-date for long. Thus, by confining myself to basic and verified information, I hope to stay on relatively safe ground.*

---

*Reference to some updated books is given in Lectures VIII and III. The reader may also consult the series *Stars and Stellar Systems*, Volumes I to IX, G. P. Kuiper, ed., The University of Chicago Press, Chicago, 1955–1975, as well as the excellent introduction to the subjects of nucleosynthesis and Stellar evolution; *CNO Isotopes in Astrophysics*, J. Audouse, ed., Reidel, Boston, 1977. Another important reference is *Physical Processes in the Interstellar Medium* By L. Spitzer, Jr., Wiley-Interscience, N.Y., 1978.

My next purpose is twofold: (1) *to present new observational evidence provided by advanced new techniques in astronomy and astrophysics*; (2) *to demonstrate the all-embracing, unified role of gravitation in generating structures, asymmetries, recording, clocks, synthesis, and evolution.*

Additional relevant matter can be found in Lectures VI to IX. Lecture III contains a brief account of the Newtonian and the Einsteinian theories of gravitation. Hence these lectures, and especially Lecture III, should be viewed as inseparable from this lecture.

In selecting the topics to be discussed next, I have been guided by the central ideas presented in the *Introduction* and in Lectures VI, VII, VIII and IX. For this reason I see it fit to begin by a brief discussion of the unified role gravitation plays in generating structures and asymmetries *on earth*. It is by discussing this subject first that we may later appreciate the all-embracing role gravitation plays in generating the large-scale structures, asymmetries, recording, clocks, and syntheses in astrophysical and cosmological systems.

I.1 GRAVITATION, ASYMMETRY AND STRUCTURE

I.1.1 A Fallacy Associated with Current Equilibrium Theories

The forces responsible for wearing down all terrestrial land masses are gravity and (gravitationally-induced) solar energy (see §I.2. below). Radiation energy from the sun evaporates sea water; and, using the atmosphere as its agent, the solar energy causes circulation of water vapor, part of which is (gravitationally) precipitated as rain and snow. Gravity attracts the water toward the sea, much of it by circuitous routes. This hydrolic cycle performs a vast amount of work (Fig. VI.2).

Water and air react with rock materials, changing them chemically and breaking them into small pieces. Running water moves the loose rock debris and in the process causes further breakup of bedrock. Winds and waves join the attack on the lands. All these activities are included in the process of *erosion*, which sculptures the lands and tends to reduce mountains toward sea level.

Thus, at first sight, one may conclude that gravity, aided by gravitationally-induced solar heat, strives "to establish statistical equilibrium", or "maximum disorder", by reducing structure, form, and regularities into shapeless, structureless, formless uniformity, i.e., that "nature strives to establish (statistical) equilibrium". This, indeed, is the claim made in current textbooks in physics and philosophy.

But these are false impressions, for they select and analyse only a single group of processes, ignoring the rest of nature. And this, indeed, is the great fallacy associated with the current theories of thermodynamics and statistical mechanics (*Introduction*, Assertion 9 and below).

The fact is that the processes of erosion and weathering are opposed by other (gra-vitationally-induced) processes, which generate structure, order and non-equilibrium conditions. There is abundant evidence that the lands have been elevated repeatedly, on a large scale, by gravitationally-induced forces. These forces rejuvenate the structure of the lands by forming mountain chains, broad upwarps, sea-floor spreading and structuring, volcanic and metamorphic structuring, repeated folding, structured strata, etc. Moreover, the very process of erosion and weathering creates new structures; by sedimentation, by river meandering, by valleys, lagoons, springs and glaciers (see below).

The two opposing processes of structuring and erosion are not haphazard, nor spontaneous or statistical; they are fully controlled by gravity, and by gravitation-ally-induced processes in the solar system and beyond (Lectures VI and VII). *Most important, they do not tend towards equilibrium conditions*, neither on earth, nor in the solar, stellar, galactic, or cosmological structures (see below).

These processes will therefore be given special attention in this preliminary lecture.

I.1.2   Gravitationally-Induced Sedimentary Structures

The study of sedimentary begins by an analysis of the chemical materials that structure the various rock sediments. It then evaluates the structure of the stratified rocks, in space and time, as well as the geological processes recorded in them. This broad study is called *stratigraphy*.

A fundamental rule in this study, known as *"the law of superposition"*, is that in a sequence of layered rocks, as they were laid down, any layer is older than the layer next above it. This seems elementary for rocks that have not been disturbed; but more difficult in structured strata that have been overturned, or even completely inverted.

The environments in which structured sediments accumulate vary widely in geological times, in the size and geography of areas they occupy, and in their climatic conditions. Sedimentary environments are essentially *continental* (on land), or *marine* (on the floors of seas and oceans). A transitional *coastal* environment is formed by beaches, lagoons, estuaries and deltas. There are the *water-lain muds, sands and gravels* of rivers and lakes; there are the *rock screes, evaporite salt* deposits, the *wind-blown (aeolian) sands of desert, and the ice-borne (glacial) boulder clays* of high mountains and polar regions. Although fragmental (clastic) *silts, sands and shingles* from the bulk of structured sediments laid down in coastal environments, conditions may permit local accumulation of organic material, such as *peat, shell banks*, and *coral reefs*. But at this point I want to stress only two points:

1) Accumulation of structured sediments may be dislodged and *redeposited by strong gravity currents*.

2) Covering more than two-thirds of the earth surface, structured sedimentary rocks and sediments provide the main *geological record* of past geographies, shifting coastlines and changing climates. Because fossils are restricted to them, they yield much information on the changing patterns of plant and animal life, and have supplied the principal means of *recording geological times and biogeochemical evolution.*

Most sequences of consolidated sedimentary structures show *distinct layering,* known as *bedding, or stratification,* caused by slight alterations in composition, or by changes in rock type. Various modes of stratification, coupled with other bedding-plane features, including *"fossil rainpits", "suncracks",* and *"water ripple marks",* provide records of the nature of ancient sedimentary environments. Thus, ancient *patterns of wind and water flow can be deduced from such geological structures.* For example, *asymmetric structures of bedded sandstone showing a shallow ridge-and-trough* (which can be seen on many modern sandy beaches after the tide has receded), *are actually 'water fossil ripple marks' hundreds of million years old. They indicate that the sea water currents responsible for their formation flowed mainly in one direction.* Because such ripples tend to form parallel to a shoreline, their orientation in bedded rock structures provides *"recorded information"* about past geographies. Similarly, one can obtain information on the formation of sand dunes. (Dunes of different shape and size reflect the variation in sand supply, wind speed, and in the amount of stabilizing vegetation). *Most important is the fact that it is gravity which "stands behind" these structured recordings!*

I.1.3 Differentiation and Structuring by (Gravitationally-Induced) Fractional Crystallization.

Here gravitation helps directly and indirectly. As a melted magma (molten or semi-molten rock which originates underground) cools slowly, the first-formed crystals grow and *settle down* gradually. *This selective settling causes a change in the chemical composition of the remaining solution. Hence, the next minerals which crystalize (and settle atop the first layer), have a different composition. Thus, as the temperature of the mixture continues to fall, new minerals crystalize from the remaining liquid, i.e., a new layer of minerals forms on top of the previous layer, while the next mineral-species is growing and settling. As the remaining liquid cools further other minerals start to crystalize and settle to form new (gravitationally-induced) structures.*

Thus the rock materials removed by gravity-induced weathering and erosion are deposited elsewhere to form new sedimentary structures, either by settling from suspension, or by precipitation and crystalization from solution.

I.1.4 Gravitationally-Induced Volcanic Structuring

Rock formations and structured strata derived from the products of ancient volcanic eruptions, and the eroded remains of extinct volcanoes, form the landscape in many countries.

Volcanic eruptions can form many different structures. (They also form new ecosystems if they erupt, say, in the middle of heavily forested areas).

There are various types of volcanoes; from the *"fissure volcano"*, which builds up over long cracks in the crust (in this case the lava is usually runny basalt which spreads out as *thin* flows to form plateau structures), to the *"central volcano"* (which is fed by a single pipe and builds a cone with a summit crater). Basalt eruptions from central volcanoes produce thin lava flows, spreading over a wide area to form flatish volcanoes (*shield volcanoes*) shaped like upturned saucers. There are also the *"strato-volcanoes"* which are characterised by *steeper slopes* and in which short lava flows *pile up around a central crater.*

Many volcanoes have an eruptive cycle where initially gas pressure is high and activity explosive. As pressure is released by explosions the style of activity becomes milder, terminating in lava flows. *Such volcanoes are built of alternating layers of lava and ash.* (In explosive eruptions the embedded gas pulverizes the lava, producing billowed ash-filled clouds, which, then, settle to *form alternating structures of lava and ash*, as the phases of eruption change.)

As the volcanoe gets older it may progressively change its phases and modes of eruption. A volcanoe which starts life as a basalt shield volcanoe, erupting frequently, may change so that it erupts often but more explosively with pasty lava, building a strato-volcano on top of the older shield volcanoe. Thus, *as the volcanoe grows older, its internal structure becomes more complex.* The cone builds higher and higher, and greater gas pressures are needed to push magma up to the summit crater. Often it happens that lava does not reach the summit crater, but breaks out on the side of the volcanoe.

I.1.5 Structuring By (Gravitationally-Induced) Deformation

All structured sediments and most volcanic structures are initially formed as (almost) *horizontal layers*. In stable regions of the continental crust, and over much of the ocean floor, they have stayed so for many millions of years. But in unstable regions of the crust, the structured strata have been *folded, squeezed* and *fractured*, processes collectively called *deformation. The folded strata thus form new, more complicated structures.*

Folds can be all sizes, from huge arches or overfolds scores of kilometers across, to *microscopic* crinkles. *Fractures can be anything from the great dislocations which bound the plates of the earth crust, to minute displacements in single crystals.*

The way in which rocks change their shape under stress is determined by their composition, the temperature and pressure under which they are confined, the pressure of water contained in pores in the rock, and the strain-rate of compression in a given time. Depending on such factors, the rock may fracture, crush, fold, or flow plastically.

For example, a hard rock-quartzite—compressed rapidly at low, *near surface* temperatures and pressures, will *fracture*, or even be *crushed to powder! The same quartzite, deep in the interior of an active fold-mountain range, at the high temperatures and pressures that the gravitational pull of the earth generates, and under steady stress, will fold into intricate structures and shapes without cracking!*

I.1.6 Structuring By (Gravitationally-Induced) Metamorphism

Metamorphism means transformation. (Logically the term may be applied to *any profound change.*) In geology the meaning is restricted; it does not include the destruction of rock structures exposed to the weather, nor the fusion of rocks by igneous processes. Metamorphic structures have been developed from earlier *igneous and sedimentary structures*, by heat and pressure at some depth, and most effectively in the great mountain zones. *Resultant changes are in texture, in mineral composition, and in structural feature* (of the rock as a whole, *and of its crystaline structure—* see below).

The type of metamorphic structure produced depends on the composition of the original structure, the temperature and pressure, and the amount of water present. In certain situations, gravitationally-induced forces cause pre-existing structures *to crystalize, or recrystallize, into new structures.* Their constituent elements either *re-group* in new minerals, or *re-form* in larger crystals of the original minerals. Extreme metamorphism results from combined *dynamic* and *thermal* effects. Gravitationally-induced forces cause *magma to migrate toward the surface*, either following fractures, or drilling its way up by a combination of chemical attack and thermal shock.

Rapidly chilled magmas form glassy rocks. Slower cooling allows crystals to form; and, the slower the cooling, the larger and more structured the crystals. Some magmas crystallize to structures composed of approximately one mineral only, but most igneous rocks comprise several different minerals (see below).

Fluids *"rising"* from deep-seated plutons may combine with rock material in deformed sedimentary rocks, and the resultant product is indistinguishable from *granite* that has crystallized from magma. (To some extent, therefore, granite structures may be a product of metamorphic as well as igneous processes.)

Many of the metamorphic structures consist of flaky minerals, such as *mica* and *chlorite*, set in *parallel arrangement*. (These minerals cause the resultant structure to split into thin sheets, and the structures are said to be *foliated*). The commonest kinds of foliated metamorphic structures are *slate, phyllite, schist* and *gneiss. Marble* and *quartzite* are *nonfoliated metamorphic structures* (see below).

I.1.6.1 Sea-Floor Spreading and Structuring

Ocean floors are continually being added to by extrusion at the ridges of new, initially molten basaltic rock *"rising"* from the underlying mantle*. The spreading rates of new crust vary from 1 to 10 cm every year, on both sides of a ridge.

---

* E.g., the mid-Atlantic ridge

Proof of sea-floor spreading is found in the steadily increasing age of the sedimentary rocks which *"rest on"* the igneous crust with distance from the ridges and
and in *"fossil" magnetisation patterns* in the basaltic rocks.

As upwelling molten-basalt solidifies on reaching the sea-floor along the ridges
it is magnetised in the direction, north or south, of the earth's prevailing magnetic
field. Thus the evolving crust retains — *like a tape-recorder* — *a magnetic structure*
(or pattern) *identical on either side of the ridge.* These structured patterns can be
dated by reference to a verified time-scale of geomagnetic reversals (which occur at
irregular intervals of a few hundred thousand years).

The ridges formed by sea-floor spreading are, themselves, *new structures,* 2–4
km in height and up to 4000 km wide. They form a nearly continuous submarine
mountain structure over 40000 km in length, appearing above sea level in places
like Iceland. They are broken and repeatedly offset by innumerable *'transform faults'.*
(Deep, furrow-like *trenches* hundreds of kilometers long, tens of kilometers wide,
often flanked by chains of volcanic islands, border some oceanic regions).

I.1.7 Analysis of Metamorphic Structures

As an example, let us start with the analysis of the structure of marble (which is
a metamorphic structure formed mainly by recrystallization of limestones and dol-
omites, traversed by filled fissures and cracks or interbedded with mica schists,
phyllites, gneisses, granites, etc.) *Here gravity-induced earth movements often shatter
the crystal structure, producing fissures afterward filled with veins of calcite. It is in
this way that the beautiful brecciated or veined marbles are produced.* Sometimes the
broken fragments are rolled and rounded by the flow of marble under pressure, and
pseudoconglomerates result.*

---

* Bands of calc-silicate rock may alternate with bands of marble, or form nodules and patches, sometimes producing interesting decorative effects. The black and gray colors in marbles often result from the
presence of fine scales of graphite.

Even "the purest" of the metamorphic marbles contain mixtures of other minerals. The commonest
are quartz (SiO_2) in small rounded grains, scales of colorless or pale yellow mica, dark shining flakes
of graphite, iron oxides and small crystals of pyrite. The silicates, if present in any considerable amount,
may color the marble; e.g., green from green pyroxenes and amphiboles, brown from garent and vesuvian-
ite, yellow from epidote, chondrodite and sphene. *The presence of these minerals in the crystal structure
represents the undifferentiated mixture in the original medium which had differentiated during metamorphism
to form the new structures.* (In some marbles the structural bedding of some *sediments* were incorporated
in the final structure).

In addition to the various *macro-structures* described above we can easily detect distinct *micro-
structuring.* In marble, for instance, the principal mineral is calcite which varies in light transmission,
hardness and other properties in *various directions.* (Hence calcite *crystals transmit more light in one direction than in others.*)

Note also that slabs prepared for uses in which translucency is significant are cut parallel to that
direction. Note also that the bending of marble slabs is attributed to the *directional* thermal expansion of
calcite crystals on heating. Moreover, the basal faces of calcite crystals are less rapidly soluble than those
at right angles.

In general we say that a structure is isotropic if it is identical in all directions in space, (i.e., when it lacks any preferred internal directionality). *Planar structures*, for instance, show-up well in most *mica schists* (where the majority of mica flakes may be *parallel to a given plane*—the plane of *foliation, or schistosity*). In *linear structures* needle-shaped crystals show *preferred paralelism to a line*.

The origin of anistropic metamorphic structure is twofold:

1) The structure may be a *relic* after *premetamorphic anistropic structures* like sediments with deposition structure and lavas. Earlier *flow structures* may preserve their *anisotropic structure* as relic during metamorphic recrystallization (even if not exposed to anisotropic stress).

2) Anisotropic structure develops in a recrystallizing metamorphic medium *under anistropic stress. The stress may be very mild, like that of the gravity vector, or strong, like that produced by gravity-induced geological movements!* The stress causes strain of the medium (cataclastic with fractures, and plastic without fractures) and consequent rearrangement of mineral grains into various types of *anisotropic structures*.

Parallelism of mineral grains in metamorphic rocks with anistropic structures may be of *"dimensional orientation"* or *"lattice orientation"*.

Dimensional orientation refers to the *external* shape grains, whereas lattice orientation refers to the *internal* structure of minerals.*

Minerals with equidimensional outline such as those belonging to the regular crystallographic system may show preferred lattice orientation which can be detected only by accurate microscopic studies.**

Lattice structures are determined experimentally by *x-ray analysis* which also reveals the crystal structure. The atomic distances may then be calculated from the data (when x-rays are directed at a crystalline material they are diffracted by the planes of atoms or ions within the crystal. The diffraction angle depends on the wavelength of the x-rays and the distance between adjacent planes).

It should be stressed, however, that noncrystalline, amorphous (literally, "without form") materials have structures too. The structure of liquids has much in common with the structure of crystals; their densities and therefore their *packing factors*, are within a few percent of each other. A liquid is usually slightly less dense than a crystal. (However, this is not a general rule).

---

* Quartz grains in schists and gneisses may, for example, have lenticular shapes unrelated to the internal crystal structure. Such quartz lenses are often dimensionally oriented parallel to the schistosity plane.

** The repeating three-dimensional pattern in crystals is due to an atomic coordination within the material (*internal orientation*); and this pattern *sometimes controls the external shape of the crystal* (dimensional orientation). The planar surfaces of gems, quartz (SiO_2) crystals, and even ordinary table salt ($NaCl$) are all external manifestations of internal crystalline arrangements.

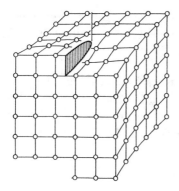

A simple lattice imperfection in crystals.

Lattice imperfections are found in most crystals. In some cases when *missing* atoms, *displaced* atoms, or *extra* atoms are involved, there may be *point defects*. The *line defect* involves the edge of an *extra plane* of atoms. Finally, imperfections may be in *boundaries*, either between adjacent crystals, or at the external surfaces of the crystal. *Such imperfections influence many of the characteristics of materials, such as mechanical strength, electrical properties and chemical reactions!*

I.1.8 Asymmetry, Structures and Gravitation: Is there a Fundamental Connection?

The great bulk of terrestrial solid matter is crystalline, and therefore the properties of crystals are to some extent also the properties of ordinary solid materials. Few things in nature are more perfect than a crystal in which immense numbers of atoms or molecules, all identical, are stacked in (almost) complete symmetry. Rarely is even one atom in a thousand out of line. Yet, surprisingly, many of the most important properties of crystals are due to the few odd places where the crystal *"goes wrong"*! In fact many crystal structures could not have grown at all without having *asymmetric* "imperfections" in them! *What causes this asymmetry?*

Only rarely in nature do actual crystal structures present *"perfect symmetry"*! *It usually happens that the crystal is so placed with respect to the vector of gravity that there will be a more rapid deposition of material on one part than on another!!!* (Only when a crystal is *"freely suspended"* in the mother-liquid under conditions of *weightlessness*! and the material for growth is supplied *at the same rate on all sides* may a perfect symmetry result.)*

The gravitational field is therefore a most powerful factor in the evolution of all asymmetric crystal structures, from metamorphic rocks to prebiotic or biotic crystal structures!!! In fact we know today that in a crystallizing suspension of organic spheres

---

* Investigations aboard SPACELAB 3, for example, include measurements of the effects of residual, transient, forced convection and growth kinetics on the homogeneity, defect structure, and external morphology within solution-grown crystals (NASA Application Notice, Dec. 14, 1977).

in water the earth's gravitational field produces an elastic deformation that is readily observed through its asymmetric effect on the evolving crystal structure! Moreover, it should also be stressed here that these very processes play a most important part in crystallized virus systems!!!
Consequently, gravity provides a most important biological function by excluding foreign particles such as antibodies from a crystal structure!!! (see also Assertion 9, *Introduction). The evolution of life on earth is therefore strongly tied up with gravitation! In fact, without gravitation, there could be no life.* This topic will be taken up again in Lecture IX (§2.3.1).*

In the rest of this lecture we discuss the various roles gravitation plays in astrophysical structures.

For that purpose we first introduce a number of technical terms starting from the concepts of *luminosity, surface temperature, H—R diagram* and *stellar evolution*.

I.2 STELLAR STRUCTURES AND THE HERTZSPRUNG-RUSSELL DIAGRAM

I.2.1 Luminosity, Surface Temperature and Stellar Evolution

We must first distinguish between two definitions of luminosity, namely: *apparent luminosity, l,* which is the rate (a directly measurable quantity) at which radiant energy from a star or a distant galaxy, impinges on a unit area of the earth's surface,** and *absolute luminosity,* L, defined as the rate of total radiant energy dissipated in outer space from a given galactic or stellar object. Then we have

$$L = 4\pi r_L^2 l \tag{1}$$

where r_L is the *luminosity distance* to the light source at rest with respect to earth. (Eq. (1) is valid in Euclidean space).
The various gravitationally-induced states a radiating star can evolve into are conveniently described by a diagram in which absolute luminosity is plotted against surface temperature, T_s. This so-called *Hertzsprung-Russell (H—R) diagram* is shown in Fig. I–1. Only portions of the diagram are occupied by actual stars, with the majority of typical stars, like the sun, lying on, or near the so-called *main sequence,* which runs diagonally from lower right to upper left. The more massive a typical star is, the higher its T_s and the higher its position on the main sequence.

The H–R diagram is useful in following the gravitational evolution of individual stars. It is only to the casual observer that the stars appear to be of fixed brightness.

---

* *Note Added in Proof:* The predictions made here & in the Introduction, Lectures VII and IX, have now been partially verified by Prof. H. Eyal–Gilladi of the Hebrew Univ. ("Ha-Aretz", Nov. 27, 1980).
** In astronomy it is called "power per unit mirror area," and denoted $l = p/A$.

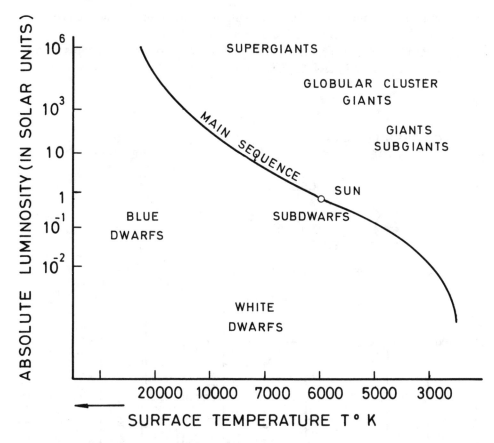

Fig. I.1 Gravitational Structuring and H—R Diagram

This diagram demonstrates the fact that gravitation controls the evolution of all stars according to their mass (and hence according to their gravitational pull and surface temperature). Most stars are in the main sequence. Unlike this schematic representation, in the actual diagram the branches are not narrow and well defined, the spread being due to errors in measuring absolute luminosities and to actual differences in structure and chemical composition among stars belonging to the same class.

Any main-sequence star follows a track which depends on *its initial mass* (and hence on its internal temperatures). In the course of the evolutionary process, many stars eventually move to the lower left of the H–R diagram, where they contract into tiny "white dwarfs"—stable iron bodies supported by the pressure of cold degenrate electrons. Collapsing stars may become supernovae, neutron stars or black holes (Figs. I.4, and VIII.2).

In fact, most of them—even the sun—vary to some extent, and some, such as the cepheids, actually fade and brighten periodically (see Fig. I.12). Generally, each star may, in its unique way, be rotating about its axis, or orbiting round a companion, or pulsating or erupting, often doing all four at the same time in a quite complex manner. These gravitationally-induced processes are monitored with great interest by astronomers.

The evolutionary changes in a star's absolute luminosity and surface temperature can be calculated theoretically and shown as an evolutionary track on the H—R diagram. Results indicate that the evolutionary track followed by a main sequence star *depends on its initial mass, i.e., it depends on the gravitational field produced by each star*!

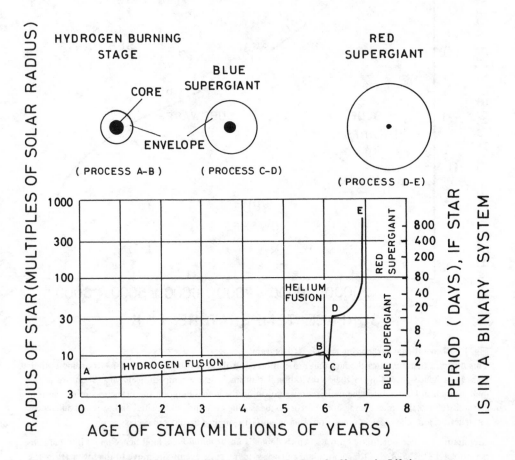

Fig. I.2 **The Stronger the Gravitational Field in the Core of a Star the Shorter its Lifetime.**
The initial mass of the star determines its evolution and life span. Generally, the *larger* the mass of a star, the *shorter* its lifetime. Further evolutionary stages are described in Figures VII-1, 2 and I.3. For the 'age of the universe' see Fig. III.4 and Lecture VIII. In this example the mass of the star is 20 solar masses. As the mass increases the rate of synthesis in the stellar core-increases. Hence the rate of hydrogen fusion in the sun is much slower than the value given for this massive star.

During the process A—B, hydrogen nuclei fuse to form helium in the coreof the star and its radius increases very slowly (initial composition at A: 70% H, 27% He and 3% heavier elements). Core temperature during this stage is about $2-3 \times 10^7 °K$ (Figures VII-1, 2). By point B, all the hydrogen in the core has been consumed and the core starts to contract—(process B–C), whereby its temperature increases. This heating causes the stellar envelope, which is still rich in hydrogen, to expand (C–D), forming a *blue supergiant*. At point D the core temperature is about $10^8 °K$ and helium nuclei begin to fuse, forming carbon and some oxygen (see Figures VII-1, 2). The helium-burning stage passes through several giant phases (*red supergiants*), process D–E

I.2.2 Main-Sequence Stars and Age

Throughout most of their lifetime, stars remain near the main sequence, moving up along it as the hydrogen in their core is depleted by conversion into helium (Figs. VI.1 and VI.2). When all the hydrogen has been used up, they veer to the right and become so-called "giants" (Fig. I–2).

Their evolutionary tracks then become quite complex and may involve one or more excursions into regions of instability. Typically, hydrogen burning releases about 6×10^{18} erg/gr of radiation energy (Fig. VII.2), a figure which can be employed to calculate the age of the star at the turn-off point. Thus, star clusters* vary from a-bout 10^6 years, the age of very massive stars, to about 10^{10} years, the age at which stars about as massive as the sun reach the turn-off point. This latter result is highly significant, since it allows an independent age comparion with time scales predicted by general relativistic cosmological models or by other empirical methods (e.g., Hubble's inverse constant).** The turn-off point in the example described in Fig. I.2 is reached after 6×10^6 years (point B).

I.2.3 Iron White Dwarfs and Chandrasekhar Limit

In all evolving stars, nuclear and gravitational energies are transformed into *outward*-radiating photons and neutrinos. The star subsists on its resources through-out a series of (gravitationally-induced) nuclear-fuel burning stages, each charac-terized by a different temperature (Fig. VII.2). During these gravitationally-in-duced stages the surface temperature and the size of the star change. A normal stable star adjusts its size like a *"thermostat"*, whereby the rate of energy release from a particular nuclear reaction just suffices to offset the radiant energy "loss" into unsaturable space (plus the energy needed for adjusting the pressure to balance the different gravitational force). Inside cores of highly evolved massive stars the temperature may reach 10^{10}°K, enough to convert matter, through a chain of exo-thermic nuclear reactions, into the most *stable* element–iron. In a typical star, tem-perature gradients between interior and outer layers are about ten million degrees C over a radius of a hundred million km i.e. ~ 0.1°C/km. (For comparison, in the earth's core the temperature drops 4200°C over a radius of about 6000 km).

Following such gravitationally-induced processes, the tracks of many stars eventually move to the lower left of the H—R diagram, where they become tiny *white dwarfs*—stable iron bodies supported by the pressure of cold degenerate electrons. White dwarfs are aged stars that have exhausted their nuclear fuel, cooled and contracted to a small size (of the order of the earth). When the temperature is sufficiently low, the electrons are frozen into the *lowest* available energy levels. Ac-

---

*Assuming that most of the stars in the cluster were formed at roughly the same time and are similar in composition.

** See Lectures VI, and III.

cording to the Pauli exclusion principle, there will be two electrons in each level (because of the two spin states available). Thus using Newtonian physics and quantum-mechanical computations for an equation of state and a polytrope Chandrasekhar was able to estimate the maximum stable mass and size of a white dwarf. This is called the *Chandrasekhar limit*, and equals $1.26M_\odot$ to $1.4M_\odot^\dagger$. The diameter of a white dwarf with this maximum mass is only about 8000 km. Theoretical considerations demonstrate that this maximum is a point of transition from stability to *instability*, so stable iron white dwarfs can exist only for masses *less* than about 1.4 times the mass of the sun (Fig. I.4).

The gravitationally-induced evolution of a white dwarf may be smooth and un-eventful so long as its mass does not exceed the Chandrasekhar limit, for then it can cool off slowly and approach the *final black dwarf state*. During that approach *the gravitational force of contraction would be balanced by the degenerate electron pressure*, and the entire star would approach the low *temperature of intergalactic or inter-stellar space* (now at a minimum of $2.7°K$, depending on location—for details see Lecture VI, section 2.1).

Stars *above* the Chandrasekhar limit must *lose mass* before they become stable white dwarfs. Thus stars that begin their gravitational evolution above that limit on the main sequence and end as white (and, later, black) dwarfs *must be mass emitters*!

The rate of mass loss depends on the evolutionary stage of the star, on its ro-tational state, on whether it is a member of a binary system or not, etc. However, other interesting processes may take place inside stars whose mass is *above* the Chandrasekhar limit. These are briefly described next.

I.3 SUPERNOVAE, GRAVITATIONAL COLLAPSE, NEUTRON STARS, PULSARS

What happens when a star, *whose mass is above the chandrasekhar limit*, reaches the end of its thermonuclear evolution and begins to cool down? At this stage nuclear reactions no longer supply the energy required to maintian the high internal pressure, gravitational contraction sets in and the internal structure starts to *collapse*. One *possibility* is that the core of the star becomes so heated during the collapse that the star explodes, becoming a *supernova*. Of all supernovae that have been recorded, none has been studied in more detail than the famous explosion of July 1054, which formed an expanding luminous cloud known as the *Crab Nebula*. As reported at the time by Chinese astronomers, the new star whose birth they witnessed, grew in brightness from day to day and in a few days outshone every other star, after which it began to fall off gradually. On the basis of these reports, one can reconstruct

---

† depending on the detailed methods used in the computations. (See also Lecture VII).

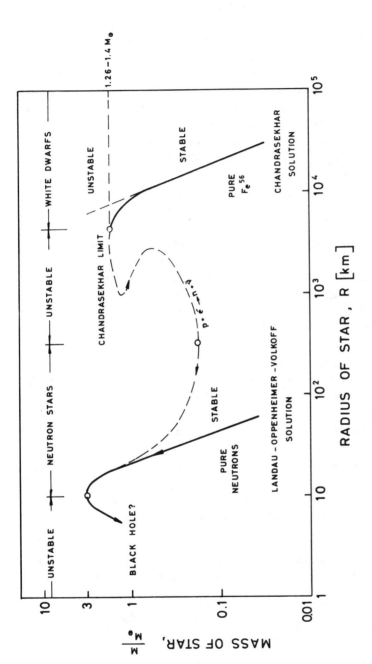

Figure 1.4 Gravitation and Superdense Structures.

The larger the mass, the smaller the final radius of a stable star. Shown are regions of stellar stability and evolution from iron white dwarfs to neutron stars. Solid straight lines represent the extrapolated non-relativistic expressions of Landau-Oppenheimer-Volkoff and Chandrasekhar, respectively. Dashed line represents the interpolating solutions of Harrison, Thorne, Wakano, and Wheeler, which take into account the shift in chemical composition from pure Fe^{56} to pure neutrons. Arrows indicate the direction of increasing central density. Transitions between stability and instability occur at the maxima and minima of M (indicated by circles). A typical density of a white dwarf is about 10^6 gr/cm³ and its radius is about 1 percent of the solar radius for about 1 solar mass. If the mass exceeds 1.4 times that of the sun, the superdense star cannot be a white dwarf. If the mass ratio exceeds 3, the superdense star cannot be a neutron star, in which case it may become a *black* hole. See also Figs. VIII.2 to 4.

a light curve characteristic of similar events which the powerful telescopes of today show, from time to time, in far away galaxies.*

Today, nine hundred years after the explosion, the boundary of the Crab Nebula has moved out about 3 light years from its center. The pattern of the expanding cloud is not uniform; inside it one can detect waving "tentacles" involving magnetic lines of force, centered on which electrons move in circular and spiral tracks radiating strong radio waves, visible light and intense X-rays into space. The source of the magnetic field was in the star itself before it exploded. In fact there is no star, so far as we know today, which does not have such a field associated with it. When the explosion occurs, the magnetic lines of force embedded in *ionized* matter follow the matter in its outward motion. Consequently, the motion of matter affects the magnetic field and the field, conversely, affects the motion—a highly complicated case of *dynamic coupling*, whose complexity is aggravated by the fact that supernova phenomena involve turbulent motions. Turbulence is a difficult empirical aspect of fluid dynamics and *magnetohydrodynamic turbulence* is a particularly difficult topic of that science. Nevertheless, by observing the pattern of brightness of supernova radiation, its absolute luminosity, its spectral composition and its direction of polarization, astrophysicists gained an idea of the structure of the tentacles, the magnetic lines of force, the electron orbits and the total energy released in the explosion (of the order of 10^{49} ergs).**

From laboratory experience and from processes observed in solar flares, or in interplanetary space, astrophysicists know that instabilities associated with ionized gas (plasma) and magnetic fields lead to *acceleration of protons* to very high energies. Such processes may well occur with much greater violence in supernova explosions. The latter may thus be important sources for the observed flux of *cosmic rays* reaching the earth. In this context it may be mentioned that the heavy elements in the crust of the earth and the other solar planets, (which are atypical in the world accessible to our observations, where about 97–99% of all matter consist of two elements only: *hydrogen* and *helium*), may well have *originated in an early supernova explosion*.*** Such a hypothesis is not ruled out in the light of available age data, including those

---

* It is this extended reach of the modern telescope which makes systematic observation of such events possible. Within our own galaxy—the Milky Way—they may occur about hundreds of years apart on the average.

** Type I supernova, easily recognizable by means of their light curves and/or spectra, have been successfully used as distance indicators (see § I.7). Members of this class may be detected up to distances of 500 Mpc. In this range they are better distance-indicator candidates than the galaxies themslves [Barbon, R., Capaccioli, M., and Ciatti, F., *Astron, Astrophys.* 44, 267 (1975)].

*** Most cosmic rays consist of high-energy protons, although α-particles and heavier nuclei up to iron are normally present. Heavier particles, up to the transuranic elements, may also be found. The mean energy of a cosmic-ray proton is about 2 Bev but individual particles may reach 10^{11} to 10^{13} Bev (i.e. 10^8–10^{10} ergs). By contrast, the most powerful accelerators of this generation would produce particles of less than 10^3 Bev. The flux of cosmic rays reaching us is isotropic and constant in time (except below

of our own galaxy and of the moon and the earth, as well as of time scales supplied by standard cosmological models and by evolution models for massive stars. In fact, it is generally held that the *solar planets* formed about 4.6 billion years ago through aggregation and condensation of dust, debris, meteors and gas that had been previously dispersed in interstellar space.** Interacting later with the protosun to form a large disk-shaped system, the original bodies have since *increased* their size through "sweeping" to produce the present planets. (Some of the debris have not yet been "swept up". They keep orbiting in the solar system, forming, *inter alia*, part of the *asteroid belt*).

There exist, however, other theories as well, concerning the origins of the solar system and of the heavy elements. The latter may also be associated with the onset of expansion of the universe, a process which bears a certain similarity to supernova phenomena and gravitational collapse.

Gravitational collapse occurs as the temperature in the iron core approaches about $8 \times 10^{9}\,^{\circ}K$ and is caused by endothermic (energy-consuming) breakdown or photo-disintegration of the atomic nuclei, (mainly by the correspondingly high energy of γ-rays). The initial energy input is provided by the core, but so much of it is absorbed by the disintegration process (see Fig. 6a), that the *pressure inside the core decreases rapidly*. This, in turn, induces contraction of the core, and an *inward fall of outer layers) releasing gravitational energy*, which in turn, is absorbed by the core

about 10 Bev, where the solar system exerts some non-uniform time effects). In interstellar space their density is about 1 eV/cm^3 (or 10^{-12} erg/cm^3), which is comparable with that of starlight. (It is a matter of definition whether one includes in cosmic rays such elementary particles as electrons, X-rays, γ-rays, and neutrinos).

** Average chemical composition of ordinary and carbonaceous meteorites and of lunar soil samples are shown in the following table (in percentages).

Component	Ordinary Chondrites	Carbonaceous chondrites	Lunar samples
silicates	75–86	76–90	98–100
Water	0.2–0.3	1.21	0
Free metals	8.3–19	0.1–3.5	0–1
Carbon	0	0.1–3.8	0
Nitrogen	0	0.01–0.3	0

These values were compiled by J.A. Wood [in *The Moon, Meteorites and Comets*, B.M. Middlehurst and G.P. Kuiper, Eds. (Univ. of Chicago Press, Chicago 1963), p. 337]. the values for carbon and nitrogen were given by E.K. Gibson, C.B. Moore, and C.F. Lewis [*Geochim. Cosmochim, Acta 35*, 599 (1971)]. The range shown for the lunar samples encompasses the mean compositions for the eight Apollo and Lunar sites, as compiled by S.R. Taylor [*Lunar Science: A Post-Apollo View* (Pergamon, Elmsford, N.Y. 1975)] and D.R. Criswell [in *Proceedings of the Second Princeton Conference on Space Manufacturing Facilities* (American Institute of Aeronautics and Astronautics, New York, 1977)].

for further disintegration, and so on. This so-called *implosion* continues until the density in the core becomes so high that a bounce occurs.* This in turn generates inward-directed *strong shock waves* and *very intense high-energy neutrino fluxes*, as a result of which the regions surrounding the core undergo very rapid heating. Since the outer layers still contain such elements as H^1, He^4, C^{12}, O^{16}, etc. (Fig. 6a), the temperature buildup promotes *very rapid exothermic thermonuclear burning*. This in turn releases so much energy in a very short time that it blows off the outer layers of the star and converts it into a luminous expanding supernova.

A supernova explosion may blow off enough matter for its mass to drop below the Chandrasekhar limit. It is widely believed that in this case the highly compressed *remnant* does not end up as a white dwarf, but rather becomes a superdense *neutron star* consisting almost entirely of *degenerate neutrons*. To understand how the latter are formed from degenerate electrons plus heavy nuclei we must evaluate the changes in the Fermi energy of the electrons as the density of the core increases. At densities greater than 10^6 gr/cm^3 complex nuclei can no longer exist because the Fermi energy of the degnerate electrons is high enough to allow the electrons to *penetrate* the *nuclei* and transform the protons to neutrons according to

$$p + e^- \rightarrow n + \nu \tag{2}$$

The neutrinos generated by this reaction escape the star and dissipate in space. Here, the nuclear energy balance favors breakdown of the nuclei with emission of neutrons, which cannot (according to the Pauli exclusion principle) decay into protons and electrons by the reaction

$$n \rightarrow p + e^- + \nu \tag{3}$$

Actually, things are not so simple. Detailed calculations by Harrison, Thorne, Wakano, Wheeler, Zeldovich, Novikov, Cameron, and others, show that the actual behavior may be represented schematically by the dashed line in Fig. I.4, and moreover that the maximum stable mass of a pure neutron star is given by**

$$M_m \cong 3M_\odot \tag{4}$$

This stable mass is known as the *Landau-Oppenheimer-Volkoff limit*.

A photon emitted from the surface of such a neutron star is red-shifted by about $z = \Delta\lambda/\lambda = 0.13$, indicating that general relativity comes into play in massive stable neutron stars. Calculations by Chiu and Salpeter§ show that if such stars have a core temperature of, say, 2×10^9 °K, their surface temperature would be 1.2×10^7°K, and their rate of energy loss to space around 4×10^{39} erg/sec for neutrinos, and 2.3×10^{37} for photons.

---

* See, however, the discussion on black holes below.
** Depending on the model employed, M_m may vary from 0.4 $M_\odot$ to about 3 $M_\odot$.
§ *Phys. Rev. Letters, 12*, No. 15, 414 (1964).

Neutron stars of low mass are, therefore, much like white dwarfs of the same order of mass, except that they are about a thousand times smaller in size and neutron-degeneracy pressure replaces electron-degeneracy pressure, and their density is about a billion times higher! (Fig. I.4).

The discovery (in 1967) of *pulsars*, stars that emit radiation in *regular pulses* at intervals ranging from a few seconds down to a small fraction of a second, first led Gold to suggest that they are actually *rotating neutron stars*. Gold's idea was confirmed by observations and is generally accepted today. Pulsars are left behind after a supernova explosion* and may begin life with a *very high rotational frequency* (perhaps as high as 10^4 sec^{-1}). Subsequently the rotational speed *slows down*[†], since the star loses energy through electromagnetic braking effects, electromagnetic radiation and perhaps even gravitational radiation.

To account for the observed pulses it is necessary to assume that the neutron star lacks circular symmetry about the axis of rotation, as would be the case if its

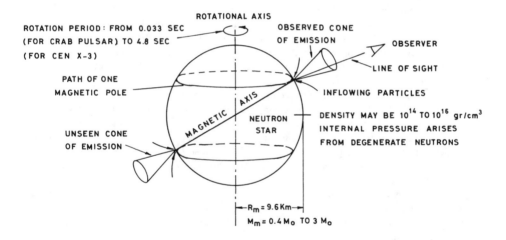

Fig. I.5 **Gravitational Clocks: Model of Pulsar Star showing the rotating neutron star with emission regions near the surface of the magnetic poles.**
The emission processes near the magnetic poles are regarded as analogs of the auroral zones on earth (although their detailed mechanisms are totally different). The pulsars are pulsed radio sources, but may also emit optical gama or X-ray radiation. The pulses repeat very exactly, with a period of 0.033 second for the fastest known rotating pulsar (the Crab Pulsar), to about 4 seconds for the slowest. The dipole field lines give a preferred direction to each part of the emission region producing the linear sweep of polarization observed in radio pulses. Due to dissipation of energy, some pulsars slow down with time (about 10^{-5} second/year for the Crab pulsar). (See also Figure I.6.)

* About a year after their original discovery, one was also found in the *Crab Nebula* and one in the *Vela Remnant*—verified sites of supernova explosions.
† On the other hand, some pulsars display restless behavior and even accelerate (see below).

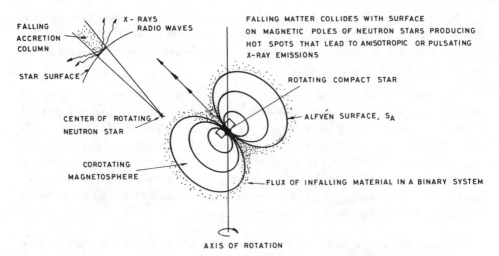

Fig. I.6 Gravitational Clocks: Pulsating sources produced by accretion of matter (see also Fig. I.9) onto the surface of a rotating compact star.

The combined effect of rotation and magnetic poles offset from the rotation poles can lead to *pulsatory x-ray and radio emission.* Gravitational energy is converted here into observed radio and X-radiation. Compact star may be a degenerate dwarf or a neutron star, but for each the detailed process is different. Far from a compact rotating magnetic star, the stellar magnetic field does not appreciably affect the flow dynamics of accreting matter; but close to the star, the magnetic field dominates flow dynamics, causing infalling matter to fall on *"hot spots"* through *"magnetic funnels".* The location of the *Alfvén surface,* S_A (where transition between the two flow regimes occurs) depends on the flow pattern of the accreting matter beyond it; the rotation period of the compact star, the mass accretion-rate, and the stellar magnetic moment. In close binary systems, three distinct flow patterns are possible: (i) approximately *radial inflow,* which occurs when the accreting matter has insufficient angular momentum for centrifugal forces to halt its free fall; (ii) *orbital motion* with *possible disk formation*,* which occurs when the accreting matter has sufficient angular momentum for the centrifugal forces alone to halt the matter outside S_A; (iii) *streaming motion,* which, if supersonic, forms a collisionless bow shock; it may be relevant in the case of accretion fed by the stellar wind from a close binary companion (Fig. I.9).

Whether one gets essentially radial inflow or orbital motion near S_A depends on the velocity and angular momentum of the streaming matter. For most geometries, two pulses of different intensity can be expected for each revolution of the compact star. However, if the rotation axis of the star coincides with the line of sight, or the magnetic axis, there will be *no pulses,* while if the rotation axis is perpendicular to the line of sight, or magnetic axis, there will be two *identical* pulses for each revolution.

Considerable observational evidence indicates that Herc. X–1 is an accreting magnetic neutron star in a binary system. This is the best known object after the Crab pulsar. It is an eclipsing (1.7 days), periodically pulsating X-ray source (1.25 sec). Its binary nature is reflected in Doppler shifts in the period of pulsation, in phase with the occultation period (see Fig. I.7 for explanation). It also shows optical behavior of the same pattern.

* If matter flattens and forms a disk, viscous forces in the disk transport angular momentum outward, allowing gradual inflow of matter toward the compact star; the inner edge of the disk is located at $S_{A'}$ where the flow of matter becomes dominated by the stellar magnetic field.

magnetic poles are offset from its rotation poles (Figs. I.5 and I.6). There are, broadly speaking, two schools of thought concerning radiation from pulsars. According to one view it originates near the surface at the magnetic poles of the star as shown in

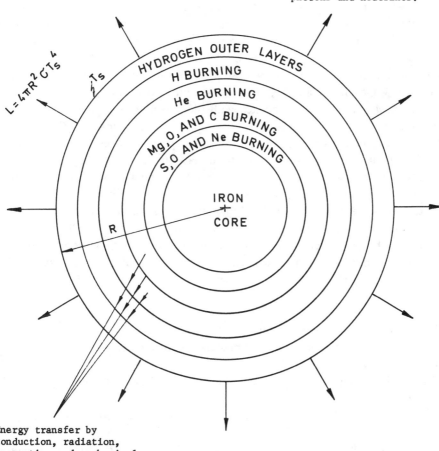

Radiation energy lost
in unsaturable outer
space in the form of
photons and neutrinos.

Energy transfer by
conduction, radiation,
convection and mechanical
means (such as hypersonic
jet streams, expansions,
explosions, or pulsations).
Temperature gradients may
normally be 0.1°C/km,
decreasing from center to
surface.

Fig. 6a: **Gravitation, Structure and Synthesis of the Chemical Elements:**
Schematic representation of nuclear "burning stages" in a highly evolved star.

During the various stages of stellar evolution the rate of energy release inside the star equals that of (ir-reversible) energy dissipation in outer space. In the interiors of highly evolved stars the temperature may reach $10^{10°}$K, enough to convert matter into the most stable element—iron. Iron, however, may undergo endothermic decomposition, and this may cause instability, explosion, or implosion which leads to gravitational collapse (Fig. I.4). Iron stars may also cool off peacefully and contract into white dwarfs and finally into black dwarfs. See also Figs. I.1 and I.4.

Fig. I.5. According to the other it involves streams of plasma, originating in non-thermal processes at unknown surface locations and flowing out in a light cylinder; the neutron star is assumed to have a co-rotating magnetosphere reaching out to a circle at which the peripheral speed is close to that of light, and the plasma thus accelerates to relativistiç speeds and is flung out of the magnetosphere. In this model there is no indication of what happens beyond the light cylinder and how the magnetic field is oriented. By contrast, the model shown in Figs. I.5 and I.6 is very clear about that and also accounts for the observed variation in the plane of polarization of the radio waves during a pulse. In this model the radiation cones can be identified with the *"magnetic funnels"* associated with the two magnetic poles. The emission processes near these poles are regarded as analogs of the *auroral zones on earth* (although their detailed mechanisms are totally different). It is estimated that in the case of the Crab pulsar, ions accelerate to at least 10^7 Bev per electron charge; this acceleration is assumed to take place very close to the neutron star, probably within about 1 km. of the surface. Since the ions and electrons follow curved magnetic field lines, they are copious emitters of high-energy *gamma rays* (of the order of 10^3 Bev). As the Crab and other pulsars produce optical and X-ray radiation, it is assumed that the key mechanism operative in the conversion of gamma-rays into X-ray, optical and radio emission is *pair creation* in the intense magnetic field.

*   *   *

Some questions which need further clarification are listed below:

1) How does a single neutron star manage to convert its rotational energy into energy of relativistic particles?

2) Detailed mechanism of mass transfer (accretion) in close binary systems (See section I.4 below).

3) The behavior of the pulse frequency in the period following a steep *speedup*, or *macroglitch*.

4) The behavior of compact stars between 2×10^{14} and 10^{16} gr/cm^3; the presence of superfluid neutron and protons; crust- and core-quakes; hydrodynamic instabilities associated with fluid neutron cores, liquid and crystalline neutron matters, etc.

5) Detailed mechanisms of formation of random and regular pulses.

I.4 X-RAY ASTRONOMY, BINARY X-RAY SYSTEMS, AND GRAVITATIONAL CLOCKS

I.4.1 Recent Discoveries

The most exciting area of space astronomy research today is the high-energy field. Recently, the *High-Energy Astronomy Observatory* (*HEAO 2*) X-ray telescope spacecraft (called also the *Einstein* Observatory) returned the first images ever obtained of X-ray sources in space from an orbiting satellite.

First target imaged by the telescope was *Cygnus X-1*, an X-ray source believed also to be a *black hole*. Cygnus X-1 is a binary star system (see below), located about 6,000 light years from earth (see Table I.1 for details and §I.5 and Lecture VIII for a discussion of black holes).

Binary star systems consist of a *superdense star* orbiting closely round *a larger 'normal' star*;—bound systems which operate like huge clocks in the depths of space. Many such systems have been discovered in the past decade, especially by the Small Astronomy Satellite (*SAS-1*), designated *UHURU* (the Swahili word for freedom), and by *HEAO 1*. These two observatories returned only numerical data, but these early data were enough to revolutionize many of our physical concepts while, at the same time, have verified some possible large-scale mechanisms predicted by Einstein's general theory of relativity (see below and in Lecture III).

Previous discoveries, through rocket-borne instruments, of extrasolar sources of X-radiation, demonstrated the potentialities of satellite instrumentation. The UHURU project, entirely devoted to X-ray observations, was expected to lead to detection of fainter sources with finer resolution and positional accuracy. What was in fact achieved was a revolution in our understanding of the nature of important cosmic objects, with X-ray observation becoming one of the most active branches of modern astronomy. (In 15 months, UHURU gathered enough data to compile a detailed catalogue of 125 X-ray sources. Many more have been detected since).

The presence of *pulsating* X-ray sources in *binary systems* was the first indication of the important role of *mass accretion* as the energy source for generation of the very large X-ray fluxes observed in these objects. Thus, the presence in these systems of white dwarfs, neutron stars, and perhaps even black holes, supply us with an insurmountable, ready-made, physical laboratory.

I.4.2 The Nature of X-ray Sources

The first indication of the nature of X-ray sources came in 1966, when the strongest of them, *Scorpius X-1*, was identified *optically* with a faint blue starlike object that resembled an old *nova*—an *exploded star*. (Earlier it had been established that old novae are *close binary systems* in which one of the stars is a *white dwarf*. In such a system gaseous matter is *transferred* from the normal companion to the small white dwarf, leading to an explosion in the outermost envelope of the latter). This discovery supported the assumption that the X-ray sources are binaries and that the X-rays are emitted from the normal companion star through an extremely heated cloud consisting of gas captured by a white dwarf, a neutron star, or even a black hole. They are generated by the enormous *acceleration* and *viscous heating* of the accreted matter during its fall toward the compact star, in the course of which its temperature may rise to 10^{7}°K. New evidence reported by the UHURU team focused attention on the *pulsating X-ray sources*; *Centaurus X-3* and *Hercules X-1* were found to pulsate precisely every 4.84 and 1.2 seconds, respectively (Table

TABLE I.1: Properties of Some Binary X-ray Sources shown in Figs. I.7, I.8 and I.9*

Name of X-Ray Binary and its Catalogue Number	Binary Period (Days)	Eclipse and Pulsation Periods of X-ray Emission	X-ray Luminosity ($\times$ Sun's Luminosity)	The *Invisible* Companion	The *Visible* Primary Star	Distance (Light Years)	Remarks
Cygnus X-1 (3U 1956 + 35) Black Hole? (Cyg X-1)	5.6	Random variation on time scale of between .001 – 0.005 and 60 sec. No Eclipse.	10^4	$M \geq 8 M_\odot$	Ninth-Magnitude Blue Supergiant; $M > 20 M_\odot$ Image taken also by HEAO-2 X-ray telescope	6,000	Identified also as a weak radio source which, in turn, led to optical identification with a spectroscopic binary of 5.6 days period.
Centaurus X-3 (3U 1118–60) (Cen X-3)	2.087	X-Ray *Eclipse*: .488 Day; Pulsation Period 4.84239 seconds.	10^4	$M \sim 0.6 - 1.1 M_\odot$?	13.4-Magnitude Blue Giant; $M > 16 M_\odot$	25,000	
Hercules X-1 (3U 1653 + 35) (Her X-1)	1.7	X-Ray *Eclipse*: .24 Day; Pulsation Period 1.23782 seconds.	10^4 Around 10^{12} – gauss field on its surface	$M \sim 1.3 - 2.5\ M_\odot$	Varies with 1.7-Day period between 13th and 15th Magnitude: $M \sim 2 M_\odot$	16,000 Age may exceed a billion years.	35-Day variations in light curve. Doppler shift in pulsation period, in phase with occultation.
SMC X-1 (3U 0115–37)	3.89	X-Ray *Eclipse*: .6 Day.	2×10^5	$M \sim 2 M_\odot$	13.4 Magnitude Blue Supergiant $M > 25M$	190,000	An X-Ray outside our galaxy.
VELA X-1 (3U 0900–40)	8.95	X-Ray *Eclipse*: 1.7 Days : Slow Flares on Time Scale of Hours.	10^3	$M = 1.6 - 2.1 M_\odot$?	Seventh-Magnitude Blue Supergiant : $M > 25 M_\odot$	8,000	Weak Radio Source. Supernova remnant.

TABLE I.1: (continued) Properties of Some Binary X-ray Sources

Name of X-Ray Binary	Binary Period (Days)	Eclipse and Pulsation Periods of X-Ray Emission	X-Ray Luminosity (× Sun's Luminosity)	The *Invisible* Companion	The *Visible* Primary Star	Distance (Light Years)	Remarks
SS 433 (4U1908 + 05)	164 days		opposing 2 jets?	$\sim 1M_\odot$? neutron star?	14 Magnitude	13,000	Rotating jets?
3U 1700–37	3.412	X-Ray *Eclipse*: 1.5 Days.	10^3	$M = (1.3 \pm 02)\, M_\odot$?	6.7 Magnitude Blue Supergiant; $M > 30 M_\odot$	9,000	Weak Radio Source.
Circinus X–1 (3U 1516–56)	Probably Longer than 15.	X-Ray *Eclipse*: Duration longer than a day; rapid irregular variation similar to that of Cygnus X–1.	?	?	Optical component not yet found.	?	
Cygnus X–3 (3U 2030 + 40) (Cyg X–3)	4.8	Sinusoidal Variation with 4.8 hour period.	At least 10^4. Peaks of more than 4×10^4.	$\sim 1M_\odot$(?)	No visible star; infrared source with 4.8 hour period.	At least 25,000	Source of intense Radio-Wave bursts.
4U 1538–52		X-Ray *Eclipse*: 3.73 days.	2×10^{38} ergs^{-1}	neutron star?	Supergiant		

* 3U refers to the third UHURU Catalogue and 4U to the fourth. SMC stands for Small Megellanic Cloud, a companion galaxy to our own (Fig. I.8). [See also Avni, Y. and Bahcall, J. N., *Astrophys. J.*, *197*, 675 (1975); Wheeler, J. C. and Shields, G. A., *Nature 259*, 647 (1976); D. A. Schwartz, *et al*, *Nature, 275*, 517, (1978); B. Margon, H. C. Ford, *et al*, *Astrophys. J. Letters, 233*(2), L 63 (1979); 230(1), L 41 (1979)].

I.1). Comparison with the pulsars (already verified as *neutron stars*) was inevitable. After a few months, a *systematic variation in intensity* was detected (2.087 days for Cent. X-3 and 1.7 days for Her. X-1).

I.4.3 The Nature of X-ray binaries and Gravitational Clocks

The physical picture is now clear: *the intensity variations result from the fact that during each revolution the compact star disappears for some time behind its large companion* (in eclipsing X-ray binary systems shown in Fig. I.7). *Moreover, the correlation between the observed orbital velocities (spectral lines) and the intensity variation is perfect!*

In five of the X-ray binaries Listed in Table I.1 the visible companion is a massive *blue supergiant*, a rare star in our galaxy. Their number *in our galaxy* is estimated to be about 10^3 in a total of some 10^{11}. Thus the probability of even one X-ray source being randomly associated with such supergiants is very small.

These X-ray binaries are all located close to the central plane of our galaxy (see Fig. I.8 for direction and Table I.1 for distances). From the fact that three of these X-ray binaries are located less then 10,000 light-years away one can estimate that

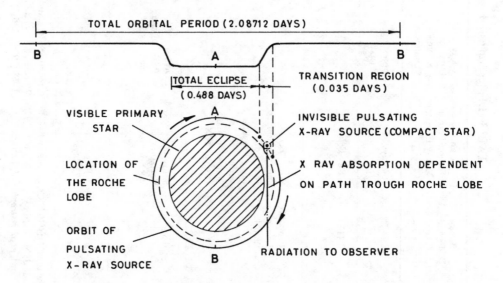

Fig. I.7 **Gravitation and the Structuring of Binary Clock Signals:**
Schematic representation of an eclipsing binary X-ray system.

A model of *Centaurus X-3* based on variations in X-ray intensity measured by the UHURU satellite. Duration of X-ray eclipses are listed in Table I.1. The measured properties of the *visible* primary star (usually a blue supergiant) can tell us the mass of its invisible (superdense) companion. The Roche lobe is defined in Fig. I.9. See also Fig. VIII.1.

about 50 such systems may exist in the galactic disk—an extraordinarily high percentage. How can this be explained? The presence of neutron stars (and possibly of black holes) in the X-ray binaries means that in these systems a *supernova explosion* must have taken place. Why was the entire system not disrupted by such an explosion? As a result of the calculations made by such theorists as R. Kippenhahn and A. Weigert in Germany, B. Paczynski in Poland, M. Plavec in Czechoslovakia, C. de Loore and J.-P. de Grève in Belgium, and D. Arnett in the U.S., we now have fairly conclusive answer to this question. For example, we know that a large amount of mass can be transferred from one star to another in the evolution of close binary systems. This mass transfer is related to the *limited space* available during stellar envelope expansion (process C-D in Fig. I.2). Thus, in most close binary systems the normal star *overflows its Roche lobe space* (a pear-shaped space named after the 19th century French mathematician E. A. Roche, who first recognized its importance—see Fig. I.9). Expansion of the envelope during the C-D process to a *red giant* may increase the radius up to several hundred times. As soon as one of the components reaches a volume larger than that of its Roche lobe, the gas outside the lobe *overflows* toward the companion star. The more massive component of a binary evolves faster, and, hence, is the first to overflow. Moreover, the ensuing loss of mass accelerates the expansion of the envelope of the mass-losing component. This mass transfer terminates when almost the entire hydrogen-rich envelope has been lost to the companion. The younger star, which may not have yet finished its hydrogen-burning stage, therefore has a longer hydrogen-fusion period.

On the other hand, the computations show that the companion who lost the mass continues to evolve through burning of helium, neon, oxygen, silicon, toward iron and its final *collapse to a neutron star* (or a *black hole*), giving rise to a supernova. (In the *less massive* binary systems, the helium star is likely to continue losing mass and might end as a *white dwarf*). *As a result of the explosion, the binary period is extended and the velocity of the center of gravity of the system is accelerated to several dozen kilometers per second* (the so-called *sling effect*). *In this way, a "runaway" binary system can be formed, moving away from its place of birth.* Much later, the star with the added mass ends its hydrogen-fusion stage and becomes a blue supergiant that, in turn, begins to lose mass at a moderate rate in the form of stellar wind. Fig. I.9 shows the details of this later process.

I.4.4 Gravitational Clocks and The Rate of Mass Transfer

For an X-ray source to be formed, the rate of mass transfer to the collapsed star may not be very large (otherwise the resulting *thick envelope* around the collapsed star may be *opaque* to X-rays). Blue supergiants with radii about 20 to 30 times the radius of the sun are observed to be losing mass from the outer parts of their atmosphere by the prevailing high *radiation pressures*, the observed rates of loss being of the order of a $10^{-6}M_\odot$ per year. The compact star in its orbit plows through this outflowing wind and captures only about 0.01 or 0.1 percent of it.

These facts explain why the massive X-ray binaries almost always include blue supergiants (Table I.1). *Unevolved* massive stars do *not* generate a sufficiently strong stellar wind to account for the observed strong X-ray sources, whereas stars that have evolved *beyond* the blue supergiant stage will have begun to overflow their Roche lobe, and, therefore, *extinguished* any X-ray emission from a compact companion, by the ensuing large rates of mass transfer. Stars *less* massive than about 15 or 20 solar masses, do *not* evolve into blue supergiants, and, therefore, never possess stellar winds that can turn a possible collapsed companion star into an X-ray source. Age calculations show that X-ray emission at the blue-supergiant stage can only start several million years *after* the supernova explosion. This explains the complete absence of an expanding supernova envelope around any of these systems (since such envelopes must have disappeared in space during that period). Figs. I.7 and I.9 explain these mechanisms and the conditions under which binaries become gravitational clocks.

I.4.5 Transient X-ray Sources

In eccentric-orbit models a compact object star is in an eccentric orbit around the more massive star. If the compact companion accretes mass from the stellar wind of the normal star, the accretion rate becomes *time dependent*, so that the resulting X-ray emission would show a *maximum* (in the absence of absorption extinction) and a *minimum*. The other possibility for describing transient X-ray sources is of course to assume that the accretion arises from direct mass transfer resulting from *the contraction of the Roche lobe*. Such eccentric orbit binaries may be formed when there is a supernova explosion in a close binary system.

X-ray transients are characterised by a rapid rise, over a few days, to a maximum emission, which may last for up to about four months and then smoothly and progressively decreases. The decrease is approximately that expected by reduced accretion rate with diminishing stellar wind intensity as the primary-compact object distance increases.

Now, a supernova explosion in a close binary system leaves a collapsed compact object, which may or may not remain bound. If mass transfer has already taken place, then the supernova occurs in the less massive star, and isotropic mass loss from a symmetric explosion cannot by itself produce an unbound "runaway". However, the system may be disrupted by an asymmetric explosion, with the formation of a "runaway" collapsed object which, if a newborn star, might be observable as a radio pulsar. In addition momentum imparted to the unexploded star by the supernova shell might contribute to the disruption of the system (Fig. VIII.1).

I.5   BLACK HOLES

The physics and thermodynamics of black holes form a subject of considerable interest. But due to the physico-philosophical difficulties involved we confine our-

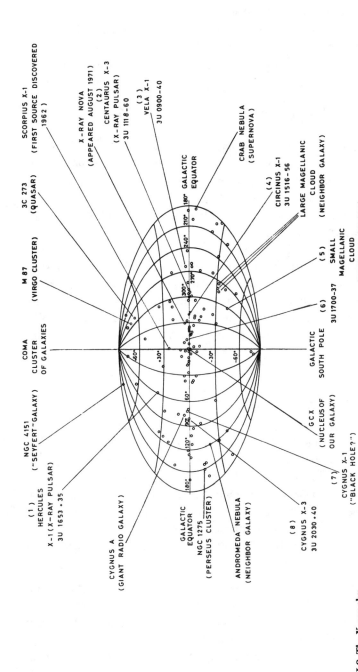

Fig. I.8 The X-ray sky

Eight X-ray sources are identified on this map. The map is an *equal-area projection in galactic coordinates*, which means that the central plane of the galaxy is aligned with the equator of the coordinate system and the *galactic center* is at zero degrees latitude and longitude. The concentration of the X-ray sources along the galactic plane, particularly toward the center, indicates that most of them lie *within the galaxy*.

Note the location of the *Virgo cluster*, the cluster nearest to our local group. Virgo contains about 2500 galaxies, each, in turn, contains around a thousand billion solar masses. It is about 55 million light-years away and it recedes from us at a velocity of about 1100 Km per second. The *Perseus cluster* (NGC 1275) contains about 500 galaxies and it recedes from us at a velocity of about 5400 km per second (thus according to Hubble's linear relation its distance is about $55 \times 5400/1100 = 279$ million light-years away).

Discoveries made during the last decade show that both the *Crab nebula* and the *Vela X-1* contain each a *pulsar* that is now identified as a *rotating neutron star* (the remnant of a *supernova explosion*).

selves, in this lecture, to simple preliminary concepts. (The subject is taken up again in Lecture VIII).

This section is subdivided into two topics:
1. The possibility of detecting a black hole in a binary system (e.g., the case of Cygnus X).
2. The possibility that black holes are at the center of violently active galaxies (e.g., BL Lacertae, Seyfert and N-galaxies as well as quasars, or even our own galaxy).

Two general remarks will be introduced at this early stage:

a) Black holes may result from gravitational collapse as was previously described. They form a hole in space (or a "frozen star") with a definite edge, over which anything can fall in—but nothing can escape, because of a gavitational field so strong that even radiation is *irreversibly* trapped and held by it.

In it the gravitational forces become stronger than the forces that hold the structure of the atomic nucleus, causing it to collapse.

b) Black holes, *if exist*, provide *thermodynamic sinks* for radiation, gas, dust or even for entire stars. Hence, they constitute an *additional* thermodynamic sink to that supplied by the *expanding* intercluster space (which *cannot be saturated* with radiation because of the very expansion of intercluster space—see Lecture VI). Theoretically they thus constitute *a most direct coupling between gravitation and thermodynamics.*

I.5.1 Black Holes in Binary Systems

The discovery of the X-ray binaries has produced evidence about stars whose enormous energy production results from the simple process of matter falling in an extremely strong gravitational field, a process in which as much as 40 percent of the rest energy of the matter can be radiated away into space! *That process is some two orders of magnitude more efficient than that of nuclear fusion!* The demonstrated existence of these processes might be helpful in gaining an understanding of energy generation in other abnormally luminous objects such as quasars, or in the nuclei of violently active galaxies (see below). It has also produced the first indication for the possible detection of black holes.

But the present evidence that Cygnus X is a black hole of Mass $>8M_\odot$ is not conclusive. It is, however, sufficient to warrant some theoretical considerations of the implication of such an object for the later stages of stellar evolution. In theory it *may* be formed by implosion of a star of Mass $30M_\odot$ as previously explained, but, alternatively, it *may* be formed by accretion on to a neutron star. For the second alternative one might envisage that the formation of a black hole began as a less massive neutron star that grew to its final mass *through accretion*, during which

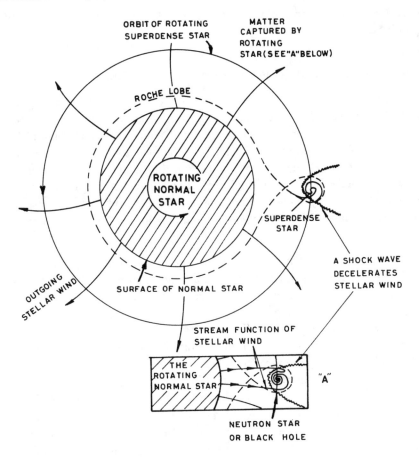

Figure I.9. **Destruction of Gravitational Clocks By Mass Transfer; Roche lobe, bow-shaped shock wave and mass transfer in a binary X-ray source.**
The orbiting superdense star may be a rotating white dwarf, a neutron star, or, perhaps, a "black hole." The superdense star is much smaller in size than the normal companion. The latter loses mass in the form of supersonic "stellar wind" (somewhat similar to the "solar wind"). The rotating superdense star, with rotating magnetic and gravitational fields, forms an obstacle in the larger star's "stellar wind," and a bow-shaped shock wave forms around it by the action of the fields. Matter flowing through this shock front is condensed and decelerates. Part of it decelerates enough to reverse its direction of flow on the forward moving section of the orbiting superdense star (see window "A") and co-rotate with the superdense star in the manner described in Figure I.6. Only about 0.1 % of the stellar wind is captured by the star to produce the observed X-rays. The *Roche lobe* is defined as the space *beyond which* an expanding normal star, in a binary system, begins to *transfer matter to its companion star.*

In the configuration shown the normal star has expanded to the stage where it begins to overflow its Roche lobe. The matter outside this space flows toward the superdense star at a comparatively low speed.

An *"accretion disk"*, or infalling *"magnetic funnels"*, may be formed, depending on the conditions explained in Figure I.5. However, a pulsar X-ray source can only be formed with infalling "magnetic funnels" of neutron stars whose axis of rotation is inclined with respect to the axis of the magnetic field. *Periodic modulation of the X-ray intensity is impossible in an accretion disk.* Therefore, the random variation of X-ray intensity in the case of Cygnus X-1 (see Table I.1) has its origin in such a disk. See also Fig. I.7.

it eventually collapsed into a black hole, i.e., it collapsed when its mass passed the M_{max} limit.*

I.5.2 Black-hole sinks at the center of violently active galaxies

In the nucleus of a galaxy the stars form a dense cluster in which *star–star collisions* may become sufficiently energetic and inelastic that the centers of the colliding stars coalesce. Such supermassive stars were first conceived by Hoyle and Fowler (1963)** as an explanation for "explosions" in the nuclei of some galaxies. Shortly thereafter, when quasars were discovered (1963), they suggested that supermassive stars are responsible for the quasar phenomenon. But whether galactic explosions or quasars are driven by such stars remains a subject of controversy in astronomical circles even today. Hence we avoid taking a position on this issue *except for the following remarks.*

Star–star collisions in a dense nucleus *may* indeed lead to stellar coalescence and the gradual building up of supermassive stars. It *may* also be possible that the entire conglomerate of stars, supermassive stars, gas and radiation in a galactic nucleus becomes so dense that it collapses to form a single black hole with a gigantic mass. *What, in principle, can happen then*? Can the idea of a supermassive black hole in the center of a galaxy explain *otherwise* unexplainable observations? Many astronomers are taking seriously the idea that black holes are at the center of quasars and a whole range of highly energetic galactic objects—at least, they claim, this is the most powerful and *"economical"* model that can be constructed within the framework of *"conventional" physics.* Indeed, in all of the unusually energetic galactic objects listed below there is good evidence that the predominant source of energy *is at the center.* (These objects are often called galaxies with *active nuclei.*) But whatever it is that is active is presumed *to be the same in each case*! While there is vigorous debate over the nature of this active center, there is virtually *unanimous consensus* that *the same process* is at the heart of all the active galaxies, and many other, less active galaxies. Presumably even our own galaxy has a source of the same type at the center, which is in continuity with the others, although less luminous.

Thus even skeptics agree that the same process, at different scales, produces the great variety of phenomena in galactic nuclei. Taking into account the great difference between the luminosities of the relatively faint nucleus of our galaxy and the luminosity of the brightest objects listed below, the process may have a range of energies differing by more than 10^6.

The black-hole theory is then claimed to explain the following astronomical phenomena, employing the smallest number of presuppositions:

---

* However, even very massive stars may form only neutron stars of $M = 1.2$–1.4 $M_\odot$ *regardless* of the total *initial* mass of the star. Most of these stars must explode, leaving only a neutron star remnant.

** Hoyle, F. and W.A. Fowler, *Nature, 197,* 533 (1963).

1. Quasars:

If quasars are *cosmological* objects (see, however, the remarks in §VIII.1), then their luminous power is so great that black holes with masses between 10^7 and 10^9 times the mass of the sun would suffice to produce the energy that is emitted from them. Thus, according to this simple model, *quasars are supermassive black holes in the center of giant galaxies, with energy supplied by the capture of gas, or even entire stars, from their surrounding.**

2. BL Lacertae objects

These objects appear to be an extreme form of quasar and would require black holes just as massive as quasars would. Like quasars they appear much like stars upon first observation, and emit most of their energy as visible light. They were discovered in 1969.

3. Radio, Seyfert and N-galaxies

Galaxies that produce most of their energy in the form of radio waves also require supermassive black holes to explain their output.

Similar conclusions apply to Seyfert galaxies, which are violently active galaxies slightly weaker than the quasars. N-galaxies, which are similar to Seyferts and quasars in many properties, have active energy sources at their centers that produce bright blue light in an otherwise red galaxy.

4. Giant elliptical galaxies

Some of the most impressive but circumstantial evidence in favor of the black-hole theory comes from relatively nearby galaxies such as the M-87 in the cluster Virgo, which is about 55 million light-years away (see Fig. I.8 and Table I.4). It has a compact radio source at its center. Recent measurements indicate that its bright center is less than 300 light-years in diameter and that stars which move around it have anomalously high velocities. Stars with such large velocities could only be held captive by an *extremely large mass* (around $6 \times 10^9 M_\odot$) at the core of the galaxy. With both the mass and the luminosity of the core of M-87 determined, it was possible to determine whether the core was brighter or darker than the rest of the galaxy. The results show that the mass-to-light ratio at the core is ten times greater than the rest of the galaxy, i.e., the core is indeed supermassive and is very small compared to the size of the galaxy (which is 300,000 light-years in diameter).

---

* The brightest quasars are about 100 times brighter than the giant elliptical galaxies, which are the brightest "normal" galaxies known (see below).

The appeal of the black hole theory is that it has, in principle, three attributes that seem to be precisely what is needed to explain the aforementioned observations:

a) Black holes can produce the enormous energy required, i.e., it is generally assumed that near the black hole matter forms *a rotating disk* (an accretion disk as described in Fig. I.9), which feeds the black hole (see also Lecture VIII).

b) Most highly resolved radio measurements show that galactic nuclei are less than 30 light-years across. The nuclei are therefore 1 millionth or less the size of a giant elliptical galaxy. Black holes, being the most compressed form of matter known, would be appropriately small. Moreover, rapid fluctuations observed in the output of active galaxies, particularly BL Lacertae and the most violent quasars, indicate *the size of the core may indeed be only light-months or-days across*!

c) The stability of galactic nuclei to rotations is a third requisite that black holes seem to fulfill. Observations show that the direction of the radio structure does not necessarily coincide with the axis of rotation of the galaxy as a whole. Because the observed radio structure *does not rotate* with the rest of the galaxy, the energetic core must be massive enough to resist this motion. (The preferred direction in space may be provided by the axis of a *spinning* supermassive black hole at the center of the galaxy.)

In comparison, *the alternatives to the black-hole theory* are rather limited. One possibility is that the core of active galaxies has a dense cluster of stars providing the requisite energy through frequent *supernova explosions*. Another possibility is that the core has not quite collapsed to the characteristic diameter of a black hole (see Lecture VIII). Such objects are usually called *spinars* or *magnetoids*, because it is postulated that rotating internal magnetic fields would stop their further contraction. Nevertheless, most astronomers estimate that spinars would subsequently collapse into black holes, perhaps within only about one-million years. Thus they might be more accurately understood as possible intermediate stages leading to a black hole.

Finally we note that the gravitationally-induced structures described above are all characterized by:

1) *Centreity.*
2) Large association of component elements "closed on themselves" *to form a new structure with a higher degree of complexity.*
3) *Evolving, time-asymmetric structures.*
4) Directed heterogeneity in which the aggregated elements are related and interconnected according to a certain pattern—*a pattern organized by the gravitational field*!

We shall return to these concepts in § IX.2.8.2.

I.5.3 On the Production of Heavy Elements in Nature

The rapid neutron capture process is known to be capable of forming heavy elements, and is believed to take place during supernova explosions†. However, estimates indicate that a *supernova explosion may not always be sufficient to explain the synthesis of heavy elements*§ Because of these limitations Chechotkin and Kowalski** have suggested that the production of heavy elements may also take place in the matter ejected from the envelope of a neutron star during *starquake processes*. They considered the case where instabilities cause the ejection of matter from the neutron star surface. Generally speaking it is now believed that the discovery of pulsars, and their interpretation as neutron stars, has provided a most promising mechanism for the nucleosynthesis of heavy elements. However, to obtain the observed density of heavy elements by such mechanisms one may have to *assume* the presence of about 10^9 neutron stars in our galaxy!* In this way neutron stars may enrich with heavy elements the surfaces of stars in their neighborhood. But confirmation of such mechanisms is needed before one can arrive at final conclusions.

I.6 GAS, DUST AND FORMATION OF STARS IN OUR GALAXY

I.6.1 Dimensions of the Milky Way

The main body of our galaxy is a disk with an overall diameter of about 100,000 light-years. The galaxy is remarkably thin: only about 1000 light-years thick in the vicinity of the solar system. The entire disk rotates, but not like a solid wheel, i.e., its *central* portions rotate *faster* than the periphery (cf. Fig. I.10). The observational eveidence for galactic rotation is that stars in the neighborhood of the solar system move in almost circular orbits around the calactic center at an average rate of about 300 km/sec.** The distance from the solar system to the galactic center (30,000 to 33,000 light-years) is so great that it would take us some 250×10^6 years to complete a *single* circuit (some call this period *a cosmic year*). Hence, since its formation (less than 10^{10} years ago) it has not rotated more than 100 revolutions.***

The mass of the entire Milky Way system is estimated to be 10^{11} solar masses, with about 50×10^9 solar masses at its nucleus. *The principal gravitational force*

---

† Seeger, P.A., Fowler, W.A. and Clayton, D.D. *Astrophys. J. Suppl. 11,* 121 (1965).

§ Ivanova, L.N., Imshennik, V.S., and Chechotkin, V.M., *Astrophys. Space Sci., 31,* 497 (1974).

** Chechotkin, V. M. and Kowalski, M., *Nature. 259,643* (1976).

* Peebles, P.J.E., *Gen. Rel. and Grav. 3,* 63 (1972).

** Some estimates are closer to 250 Km/sec. The galactic center, according to some estimates (q.v. the footnotes to § VI.2.1) is moving at about 600 Km/sec with respect to the universe as a whole. (Net velocity of sun and earth with respect to the universe as a whole is about 400 Km/sec—but in different inclination to the galactic disk).

*** It should also be stressed that the galactic center of our galaxy is moving relative to the Andromeda galaxy at about 80 Km/sec (approach). However, distances between our local group of galaxies and all other clusters of galaxies are increasing by the world expansion—see Table I.4 below for details.

that controls the galactic system is produced by the massive galactic nucleus. Spiral arms are extended out of the nucleus. They contain stars, interstellar gas and cosmic dust. The highest dust concentrations are found mainly along the central galactic plane in the *inner parts of the arms* (closest to the central nucleus).

Figure I.10
Center: NGC 224 Great Nebula in Andromeda. Messier 31. Satellite nebulae NGC 205 and 221 also shown. (48-inch Schmidt photograph from Mount Wilson and Palomar Observatories). Note the rotation of the spiral arms.

Inset (low left): A spiral galaxy in *Coma Berenices,* seen edge-on (200-inch photograph from Mount Wilson and Palomar Observatories). This is the shape an extragalactic observer, seeing our galaxy edge-on, may photograph.

I.6.2 Continuing Star Formation

The stars commonly found in the spiral arms are, almost without exception, relatively *young*. They are between 10 to 25 million years old (i.e., less than 1% of the age of the sun or the earth). Thus, *continuing star formation* is associated with the *spiral structure*. A few assumptions have been suggested to explain this phenomenon. One postulates that phenomena of *shock and compression* are responsible for such features. It *assumes* that a gravitational potential wave of a spiral form rotates within each spiral galaxy. When it passes through the interstellar medium, dense concentrations of dust and gas are produced (which thus become detectable as spiral arms). Therefore, *gravitational contraction* becomes the *major cause* for star formation in the spiral arms. Ultimately, many of these young stars may become blue-white supergiants, rich in the ultraviolet radiation capable of exciting the (observed) bright emission nebulas. However, according to another assumption, large-scale *magnetic fields* play a major role in the formation of spiral arms and protostars.

Supernova explosion may also affect the surrounding interstellar medium. Large amounts of gas enriched with elements heavier than helium are thus added to the interstellar medium. Simultaneously, tremendous amounts of kinetic energy are transmitted to the medium in the form of explosive shock waves. Indeed, several highly luminous and apparently young stars were found close to the Gum Nebula (the remains of a supernova explosion that took place 11,000 to 30,000 years ago, only 500 light-years from the solar system—the Crab Nebula is 7,000 light-years from us).

One may therefore assume that supernova explosions contribute to the triggering of star birth in the spiral arms.

I.6.3 Clouds and Globules of Dust

Interstellar clouds have been studied in different regions of the electromagnetic spectrum: X-ray, ultraviolet, visible, infrared and radio wave lengths. Such studies have yielded much information about the interstellar medium. The principal constituent is hydrogen (the ion is detected by the Balmer recombination lines* and the neutral atom by the radio wave-length of 21 centimeters that can be readily observed by radio telescopes. Molecular hydrogen is detected by means of ultraviolet observations from rockets). Additional constituents include: He, N, C, O, CH, CH^+, CN OH^-, NH_3, H_2O, H_2CO (formaldehyde), CO, CH_3OH (methyl alcohol), etc.

Also present are *cosmic dust grains* (with a diameter of the order of 0.0005 millimeter, (i.e., 1/2 a micron). Present conceptions are that these grains are a mixture of such elements as iron, silicon and carbon. Some may have been ejected into interstellar space by supernova explosions.

---

* Observed after a free electron is captured by a positively charged hydrogen nucleus.

Radio telescope studies of formaldehyde and other molecules found in dark galactic clouds have shown that the temperatures inside these clouds may be as low as 5°K or even less. *The cold grains may therefore provide cold surfaces where atoms can combine to form the observed molecules.* The grains may also filter out most of the ultraviolet radiation (which would inhibit the formation of molecules).

Within 1000 light-years of the solar system there are about a hundred fair-sized *globules of dust*, and about a dozen *large clouds* (a typical large cloud has a radius of some 12 light-years; the mass of the dust in it about 20 solar masses and it also contains much gas). Both dark clouds and globules of dust may also be associated with the gradual collapse into young protostars.

Other appropriate conditions for the *continuous formation of stars* exist in the galactic nucleus, or close to it where interstellar gas and dust intermingle with young stars.

There is another interesting class of young stars called *T Tauri*; faint stars that vary irregularly in their energy output, and are either rotating rapidly, or continuously ejecting mass. Such stars are most often found *in groups*, generally *near or within dark nebulas*.

The *oldest* star clusters and individual stars in our galaxy are found at large distances from the central plane of the disk. This fact would seem to imply that the first condensations began to form shortly after the galaxy became a separate unit in the expanding universe. Later, the galactic gas was attracted more and more toward the central plane of the galaxy, *leaving behind the first protostars. Younger stars* started to form in the central plane, as the galaxy became more flattened and concentrated, and gradually developed its present spiral arms and rotational motion.

So much for preliminary concepts about gravitationally-induced evolution. We shall return to this subject in Lecture IX, and, in the meantime, we shall turn attention to some "technical problems".

●I.7   HOW ARE COSMIC DISTANCES MEASURED?

What does one mean by *"cosmic"* distances? What the word *"measurement"* stands for in *"cosmic distances"*? And, in particular, what *precision* is being involved?

In Lecture III we review the various tests of the theory of general relativity, in which solar-system tests are confronted today by certain limitations on the precision of measurement. A summary of these limitations is given in Table I.2. (p. 118).

The best resolution ever achieved in astronomy is by VLBI (Very-Long-Baseline interferometry) radio telescopes. With the aid of atomic clocks the observations made simultaneously by radio telescopes thousands of kilometers apart (or by using as baseline the distance from the earth to an artificial satellite in orbit around the sun), can be combined to yield sharp images of some of the most distant objects in the universe (e.g., the interesting jet structure in the galaxy NGC 6251, 400,000,000, light-years away, was obtained recently by this method.)

There are a few methods to estimate cosmic distances. These are:

1) *Kinematic methods.* These include the VLBI radio telescopes, which may be useful up to distances of 10^8 pc* by interferometric radio observations, using as baseline the distance from the earth to an artificial satellite in orbit around the sun.

2) *Main-sequence photometry* (up to about 10^5 pc).

3) *Variable-stars method* (classical Cepheids and cluster variables) (up to about 4×10^6 pc).

4) *Novae, brightest stars, globular cluster, etc.* (up to about 3×10^7 pc).

5) *The brightest galaxies method* (up to about 10^{10} pc).

The last four methods are "indirect" in the following sense: if the *absolute luminosity*, L, of a radiation source can be estimated (by employing other astrophysical methods), one can determine the (luminosity) distance to it, r_L, by measuring its *apparent luminosity* l, and using eq. (1), i.e.,

$$r_L = (L/4\pi l)^{1/2} \qquad (1a)$$

Therefore, the main difficulty in these methods is associated with proper astrophysical methods to estimate L. However, before we review these difficulties, it would be of interest to say a few words about the kinematic methods.

●I.7.1 Kinematic Methods

These methods do not require the use of eq. (1). Distances to some of the nearest stars (a few thousands), can be determined from the shift (trigonometric parallax) in their apparent positions, caused by the earth's revolution about the sun.

A most important kinematic method is the "moving cluster method". For instance, the "proper-motion" distance to *Hyades* (about 41 pc), which contains about 100 stars that move through our galaxy with almost equal and parallel velocities, is found from its *radial* velocity (as determined from the Doppler shift of its spectra —see eq. 2 below for details)—the velocity components transverse to the line of sight (expressed as the product of the distance to the cluster times the "proper motion", i.e., in radians per unit time) and the speed of light. (Other methods involving statistical analysis of "proper motions", assumptions about the distribution of the stars

---

* 1 pc ≡ 1 parsec = 3.2615 light years. 1 light year = 9.4605×10^{12} km. In kinematic methods the star's distance in parsecs is $1/\pi$, with π expressed in seconds of arc. (Thus, a trigonometric parallax of 0.03" corresponds to about 30 pc).

TABLE I.2: Present Precision Limitations in Solar Measurements

(see also Lecture III, §III.1 and Ref. 129, p. 347)

Quantity to be measured	Test related to the effect of (an example)	Precision of measurement in the 1980's
Distances between two bodies in the solar system.	Perihelion shifts for Mercury, Mars, etc.; Relativistic time delays for radio waves from earth to Venus; Periodic relativistic effects in earth-moon separation.	0.3 km—by bouncing radar signals off another planet (e.g., Venus); 10 meters—by round-trip radio travel time to a radio transponder (on another planet, e.g., Venus, or in spacecraft); Less than 10cm—by laser ranging of the distance earth-moon, or by VLBI
Difference in lapse of time in solar system.	Clock on earth vs. clock in orbit about sun $\Delta t/t \sim 10^{-8}$	Stability of a hydrogen maser clock $\Delta t/t \sim 10^{-13}$ for t up to one year.
Angular separation of radiation sources (e.g., galactic jet structure)	Jets in the galaxy NGC 6251	$\sim 1''$ with optical telescope; ~ 0.0001 arc-second with VLBI radio telescopes.

velocity angles, radial velocities, and ratios of absolute luminosities are available and may be employed at distances beyond 200 pc. However, these methods are intrinsically inaccurate if the sample of stars studied does not have the assumed distribution of angles—between stars' velocities and the line of sight).

A similar method is the "expanding cluster parallax". E.g., the distance to the water masers in Orion is estimated from radial velocities with respect to the center of their expansion (see also p. 185).

●I.7.2 The Method of Main-Sequence Photometry

If the distances of some nearby stars are known by one of the kinematic methods, one can employ eq. (1a) to evaluate their corresponding *absolute* luminosities. A large portion of nearby main-sequence stars were found to obey a strict *relation* between absolute lumosity, L, and their corresponding *spectral type* (the spectral type represents the surface temperature of the radiating source and is usually denoted

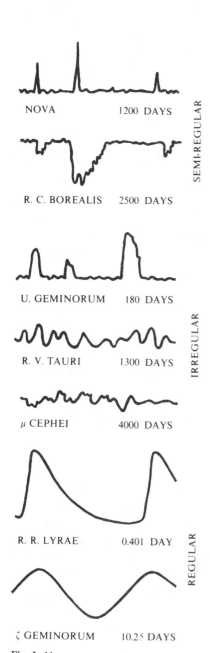

*(Upper right) Group of five galaxies with unusual con-
necting clouds, in Serpens. These connecting clouds are
evidence of the mutual gravitational interactions among
these galaxies. (From Mount Wilson and Palomar Observatories).*

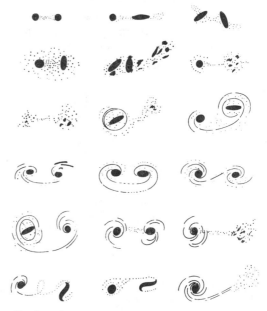

Fig. I. 11

*(Left' Typical light curves of variable
stars, (From W. Becker, Sterne und Stern-
systeme. Dresden: T. Steinkopf, 1950, pp.
105, 125) cf. §I.7.3.*

*(Lower right) A schematic diagram taken from Zwicky
showing the various types of bridged galaxies. (From
Handbuch der Physik, S. Flugge, ed., vol. 53. Berlin:
Springer, 1959)*

by one of the letters O, B, A, F, G, K, M, R, N, S, with O very hot and S the coldest)*. The relation between surface temperature and absolute luminosity was (independently) discovered by E. Hertzsprung and H. N. Russell during the decade 1905–1915 (see Fig. I.1)

Thus, if we know the surface temperature of any main-sequence star (from its spectra), we employ the H–R diagram to estimate its L, and, by measuring also the apparent luminosity l of this star we can calculate its distance. However, this method works best only for the lower part of the main sequence (low temperatures) where the Hertzsprung-Russell relation is well-defined (minimum scatter of data).

Present knowledge of cosmic distances depends on the precision by which kinematic methods evaluate the distance to the *Hyades*. The Hyades belong to one of the 650 *open clusters* known in our galaxy. Stars in open clusters, as well as most nearby stars like the sun, generally belong to *Population I*, which is characterized by *high metal content and relative youth*. Their location is limited in our galaxy to the *spiral arms* (see §I.5). Each of the open clusters contains 20 to 1000 stars.

There are about 130 catalogued *globular clusters* in our galaxy, (such as M13 in Hercules), each containing 10^5 to 10^7 stars. They belong to the so-called *Population II* (characterized by *lower metal content and greater age*, they pervade the whole galaxy).

The main-sequence method is especially useful when applied to clusters of stars (but it works best with the open clusters). In this case all stars have the same distance from earth and the main sequence stars can be picked out by plotting apparent luminosity versus spectral type for a large sample of cluster stars. This procedure was carried out with Hyades stars *to calibrate the H-R main-sequence relation* (taking the distance to the Hyades as 40.8 pc as evaluated by the kinematic method).

●I.7.3 The Variable-Stars Method

Most stars, including the sun, shine steadily over very long periods; the sun's luminosity has been changed only slightly over the last millions of years. However, *there are about 10,000 catalogued stars whose apparent luminosities fluctuate considerably over short periods of a few hours, days, weeks, and months.* Astronomers found that some of the short-period variables, of the kind known as cepheids, obeyed an empirical law which connected their periods with their absolute luminosities; *the longer its observed period, the greater the absolute luminosity!***

---

* Some students prefer to memorize this by the mnemonic 'poem': "Oh, Be A Fine Girl, Kiss Me Right Now, Sweetheart". The code-order (not the 'poem') is frequently used to express surface temperatures in the H–R diagram.

** First discovered by *Henrietta Swan Leavitt* in 1912, using 25 Cepheids in the small Magellenic cloud. Absolute luminosity was determined from kinematic measurements of distance for the nearest cepheids and the inverse square law, eq. 1a.

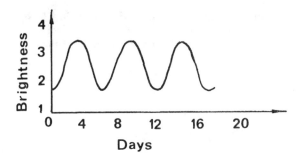

Fig. I.12 **The variation in brightness of the variable star Delta Cephei whose period is about 6 days.**

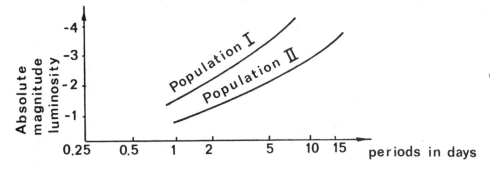

Fig. I.13 **The period-luminosity (P–L) relation for Cepheid variables;**
The longer the observed period of a Cepheid, the larger its absolute luminosity. Thus, a measurement of
its period enables its absolute luminosity to be estimated; then a comparison between its real (absolute)
and observed (apparent) brightness yields its distance by using eq. 1a.

Thus, a measurement of a Cepheid's period enables its absolute luminosity to be
estimated; then, a comparison between its real (absolute) and observed (apparent)
brightness yields its distance by using eq. 1a. Therefore, it became possible to use
the Cepheids as *"standard candles"* in outer space.

This method has given trouble in the past because there are actually different
period-luminosity (P-L) relations for different classes of variables, but these difficulties have by now been mostly overcome, as briefly explained below.

There are two families of variable stars: the *RR Lyrae stars* (or "cluster variables") and the *Delta Cephei stars* (or *"classical Cepheids"*). The former ones have
periods ranging from a few hours to a day, and belong to *Population II*, while the
Delta Cephei have periods ranging from 2 to 40 days, and belong to *Population I*
(but there is another type of variable stars, called the *W Virginis stars*, which belong
to *Population II*, but have periods like the classical Cepheids; and these stars were
confused with Cepheids before *Walter Baade,* an American astronomer, distinguished, in 1952, between the two different populations (Fig. I.13).

Table I.4: Recession Velocities Associated with Selected Clusters of Galaxies
(arranged in order of increasing recession velocity, cf. Fig. I.14)
(for identification see also Figures I.8 and I.16.)

Cluster name	Estab. No of Galaxies in the cluster	Recession velocity, cz, in km/sec
Virgo*	2500	1150
Pegasus I	100	3800
Cancer	150	4800
Pisces	100	5000
Perseus	500	5400
Coma	1000	6700
Hercules		10300
Pegasus II		12800
Cluster A	400	15800
Ursa Major I	300	15400
Leo	300	19500
Cor. Bor.	400	21600
Gemini	200	23300
Boötes	150	39400
Ursa Major II	200	41000
Hydra		60600

* The brightest galaxy in this cluster is NGC4472. It also contains the large eliptical galaxy M87 in which there are about 2000 globular clusters that reveal a sharp cutoff in their luminosity distribution (see Table I.5).

Distances within the Milky Way were unaffected by Baade's discovery—for they had been determined from Population II, and so are in fact correct. Unfortunately, the variable stars observed in other galaxies are of Population I (a fact unknown till 1952). This means that their absolute luminosity had been underestimated before Baade made his discovery. In consequence their distances had also been underestimated.

Baade's discovery was a dramatic one for it multiplied cosmic distances about 500%, thus greatly increasing the scale of the universe accessible to our observations. One result was that our galaxy was no longer considered a giant among other galaxies of the universe. It lost its privileged status, and shrank to a modest average size that characterizes all other galaxies. As Copernicus dethroned the Earth, and Shapley the Sun, so Baade dethroned the Milky Way. Thus, the traditional *"geocentric picture of the Universe"* received its last blow in 1952 and has been completely discredited since. But this significant discovery was preceded by other dramatic events that are briefly described next.

●I.7.3a A Few Historical Events that Revolutionized Astronomy

In 1905 Agnes Clerke, a historian of astronomy, wrote: *"The question whether nebulae are external galaxies hardly any longer needs discussion. It has been answered by the progress of research. No competent thinker, with the whole of the available evidence before him, can now, it is safe to say, maintain any single nebula to be a star system of co-ordinate rank with the Milky Way."*

It was a philosopher who first predicted that the Milky Way is a rotating island of stars; in order to explain the flattened form of the Milky Way, *Immanuel Kant* (1724–1804), the founder of critical idealism, wrote in 1755:

"... the whole nest of (the fixed stars) are striving to approach each other through their mutual attraction ... ruin is prevented by the action of the centrifugal forces ... "

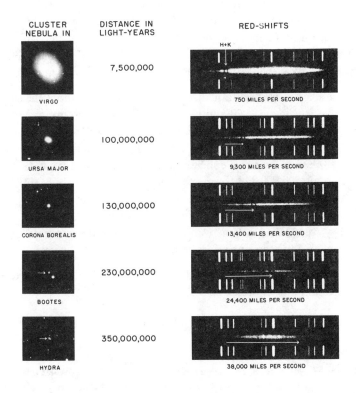

CLUSTER NEBULA IN	DISTANCE IN LIGHT-YEARS	RED-SHIFTS
VIRGO	7,500,000	750 MILES PER SECOND
URSA MAJOR	100,000,000	9,300 MILES PER SECOND
CORONA BOREALIS	130,000,000	13,400 MILES PER SECOND
BOOTES	230,000,000	24,400 MILES PER SECOND
HYDRA	350,000,000	38,000 MILES PER SECOND

Figure I.14 **Spectral Red-Shifts vs. Distances of a few well-known Galaxies.**
Relation between red shift (velocity) and distance for extragalactic nebulae. Arrows indicate shift for calcium lines H and K. The distances given in this photograph are based on the old Hubble constant. They should be multiplied by approximately 5 to agree with the presently accepted value of the Hubble constant. (From Mount Wilson and Palomar Observatories)

Then he asks, for the first time, how a very remote galaxy would appear to us. His answer:

"Circular if its plane is presented directly to the eye, and elliptical if it is seen from the side or obliquely. The feebleness of its light, its figure, and the apparent size of its diameter will clearly distinguish such a phenomenon when it is presented, from all the stars that are seen single . . . this phenomenon . . . has been distinctly perceived by different observers [who] . . . have been astonished at its strangeness . . . Analogy thus does not leave us to doubt that these systems [planets, stars, galaxies] have been formed and produced . . . out of the smallest particles of the elementary matter that filled empty space."*

Star counts were begun in the 18th century by *W. Herschel* who propsed a disk-like model for the Milky Way. Later, in 1918, *H. Shapley* made use of distances calculated by the Cepheid-variable method and modified Herschel's model by placing the sun near the edge of the galaxy instead of at the center where it had always previously been placed.**

In 1926–7, *Lindblad and Oort* discovered from motions of the stars that our galaxy is rotating, thereby confirming Kant's prediction. At that time the distance from the solar system to the galactic center (Sagittarius, see slso Fig. I.8) and the sun's period for a complete revolution (about 10^8 years) were already known and it allowed an estimation of the mass of the Milky Way (just as one determines the mass of the sun from the earth's orbital period and the distance of the sun). The resulting mass of the Milky Way was 10^{44} grams, or about 10^{11} solar masses.

But another, even more dramatic discovery, was made at about the same time by *Edwin P. Hubble*. Starting in 1923 and using the 100-inch telescope at Mount Wilson (California) he managed to detect Cepheids in some of the spiral nebulae, including Andromeda (NGC 224 (M31)—see Table I.5). which is visible with the naked eye. Using the P-L relation for the Cepheids he calculated their distance. His results: a distance of about 800,000 light-years for the Andromeda nebula (now known to be 2,200,000 light-years), and similar values for some other spiral nebulae. These results proved that these nebulae are hugh stellar systems outside our own galaxy. Henceforth they were called galaxies, and, consequently, the concept *"extragalactic distance"* became a common one in astronomy.***

Hubble's greatest discovery, namely that all clusters of galaxies recess from each other, was based on his linear relation

$$U_r = H_0 r \tag{2}$$

* Taken from C.W. Misner, K.S. Thorne and J.A. Wheeler, *Gravitation*, W.H. Freeman and Comp., San Francisco, 1973, pp. 756, See also Kant I, in *Universal Natural History and Theory of the Heavens*, Univ. of Michigan Press, Ann Arbor, 1969.

** More recent studies revealed that Shapley's calibration contained errors due to the neglect of interstellar absorption.

*** Calibrating the Cepheid P–L relation was also carried out by Hertzprung, Russell, R. E. Wilson, A. Sandage and others.

TABLE I.5: Distances and Radial Velocities of Selected Galaxies*

Name and/or code of galaxy	Type**	Distance from solar system in 10^6 parsecs***	Apparent Magnitude m_{pg}	Radial Velocity, cz (obs.) in (km/sec)
The Nearest Neighbors in the Local Group (from 1.6 to 17 Milky Ways' diameters away).				
LMC****	Ir or SBc	0.049	0.86	+ 280
SMC****	Ir	0.058	2.86	+ 167
Ursa Minor	dE	0.077	?	?
Draco	dE	0.08	?	?
Sculptor	dE	0.09	10.5	?
Fornax	dE	0.13	9.1	+ 40
Leo I	dE	0.23	11.27	?
Leo II	dE	0.23	12.85	?
NGC6822	Ir	0.52	9.21	− 40
Other Members of Local Group (all about 21 to 25 Milky Way's diameters away)				
NGC224 (M31) (Andromeda Nebula)	Sb	0.65	4.33	− 270†
NGC205	E6p	0.65	8.89	− 240
NGC221 (M32)	E2	0.65	9.06	− 210
NGC147	dE4	0.65	10.57	?
NGC185	E0	0.65	10.29	− 340
NGC598 (M33)	Sc	0.74	6.19	− 210
ICI613	Ir	0.74	10.00	− 240
Miscellaneous Bright Galaxies				
NGC3031 (M81)	Sb	2.0	7.85	+ 80
NGC3034 (M82)	Scp	2.0	9.20	+ 400
NGC5236 (M83)	Sc	2.4	7.0	+ 320
NGC5128 (Cen A)	EOp (R)	~ 4.0	7.87	+ 260
NGC4736 (M94)	Sbp	4.3	8.91	+ 350
NGC5055 (M63)	Sb	4.3	9.26	+ 2600
NGC5194 (M51)	Sb	4.3	9.26	+ 550
NGC5457 (M101)	Sc	4.3	8.20	+ 400
Messier Galaxies in the Virgo Cluster (Visual Magnitudes)				
NGC4486 (M87)	EOp(R)	15 ± 5	9.3	+ 1220
NGC4594 (M104)	Sb	15 ± 5	8.1	+ 1020

* For identifications see also Fig. I.8. Average number of stars in a typical galaxy equals 10^{11}. Distances, types, and magnitudes are mostly taken from the compilation of S. van den Bergh, *Observers Handbook of the Royal Canadian Astronomical Society*, 1971.

** "Type", E. denotes "elliptical", with E0, E1, ... increasingly flat; S. denotes "spiral", with S0, Sa, Sb, Sc, ... increasingly open; SB denotes "barred spiral", with SB0, SBa, SBb, SBc ... increasingly open Ir denotes "irregular"; p denotes "peculiar"; d denotes "dwarf"; R denotes "strong radio source".

*** 1 parsec = 3.2615 light years.

**** Large and Small Magellenic Clouds, respectively.

† About ⁻80 km/sec when calculated for the galactic center of our galaxy.

where H_0 is a constant, U_r the radial velocity of the radiation source and r the distance to it. He arrived at this great discovery by measuring distances by means of the Cepheid variables and combining these results with radial velocities deduced from characteristic spectral lines emitted by the observed radiation source (Fig. I.14). To arrive at eq. 2 he employed the *Doppler principle*, which for non-relativistic velocities (see Lecture III), can be written as

$$z = \frac{\lambda - \lambda_L}{\lambda} = \frac{U_r}{c} \tag{3}$$

where z is the red shift factor, λ the moving source wavelength of a characteristic spectral line, λ_L that of the same line measured at rest in a laboratory and c the velocity of light in vacuum. Hubble's observations (which have since been confirmed to extremely great distances) demonstrated *systematic, isotropic galactic recession* (i.e., *expansion* of the universal system of clusters of galaxies). This topic is further discussed in Lecture III and VI.

Hubble was forced to make his discovery in a series of steps. Most extragalactic distances are too large for Cepheid variables to be detected inside galaxies. (Observations of Cepheids can take us out to about 14×10^6 million light-years. At greater distances, the Cepheids become too faint to be seen individually, and more "indirect" methods have to be used—as explained below. RR Lyrae stars can be em-

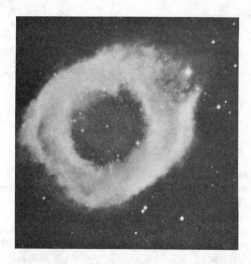

Figure I.15 **The expanding envelope of NGC 7293 planetary nebula in Aquarius.**
The expansion of the envelope and the internal motions are clearly indicated. (200-inch photograph from Mount Wilson and Palomar Observatories)

ployed only for shorter distances, i.e., up to Magellenic clouds, Ursa Minor, Draco and the Sculptor systems.) Hubble's first step was, therefore, confined to the nearest galaxies where Cepheid variables are detectable—the group now called the "local cluster" or the "local group" (see Table I.5), to which our galaxy belongs.

The Cepheids are therefore not bright enough to be used to determine the distance of the nearest cluster of galaxies, *the Virgo cluster.*

For his next "distance candles" Hubble used *supergiant stars,* which are intrinsically brighter than Cepheids. These stars are detectable individually in some distant galaxies whose Cepheids are invisible. Hubble's second step brings us to the next method.

●I.7.4 The Novae, Brightest-Stars, and Globular-Clusters Methods

This class of distance indicators is used to reach beyond our local group and it includes the following astrophysical methods:

●I.7.4.1 The brightest stars indicators

Stars of each galaxy in the local group—about fifteen to twenty galaxies—have been found to have a well-defined *maximum* absolute luminosity, of about $M_v = -9.3$ magnitude (distances and absolute luminosities were calculated *via* the Cepheids). Hence, they can be used as distance indicators out to about 10^8 light-years. However, at larger distances it becomes difficult to distinguish between them and *non-stellar emission regions* (Hubble's 1936 calibration of the distance scale via supergiants, etc., was in error, probably because he confused such objects and brightest stars).

●I.7.4.2 Globular clusters as distance indicators

The globular clusters in our galaxy have absolute luminosities that are typically about $M_v = -8$. But they vary considerably about this mean. However, recent studies of the 2000 globular clusters in the *M87 galaxy* of the *Virgo cluster* reveal *a sharp cutoff* in their luminosity distribution. Sandage suggests* that the absolute luminosity of this phenomenon in M87 be identified with the known absolute luminosity of the brightest globular cluster B282 of the nearby *Andromeda galaxy.* This places *the Virgo cluster at 1.7×10^7 pc,* or about 55 *million light-years away.* A more recent estimate** puts this distance at around 65 million light-years. This method may prove to be more reliable than the other methods in this class.

●I.7.4.3 The Novae method

The phenomena of novae take place in a typical galaxy at a rate of 40 per year. They are characterized by a sudden flare—an increase by four to six orders of magnitude

---

* A. Sandage, *Astrophys. J., 152,* 1149 (1968); *Observatory, 88,* 91 (1968).
** G. de Vaucouleurs, *Astrophys., J., 159,* 435 (1970).

in the value of luminosity of a star. They have been used as distance indicators since 1917, when a nova was detected in the spiral nebula NGC6946. The brightest novae reach $M_v \cong -7.5$, so they can, in principle, be used as distance indicators out to about 30 million light-years. However, they tend to exist in the bright central regions of galaxies, and, consequently, it is difficult to measure their apparent luminosity.

●I.7.4.4 **Non-stellar objects as distance indicators**

Included in this type of indicators are *large clouds* of (interstellar) hydrogen that are *ionized* and made *luminous* by the presence of *very hot stars* (the H-II regions). Since their size may reach hundreds of parsecs, their angular diameters are employed to estimate their distances out to about 300,000,000 light-years.

●I.7.5 **Brightest Galaxies as Distance Indicators**

Hubble checked the method described in §I.7.4.1 by using it on the local group, whose distances were already known from their Cepheid variables, and then applied it to more distant galaxies. In this way he was able, in the 1930's to extend the range of "measured" distances from 1 million light-years to 10 million light-years. Forty, fifty years ago this was the maximum distance Hubble could "measure" using the brightest stars as distance indicators. Forced to invent another method he turned to the galaxies themselves as distant indicators. He simply assumed that *they all have the same absolute luminosity*; But he backed this assumption by empirical results which demonstrate the maximum error involved. His measurements of *apparent* luminosities of galaxies in the nearest cluster Virgo demonstrated that their *absolute* luminosities differ from one another by at most a factor of 10. (The assumption that all galaxies have an absolute luminosity in the middle of this range of values then leads to an error in their distance of about a factor 3 in unfavorable cases.) Today we know that he somewhat underestimated the error involved. However, when one deals with *clusters* of galaxies one can reduce the error by restricting the observations to the *brightest* galaxies in the cluster. And so Hubble did.*

Hubble's procedure was validated when it was found that use of the brightest galaxies as distance indicators gave a good linear relation between (luminosity) distance r and z for 10 clusters with z ≪ 1 (eq. 3). These brightest galaxies are usually the elliptical galaxies known as type E in Hubble's classification scheme (see Table I.5).

Using the brightest galaxies as distance indicators Hubble was able to explore the universe out to the *cosmological* interesting distances of 500 million light-years — a region that contained about 100,000,000 galaxies. According to Hubble, the average distance between galaxies (remember that his distance-scale was considerably modified later by Baade as explained in § I.7.3), is about a million light-years. This does not mean that the galaxies are dispersed uniformly in space with a given spac-

---

* E. Hubble, *Astrophys. J.*, *84*, 270 (1963).

ing between them. The actual picture shows considerable clustering, ranging from pairs of galaxies through clusters with fifteen or twenty members like our group, up to clusters such as the one in Virgo containing about 2500 galaxies.

It is not yet clear whether these clusters are gravitationally bound systems. Frequently, the group members are not sufficiently massive to bind the clusters gravitationally. Yet, there may be additional, yet undetected, gas and faint stars between the galaxies to provide the necessary 'missing matter'. This is also a central problem in relation to the overall ('smeared-out') density of matter in the universe, a quantity that plays a key role in differentiating between cosmological models (see § III.2..).

The brightest E galaxy in the Virgo cluster is NGC4472 (M49) with absolute (luminosity) magnitude of around -21.7 (determined by using the globular clusters in M87 (see Table 5) to give the distance to the Virgo cluster). Therefore, if we assume that all bright E galaxies have absolute magnitude of about -21.7, then they can be used as distance indicators out to a distance of about 10^{10} pc. However, as one looks out to greater and greater distances, one tends to select increasingly rich clusters of galaxies for study, and if there is no absolute upper limit to galactic luminosity, the brightest galaxies in these clusters will have greater and greater absolute luminosity. If one mistakenly assumes that these distant galaxies have the same absolute luminosity as NGC4472, one will *underestimate* their true luminosity distance. This is the so-called *Scott effect* which is still a matter of controversy. Nevertheless, the distances to clusters at greater distances than that of the Virgo cluster are determined today by assuming that their brightest E galaxies have the same absolute luminosity as the brightest galaxy NGC4472 in the Virgo cluster.

Figure I.16 **The Virgo cluster of galaxies**
The Virgo Cluster of galaxies at a distance of about 17 megaparsecs. Its speed of recession is 1150 km/sec. and it contains about 2500 galaxies. (Yerkes Observatory photograph) cf. Table I.4

Type I supernova have also been used as distance indicators (up to 500 Mpc). In this range they are better candidates than the galaxies themselves [Barbon, R., et. al. *Astron. Astrophys.*, *44*, 267 (1975)].

●I.8 NEUTRINO ASTRONOMY AND ASTROPHYSICS

Within the last few years it has been demonstrated in experiments that there should be two types of neutrinos and two types of antineutrinos, i.e., the so-called electron neutrinos v_e, $\bar{v}_e$, and the muon neutrinos v_μ, $\bar{v}_\mu$. Since the rest masses of these particles are zero, their equilibrium concentration and the corresponding energy density and pressure are rather close to the values that are applicable to quanta of electromagnetic radiation. All stars are utterly transparent to neutrions; any neutrino created immediately leaves the star carrying its energy with it to unsaturable space. (A layer of lead several light-years thick is required to stop their dissipation into space). Neutrino-energy losses to space become important in determining the evolution in density and temperature of carbon-oxygen cores in advanced stars (q.v. Figs. VII.2 and VII.3).

The main characteristic of the neutrino is the weakness of its interaction with all other particles, as well as with other neutrinos. While photons are fully neutral, neutrinos are neutral only in the sense of electric and baryonic charge. According to contemporary views, neutrinos are helical particles, i.e., for a given direction of the momentum, the neutrino spin can only be antiparallel to that momentum. The antineutrino spin, accordingly, must be parallel to its momentum (§II.13).

Thermodynamic equilibrium between neutrinos and other particles can occur only in an unbounded, uniform distribution of matter, i.e., in the "big bang" evolving world. In this case the large mean free parth of the neutrino has no significance since no net transfer of neutrinos is experienced by any expanding mass. Thus at the early stages of the world expansion, when the temperature was above 10^{11}°K, the rate at which thermodynamic equilibrium is approached is high compared with the rate of decrease in density and temperature. Only in such catastrophic events as supernova explosions may some of the neutrinos and antineutrinos be unable to escape to outer space. It is not only the short time scale of the explosion that may prevent neutrino escape; it is also due to the extremely high values of matter density, the high electron temperature and the neutrino energy, which prevail in the cores of "advanced" stars (Fig. VII.2).

●I.8.1 Neutrino Cooling

There is an important difference between the creation of electron-positron pairs e^+ e^- and the creation of neutrino-antineutrino pairs in galactic masses. While both processes are accompanied by an energy consumption and reduction of pressure (compared with the pressure which would have existed without the creation of the pairs) only the neutrinos escape to outer space. At equilibrium (at a given T and ρ), e^+ e^- are created and annihilated at the same rate. Therefore, they do not cool the system immediately after their creation as do the neutrinos escaping to space. For instance muon neutrinos can escape easily from a neutron star, thus becoming the important factor in its *cooling* immediately after it is created. There is a great variety of neutrino processes in "advanced" galactic masses, including in a nominal way, the so-called *Urca Process* (the analogy between these neutrino processes and the gambling operation at the "Casino de Urca" near Rio de Janeiro was suggested by George Gamow who coined this term. There also, no matter how you played the game, you always seemed to lose). The simplest example of the Urca process is when a proton captures an electron to form a neutron and a neutrino and the neutron subsequently decays back to a proton and an electron plus antineutrino. Consequently, after each cycle one ends up with the same composition but has lost the energy dissipated in unsaturable expanding space by the escaping neutrinos.

●I.8.2 Neutrino Cooling and Entropy

There is a threshold density active in the Urca Process. For the same composition Urca energy losses to cold space would be drastically enhanced by density variations associated with convection in the interiors of stars. The larger the density deviation from the threshold for electron capture, the stronger are the Urca losses by neutrinos dissipating in outer space. Thus, during a fluctuation involving a compression followed by expansion e^+ e^- are created but they disappear during the expansion phase, giving back their energy. Thus, when e^+ e^- pairs are considered, the local entropy of the system remains invariant; not so for neutrino pairs! As far as outer space keeps expanding its neutrino concentration remains negligibly small compared with the equilibrium concentration, and the Urca Process increases the dissipation of neutrinos in space thereby causing the entropy to increase.

We also stress here that neutrino radiation is characterized by viscous dissipation and viscous stress similar to those found in fluid dynamics (see §VI.7).

The rate of neutrino losses (du_ν / dt) to unsaturable space determines the entropy change dS / dt according to

$$\frac{dS}{dt} = -\frac{1}{\rho T} \cdot \frac{du_\nu}{dt} \tag{4}$$

where

$$\frac{du_\nu}{dt} = 4.3 \times 10^{15} T^9 \text{ erg sec}^{-1} \text{ cm}^{-3} \tag{5}$$

and T is the temperature in units of $10^{9}°$K subject to the condition $T > 6$. For stable configurations, which depend on S as a parameter, the rate of entropy change determines the rate of evolution. However, for supermassive stars ($M > 10^4 M_\odot$) the gravitational self-closure takes place before the temperature rises high enough to enable neutrinos to carry away a substantial portion of the stellar mass.

Finally it should be stressed that the earth moves through the isotropic neutrino medium in interstellar space. *Consequently, all motions, clocks, radiation, irreversible processes, etc., can be measured with respect to this (expanding) isotropic frame of reference* (see §VII.7)! *This weak interference with local processes is not without other fundamental consequences.*

●I.8.3 Neutrino Telescopes

While only a small fraction of advanced stars manifest themselves as supernova explosions, all such collapses emit intense pulses of neutrinos and anti-neutrinos, both electronic and muonic. Unlike electromagnetic radiation, neutrinos find all parts of the galaxy transparent, even the galactic center. Therefore, a few astronomers have launched a search for neutrino pulses.

A few laboratories have been set up for this purpose. The most famous ones are Davis' ^{37}CL Neutrino Observatory in the Homestake Gold Mine in Lead, South Dakota, and the Mont Blanc Tunnel Laboratory, 5000 miles away. The most sensitive search to date for extraterrestrial neutrinos has been performed by R. Davis and his collegues. Using the reaction

$$\nu_e + {}^{37}\text{CL} \rightarrow {}^{37}\text{Ar} + e^- \tag{6}$$

they set a limit that implies that $< 10^6 \ \nu_e/\text{cm}^2/\text{sec}$ arrive at earth from $\text{B}^8 \rightarrow \text{Be}^8 + e^- + \nu_e$ decays in the sun (use is made here of the neutron-rich chlorine-37 which make up about 25% of all chlorine atoms. Its nucleus contains 17 protons and 20 neutrons. If one of these neutrons absorbs a neutrino, it becomes a proton and emits an electron. The nucleus will then have 18 protons and 19 neutrons and will be argon-37. But liquid chlorine — which is denser than chlorine gas — is highly corrosive. Instead liquid carbon tetrachloride, or liquid perchloroethylene may be used. Neutrino absorption forms free argon atoms out of bound-to-a-molecule chlorine atoms. Free argon atoms are then formed as tiny bubbles of gas that rise in a big experimental vessel and are collected at its top. Argon-37 is radioactive and its presence can therefore be detected even in small concentrations. One way of doing this is to wait a few weeks to allow as much as argon-37 as possible to accumulate, then flush the tanks with helium gas that sweeps out the argon-37 atoms for analysis. This system may therefore be called *neutrino telescope*. Solar neutrinos can reach the tank of perchloroethylene without trouble even though it is a mile deep. Cosmic-ray effects are therefore shielded off and all that is left is the effect of the trace radioactivity in the earth surrounding the neutrino telescope).

Unlike photons the neutrinos from the center of the sun reach us directly. Therefore, one may be able to deduce, in more details, the nature of the fusion reactions going on there. The actual energies of the neutrinos formed depend on the kinetics of the fusion reactions. From the energy spectrum of the neutrinos observed one may hope to deduce the exact reactions in the sun's center and from that the internal temperature and perhaps other characteristics as well. Since these neutrinos have an effective mean energy of ~ 10 Mev, we can use this result to determine a limit of <1/day on the rate of neutrino pulses (10^{11} v_e/cm^2) generated from other stellar sources (for a collapsing star of mass greater than about 1.2 M$_\odot$ located at the galactic center about 10^{52} ergs are radiated away in the form of electronic neutrinos and antineutrinos. The average neutrino energy predicted by various astrophysical models of collapsing stars varies from 10 to 40 *Mev*. The neutrino flux at the earth for the aforementioned collapsing star is about 10^{11} neutrinos per square centimer. Note that in advanced stars neutrinos are being formed in such numbers as to make the star catastrophically leaky. The escaping neutrinos can carry off all the internal store of energy of the star in a short period—of the order of a single day).

Neutrino astronomy may even tell us more general facts about the universe. Intercluster space and all galactic bodies contain huge floods of neutrinos moving through every part of them. These floods represent records of events that have taken place throughout the universe's history. Neutrino astronomy should eventually tell us whether the present universe is dominated by neutrinos or by matter. It may be only because we have not yet detected the neutrino background that matter seems to dominate.

In closing this section I mention two additional topics:

1) Future temperatures of the earth may be calculated from the upper limit of the solar neutrino flux (the earth's temperature history can be investigated experimentally by means of Urey's method of isotopic paleotemperature analysis. Solar luminosity can be calculated from the upper limit of the solar neutrino flux).

2) Other modes of neutrino detection have been proposed. For instance the one employing the reaction

$$\bar{v}_e + p \rightarrow n + e^+ \tag{7}$$

●I.9 THE EMERGENCE OF GAMMA-RAY ASTRONOMY

Gamma-rays suffer negligible absorption or scattering as they travel in space; hence they may survive billions of years and still reveal their source. Studies of the spatial, temporal, and energy distribution of cosmic gamma-rays can, therefore, provide valuable new information for resolving some of the major problems in astrophysics and cosmology.

Many potential sources of cosmic gamma-rays exit. One important task of present-day astronomy is therefore, to identify them.

White dwarfs, neutron stars and black holes in binary systems provide several possible sources for energy in gamma-ray pulses. The observed sudden changes in the rotational speed of a neutron star ("glitches"), "starquakes", volcanic activities or other sudden changes may release gamma-ray bursts. Mass transfer in a binary system is another potential source. A single proton falling onto a white dwarf can release one *Mev*, and one falling onto a neutron star, or into a black hole, can release as much as 100 *Mev*. Blobs of plasma balanced in equilibrium over the magnetic funnels of the neutron star (Figs. I.6 and I.9) may accelerate downward during a glitch and release gamma rays in the process. Other types of flare models may be proposed (e.g., when low-energy photons are boosted in energy by extremely energetic electrons in flares from normal stars or in encounters between relativistic dust particles and photons).

During the past few years the number of point gamma-ray sources identified has increased frome one (the Crab Nebula and pulsar) to at least ten. Several of these new sources are identified with radio pulsar, but many have not yet been identified. The results to be obtained with the newer instruments aboard the gamma-ray observatories now contemplated will probably yield a better view of the galactic gravitational structure, magnetic fields, and cosmic rays.

Indeed, we may now expect that gamma-ray astronomy will also be applicable to problems of cloud formation, the confinement of galactic cosmic rays, the galactic halo, and the diffuse cosmic gamma-ray background and its relation to cosmological problems. Observations of gamma-ray emissions from supernova ejecta and the accumulated background of the universe may also make it possible to prove that supernova eject new nuclei, measure the supernova yield, determine the supernova structure in more details, learn more about the average density of the universe, and further evaluate the evolving world around us.

●I.10 EXPLORATION OF EXTRA-SOLAR SPACE BY UNMANNED SPACECRAFTS

Launched in the early 70's, *Pioneer 11* has a good chance to emerge from the solar system into the depths of extra-solar space around 1993. After grazing Saturn in 1979, Pioneer 11 is now heading in the same direction the entire solar system is moving through the galaxy, but faster and ahead of debris from our sun and planets.

Pioneer 10, on the other hand, is headed the opposite direction, where the solar system had been. Voyagers 1 and 2, the two spacecrafts that produced the spectacular colored imagery of Jupiter in 1979 and discovered the active volcanoes on the ruby-red Jovian satellite Io, were targeted to follow Pioneer 11 past Saturn in 1980 and 1981. *Voyager 2* will pass near Uranus in January 1986 and near Neptune in August 1989, if its longevity matches that of the Pioneers.*

---

* In mid-November 1980 *Voyager I* has indeed passed near Saturn, producing excellent color images of its rings and satellites. It is scheduled to leave the solar system about 1990 and proceeds in the direction of the constellation Ophiuchus.

Pioneer 11 was originally designed with solar panels, whose power output and lifetime would have been enough for a mission up to Jupiter. *Solar* energy would *not* have done the job for longer missions and NASA should be credited for switching to *radioisotope thermoelectric generators* that can deliver the energy and longevity required for much greater ranges. With increasing competence in telecommunications, the range of such space explorations now begins to overlap with some astronomical-cosmological studies. Thereby, it gives the whole subject of space science a greater degree of credibility and testability.

Indeed, the capability of telecommunications in 1983 is a factor of 150,000 better than that used with the 1965 Mariner mission to Mars. And it allows reliable detection of signals less than 10^{-18} watt per square meter arriving from distances greater than 1700 million kilometers (where they have been produced by a transmitter with a power of only 10 to 30 watts). One of the triumphs of the Pioneer 11 mission is the ability of the TRW designed and built spacecraft to transmit a readable signal to earth (170 decibels below 1 milliwatt) over a 1700-million-kilometer line-of-sight distance (4 times weaker than the Jovian distance for which it had been originally designed).

Another feat with Pioneer 11 was the use of a NASA technique for electronically tying together the 210-ft. and 111-ft. deep space network dishes at Goldstone, California. Together with other improved techniques, it is possible today to fix the position of spacecrafts over the many millions of kilometers flight ranges, with an accuracy of about 1km.*

The results obtained from these missions leave little room for romantic speculations about the possibility of life in the solar system outside earth. Yet, those who have yearned for evidence that some forms of life exist on other bodies of the solar system should not belittle the physical evidence that has been accumulated about the fine structure of the solar system. After all, we are the first generation that has escaped from the gravitational pool of the earth, the first to send computer-controlled robots to explore the entire solar system and beyond, and we shall probably be also the first to understand how the solar system had been formed, how life first evolved in it, and how the whole system has interacted with the vast extra-solar space. The Pioneer, Voyager, and other forthcoming missions move us rapidly toward these goals and beyond.

---

* On March 3, 1982 Pioneer 10 completed 10 years of space-flight operation and continued to be the most remote man-made object in history. With present deep space network capability, it will be feasible to receive its error-free data until at least 1990. By then it will have passed through the outer boundary of the heliosphere and would, at last, enter the "actual" interstellar medium discussed in lecture VI (see also page 282 and Fig. VI.1).

FROM "CONSERVATION" IN CLASSICAL PHYSICS TO SOLITONS IN PARTICLE PHYSICS

Conservation laws result generally from the existence of exact symmetries.

Jean-Pierre Vigier, "Possible internal subquantum motions of elementary particles" in *Physics, Logic, and History* W. Yourgrau and A. D. Breck, eds. Plenum Press, New York, 1970, p. 192.

The general concept of conservation in nature has not yet been well understood. On one hand, all evidence supports the idea of conservation of certain things and the growth or decline of some others. For the former we speak about *conservation* equations, for the latter on *balance* equations. On the other hand, conservation, in its broadest sense, includes unresolved physico-philosophical problems.

To begin with, one may mention the powerful concepts of 'symmetry' and 'asymmetry'. Here, one conclusion which emerges from Lecutres V, VI, and IX may be called *"The Conservation Of Symmetry Or Asymmetry"*; that is to say, *once a symmetric mathematical formalism starts* (e.g., by using time-symmetric concepts and/or equations), *that symmetry remains therein irrespective of any mathematical funambulism*! The same rule applies to asymmetry.

Can symmetry or asymmetry 'be broken'?

The answer to this question is categorically negative. It is only by the imposition of incorrect, or inconsistent mathematics (or concepts) that physicists and mathematicians have been able to violate (or 'break') symmetry or asymmetry. For example, time-symmetric differential equations 'become' asymmetric by a 'combination with' *initial conditions* and an integration in an asymmetric time direction (§IX. 2.4.1), or by *a priori* selecting a time-asymmetric probability, and imposing it on a time-symmetric formalism (Lecture V).

But the concept "conservation" involves other subtleties, especially in the general theory of relativity and the physics of time *asymmetries*. In classical physics we normally assume the conservation of (linear and angular) momentum, mass and certain energy expressions and the creation (or consumption) of kinetic and dissipative energies. The same applies to entropy, chemical species, population dynamics, electric charge, baryon number, lepton number, soliton population, etc. (§II.12, II.13).

According to the general theory of relativity (the theory of gravitation—see Lecture III), the simple conservation methods of classical physics do not hold true. It is only when the gravitational field is included (in quite a subtle way) in the general set of certain field equations, that the orthodox idea of conservation can be restored mathematically. Nevertheless, in actual practice, the conservation equations of classical and special relativistic physics hold with sufficiently good accuracy for the purpose of most calculations. Moreover, by "saving" the conservation equations, physicists have been led to important discoveries (q.v., the most instructive example of the discovery of the neutrino, §II.13).

II.1   AIM AND SCOPE

The next two lectures are primarily intended to provide minimal prerequisites for the main lectures by way of simple derivations and physical interpretations of some important balance equations in classical physics and of certain basic equations in special and general relativistic cosmology. Therefore, they may be read independently of the main body of the text, and, as such, may also prove useful in *introductory courses of physics or engineering science.*

In what follows, I present first a highly simplified, yet unified mathematical procedure, which may serve as a common basis for the derivation of different conservation and balance equations. This simplified procedure leads to *a single general differential equation* that is suitable for the description of the distribution of *any macroscopic quantity over space and time in classical physics.* It is presented here for a moving control volume. Using this single equation as a starting point, our most important mathematical tools in classical physics are most easily derived [including the equations of *continuity* (or mass conservation); the equations *of motion* (including the *Navier-Stokes* and *Euler equations*); the equations *of kinetic energy and of general energy* in terms of *internal energy and enthalpy,* including the case of a *multicomponent reacting mixture* characterized by *temperature gradients* (unsatisfactorily presented in most textbooks even today)]. The *chemical species, entropy* and *number-density balance equations are* derived in the same manner.

The approach adopted here is intended to meet a growing need in undergraduate and graduate education for more emphasis on insight into unified physical meanings than on mathematical techniques. Indeed, present university students—who would probably reach peak productivity in the 1990's—should be proficient in subject matter that would retain its universality, validity and usefulness for the coming decades. Accordingly, their university "diet" should be more and more unified in substance and methodology.

While the introductory material is not intended for advanced mathematical-physical analysis, it still places considerable emphasis on material which is sufficiently basic to cut across a number of traditional disciplines in current university education.

An incidental idea behind the present approach consists of exposing the reader to a variety of notation methods. My experience shows that, in the long run, this not only helps the student cope with diverse textbooks and articles, but, also, enables him to gain better physical insight into the problems at hand. Accordingly, some quantities are presented here in alternative equivalent forms. For instance, the gradient of a scalar macroscopic property ψ is written down as

$$\operatorname{grad} \psi \equiv \mathbf{V}\,\psi \equiv \frac{\partial \psi}{\partial x_i} \qquad \text{(a vector)}$$

and its convective terms as

$$\mathbf{u} \cdot \operatorname{grad} \psi \equiv \mathbf{u} \cdot \nabla \psi \equiv V_i \frac{\partial \psi}{\partial x_i} \qquad \text{(a scalar)}$$

where $\mathbf{u}$ represents a velocity vector and Latin indices run through the three values 1,2,3, with x^1, x^2, x^3 representing the Cartesian components of the *position* vector $\mathbf{x}$, and *repeated* indices are to be summed up only over the spatial coordinates (*Einstein's summation convention*), which will subsequently be employed in summation over the four values 1, 2, 3, 0. (x^0 denoting the *time component* whenever *Greek indices* occur in the formulation).

Accordingly the *spatial derivatives* of a *vector* property, such as the velocity, $\mathbf{u}$, are written down as

$$\operatorname{div} \mathbf{u} \equiv \mathbf{V} \cdot \mathbf{u} \equiv \frac{\partial V_i}{\partial x_i} \qquad \text{(a scalar)}$$

and the *gradient* of a vector macroscopic property as, e.g.;

$$\operatorname{grad} \mathbf{u} \equiv \mathbf{V}\mathbf{u} \equiv \frac{\partial V_i}{\partial x_k} \equiv \begin{bmatrix} \dfrac{\partial V_x}{\partial x} & \dfrac{\partial V_y}{\partial x} & \dfrac{\partial V_z}{\partial x} \\[2ex] \dfrac{\partial V_x}{\partial y} & \dfrac{\partial V_y}{\partial y} & \dfrac{\partial V_z}{\partial y} \\[2ex] \dfrac{\partial V_x}{\partial z} & \dfrac{\partial V_y}{\partial z} & \dfrac{\partial V_z}{\partial z} \end{bmatrix} \qquad \text{(a second-rank tensor)}$$

Here, the array in parentheses is applicable *only* when the velocity is given in terms of its Cartesian components V_x, V_y and V_z. In this expression, when rows and columns are interchanged we denote the result as

$$\widetilde{\operatorname{grad} \mathbf{u}} \equiv \mathbf{V}\mathbf{u}^{T} \equiv \frac{\partial V_k}{\partial x_i}$$

which is also called the *transposed tensor* of grad $\mathbf{u}$. The notation for the various *time derivatives* and other tensors will be introduced in due course.

II.2 LIMITATIONS OF THEORY

II.2.1 Micro- vs. macro-analyses

To understand the behavior of a medium of matter and/or radiation we have, in principle, to study that of all discrete "particles" (molecules, atoms, electrons, photons, neutrinos, etc.) which make it up. Complete information on the motion of a system is obtainable by constructing (and integrating) the appropriate differential equations for *each* of the individual particles involved. Unfortunately, even if these equations lend themselves to integration in general form—we would still be unable to assign the exact conditions required for stating the *initial* coordinates and velocities of all the particles involved. (To say nothing of the fact that most gases in the lower atmosphere contain more than 10^{20} particles per cubic centimeter!). We may thus be tempted to conclude that the complexity of the problem at hand increases with the number of particles, and give up our attempt to characterize the system. However, our situation is not nearly as bad since observations show that when the number of particles is "very large", new types of "*regularities*" seem (to us) to supervene—the so-called "average" laws representing *macroscopic* systems. At this juncture it should be noted that while motion of *macro*scopic systems may obey the same dynamical laws as that of *microscopic* ones, *the reverse may not* always be valid. Traditional thermodynamics tells us that the dynamical laws appear *reversible* on *microscopic* scales involving single particles (cf. Time-Reversal Invariance in Lecture IV), but *irreversible* on *macro* scales. This, however, is not always the case (q.v. § IV.11). Yet traditional "macroscopic" theory confines itself to "continuous" systems whose infinitesimal change in volume dV is such that it still contains a "very large" number of particles. For the limited purpose of this lecture we make the same assumption.

There are other factors which restrict the validity of such macro theories. Perhaps the most important of these is related to the *mean free path*, λ, which represents the density of the medium considered* and varies with its local properties. When we examine a certain volume of a fluid (or introduce a macroscopic [solid] instrument of given size into it), the question arises, to what extent the theory remains valid if a characteristic scale Δx of the system (or of the instrument) is continuously reduced.

---

* In kinetic theories, the mean free path is a statistical average distance traversed by material particles between collisions (generally at a given temperature and pressure). For instance, in atmospheric air at ordinary temperature

$$\lambda = 1.26 \sqrt{\gamma} \left(\frac{\mu}{\rho a} \right)$$

where $\gamma = C_p/C_v (= 1.4)$ and μ, ρ and a are, respectively, viscosity, density and the speed of sound. The mean free path in air varies accordingly from about 10^{-7} ft. at sea level to about 1 ft. at an altitude of 80 miles (i.e., the smaller the λ, the denser the fluid).

To give a quantitative answer to this question we usually introduce a dimensionless criterion, defined as

$$K = \frac{\lambda}{\Delta x}$$

and called the *Knudsen number*.

II.2.2 On some Other Limitations of Classical Continuum Physics.

There are many macroscopic phenomena vital to technological applications and scientific curiosity that cannot be explained or predicted in rational manner within the scope of classical continuum physics. To cite a few : The unresolved problems of *liquid crystals, fracture mechanics, turbulence, bubble & drop phenomena, emulsion and suspension mechanics, dynamics of granular, porous and composite materials, non-local thermodynamics,* etc. A specific example : In the vicinity of a crack tip in a solid body the importance of the crack tip radius in the stress distribution is well-known, yet it has not yet been satisfactorily solved.

Internal time effects become responsible to various failures of theory and accuracy of measurements. They arise from the *travel time* of the disturbance from one sub-body to the neighboring one. Most physical phenomena we deal with arise from the response of bodies to disturbances imparted to them by external effects. But when the minimum desired *period* (alternatively a *frequency*) of a disturbance becomes *comparable* with the *internal transmission time* (or *associated frequency*), the individual response of sub-bodies must be taken into consideration. In dynamical systems this is related to a proper relationship between *wave length* and a characteristic *internal length*.

Accurate experimental measurements on lattice dynamics have verified *the failure of continuum theories in the regions where the wave length is comparable to atomic distances in a lattice.* For instance, classical elasticity gives a constant phase velocity $C*$ for plane longitudinal waves in an isotropic elastic solid, while the experiments show that $C*$ depends on the wave length.

These limitations indicate *the necessity to construct new continuum theories.* Recently a few such theories were proposed by several workers, notably by Eringen, Truesdell, Coleman, Kroner, Kunin, Edelen, Kestin, Rice, and Chu (for a review see my MDT, pp. 121–142; CRT, pp. 107–118; 275–343; 483–511).

At the present time there does not exist any generally acceptable formalism for the extension of thermostatics to the description of irreversible processes in solid continua. Difficulties begin at the very foundations of the subject. Processes in strained solid continua generally involve non-uniform fields, impurities, etc., which transverse a sequence of "nonequilibrium states", that may also include inelastic deformations.

Therefore, in the following introductory notes we confine ourselves to the domain of "*simple fluids*" under such conditions where the formalism is verified by measurements. (For more advanced studies on these topics the reader is referred to my MDT and CRT.)

II.3 THE GENERAL MACROSCOPIC EQUATION

Consider a control volume V moving with "*center of mass velocity*," **u**, as shown in Fig. II-1 (see also § II.4 below):

The basis of our derivation is the general balance statement for a macroscopic property;

$$\begin{bmatrix} \text{Time rate of increase} \\ \text{of a property within} \\ \text{volume } V. \end{bmatrix} = \begin{bmatrix} \text{Net inflow of property} \\ \text{across surface A per} \\ \text{unit time.} \end{bmatrix} + \begin{bmatrix} \text{Net generation} \\ \text{rate of property} \\ \text{within volume } V \end{bmatrix}$$

(1)

which is assumed valid everywhere in the fluid. We now assign the symbol ψ to the general macroscopic property per unit volume. In the following we would be interested in evaluating "ψ-properties" such as:

$\psi_1 = \rho$ *total density* (total mass per unit volume),

$\psi_2 = \rho\mathbf{u}$ *linear momentum* (per unit volume) (vector),

$\psi_3 = \rho u$ *internal energy* (per unit volume), where u is internal energy per unit mass,

$\psi_4 = \frac{1}{2}\rho|\mathbf{u}|^2$ *kinetic energy* (per unit volume),

$\psi_5 = \rho_k$ *species density* (mass of chemical component k, per unit volume),

$\psi_6 = \rho s$ *entropy* (per unit volume), where s is entropy per unit mass,

$\psi_7 = f$ *number-density function* (per unit volume),

and others.

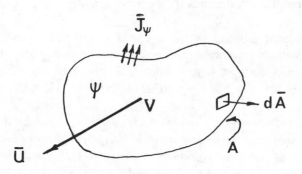

FIG. II.1: **Notation for a moving control volume**

Before proceeding to formulate equation (1) in mathematical symbols, we must familiarize ourselves with the three *physical versions of the differentiation operator with respect to time*, namely the *substantial operator* D/Dt (or the *operator following the motion*), the *partial* (or local) *operator*, $\partial/\partial t$, and the *total operator*, d/dt. In visualizing the three versions in practice, we shall use the illustration of an observer seated in a boat in a flowing stream, for whom the time derivatives of ψ is represented by the incidence of fish encountered by him.

(i) Substantial Time Derivative, $D\psi/Dt$

This is the time derivative for an observer *co-moving* with the fluid medium at the velocity of the latter, $(\mathbf{u})$—in our illustration, one whose boat is adrift. This particular observer is at rest relative to his immediate vicinity*.

(ii) Partial Time Derivative, $\partial\psi/\partial t$

This is the time derivative for a *fixed* (local) observer—one whose boat is *at anchor*. Clearly, when the flow velocity $\mathbf{u} = 0$,

$$\frac{D\psi}{Dt} = \frac{\partial\psi}{\partial t} \tag{2}$$

In general, the two operators are related through:

$$\frac{D}{Dt} \equiv \frac{\partial}{\partial t} + \mathbf{u} \cdot \text{grad} \equiv \frac{\partial}{\partial t} + \mathbf{u} \cdot \mathbf{V} \tag{3}$$

or, as an example, by

$$\frac{D\psi}{Dt} = \frac{\partial\psi}{\partial t} + V_x\frac{\partial\psi}{\partial x} + V_y\frac{\partial\psi}{\partial y} + V_z\frac{\partial\psi}{\partial z} \equiv \frac{\partial\psi}{\partial t} + V_i\frac{\partial\psi}{\partial x_i} \tag{3a}$$

where V_x, V_y and V_z are the Cartesian x-, y- and z-components.

The physical difference between the operators may also be illustrated by the corresponding concepts of "equilibrium" and "steady state." "Equilibrium" along a streamline (with respect to ψ alone) is normally given by the condition**

$$\frac{D\psi}{Dt} = 0 \tag{4}$$

which in the above analogy means that the incidence of fish is constant throughout the stream. On the other hand,

---

* In astrophysical and cosmological applications, for instance, a co-moving observer is also frequently in a state of *"free fall"*—which, under the principle of equivalence (see Lecture III) means absence of a gravity force.

** Mathematically speaking, $D\psi/Dt$ may be zero at a given point without attaining equilibrium conditions along the streamline.

$$\frac{\partial \psi}{\partial t} = 0 \quad \text{(steady state)} \tag{5}$$

means that ψ does not change with time for a given point in space, but may vary from point to point. Therefore $(\partial \psi / \partial t) = 0$ does not necessarily imply $(D\psi / Dt) = 0$.

Accordingly, steady state is characterized by

$$\frac{D\psi}{Dt} = \mathbf{u} \cdot \text{grad}\, \psi \left(= V_x \frac{\partial \psi}{\partial x} + V_y \frac{\partial \psi}{\partial y} + V_z \frac{\partial \psi}{\partial z} \right) \tag{6}$$

(iii) Total Time Derivative, $d\psi/dt$

In a spatial system of coordinates, it is defined by

$$\frac{d\psi}{dt} = \frac{\partial \psi}{\partial t} \frac{dt}{dt} + \frac{\partial \psi}{\partial x} \frac{dx}{dt} + \frac{\partial \psi}{\partial y} \frac{dy}{dt} + \frac{\partial \psi}{\partial z} \frac{dz}{dt} \tag{7}$$

with Eq. (3) as a particular case, where

$$V_x = \frac{dx}{dt}; V_y = \frac{dy}{dt}; V_z = \frac{dz}{dt} \tag{8}$$

denote the components of the fluid velocity. This derivative concerns our observer when he *starts the motor* of his boat, and cruises up-, down- or cross-stream, with (8) referring to the components of the *boat* velocity.

In the light of the above, in analyzing a *moving control volume*, we must use the derivative following the motion, $D\psi/Dt$, expressed as

$$\frac{D}{Dt} \int_V \psi \, dV \tag{9}$$

The net *inflow* across the surface A is therefore given by the closed surface integral of the flux

$$- \oint \mathbf{J}_\psi \cdot d\mathbf{A} \tag{10}$$

which covers all possible inflow-outflow fluxes along the interface. [$(d\mathbf{A})$ is normal to and directed outwards from the surface; hence the net *inflow* is associated with the *negative* sign]. The scalar product in Eq. (10) represents the first right-hand term in Eq. (1).

Here $\mathbf{J}_\psi$ is the surface flux of property ψ, as determined by a *moving* observer! Its value is usually different for a stationary one!

For the generating term in Eq. (1) we simply write

$$\int_V g_\psi \, dV \tag{11}$$

where g_ψ is the net rate of generation of ψ per unit volume. Substituting (9), (10) and (11) in Eq. (1) we obtain

$$\frac{D}{Dt} \int_V \psi \, dV = - \oint_A \mathbf{J}_\psi \cdot d\mathbf{A} + \int_V \dot{g}_\psi \, dV \tag{12}$$

which is the identical mathematical counterpart of (1). Integral differentiation yields

$$\frac{D}{Dt} \int_V \psi \, dV = \int_V \frac{\partial \psi}{\partial t} \, dV + \oint_A \psi(\mathbf{u} \cdot d\mathbf{A}) \tag{13}$$

where the volume integral on the right represents the net change of property ψ in V, if V is stationary (i.e. $\partial/\partial t$ represents the partial time derivatives as determined by an observer *fixed in space*—that is, with respect to the inertial coordinates employed in measuring $\mathbf{u}$). The surface integral accounts, therefore, for the difference between a co-moving and a fixed observer in counting the net rate of change in ψ.*

Substituting (13) in (12) and resorting to Gauss' theorem

$$- \oint_A \mathbf{J}_\psi \cdot d\mathbf{A} = - \int_V \operatorname{div} \mathbf{J}_\psi \, dV \tag{14}$$

$$\oint_A \psi(\mathbf{u} \cdot d\mathbf{A}) = \int_V \operatorname{div}(\psi \mathbf{u}) \, dV \tag{15}$$

we obtain

$$\int_V \left[\frac{\partial \psi}{\partial t} + \operatorname{div} \mathbf{J}_\psi + \operatorname{div}(\psi \mathbf{u}) - g_\psi \right] dV = 0 \tag{16}$$

Subject to the continuum hypothesis, (16) remains valid for an infinitesimal volume dV containing a large number of particles, i.e. the integrand of (16) is itself zero, yielding the final working form of the *general property equation*:

$$\boxed{\frac{\partial \psi}{\partial t} + \operatorname{div} \mathbf{J}_\psi + \operatorname{div}(\psi \mathbf{u}) - g_\psi = 0} \tag{17}$$

---

* The limitations associated with this procedure, as well as with eq. (13), are not to be discussed here. Our procedure is based on eq. (16) which one may obtain by other, more advanced procedures.

II.4   CONTINUITY EQUATION (TOTAL MASS CONSERVATION)

In this simple case the property in question is the total density ρ, defined as

$$\rho = \sum_{k=1}^{n} \rho_k, \tag{18}$$

where ρ_k is the local density (or mass *concentration*) of the chemical component k (i.e. the mass of species k actually present, per unit volume), and n the total number of chemical species which make up the fluid medium ($k = 1, 2, ..., n$). For any material and/or radiation media, ρ should be subject to an appropriate equation of state.

Conservation of total mass (except for nuclear reactions), requires that $g_\psi = 0$. Since the fluid element under consideration is co-moving with the local mass stream, there is no *total* mass transfer between one element of the fluid to another. This fact is clarified by defining **u** as the *center-of-mass* velocity (see Fig. II.1 above), namely

$$\mathbf{u} = \frac{\sum_{k=1}^{n} \rho_k \mathbf{u}_k}{\rho} \tag{19}$$

where $\mathbf{u}_k$ is the *local* mass velocity of k with respect to an inertial frame of reference.

Under these conditions, the general property equation (17) reduces to

$$\boxed{\frac{\partial \rho}{\partial t} = - \operatorname{div}(\rho \mathbf{u})} \tag{20}$$

which is known as the *continuity equation*. The physical interpretation of $\operatorname{div}(\rho \mathbf{u})$ is, therefore, rate of change in mass per unit volume (see also below). In (x, y, z) coordinates, (20) takes the form

$$\frac{\partial \rho}{\partial t} = - \left[\frac{\partial}{\partial x}(\rho V_x) + \frac{\partial}{\partial y}(\rho V_y) + \frac{\partial}{\partial z}(\rho V_z) \right] \tag{21}$$

or, in more convenient form,

$$\boxed{\frac{D\rho}{Dt} = - \rho \operatorname{div} \mathbf{u}} \tag{22}$$

using the mathematical identity

$$\operatorname{div} \rho \mathbf{u} = \rho \operatorname{div} \mathbf{u} + \mathbf{u} \cdot \operatorname{grad} \rho$$

and D/Dt as defined at the outset. In (x, y, z) coordinates (22) reads

$$\frac{\partial \rho}{\partial t} + V_x \frac{\partial \rho}{\partial x} + V_y \frac{\partial \rho}{\partial y} + V_z \frac{\partial \rho}{\partial z} = - \rho \left[\frac{\partial V_x}{\partial x} + \frac{\partial V_y}{\partial y} + \frac{\partial V_z}{\partial z} \right] \tag{23}$$

In Eq. (23) the mathematical concept *divergence* (space derivatives of a vector or a tensor) acquires a physical meaning. When the space derivatives of **u** do *not* vanish, we have a *compressible* fluid, whose density changes along its path of motion as $-1/\rho \, (D\rho/Dt)$. What is, however, the physical meaning of div $\mathbf{u} = 0$?

According to definition (4), the density of the fluid does *not change along its path of motion and in time* if, and only if;

$$\frac{D\rho}{Dt} = 0 \qquad \text{(incompressible fluid)} \tag{24}$$

which is equivalent to stating that

$$\text{div } \mathbf{u} = 0 \qquad (\textit{incompressible fluid}) \tag{25}$$

● II.5 CONSERVATION OF LINEAR MOMENTUM AND GRAVITY

In this case the general macroscopic property per unit volume is

$$\psi = \rho \mathbf{u}, \tag{26}$$

which is the momentum per unit volume (a vector). The largest unknown in this case is the flux of momentum. To overcome this difficulty we substitute $\psi = \rho \mathbf{u}$ in the third term of Eq. (17), which thus becomes div $|\rho \mathbf{uu}|$.* Obviously the dimensionality of div $\mathbf{J}_\psi$, in this case, should be the same as that of div $|\rho \mathbf{uu}|$. Now, $\rho \mathbf{uu}$ is a *tensor of second rank*, which in Cartesian coordinates reads

$$\rho \mathbf{uu} = \begin{bmatrix} \rho V_x V_x & \rho V_y V_x & \rho V_z V_x \\ \rho V_x V_y & \rho V_y V_y & \rho V_z V_y \\ \rho V_x V_z & \rho V_y V_z & \rho V_z V_z \end{bmatrix} \tag{27}$$

with the first row containing the three x-convective components of the momentum, the second row the y-components and the third row the z-components.** We see that $\mathbf{J}_\psi$, in this case, should, on the one hand, have the property of a flux of momentum (per unit area normal to the flux in question), and on the other be a second-rank

---

* The physical interpretation of div $|\rho \mathbf{uu}|$ is analogous to that of div $|\rho \mathbf{u}|$, except that $\rho \mathbf{uu}$ is a *tensor quantity* (cf. eq. 27) while $\rho \mathbf{u}$ is a *vector*; therefore, div $|\rho \mathbf{u}|$ represents the rate of change of mass (a scalar) per unit volume, while div $|\rho \mathbf{uu}|$ represents the convective rate of change of momentum (a vector) per unit volume.

** In one-dimensional flow, div $|\rho \mathbf{uu}|$ reduces to $\partial/\partial x (\rho V_x V_x) \equiv$ the net rate at which ρV_x enters and leaves through the $dzdy$ planes of volume element. Hence, the dimensionality of the last expression is $\rho V_x -$ momentum per unit time per unit volume.

tensor, like $\rho\mathbf{uu}$ (i.e., with a total of nine components). This tensor is known as the *pressure tensor* $\mathbf{P}$ (or sometimes as the *stress tensor*). Thus, we substitute $\mathbf{P}$ for $\mathbf{J}_\psi$ in eq. (17) and obtain the most general *momentum equation*

$$\frac{\partial(\rho\mathbf{u})}{\partial t} \qquad + \operatorname{div}\mathbf{P} \qquad + \operatorname{div}(\rho\mathbf{uu}) \qquad - \Sigma_k \rho_k \mathbf{F}_k = 0 \qquad (28)$$

nonsteady	pressure	momentum	source terms
state terms.	and viscous	convective	(due to external
	terms.	terms.	forces).

which is written here for a *fixed observer*. In this case (unlike that of total density), *momentum can be generated* by $\mathbf{F}_k$ forces, such as *gravity* or an *electromagnetic field*!! As for the latter it may act only on certain ion species k in the fluid, while the other species remain unaffected. Therefore, in general, the terms $\mathbf{F}_k$ should carry the subscript k (force per unit mass of k). Hereinafter we will consider only *conservative forces*, which can be derived from the gradient of a potential independent of time. For the source of momentum *due to gravity*, the last term in eq. (28) must be replaced by $\rho\mathbf{g}$, where $\mathbf{g}$ is the gravity vector.* Now, writing

$$\frac{\partial(\rho\mathbf{u})}{\partial t} = \rho\frac{\partial\mathbf{u}}{\partial t} + \mathbf{u}\frac{\partial\rho}{\partial t}$$

and resorting to the identity

$$\operatorname{div}|\rho\mathbf{uu}| = \rho\mathbf{u}\cdot\operatorname{grad}\mathbf{u} + \mathbf{u}(\operatorname{div}\rho\mathbf{u}),$$

Eq. (28) takes the form

$$\rho\frac{\partial\mathbf{u}}{\partial t} + \mathbf{u}\frac{\partial\rho}{\partial t} + \operatorname{div}\mathbf{P} + \rho\mathbf{u}\cdot\operatorname{grad}\mathbf{u} + \mathbf{u}\operatorname{div}|\rho\mathbf{u}| - \rho\mathbf{g} = 0,$$

$$\text{I} \qquad\quad \text{II} \qquad \text{III} \qquad\quad \text{IV} \qquad\qquad \text{V} \qquad\qquad \text{VI}$$

where terms II and V cancel out on account of the continuity equation. Combining terms I and IV to form the substantial derivative $D\mathbf{u}/Dt$, we finally obtain a simplified equivalent of eq. (28):

$$\boxed{\rho\frac{D\mathbf{u}}{Dt} = -\operatorname{div}\mathbf{P} + \rho\mathbf{g}} \qquad (29)$$

which is clearly of the form

$$[\text{mass}] \times [\text{acceleration}] = \text{sum of forces} \qquad (30)$$

i.e., a statement of *Newton's Second Law*. The physical interpretation of eq. (29) in terms of a velocity field will be given below. At this point I only stress the fact that *the gravity vector $\mathbf{g}$ in eq. 29 is a source of motion and acceleration*!!

---

* cf. eq. 2 in §III.1.1

● II.6   THE NAVIER-STOKES EQUATIONS AND GRAVITY

To determine the velocity distribution we must specify *how the various components of* **P** *depend on the velocity field, and its relevant derivatives.*

The general form of **P** is obtainable subject to inclusion of an *"internal friction"* effect (governed by the Second Law of Thermodynamics, see below) which occurs between adjacent fluid particles that move with *different* velocities, thus generating *relative* motion within contiguous parts of the fluid. Accordingly, the nine components of **P** must, *inter alia*, depend on the space derivatives of the velocity components.* For this purpose we resolve **P** into *hydrostatic* and *viscous* contributions, namely

$$\mathbf{P} = \tau + p\mathbf{I} \tag{31}$$

where **I** is the unit matrix, p the *hydrostatic pressure* (a scalar), and τ the *viscous stress tensor.* The above resolution of the pressure tensor **P** is feasible here and enables us to differentiate between the hydrostatic pressure (which for compressible fluids *at rest* may be identified with that of classical thermostatics), and the viscous fluxes of momentum represented by the nine components of the *viscous stress tensor* τ (which are, in turn, related to the velocity field, as will be shown later). τ vanishes for an *ideal fluid*, since the latter is *defined* as one *without internal friction* (viscosity effects) *and without heat exchange between its parts.*

The concept of hydrostatic pressure may be retained *only when we assume*, as we normally do, that *even when the fluid is in motion,* a *"local thermodynamic equilibrium"* (LTE) prevails.† More specifically, hydrostatic pressure in a moving fluid is assumed such that a surface element within it *always* undergoes a stress *normal* to itself, and this stress is *independent of the orientation.* Thus, when an element of viscous fluid moves relative to its neighbors, *we assume* that it undergoes normal stresses, which in Cartesian coordinates read

$$P_{xx} = p + \tau_{xx}; \qquad P_{yy} = p + \tau_{yy}; \qquad P_{zz} = p + \tau_{zz} \tag{32}$$

τ_{xx}, τ_{yy}, and τ_{zz} being the components of the viscous stress tensor *normal* to the yz, zx, and xy planes, respectively. In addition, *we assume*, throughout the following treatment, that the τ (hence also **P**), *is symmetric*, i.e.,

$$\tau = \tau^{\mathbf{T}} \text{ (or } \tau_{ij} = \tau_{ji}), \tag{33}$$

a requirement which is valid for *"non-polar, inelastic fluids"* in which *torques* can

---

* Other effects such as *relaxation times* due to *fading memory* can also have a marked effect. For a brief discussion see Lecture IV Section 16.

In assuming that the equilibrium term in **P** is a scalar, the theory is restricted to *non-elastic fluids* (cf. also the subsequent discussions).

† This postulate should always be questioned and verified.

only derive from the action of direct forces.[++] (We can also *assume* that *angular momentum is conserved* and *deduce* the *symmetry* (33) from this assumption.)[*]

Furthermore, we have assumed that the fluids are *isotropic*, which means that there is *no internal sense of orientation* within the fluid (i.e. a functional relation between stress and deformation strain must be independent of the alignment of the coordinate system *chosen* by the analyst).

For astrophysical and cosmological applications we must also take into account viscous stresses generated by *high-pressure radiation* (see Lectures VI and III).

We can now substitute the expression $\mathbf{P} = p\mathbf{I} + \tau$ in the equation of motion (28 or 29) and obtain

$$\frac{\partial(\rho\mathbf{u})}{\partial t} = - \operatorname{div}|\rho\mathbf{u}\mathbf{u}| - \operatorname{grad} p - \operatorname{div}\tau + \rho\mathbf{g}, \tag{34}$$

or

$$\rho\frac{D\mathbf{u}}{Dt} = - \operatorname{grad} p - \operatorname{div}\tau + \rho\mathbf{g} \tag{35}$$

For an *ideal fluid*, τ vanishes and (35) reduces to

$$\rho\frac{D\mathbf{u}}{Dt} = - \operatorname{grad} p + \rho\mathbf{g} \tag{36}$$

which is the *Euler equation*. Note again that the gravity vector $\mathbf{g}$ is the source of acceleration. This fact explains many phenomena in nature, e.g., the *atmospheric circulation*, the *hydrological cycle*, etc. (Lecture I, VI and VII).

If the (local) relative-velocity gradients are "moderate," we can *assume* that, as a first approximation, viscous momentum transfer depends only on the *first* derivative of the velocity components. This is a *linear* approximation which is verified by observation in many fluids, and known as *"Newtonian."*[**] Neglecting relaxation phenomena, there can be no forms in τ which are independent of $\partial V_i/\partial x_k$ ($i, k, = 1,2,3$), since by definition, τ vanishes when $\mathbf{u} = $ constant everywhere in the fluid. This result can be demonstrated with a fluid in a state of *mechanical equilibrium* (without internal friction or relaxation effects) in a cylindrical vessel rotating at

----

[++] A *polar fluid* is one capable of *transmitting stress couples*, and thus subject to *body torques*.

[*] cf., e.g., Aris, R., *Vectors, Tensors, and Basic Equations of Fluid Mechanics*, Prentice-Hall, N.Y. 1962 pp. 102. It should be noted that *it is equally valid to assume symmetry of τ and deduce conservation of angular momentum!* Fluids obeying eq. (33) are also called *"structureless"* fluids. In these circumstances (33) may be viewed as a highly *restrictive assumption* (For a theory of polar fluids with asymmetric viscous stress tensors and applications to *lubricity* problems see Gal-Or and Zehavi's paper in the Intern. (ASME co-sponsored) Congress on Gas Turbines, Haifa, 1979).

[**] In recognition of the historical fact that the *linear* relation for simple shearing motion

$$\tau_{xy} = - \mu\frac{\partial V_x}{\partial y}$$

(for the significance of the proportionality factor μ — see below) was originally put forward by Newton.

a constant angular velocity ω. Obviously, at this juncture, the vector product equals the local velocity according to $\omega x \, \mathbf{r} = \mathbf{U}$, but $\tau = 0$.

Linear combinations of space derivatives which vanish even when $\mathbf{U} = \omega x \mathbf{r}$, are given by the sum

$$\frac{\partial V_i}{\partial x_k} + \frac{\partial V_k}{\partial x_i} \qquad (\mathrm{i,j,k,} = 1,2,3) \tag{37}$$

Therefore, according to our assumptions, τ must only contain combinations of these derivatives. For isotropic fluids, the most general tensor (of rank two) which satisfies these conditions is

$$\tau_{ik} = A \left[\frac{\partial V_i}{\partial x_k} + \frac{\partial V_k}{\partial x_i} \right] + B \frac{\partial V_j}{\partial x_j} \delta_{ik}, \tag{38}$$

when A and B are independent of the velocity components, and $\delta_{ik} = 1$ if $i = k$; $\delta_{ik} = 0$ if $i \neq k$. As a matter of convenience and common practice in fluid dynamics, (38) is rewritten in another general form

$$\boxed{\tau_{ik} = - \mu \left[\frac{\partial V_i}{\partial x_k} + \frac{\partial V_k}{\partial x_i} - \frac{2}{3} \frac{\partial V_j}{\partial x_j} \delta_{ik} \right] - \kappa \frac{\partial V_j}{\partial x_j} \delta_{ik},} \tag{39}$$

where the new coefficients are, again, assumed *independent* of the velocity components. The first coefficient μ is called the *shear viscosity*, while κ is termed the *expansion (or volume) viscosity).** Using another notation eq. (39) is written as†*

$$\tau = -\mu[\mathrm{grad}\,\mathbf{u} + \mathrm{grad}\,\mathbf{u}^T] + [2/3\mu - \kappa]\,\mathrm{div}\,\mathbf{u}\mathbf{I}$$

$$= -2\mu\,\mathbf{D} + [2/3\mu - \kappa]\,\mathrm{div}\,\mathbf{u}\mathbf{I} \tag{40}$$

For incompressible viscous fluids, div $\mathbf{u} = 0$, and eq. (39) reduces to

$$\tau_{ik} = -\mu \left[\frac{\partial V_i}{\partial x_k} + \frac{\partial V_k}{\partial x_i} \right], \tag{41}$$

or

$$\tau = -2\mu\,\mathbf{D} \tag{41a}$$

The viscous stress tensor of an incompressible fluid is, therefore, characterized by a single viscosity coefficient. Thus, for a *Newtonian* incompressible fluid in (x, y, z) Cartesian coordinates:

Normal viscous stresses
$$\begin{cases} \tau_{xx} = -2\mu\,\partial V_x/\partial x \\ \tau_{yy} = -2\mu\,\partial V_y/\partial y \\ \tau_{zz} = -2\mu\,\partial V_z/\partial z \end{cases}$$

---

* *Also known as bulk viscosity. For a brief discussion see Chapman and T. G. Cowling, "The Mathematical Theory of Non-Uniform Gases," Cambridge, 1970. (3rd ed.).*

*† The term $\mathbf{D} = \frac{1}{2}[\mathrm{grad}\,\mathbf{u} + \mathrm{grad}\,\mathbf{u}^T]$ is called the *rate-of-deformation tensor*.

Tangential shearing stresses
$$\begin{cases} \tau_{xy} = \tau_{yx} = -\mu\left[\partial V_x/\partial y + \partial V_y/\partial x\right] \\ \tau_{yz} = \tau_{zy} = -\mu\left[\partial V_y/\partial z + \partial V_z/\partial y\right] \\ \tau_{xz} = \tau_{zx} = -\mu\left[\partial V_x/\partial z + \partial V_z/\partial x\right] \end{cases}$$

Substituting (39) in the equation of motion (35) we finally obtain the working form of the *Navier-Stokes equation*

$$\rho\left[\frac{\partial V_i}{\partial t} + V_k\frac{\partial V_i}{\partial x_k}\right] = -\frac{\partial \dot{p}}{\partial x_i} + \frac{\partial}{\partial x_k}\left[\mu\left(\frac{\partial V_i}{\partial x_k} + \frac{\partial V_k}{\partial x_i} - \frac{2}{3}\frac{\partial V_j}{\partial x_j}\right)\right] + \frac{\partial}{\partial x_i}\left[\kappa\frac{\partial V_j}{\partial x_j}\right] + \rho g_i \tag{42}$$

—a "vector equation" comprising three "components," one for each x_i.

● II.7   KINETIC-ENERGY EQUATION AND DISSIPATION FUNCTION IN GRAVITATIONAL FIELDS

In this section we derive, from the equation of motion, a balance equation for the kinetic energy. From the result we identify the term responsible for irreversible conversion of mechanical into internal energy. The latter is characterized by a dissipation function, which also plays an important role in the balance equation for internal energy (to be derived in the next section), and in the evaluation of an (entropy-free) thermodynamic arrow of time (see below).

The balance equation is derived here for the kinetic energy per unit volume, i.e. $1/2\,\rho|\mathbf{u}|^2$. For this purpose, rewriting the equation of motion (29) in the form

$$\rho\frac{DV_i}{Dt} = -\frac{\partial}{\partial x_k}P_{ki} + \rho_k\mathbf{F}_k, \tag{29a}$$

and resorting to (scalar) multiplication of both sides by $V_i(t, x_i)$, we have

$$\rho\frac{D\left[\frac{1}{2}|\mathbf{u}|^2\right]}{Dt} = -\frac{\partial}{\partial x_k}(P_{ki}\,V_i) + P_{ki}\frac{\partial V_i}{\partial x_k} + \rho_k\mathbf{F}_k\,V_i, \tag{43}$$

since

$$-V_i\frac{\partial P_{ki}}{\partial x_k} = -\frac{\partial}{\partial x_k}[P_{ki}V_i] + P_{ki}\frac{\partial V_i}{\partial x_k} \tag{44}$$

In tensor notation, eq. (43) reads

$$\rho\frac{D\left[\frac{1}{2}|\mathbf{u}|^2\right]}{Dt} = -\operatorname{div}[\mathbf{P}\cdot\mathbf{u}] + \mathbf{P}:\operatorname{grad}\mathbf{u} + \sum_k\rho_k\mathbf{F}_k\cdot\mathbf{u}, \tag{45}$$

where

$$\mathbf{P} : \operatorname{grad} \mathbf{u} \equiv P_{ki} \frac{\partial V_i}{\partial x_k} = p \operatorname{div} \mathbf{u} + \tau : \operatorname{grad} \mathbf{u} \tag{46}$$

Referring to the equation of continuity, it is readily shown*, that, for any arbitrary local property ψ,

$$\rho \frac{D\psi}{Dt} = \frac{\partial(\rho\psi)}{\partial t} + \operatorname{div}(\rho \mathbf{u} \, \psi) \tag{47}$$

Resolving $\operatorname{div}(\mathbf{P} \cdot \mathbf{u})$ into its pressure and viscous contributions, we now rewrite (45) in the form

$$\frac{\partial}{\partial t}[1/2\rho|\mathbf{u}|^2] = -\operatorname{div}[1/2\rho|\mathbf{u}|^2\,\mathbf{u}] - \operatorname{div}|p\mathbf{u}| - \operatorname{div}|\tau\cdot\mathbf{u}| \quad + p\operatorname{div}\mathbf{u} \quad + \tau: \operatorname{grad}\mathbf{u} + \sum_k \rho_k \mathbf{F}_k \cdot \mathbf{u}$$

Rate of Increase in K.E.	Net rate of Input of K.E. by Bulk Flow.	Rate of Work done by Pressure.	Rate of Work done by Viscous Forces.	Rate of Reversible Conversion into Internal Energy.	Rate of Irreversible Dissipation (loss) to Internal Energy.	Rate of Work done by External (e.g. gravity) Force.

$$\tag{48}$$

Note again the role of the gravity vector in *generating* kinetic energy.

The introduction of the term representing the *irreversible conversion* of gravitational-mechanical into internal energy (i.e., loss of kinetic energy)—is based on observations and on the *second law of thermodynamics* (Lecture IV). The latter may, therefore, be illustrated (in a restricted form) by the important inequality

$$\left.\begin{array}{l} \text{Rate of Irreversible} \\ \text{Increase in Internal} \\ \textit{Energy per unit volume} \\ \text{due to } \textit{Viscous} \\ \textit{Dissipation of Gravitational-} \\ \textit{Mechanical Energy.} \end{array}\right\} = \Phi = -\tau : \operatorname{grad}\mathbf{u} \equiv -\tau_{ki} \frac{\partial V_i}{\partial x_k} \geq 0 \tag{49}$$

For a compressible Newtonian fluid eq. (49) takes the form**

$$\Phi = -\tau : \operatorname{grad}\mathbf{u} = \mu \frac{\partial V_i}{\partial x_k}\left[\frac{\partial V_i}{\partial x_k} + \frac{\partial V_k}{\partial x_i}\right] - [2/3\,\mu - \kappa]\frac{\partial V_j}{\partial x_j}\frac{\partial V_j}{\partial x_j} \geq 0, \tag{50}$$

---

* Bearing also in mind the identity $\operatorname{div}(\rho\psi\mathbf{u}) = \rho\mathbf{u} \cdot \operatorname{grad}\psi + \psi \cdot \operatorname{div}[\rho\mathbf{u}]$.

** Some textbooks define $\mu\Phi'$ as Φ with κ neglected. Note that $(\partial v_j/\partial x_j)^2 \equiv (\operatorname{div}\mathbf{u})^2$, which vanishes for an incompressible fluid. Note also that the inequality in eqs. (49) and (50) is *postulated* by the second law of thermodynamics (eq. 77 below), *not deduced* from the formalism presented here.

where Φ is the so-called *dissipation fuction*. For an *incompressible* Newtonian fluid, eq. (49) reduces to:

$$\Phi = \frac{1}{2}\mu\left[\frac{\partial V_i}{\partial x_k} + \frac{\partial V_k}{\partial x_i}\right]^2 \geq 0 \tag{51}$$

Therefore, the second law of themodynamics dictates the *positive* nature of the shear viscosity, i.e., $\mu > 0$. Hence, the assumption of a positive value for the shear viscosity is *deduced* from the science of thermodynamics. But the second law of thermodynamics is a postulate; indeed, a postulate supported by all empirical evidence, but still a postulate.

One of the main aims of gravitism is, therefore, *to deduce* the second law of thermodynamics, the positive nature of viscosity and the positive energy dissipation *from the theory of gravitation. This, in turn, will reduce thermodynamics to an integral core of the general theory of relativity* (Lectures VI and VII).

To proceed we note that the physical meaning of $(-\tau:\mathrm{grad}\,\mathbf{u}) \geq 0$, is that all flow systems are subject to *irreversible conversion* ("*degredation*") *of mechanical into thermal energy*! This (irreversible) viscous dissipation increases the internal energy of the fluid. In certain cases, e.g., flight at hypersonic speeds, lubrication, and other processes characterized by high velocity gradients, it also entails a significant increase of temperature. We shall return to this subject in subsequent sections.

● II.8   FIRST LAW OF THERMODYNAMICS OR ENERGY CONSERVATION EQUATION

The general macroscopic property involved in deriving the first law of thermodynamics is ρu, where u is the *internal energy* per unit mass (a scalar). It is assumed to be an absolutely continuous additive property. Now, using $\psi = \rho u$ as the variable in the general property equation, (17), we immediately obtain the first law of thermodynamics

$$\frac{\partial(\rho u)}{\partial t} + \mathrm{div}\,\mathbf{J}_q + \mathrm{div}(\rho u\mathbf{u}) - g_u = 0 \tag{52}$$

where g_u represents the source terms of internal energy (to be specified below). This version of the first law is less familiar than the one obtained with the aid of eq. (47); it reads

$$\rho\frac{Du}{Dt} = -\mathrm{div}\,\mathbf{J}_q + g_u \tag{53}$$

where $\mathbf{J}_q$ is the *total* heat transfer flux due to *conduction* $\mathbf{q}$, *plus* the possible energy (or enthalpy) transfer due to *mass fluxes* of the individual species $\mathbf{J}_k = \rho_k(\mathbf{u}_k - \mathbf{u})$, namely

$$J_q = q + \sum_{k=1}^{n} h_k J_k \tag{54}$$

h_k being the specific enthalpy of component k. (The second term vanishes for a single-component fluid).

What remains now is to analyze the source term in (53). In the preceeding section we have demonstrated (cf. eq. (48)) why the terms $(-\Phi)$ and $(+p\,\mathrm{div}\,\mathbf{u})$ represent, respectively, irreversible and reversible interconversions of kinetic-mechanical and internal energy.

Therefore, in g_u, these terms should appear with opposite signs to those of eq. (48), i.e.,

$$g_u = -\boldsymbol{\tau} : \mathrm{grad}\,\mathbf{u} - p\,\mathrm{div}\,\mathbf{u} + \sum_k \mathbf{J}_k \cdot \mathbf{F}_k, \tag{55}$$

where a possible generation of internal energy, through the net work done by an external force $\mathbf{F}_k$, is added. Substituting (55) and (54) in (53), we finally obtain

$$\rho \frac{Du}{Dt} = -\mathrm{div}\,\mathbf{q} - \mathrm{div} \sum_{k=1}^{n} h_k\,\mathbf{J}_k + \Phi - p\,\mathrm{div}\,\mathbf{u} + \sum_k \mathbf{J}_k \cdot \mathbf{F}_k \tag{56}$$

The third and the last terms vanish identically when $\mathbf{J}_k = 0$. Eq. (56) is a very useful statement of the first law, which in many textbooks appears simply as

$$\boxed{du = d'q - d'w - d(\text{Kinetic Energy}) - d(\text{Potential Energy}),} \tag{57}$$

where $d'q$ and $d'w$ are "infinitesimal quantities" of heat and work, respectively. The equivalence of (56) and (57) becomes clearer if we resort to the continuity equation and write

$$p\,\mathrm{div}\,\mathbf{u} = -\frac{p}{\rho}\frac{D\rho}{Dt} = \rho p \frac{D}{Dt}(1/\rho) = \rho p \frac{Dv}{Dt}, \tag{58}$$

where v is the specific volume ($v = 1/\rho$). By these means, (56) is recast in the more familiar form

$$\boxed{\frac{Du}{Dt} = -\frac{1}{\rho}\mathrm{div}\,\mathbf{J}_q - p\frac{Dv}{Dt} + \frac{1}{\rho}\Phi + \frac{1}{\rho}\sum_k \mathbf{J}_k \cdot \mathbf{F}_k} \tag{59}$$

which is another statement of the first law for *inelastic* (but not necessarily Newtonian) fluids. It should be noted that most textbooks resort to highly elaborate and circuitous derivations in obtaining eq. (56) or (59), while the use of the general property equation is much more convenient as a starting point.

Referring to possible interaction with an unpolarized electromagnetic field, we identify the term responsible for irreversible conversion of gravitational and electromagnetic energies into internal energy as

$$b = \frac{1}{\rho} \sum_{k=1}^{n} \mathbf{J}_k \cdot \mathbf{F}_k = \mathbf{i} \cdot [\mathbf{E} + 1/c\,(\mathbf{u} \times \mathbf{B})] + \frac{1}{\rho} \sum_{k=1}^{n} \mathbf{J}_k \cdot \mathbf{g} \geq 0 \tag{59a}$$

where

$$\mathbf{F} = z_k [\mathbf{E} + 1/c\,\mathbf{u}_k \times \mathbf{B}] \tag{59b}$$

is the *Lorentz force* (per unit mass) acting on component k, and

$$\mathbf{i} = \sum_{k=1}^{n} z_k \mathbf{J}_k$$

z_k being the charge per unit mass of component k; $\mathbf{E}$ and $\mathbf{B}$ are the electric and magnetic field intensities respectively (see also Lecture VI on the thermodynamic and electromagnetic Arrows of Time).

The inequality in eq. (59a) should always contain the gravity vector $\mathbf{g}$. This inequality is based on observations and on advanced formulations of the "second law" of thermodynamics. We can therfore restate the first and second laws of thermodynamics for this case as

$$\frac{Du}{Dt} = -\frac{1}{\rho} \operatorname{div} \mathbf{J}_q - p\frac{Dv}{Dt} + \frac{1}{\rho}\Phi + b \qquad \text{(the first law)} \tag{59c}$$

$$\Phi \geq 0; \qquad b \geq 0 \qquad \begin{array}{l}\text{(the second law in a weak} \\ \text{gravitational field)}\end{array} \tag{59d}$$

These equations will be employed in Lectures VI and VII. Again we note that, in its general form, the mathematical expression $b \geq 0$ contains the gravity vector $\mathbf{g}$, plus certain subtle expressions for the so-called "diffusion fluxes" of certain solid, liquid, or gaseous elements by the differentiating and stratification acts of gravity or of gravity-induced forces, e.g., geological and atmospheric "natural convection" (q.v. Lectures VI and VII).

● **II.9   FIRST LAW AND ENTHALPY**

Recalling the definition of the specific enthalpy

$$h = u + pv = \frac{1}{\rho} \sum_{k} \rho_k h_k, \tag{60}$$

eq. (56) becomes

$$\rho\frac{Dh}{Dt} = -\operatorname{div}\mathbf{q} - \operatorname{div}\left[\sum_{k} h_k \mathbf{J}_k\right] + \frac{Dp}{Dt} + \Phi + b \quad \begin{array}{l}(b \geq 0) \\ (\Phi \geq 0)\end{array} \tag{61}$$

which is a very useful relationship in physics and engineering. It should be noted that nowhere in these energy equations (formulated in terms of u or h), does there appear a *source term* due to, say, energy released by combustion, or consumed by dissociation, or by any other reaction. The reason is that h and u are not affected by such processes. (The internal energy of, say, a constant-volume adiabatic system, remains the same during any chemical change of composition, even though its temperature may be affected.) For this reason, all the more so in view of the fact that temperature is the quantity measured in most cases, it is useful to reformulate the first law accordingly.

● II.10   FIRST LAW IN TERMS OF TEMPERATURE FIELD

It may be surprising to learn that the complete first law of thermodynamics was not formulated correctly, in terms of the temperature field, until 1959 (by H. J. Merk in the Netherlands)*—hence, its absence in most physics textbooks. In the following, we derive this equation, first on a particular level for a single-component ideal gas, and then present the derivation as due to Merk.

(a) Particular Energy Equation

For a single-component gas, $J_K = 0$. The assumption of "*ideal*" behavior means that h is independent of the pressure. Accordingly

$$\frac{Dh}{Dt} = \left(\frac{\partial h}{\partial T}\right)_p \frac{DT}{Dt} \equiv C_p (DT/Dt) \tag{62}$$

where C_p is the *specific heat at constant pressure*.* Substituting (62) in (61), we obtain

$$\rho C_p \frac{DT}{Dt} = -\operatorname{div} q + \frac{Dp}{Dt} + \Phi \tag{63}$$

(b) The Complete Energy Equation

The complete equation follows directly from the definition of enthalpy, namely

$$dh = \frac{1}{\rho} \sum_k h_k \, d\rho_k + \frac{1}{\rho} \sum \rho_k \, dh_k \tag{64}$$

From thermodynamics, we know that h_k is a function of T, p and the composition of the fluid. Hence,

---

* Merk, H. J., *Appl. Sci. Res* (A) 8, 73–99 (1959).
* Note that C_p itself is not necessarily constant, as is incorrectly implied by some textbooks, (cf. e.g., Yan, S. W. "Foundations of Fluid Mechanics," Prentice, London, 1967; Schlichting, Boundary Layer Theory), by misleading and utterly mistaken derivations (see also below).

$$d\,h_k = C_{P,k}\,dT + [l_{T,k} + V_k]\,dp + \sum_k \left(\frac{\partial h_k}{\partial p_k}\right)_{P,T} d\rho_k, \tag{65}$$

where v_k is the specific volume of component k, and

$$C_{P,k} = \left(\frac{\partial h_k}{\partial T}\right)_{p,\rho_k} ; l_{T,k} = \left(\frac{\partial h_k}{\partial p}\right)_{T,\rho_k} - v_k. \tag{66}$$

Here, $l_{T,k}$ represents a sort of partial "latent heat" under non-ideal behavior (its dimensionality is the same as that of v_k, see below).

Since enthalpy is an extensive property, the last term in (65) vanishes. By (64) and (65) we thus have

$$\frac{Dh}{Dt} = C_p\frac{DT}{Dt} + [L_T + v]\frac{Dp}{Dt} + \sum_k h_k\frac{DC_k}{Dt} \tag{67}$$

where

$$\mathbf{C}_p = \sum_{k=1}^{n} C_k\,C_{p,k} ; \qquad C_k = \frac{\rho_k}{\rho}$$

$$\mathbf{L}_T = \sum_{k=1}^{n} C_k l_{T,k} ; \qquad v = 1/\rho = \sum_{k=1}^{n} C_k v_k \tag{68}$$

We must now formulate a balance equation for DC_k/Dt. For this purpose we again refer to eq. (17), where in this case $\psi = \rho_k$, and write down the *species conservation equation*

$$\frac{\partial \rho_k}{\partial t} + \operatorname{div}\mathbf{J}_k + \operatorname{div}(\rho_k\mathbf{u}) - g_k = 0 \tag{69}$$

However

$$\operatorname{div}(\rho_k\mathbf{u}) = \rho_k\operatorname{div}\mathbf{u} + \mathbf{u}\cdot\operatorname{grad}\rho_k \tag{70}$$

Hence

$$\frac{D\rho_k}{Dt} = -\operatorname{div}\mathbf{J}_k - \rho_k\operatorname{div}\mathbf{u} + g_k \tag{71}$$

Introducing the *mass fraction* C_k and recalling the continuity equation, eq. (71) takes the simpler form

$$\boxed{\rho\frac{DC_k}{Dt} = -\operatorname{div}\mathbf{J}_k + g_k} \tag{72}$$

Finally, substituting (72) in (67) and then in (61), we obtain *the correct first law of thermodynamics in terms of the temperature field*:

$$\rho C_p \cdot \frac{DT}{Dt} = -\operatorname{div}\mathbf{q} + \mathbf{L}_T \frac{Dp}{Dt} + \Phi + Q_P$$

$$-\sum_{k=1}^{n} \mathbf{J}_k \cdot \operatorname{grad} h_k + \sum_{k=1}^{n} \mathbf{J}_k \cdot \mathbf{F}_k \qquad (73)^*$$

where

$$Q_p = -\sum_{k=1}^{n} h_k g_k \qquad (74)$$

is the heat (positive for an exothermic process) generated at constant pressure per unit volume per unit time. For an ideal gas $\mathbf{L}_T = 1$, and if $\mathbf{F}_k = 0$, and a single-component fluid is assumed, eq. (73) reduces to eq. (63), and $\mathbf{C}_p$ becomes C_p.

● II.11   ENTROPY BALANCE EQUATION

For this case the general property is $\psi = \rho s$, where s is the entropy per unit mass. Eq. (17) now reads:

$$\frac{\partial(\rho s)}{\partial t} + \operatorname{div} \mathbf{J}_s + \operatorname{div}(\mathbf{u}\rho s) = \sigma \qquad (75)$$

where σ is the internal entropy production rate (per unit volume).
    Resorting to Eq. (47), eq. (75) is recast as

$$\rho \frac{Ds}{Dt} = -\operatorname{div} \mathbf{J}_s + \sigma \qquad (76)$$

which is equivalent to eq. (3) in Lecture IV, which reads:

$$dS = d_e S + d_i S \qquad (IV\text{-}3), \qquad (76a)$$

The "second law" of thermodynamics is most commonly stated as

$$\sigma \geq 0 \qquad (77)$$

or as

$$d_i S \geq 0 \qquad (77a)$$

It will be of special interest to compare eqs. (77) with (77a), with eq. (59d) and with

---

* In most textbooks, the term containing the gradient of h_k does not appear due to the incorrect derivations mentioned above.

eq. 62 in Lecture VI. These last statements are subject to certain limitations, some of which are stressed in Lecture IV.

Similar balance equations for, say, number-density functions, moment of momentum, etc., can be easily derived from the general property equation (17). Coupled together, or solved individually with various initial and boundary conditions, the set of equations presented in this lecture forms the main basis of many problems in classical physics and engineering science. But, coupled with some equations to be derived in the next Lecture, they also form the theoretical basis of some of the most important calculations in astrophysics.

II.12   BEYOND CLASSICAL PHYSICS: SOLITONS, ANTISOLITONS AND CONSERVATION

We now move into another domain, where the subtle questions of energy dissipation and dispersion are coupled to some non-linear properties of field equations, conservation laws, and non-dispersing waves. [Most of this subject lies beyond the scope of this lecture. Yet, a few introductory remarks may best be given at this stage of the course. Further remarks will be introduced in due turn (in Lectures III, V and IX)].

The conclusion drawn from the previous section may be stated another way: *All macroscopic processes are dissipative and involve entropy increase, irreversibility and time asymmetry. Hence, all macroscopic waves are also dissipative, for they involve a decaying tendency due to the viscosity of the medium* (a manifestation of friction).

There is, however, an important class of waves that do *not* dissipate or disperse but, instead, *maintain their size and shape indefinitely.* Such waves have long been known in fluid dynamics where they are called *SOLITARY WAVES* or *SOLITONS.* What has been discovered recently is that such (nondissipative) waves also arise from some field equations formulated to unite the four fundamental fields of physics, in particular in the domain of quantum field theories formulated to describe elementary particles. Indeed, the relation between waves and particles has been known since the formulation of quantum mechanics in the 1920's (Lecture IV). During the past few years, however, a new relation between them has emerged, a relation of great importance to the advance of unified field theories [IX.2.20].

Solitons are *not* immune from dispersion; rather, they are waves in which the effects of dispersion *are exactly canceled by some compensating phenomenon.* The compensation is possible only in a certain class of waves, those whose equation of motion is *nonlinear,* i.e., those in which separation of variables and superposition is *not* possible and the propagation of waves is influenced not only by the shape of the disturbance but also by its magnitude.

By convention a solition is a disturbance where the field *increases* with the coordinate and an ANTI-SOLITON is a disturbance where the field *decreases* with the increasing coordinate. It is possible, by mathematical procedures, to *"create"* a *soliton-antisoliton pair* that "corresponds" to the *creation of a pair of particle-antiparticle in a quantum field theory*. In the reverse process, a soliton and an anti-soliton collide and *annihilate* each other "in analogy" to the mutual annihilation observed among particles and antiparticles.

Conservation may then be realized in the following sense. Since any soliton-antisoliton pair can be created or annihilated, it is only the *difference* between the total numbers of these objects that is conserved. Absolute conservation of the soliton cannot, in general, be proved unless some conditions are made.

II.12.1 The New Challenge Posed By Solitons In Particle Physics

The ramifications of the creation and annihilation of soliton-antisoliton pairs is far reaching and revolutionary in physics and philosophy, for it brings us face to face with the following set of questions and challenges:

■ Can the creation and annhilation of soliton-antisoliton pairs "correspond" to the physical creation and annihilation of particle-antiparticle pairs in a proper quantum field theory?

■ What else is there out of which to build an "elementary particle" except the geometry of a proper physical field? What else is there out of which matter is made except a soliton twist in the geometry itself? Indeed, can we express photons, electrons, positrons, quarks, antiquarks, etc., in terms of a fully-deterministic unified field theory?

■ Do the new PETRA experiments (§ IX.2.21) give us the first strong proof of *quantum chromodynamics*, or of the validity of Einsteinian physico-philosophical doctrine about the existence of a fully-deterministic unified field?

■ Indeed, what does a soliton tell us about the structure of the vacuum state (§ IX.2.20)? about the structure of matter? and about the structure of our theories?
These questions constitute some of the most fascinating topics in physics and philosophy. They are also among the most fundamental ones. To see that one must first read about fields, especially the metric field of Einstein's general theory of relativity (Lecture III). But, most important, the concepts of causality, time, asymmetry, uncertainty, gravitation and unification must first be clarified and well understood (cf. Lectures IV and IX). Hence, all we can say, at this stage of the course, is that the various balance equations that have been developed in this lecture are based on the following assumptions:

■ Strict causality and full determinism in the description of the distribution of (macroscopic) mass, momentum, energy, etc. in space and time.

■ Newtonian space and time as the stages "upon which" physical processes take place, not as the dynamic (curved) space-time of the general theory of relativity, nor as the eventual "building material" of mass, energy and momentum.

Endless types of soliton-antisolitons entities can be constructed from the balance equations presented in this lecture (see II.12.2 below). These, of course, are not of the kind used in quantum field theories. Yet, they may soon prove as highly exciting phenomena in the various domains of classical physics and biophysics (see below).

II.12.2 Solitons in Biophysics And Memory Theories?

There is some evidence, that goes back to 1956*, which indicates that the activity of populations of *neurons* may involve the existence of physico-chemical solitary waves. Moreover, the idea of solitons in *neuroanatomic structures* may be highly important in memory theories, including the evolution of ion concentrations in *cortical structures***.

Now, what do we mean by solitons in biological systems?

To begin with let us refer to the definition of diffusion and heat fluxes by equation II-54, and to the energy equation, II-73, in which the source term due to chemical reactions appears. By coupling Eq. II-73 with Eq. II-20, II-42, and II-72, one can characterize any biological phase in terms of the velocity, concentrations and temperature fields. Endless solitary waves of, say, ion concentrations, can be derived, in principle, from this set. The simplest would involve only one space coordinate (and time) in non-flow systems without temperature gradients. For a given biological system one needs to specify also the diffusion coefficients of, say, potassium and calcium ions, and the constants of the possible chemical reactions involved.

If the possible creation and annhilation of solitary waves in the extracellular space of *brain structures* is considered, additional parameters must be specified (e.g., the calcium conductance of presynaptic membrane- cf. the terminology employed in § IX.2.5.1;-the potassium and calcium equilibrium constants; and the membrane potential). Indeed, such a simplified set of equations was found suitable for the (qualitative and to some extent quantitative) prediction of *potassium and calcium ion concentrations during cortical spreading depression****. Then, by generating a local jump in the ion concentration of potassium (while keeping that of calcium at a steady-state value), one can generate *moving solitary waves in the ion concentrations of potassium and calcium.*

---

* Beurle, R. L., *Philos. Trans. R. Soc. London, Ser. B 240*, 55 (1956). See §IX.2.5.1 for terminology.

** Tuckwell, H. C. and R. M. Miura, *Biophys. J. 23*, 257 (1978); Tuckwell, H. C., *Science, 205*, 493 (1979). See §IX.2.5.1 for terminology.

*** Nicholoson, G., G. T. Bruggencate, R. Steinberg and H. Stockle, *Proc. Natl. Acad. Sci. U.S.A. 74*, 1287 (1977); Kraig, R. P. and C. Nicholson *Neuroscience 3*, 1045 (1978).

Now, when two such waves propagate from remote sources and collide, head on, they annihilate one another and the fields of ion concentrations return to their steady-state ("resting") values. Moreover, with modifications of the numerical constants involved, it was found that the result of a collision of two such solitary waves was not a return to resting values, but the emergence of two solitary waves of the same amplitude and velocity as the colliding waves. However, only one solitary wave emerged from the collision when slightly asymmetric data were employed.

This set of reaction-diffusion equations is a *nonlinear* one and it arises in many biological systems. Examples are the *Hodgkin-Huxley*[+] *equations* and the approximating equations of *Fitzhugh*[++] and *Nagumo et al*[+++] *for the propagation of the nerve impulse.* We shall return to this fascinating topic in §III.5 and in §IX.2.20. Meanwhile a final word about the powerful role of conservation is due.

II.13 NEUTRINOS AND THE POWERFUL ROLE CONSERVATION EQUATIONS PLAY IN SUBATOMIC PROCESSES (ADDENDUM).

Conservation laws are, as we have seen, very fundamental in classical physics and engineering science. Now, in bringing this lecture to a close, we shall demonstrate their role in particle physics. For that purpose we shall use the exciting history of the discovery of the neutrino.

The necessity of the existence of the neutrino was suggested by Wolfgeng Pauli in 1931. Later Enrico Fermi suggested the name neutrino and worked out a detailed theory of neutrino production. However, the initial postulation of the existence of neutrino and its predicted properties received a mixed reception from physicists. Some considered it only "*a gimmick to save the conservation bookkeeping*", *only a figment of scientific mysticism about conservation laws.* Indeed, the postulation of the neutrino made it possible not only "*to save*" three conservation laws—those of energy, momentum, and angular momentum—but to construct another conservation law as well; *the law of conservation of lepton number,* in which *the net lepton number of a closed system remains constant.**

Thus, when neutrinos and antineutrinos are taken into account, the lepton numbers are conserved in the subatomic events. In fact, the number of conservation

---

[+] Hodgkin, A. L. and A. F. Huxley, *J. Physiol. (London) 117,* 500 (1952).

[++] Fitzhugh, R. *Biophys. J. 1,* 445 (1961).

[+++] Nagumo, J., S. Arimoto, S. Yoshizawa, *Proc. IRE 50,* 2061 (1962).

[*] The lepton number of the electron is $+1$, of the positron -1, of the photon 0, of the neutrino $+1$, and of the antineutrino -1. All baryons have lepton number of 0.

laws associated with the neutrino are more than four, for they include the following list:

1) energy
2) linear momentum
3) angular momentum
4) electric charge
5) baryon number
6) electron family number
7) muon family number
8) parity (see §IV.11)

Here the conservation of lepton number is subdivided into an *electron family* and a *muon family*. Included in the electron family are the electron, the positron, the electron-neutrino, v_e, and the electron-antineutrino-$\bar{v}_e$. The electron and electron-neutrino each have an electron family number of $+1$, while the positron and the $\bar{v}_e$ each have -1.

Included in the muon family are the negative muon, the positive muon, the muon-neutrino v_μ and the muon-antineutrino $\bar{v}_\mu$. The negative muon and the $\bar{v}_\mu$, would each have a muon family number of $+1$, while the positive muon and the $\bar{v}_\mu$ each have a muon family number of -1. (The photon is a member of neither family and has an electron as well as muon family number of 0. Particles of the electron family would thus have a muon family number of 0, and vice versa. In addition all mesons and baryons have electron and muon family numbers of 0). Consequently, the conservation of lepton number has been replaced by the laws of conservation of electron family number and of muon family number.

However, these conservation laws were only valid if v_e, $\bar{v}_e$ and v_μ, $\bar{v}_\mu$ were not identical in properties. Now both sets of neutrinos are massless and uncharged. Both have *spins* of $+1/2$ or $-1/2$. Like the neutron, the spinning neutrino might generate a magnetic field which can have one of two orientations and consequently we have a neutrino and antineutrino, just as we have a neutron and antineutron.**

The experiment which demonstrated the difference between v_o, $\bar{v}_e$ and v_μ, $\bar{v}_\mu$ was first conducted in 1962 at the Brookhaven Laboratories.*** The conclusion derived

---

** It has been estimated that *the average antineutrino will travel freely through 3500 light-years of solid lead before being absorbed.* This is due to its weak interaction.

*** By smashing high-energy protons into a beryllium target an intense beam of pions was obtained, of both the positive and negative types. About 10% of the highly unstable positive pions in this beam had broken down to positive muons and v_μ while an equal percentage of the negative pions had broken down to the negative muons and $\bar{v}_\mu$. The positive pions might also produce positrons, electrons and v, but in negligible quantities. Passing through a wall of armor plate 44 feet thick the v_μ and $\bar{v}_\mu$ reached a spark chamber. About a hundred trillion neutrinos generated about 50 events, every one with a negative muon; not a single one with an electron (since only v_μ and $\bar{v}_\mu$ were entering the spark chamber, only negative muons must be formed. The different tracks produced by muons and electrons in spark chambers can be easily distinguished). By 1959 Maurice Goldhaber had also shown that neutrinos and electrons were "left-handed", antineutrinos and positrons "right-handed" (see a discussion on CPT conservation in §IV.11).

from this experiment was that there were indeed two types of neutrinos. These results not only demonstrated the existence and different properties of ν_e, $\bar{\nu}_e$ and ν_μ, $\bar{\nu}_\mu$, but also reaffirmed *the power and universal validity of conservation laws which do not involve the gravitational field. For it is only in the general theory of relativity that the* conservation laws of energy and momentum may not apply in the usual sense. But to clarify this statement we must turn now to Lecture III.

The most incomprehensible thing about the world is that
it is comprehensible.

Albert Einstein

FROM GENERAL RELATIVITY
AND RELATIVISTIC COSMOLOGY
TO GAUGE THEORIES

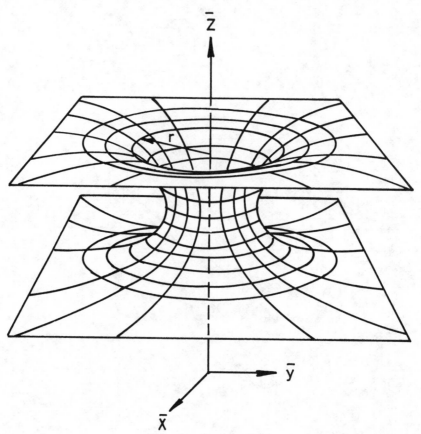

All reality is one in substance, one in cause, one in origin.

Attributed to **Giordano Bruno**

Theoretical physics is actual philosophy.

Max Born

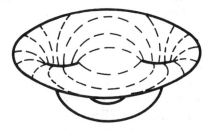

III.1 INTRODUCTION

The most important intellectual consequence of the general theory of relativity is the clarification it has brought into the relations between *science and philosophy*. In fact, the effects of the general theory of relativity were to re-establish physics on a *philosophical basis* and thereby to *elucidate* many bewildering phenomena and implications of modern research and thinking.

But the great impact of post-World-War-II technology and of American academe (Appendix VI) caused a rift which gradually opened a gap between physics and philosophy (Introduction). Today a person expert in one is usually very inexpert and ill-informed in the other. Up to World War II the subject matter of this lecture formed *a unity* (mostly owing to Einstein) — the rudiments of what we now call physics or the exact sciences being simply an integral part of a larger system of thought.

Since Einstein died in 1955, many discoveries have been made in astronomy, astrophysics and the rest of the exact sciences. These discoveries force us to *re-examine* some of our previous convictions.

Here, one task of philosophy is to give rational accounts of new evidence with a minimum of presuppositions within the framework of a unified theory. Such an attempt is undertaken in Lectures VI to IX. In this lecture I confine myself to brief remarks on a few 'accessory concepts' and formulations in general relativity and to a summary of important tests of its validity. A few cosmological concepts that have recently gained special importance are also included.

* * *

Numerous papers and books have been written during the last sixty years on general relativity and on its physico-philosophical consequences. Yet, due to the great advances in physics, astronomy, astrophysics and cosmology, only a few of them contain today valid information. It is especially due to the dramatic achievements in observational astronomy, in the last decade or so, that the new empirical evidence has provided us with a real revolution in our understanding of the physical and philosophical consequences of the theory of general relativity, without parallel in its history. These discoveries constrained both general relativistic cosmological models and wild speculations based on other theories of gravitation (the general theory of relativity is the best theory we have today on gravitation). The new observations have, therefore, transformed the great majority of these papers and books into *unreliable* sources of information. It happened almost overnight. And more students than ever now ask for *updated* material on relativistic stars and relativistic cosmology. Since a great flow of new information keeps coming in even as this lecture is being written, its contents will not remain up-to-date for long. Moreover, in a single introductory lecture I can only make a few remarks about these concepts. The main feature of this lecture is therefore restricted to the use of a 'minimum formalism' whereby the general theory of relativity is presented by focusing attention on simple symmetrical concepts and by introducing the idea of the curvature of spacetime via the easily visualised concepts of ordinary curved surfaces and Newtonian gravitation.*   ***

We start with two questions: *In what ways does Einstein's theory of gravitation differ from its Newtonian counterpart? And why is it needed at all?*

Space and time in the Newtonian theory of gravitation are the "stage" *upon which* physical processes are displayed. Here absolute, uniform time and a "universal now" (in absolute Euclidean space) were assumed to exist. The arrival of *special* relativity brought about the collapse of (absolute) Newtonian space and time. The "universal now" was replaced by the more modest "*here and now*"—the "*event*". Time itself became an integral part of a new structure: "*four-dimensional space-time*". In this continuum the concept of the "same moment" in two different places is without absolute meaning. Yet this space-time is uncurved, i.e. it is "flat."**

---

* The reader interested in further reading should be warned *against* the use of most relativistic models contained in books published *before* 1972. A few excellent books fill this lacuna: (i) *Gravitation and Cosmology: Principles and Applications of the General Theory of Relativity*, by Steven Weinberg (Wiley, N.Y., 1972); and (ii) *Gravitation*, by C.W. Misner, K.S. Thorne, and J.A. Wheeler (W.H. Freemen and Company, San Francisco, 1973).

Other books, especially designed for the general educated reader are, (iii) *Modern Cosmology*, by D.W. Sciama (Cambridge University Press, reprinted with corrections, 1972); (iv) *Space, Time, and Gravity*, by R.M. Wald, University of Chicago Press, Chicago, 1977; (v) *Cosmology and Geophysics*, by P.S. Wesson, Adam Hilger, Bristol, 1978; and (vi) *General Relativity and Cosmology*, by J.V. Narlikar, MacMillan Press, London, 1979. A classic book on this subject was written in 1933 by R.C. Tolman. It is entitled *Relativity, Thermodynamics and Cosmology* (Oxford Press, 1933), and is perhaps one of the best books ever written on relativity. However, its *thermodynamical* and *cosmological* portions could *not* stand the test of time. Updated surveys on relativistic thermodynamics and kinetic theories are available in two volumes edited by this author: *Critical Review (of Classical and Relativistic) Thermodynamics*, Mono, Baltimore, 1970, [**CRT**], and *Modern Developments in Thermodynamics*, Wiley, N.Y. 1974 [**MDT**].

Some familiarity with vector and tensor algebra is assumed for Section III.2. The concepts of stress tensor, conservation equations, conservative fields, momentum, mass density and energy fields were discussed in Lecture II.

** For instance, in such a "*flat*" continuum the sum of the angles of a triangle always equals 180 degrees and the surface area of a sphere of radius r is $4\pi r^2$, i.e., Euclidean geometry applies everywhere! Not so in *curved* space-time! Here the corresponding quantities may be greater or smaller than 180 degrees or $4\pi r^2$ (see also III.2.7).

But the special theory of relativity did not only solve physical and philosophical problems, it posed new ones. In particular, it was restricted to *inertial, nonaccelerating frames of reference in the absence of gravitational fields*. It took Einstein about a decade to solve the new problems. In his 1916 theory of general relativity the *curvature of space-time is "affected" by the distribution of masses and their motion*: at a given point in space-time ("here-now", for example) it is proportional *to the amount of mass-energy and momentum present at that point in space-time*. In short, mass-energy and momentum *curve* the geometry, or to put it more simply: *Matter acts on space causing it to curve. In turn, space acts on matter and is "coupled" to it.* Far away from masses (which produce a gravitational field), in empty space, space-time is "flat" and is subject to the special theory of relativity. Therefore, *gravitational forces find their expression in the "curvature" of space-time.*

Negligible gravitational forces mean negligible "curvature" of space-time. Or, to put it another way, the effect of geometry on matter, or, in turn, of matter on geometry, is what we mean by the concept *"gravitation"*. Consequently, general relativity tells us that the *geometrical* properties of space-time determine the evolution of the *physical* processes in space *and* time. In other words, *the fundamental concepts of science involving geometry, mathematics and physics are coupled together; there is no a priori requirement to distinguish, fundamentally, between changes in geometrical properties and changes in physical processes.*

To proceed we must now become more technical and concrete; we must slowly familiarize ourselves with *Einstein's field equations*, first with their main physical meanings, and, secondly, with their mathematical derivation and applicability. We start with the first task.

III.1.1 Einstein's Field Equations in General Relativity

Einstein's *nonlinear partial differential equations* provide us with a complete theory of gravitation, that is, with a mathematical description of gravitational fields linked with arbitrary physical systems. This general-relativistic theory of gravitation differs from its Newtonian counterpart mainly in the following respects:

i) It is based on *ten potentials* [of the *symmetric metric** $g_{\mu\nu}(\mu,\nu = 1,2,3,0)$]—instead of the *single* Newtonian potential ϕ determined by Poisson's equation

$$\boxed{\nabla^2\phi = 4\pi G\rho} \tag{1}$$

where G is Newton's constant (equals to 6.670×10^{-8} in c.g.s. units), and *forces* are determined from

$$\boxed{\frac{d^2\mathbf{x}}{dt^2} = -\nabla\phi} \tag{2}$$

---

* Mathematical definitions and physical interpretations of the metric are given later.

For a spherical body of mass M, ϕ varies with distance according to

$$\phi = -\frac{GM}{r} \tag{2a}$$

ii) In contrast to Maxwell's equations (which are linear because the electromagnetic field does not itself carry charge, whereas gravitational fields** *do* carry energy and momentum, or are generated by them) and to the Newtonian theory—in general relativity the total effect of several bodies *is not the simple sum of their individual effects,* i.e., the theory is *non-linear* and the classical type of conservation equations *may not apply.* Conservation is *conditional* on our adopting a *quasi-Minkowskian* coordinate system, such that $g_{\mu\nu}$ approaches the Minkowski metric $\eta_{\mu\nu}$ at great distances from the finite material system under study, and then follows the behavior of the *total* energy-momentum "tensor" of matter *and* gravitation and of the energy-momentum "vector" of the system (including matter, electromagnetism *and* gravitation). With the matter in this system subdivided into *widely* separated subsystems, equality of the *total* energy and momentum to the *sum* of component energies and momenta is preserved in *each* subsystem. Here the total energy-momentum "vector" is not only *conserved* but also *additive* and its most important feature from our new *thermodynamic point of view* is that *its time component is always positive,* vanishing only for matter-free empty space. This gravitational time asymmetry is of a most fundamental value (cf. Lectures VI to VIII).

iii) The *source of gravitation* (as expressed by the energy-momentum tensor) are *mass-, energy-,* as well as *pressure-* and *momentum-transfer*—see below.

iv) General relativity is usually formulated in a *geometrical* "language." The ten potentials of the metric describe the *local* properties of the space-time continuum, which acquires a curvature *in the presence of a gravitational field* (i.e. of mass, energy and momentum) but is *"flat"* (as in Euclidean geometry) in *"empty space"* remote from gravitational sources. Generally speaking, the *denser* a galactic object, the *more curved the path of a photon or a neutrino moving inside or near it*! In an extremely dense medium (e.g., the product of gravitational collapse associated with formation of a black hole)*, the curvature of space-time may be so high as actually to keep any photon or neutrino "imprisoned". The medium then acts as a (one-way) *"thermodynamic sink"* which soaks up any nearby matter and radiation, letting no signal come out beyond a certain "horizon." (see Figs. I.4 and VIII.2 to 4).

v) The discrepancies between general-relativistic and Newtonian predictions *increase* with the density of the galactic body considered. For weak gravitational fields, the two theories yield the *same* predictions (see below).

** See also the subsequent discussion of the 'principle of equivalence', inertial acceleration and terrestrial timekeeping.

* See §VIII.I on the Schwarzchild solution and §I.3; I.5.1 and I.5.3 as well as Figs. I.4 and VIII.2.

vi) Unlike the Newtonian theory, there is no obstacle in general relativity to visualizing a gas and/or radiation cloud *"filling the whole of space"*, and to describing its behavior and dynamics. This point becomes important in *relativistic cosmology* (see below) and in the theory of *cosmic time asymmetries.*

vii) General relativity is based on the *principle of equivalence* of gravitation and inertia, or on the principle of *General Covariance.*

III.1.1a The Principle of Equivalence

The principle of equivalence does not lend itself to simple formulation. In general terms, it is based on verifications of gravitational and inertial mass equivalence as demonstrated by Galileo, Huygens, Newton, Bessel, Eötvös, and Dicke, (and more recently by Cannon and Jensen[+]). Einstein concluded that as a consequence *it is possible, at any space-time point, to set up a locally inertial coordinate system* (e.g., one attached to a *very small, freely-falling* "elevator"), *such that, with a sufficiently small region about the point in question, the "laws of nature" take the same form as in unaccelerated, special-relativistic coordinate systems in the absence of gravitation.* (Note that for large, freely-falling systems in an inhomogeneous or time-dependent gravitational field, inertial forces *may not exactly* cancel gravitational forces; hence our analysis *should be restricted to small regions of space and time* in which the field does not vary significantly). This equivalence enables us, *in principle, to set up at our point a freely-falling* (locally) *inertial coordinate system in which the medium can be described by the laws of special relativity!* It is incorporated into the formalism of general relativity through the requirement that the *equations be invariant under general coordinate transformations!*

There is some vagueness here as to which *"laws of nature"* are meant. Is it *all* the laws of nature (the so-called *"strong"* equivalence), or only the laws of motion of freely-falling bodies *("weak" equivalence)?* The experiments of Eötvös and Dicke provide direct verification for the "weak" version as well as some indirect evidence for the "strong" one. But this question is still open. (See also *terrestrial timekeeping* below).

The principle of equivalence assesses the effects of gravitation on physical systems and leads to Einstein's field equations (but does *not, by itself,* determine the field equations for gravitation). In this context use may be made of the *alternative* version known as the *Principle of General Covariance*, which states that a physical equation is valid in describing general gravitational fields provided (i) *it is generally covariant*, i.e. preserves its tensor form under a general coordinate transformation $\times \rightarrow \times'$, and (ii) *it holds in the absence of gravitation*, i.e., is compatible with the laws of special relativity when the metric tensor equals the *Minkowski tensor* $\eta_{\mu\nu}$ and when the *coefficient of affine connection* (see below) vanishes. This does *not* imply Lorentz

---

[+] For details see §III.4.

invariance, nor any other invariance principle. [There exist many covariant equations which reduce to a given special-relativistic equation in the absence of gravitation. A number of gravitation theories have been proposed in addition to Einstein's field equations, but none is compatible, like them, with all sets of available observations (see §6 in III.1.2a below)].

III.1.2 Confirmations of Einstein's Theory of Gravitation

The validity of Einstein's theory has been verified by many and various tests. Best known are the solar system experiments and the cosmological ones.

III.1.2a The Solar System Experiments

Five tests have been carried out in the solar system (i.e., in gravitational fields that are to a good approximation static and (except for the fifth) spherically symmetric. These are;

1) *The deflection of light by the sun* (including the deflection of radio waves). These experiments give us the clearest and most definite tests of general relativity. [Recent results are 1.70 ± 0.10 sec. in very good agreement with Einstein's prediction, 1.75 sec. The famous results from the eclipse of May 29, 1919, are 1.98 ± 0.16 sec. (Sobral) and 1.61 ± 0.40 sec. (Principe)].*

2) *The time delay of radar echoes.* It is an effect of *space-time curvature on electromagnetic waves involving a relativistic delay in the round-trip travel time for radar signals.* In one type of experiments the reflector is the surface of a planet *(Venus or Mercury)* (Fig. III.1a). In another the "reflector" is electronic equipment on board of a spacecraft that receives the signal and transmits it back to earth. Here, again, the agreement with Einstein's prediction is very good [defining E = (observed delay)/ (Einstein's prediction) *the observational results are*: $E = 1.015 \pm 0.02$ *(Venus and Mercury as reflectors,* wave lengths 3.8 cm and 70 cm, Shapiro, *et al.,*** 1967–1970); $E = 1.00 \pm 0.014$ *(Mariner VI and VII spacecrafts as "reflectors",* wave length 14 cm; Anderson *et al.,*** (1971)].***

3) *The precession of the perihelia of the orbits of the inner solar planets.* General relativistic corrections to Newtonian theory (idealized, spherical, inverse-square law) are not the only cause of shift in the perihelion of a planetary orbit. Such non-

---

* Precision of a one-day measurement in the 1970's is about 0.1 sec. (angular separation of two quasars with radio telescope).

** *Phys. Rev. Lett.* 2 6, 27–30 (1971); *Proceedings of the Conference on Experimental Tests of Gravitational Theories,* Davies, R. W. ed. Calif. Inst. Tech., J. P. L., T. M. 33–499, Nov. 11–13 (1971), pp. 111–135 (Anderson, J. D. *et. al* and Shapiro *et. al).*

*** Precision of earth-moon separation by laser ranging is now about 10 cm., of a radio transponder (on another planet or in a spacecraft) from earth, by measuring round-trip radio travel time, about 10 meters, and of Mercury and Venus from earth (by bouncing radar signals off them) about 300 meters.

sphericities and resulting shifts are due to gravitational pulls of other planets and to deformation of the sun ("solar oblateness"; "quadrupole moment"). In addition there is an apparent perihelion shift caused by the precession of the earth's axis. Except for the solar quadrupole moment the theory predicts the values with high precision as shown in Table III.1.**

TABLE III.1

Comparisons of Theoretical and Observed Centennial Precessions of Planetary Orbits (after Steven Weinberg, *Gravitation and Cosmology,* Wiley, 1972, pp. 198). See Fig. III.1b.

Planet observed	Ellipse semimajor axis, C	Eccentricity e	Revolutions per century	Observed precession	Precession according to general relativity
	km	—	r.p.c.	[seconds per century]	[seconds per century]
Mercury	57.91×10^6	0.2056	415	(42.56 ± 0.94*) 43.11 ± 0.45	43.03
Venus	108.21×10^6	0.0068	149	8.4 ± 4.8	8.6
Earth	149.60×10^6	0.0167	100	5.0 ± 1.2	3.8
Icarus	161.0×10^6	0.827	89	9.8 ± 0.8	10.3

* Residual shift per century to be attributed to general relativity *plus* 'solar oblateness'. The corresponding shift per century produced by *Newtonian gravitation* of other planets on Mercury is 531.54 ± ± 0.68 seconds per century! The second value for Mercury is *without* solar oblateness effects.

4) *The gravitational redshift of spectral lines.* The existence of gravitational redshifts can be deduced from the *equivalence principle* by considering two experimenters in a rocket ship that maintains a constant acceleration g (the principle of equivalence requires, *inter alia*, that if a proper redshift is observed in an experiment performed under conditions of uniform acceleration in the absence of gravitational fields, then the same redshift must be observed by an experiment performed under conditions where there is a uniform gravitational field but no acceleration).

"Scientific methodology" requires us to seek experimental verification for any new theory or principle: first, *to predict* an effect, and, secondly, *to observe it!* The most dramatic verification of this type was test (I) and the discovery of the *expanding universe* (Lecture VI and, also, below). Einstein, in principle, could treat the gravitational redshift as the equivalent of experimental confirmation of his principle of equivalence. But such an attempt was conducted only in 1960 (Pound and Rebka) and, with improved precision, in 1965 *(Pound and Snider)**. It used the *Mossbauer effect* to measure the redshift of 14.4 *KeV* gamma rays from Fe^{57}. The emitter and absorber of the gamma rays were placed at rest at the bottom and top of a tower at

** Accuracy degrades rapidly as we move away from the sun, both because the smaller eccentricities make observation of the perihelia more uncertain, and because as the dimension of the ellipse (called semilatus rectum) increases, the precession per revolution and the revolutions per century both decrease.
* *Phys. Rev. B.,* *140*, 788–803 (1965).

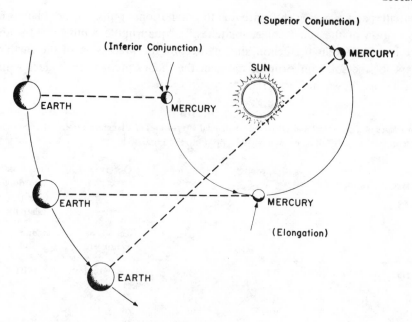

FIG. III.1a: **Geometry of the time-delay test of general relativity**. As the planets approach superior conjunction, the paths of the radar signals pass closest to the sun and experience the maximum retardation, predicted by general relativity to be about 200 microseconds

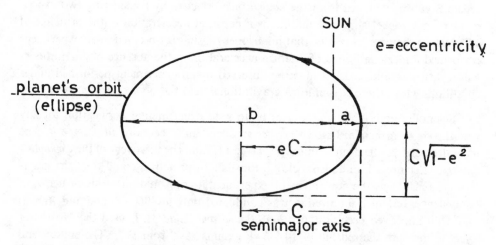

FIG. III.1b: **Perihelion is the point of a planet's orbit when it is nearest to the sun** (distance *a* is minimum) [from Greek *peri* = around, plus *helios* = sun]. Since the time of *U. J. Leverrier*, (the French astronomer) (1811–1877), it was known that there is a *significant discrepancy* between the observed motion of the Mercury perihelion and the one predicted by *Newton's theory*, even if we take into account the perturbations caused by the other planets. The first great success of Einstein's general relativity came when Einstein showed that this discrepancy *disappears* if one performs the calculations *relativistically*.

the Jefferson Physical Laboratory of Harvard University, separated by a height $h = 22.5$ meters. They measured the gravitational redshift of photons *rising against gravity* (through a helium-filled tube in a shaft). Both source and absorber were placed in temperature-regulated ovens.

The final result for the gravitational redshift was (0.9990 ± 0.0076) times the value predicted from the principle of equivalence (difference between "up" experiment and "down" experiment), i.e., the observed redshift agreed to better than 1 percent precision, with the general relativistic prediction of

$$z = \frac{\Delta\lambda}{\lambda} = gh = 2.46 \times 10^{-15} \tag{3}$$

where z is the redshift parameter defined by eq. (10) below (and also in Lecture VII), and the derivation of eq. (3) is to be given next.

Consequently, with the "*Pound-Rebka-Snider experiments*" general relativity is more established than during its early period when Einstein had tried (1911) to support this prediction by resorting to the concept of *energy conservation, applied in the context of Newtonian gravitation theory.*** Since the physical reasonings involved here are fundamentally important, albeit very simplified, I shall describe them next.

Let *a particle of rest mass m* starts from rest in a gravitational field g at point A and *fall freely* for a distance h to point B. Gaining the kinetic energy mgh, its total energy at B becomes

$$E_1^B = m + mgh \tag{3a}$$

If the particle undergoes *annihilation* at B it converts E into *a photon* of the same total energy. Let this photon travel upward (against the gravitational field) to A.

Now *if* we assume that the photon does *not* interact with the gravitational field, it will have its original energy on arrival at A and may be converted (by a suitable apparatus) into another particle which starts at A $(h = o)$ with an "excess" energy mgh that *violates the conservation of energy*, i.e.,

$$E_2^A = m + mgh \quad (h = o) \tag{3b}$$

To prevent this violation Einstein claimed that the photon *must pay back the energy by undergoing a redshift*, i.e.,

$$\frac{E^B}{E_A} = \frac{E^A(1 + gh)}{E^A} = 1 + gh = \frac{(h\nu)^B}{(h\nu)^A} = \frac{\lambda^A}{\lambda^B} = 1 + z \tag{3c}$$

where ν is the photon's frequency and λ its wavelength. Eq. (3c) means a *decrease in energy of radiation* owing to *work done against the gravitational attraction*, which, in turn, means drop in frequency or an increase in wavelength (i.e., a redshift).

** There have also been proposals to measure the gravitational redshift of light from an artificial satellite [see Kleppner, D., *et. al., Astrophys. Space Sci.,* 6, 13 (1970)].

This *gravitational* redshift has been verified by the 'Pound-Rebka-Snider experiment'. But a gravitational redshift means that a consistent theory of gravity *cannot* be constructed within the framework of *special* relativity.* The Euclidean ("flat") space-time of special relativity (see below) is, therefore, incompatible with these results and one must undertake the mathematical analysis of *curved space-time* in the neighborhood of any such process. *Consequently, space-time in the presence of a gravitational field must be curved!*

But *another conclusion*, which follows from the 'Pound-Rebka-Snider experiment', is of *fundamental importance to our thermodynamics, since it describes the second source of irreversibility in nature* (the first source is described in Lecture VI and the third in VII). This conclusion may be stated as follows:

*Gravitational redshift introduces an embedded time asymmetry into most astrophysical photon/neutrino processes, since all radiation energy dispersed away from all self-gravitating (planetary, stellar, galactic) systems undergoes a drop in energy and since in the universe accessible to our observations the net radiation energy transfer is from galactic systems into cold, unsaturable** space. Moreover, any uniform acceleration, in the absence of a gravitational field, produces the same effects.*

To support the last statement I now return to the two experimenters in a rocket ship that maintains a constant acceleration g. (I employ units in which Planck's constant and the speed of light are both unity ($h = c = 1$) unless stated to the contrary).

Let the distance between the two observers be h *in the direction of acceleration*. Suppose for definiteness that the rocket ship was at rest in some inertial coordinate system when the bottom observer sent off a photon. It will require time $t = h$ for the photon to reach the upper observer. In that time the top observer acquires a velocity $gt = gh$. He will therefore detect the photon and observe a *Doppler redshift $z = gh$*. The results here are, consequently, *identical* to equation (3c). Hence, the *principle of equivalence* requires that, if this redshift (or *gravitational* time asymmetry) is observed in an experiment performed under conditions of *uniform acceleration* (in the absence of gravitational fields), then the *same* drop in energy must be observed in an experiment performed under conditions where there is a *uniform gravitational field* (but no acceleration), i.e., *gravitational and accelerational time asymmetries are equivalent!*

Consequently, *gravitation and/or acceleration are linked to a fundamental time asymmetry in nature.* But as we shall see below the *quantitative* effect of this phenomenon is negligibly small in terrestrial processes. Yet, in principle we must regard an earth-bound laboratory as accelerating upward, with acceleration g, relative to an inertial frame of reference.

---

* Schild, A., "Time", *Texas Quarterly*, **3**, No. 3, 42–62 (1960); in *Relativity Theory and Astrophysics*, Am. Math. Soc., Providence, R. I., 1967, pp. 1–105.

** Unsaturable space is the consequence of intercluster expansion as explained in §VI.3 and Fig. VI.1.

In fact, it has been shown by Tolman that thermal equilibrium in a gravitational field does *not* correspond to constant temperature (zero gradients of temperature), but to a constant *redshifted temperature*; $T\sqrt{-g_{00}}$ = constant, where g_{00} is the time-time component of the metric (to be defined below).

Thus, the prediction of general relativity that a clock (i.e., radiation emitted from atoms with a given periodicity) in a gravitational field *runs slower* than the same clock would run if the gravitational field is absent, has been verified qualitatively and quantitatively. All other investigations have confirmed the existence of this important effect. However, the problem is usually accuracy: for atoms located on the surface of the sun, there are so many other disturbing factors which modify the effect that we hardly can attach great significance to the numerical values obtained for the redshift caused by the sun. On the other hand, measurements performed on light emitted by the companion to Sirius, for example, do give the same order of magnitude as that predicted by the theory.

In closing this section it should be stressed that one must carefully distinguish among *different types of redshifts*. Thus, there is the *cosmological redshift* discussed in Lecture VII (eq. 10), the *Doppler effect* and the *pure time dilation* in special relativity (see III.2.1 below). It is neither useful nor strictly correct to interpret the frequency shifts of electromagnetic waves from very distant sources in terms of a special-relativistic Doppler effect alone. The frequency of these waves is also affected by the gravitational field of the universe. This is a subtle problem associated with the cosmological principle (see below).

5) *Precession of gyroscope as experimental test of general relativity.* The experiments discussed thus far measure the motion of electromagnetic waves, planets, and spacecraft through the solar system. An entirely different type of experiment measures changes in the orientation of a gyroscope moving in the gravitational field of the earth. This experiment can measure what is called the 'dragging of inertial

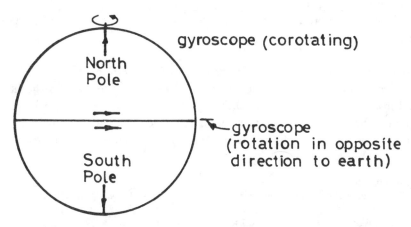

FIG. III.2: **Gyroscope in a freely moving satallite measures "dragging of inertial frames" by angular momentum.**

frames by the angular momentum of the earth'. There are three contributions to the precession:

a) *The Thomas precession* (which is important in evaluating the fine structure of atomic spectra). Yet, it should play *no role* in a gyroscope in a freely moving satellite (as the experiment is designed).

b) *Precession due to earth's rotation.* To understand it one may consider an *analogy**, whereby a solid sphere is rotating in a viscous fluid. As it rotates, the sphere drags the viscous fluid with it. Now, if in the laboratory we immerse little rods at various points in the fluid, all oriented perpendicular to the sphere's surface (rod density should equal fluid density) and observe their motion we can detect the following:

Near the poles: fluid rotates rods in the same direction as the sphere rotates.

Near the equator: near the sphere surface fluid is dragged more rapidly than at larger distances. Consequently, the end of a rod closest to the sphere is dragged by the fluid more rapidly than its far end. Therefore, the rod rotates in the direction *opposite* to the rotation of the sphere.

To put it another way: a gyroscope in a freely moving satellite is rotationally at rest relative to the inertial frame of reference in its neighborhood. It and the local inertial frames rotate relative to distant galaxies with earth's angular velocity (since the earth rotation "drags", so to speak, the *local inertial frames* along with it). Thus, near the north and south poles the local inertial frames rotate in the *same* direction as the earth does, but near the equator they rotate in the *opposite* direction**. This precessional angular velocity is in the order of magnitude of 0.1 seconds of arc per year.

c) *Precessions due to acceleration and earth's Newtonian potential.* This precession is due to motion through the earth's curved, static space-time geometry (field). It amounts, in order of magnitude, for a gyroscope in a near-earth, polar orbit, to 8 seconds of arc per year.

Both (b) and (c) precessions can be detected and equipment for this purpose has been prepared.

6). *Experimental Disproofs of Other Theories of Gravitation.*

A great number of "gravitation theories" have been proposed in the literature since Einstein published his general theory of relativity in 1915–1916. A systematic and comprehensive review of experimental evidence compared with these theories was presented by Nordtvedt in 1972***. He found that none is compatible, like Ein-

---

* Note that it is only an analogy.

** This effect is also important in black-hole physics (see the discussion of rotating black holes in Lecture VIII).

*** Nordtvedt, K. L. *Science*, 178, 1157 (1972). For a review of *"metric theories"* see also Ni, W. -T. *Astrophys. J.*, *176*, 769 (1972). For a review of *"nonmetric theories"* see Trautman, A., *Bulletin de* l'Academie Polonaise des Sciences (math., astr. phys.), 20, 185 (1972).

stein's field equations, with all sets of available observations. General relativity has emerged from each of these tests unscathed—a remarkable tribute to Einstein's intellect. Yet, a number of subtle arguments remain to be settled especially in regard to the principle of equivalence.

In closing this section I stress a number of solar-system tests that are of interest now and may become feasible in the future as technological capabilities improve.

For example: A compelling alternative to Einstein's general relativity was published in 1922 by *Whitehead***. Since it makes *exactly the same predictions as Einstein's theory for the so-called "four standard tests"* (gravitational redshift, perihelion shift, light deflection, and radar time-delay), many scientists considered it as a "*reliable alternative*", It was only in 1971 that it was realized that Whitehead's theory predicts a time-dependence for the ebb and flow of *ocean tides* that is *completely contradicted* by everyday experience***. [Whitehead's theory of gravity, which is *a two-tensor theory*, predicts that the galaxy produces velocity-independent anisotropies in the value of the Cavendish gravitational constant which appears in Newton's Law: Force $= -G_c m_1 \cdot m_2 / r^2$. Changes in G_c are also predicted by other theories of gravitation—such as the *Dicke-Brans-Jordan* (see Nordtvedt and Ni for a review)—according to which its value is determined by the distribution of matter in the universe. If correct, Whitehead's predictions would produce earth tides with periods of 12 sidereal hours and amplitudes of $\Delta g / g \sim 2 \times 10^{-7}$, where g is the "local acceleration of gravity on earth". The absence of such tides disproves this theory. Experimental results related to anomalies in earth rotation rate due to dependence of G_c on velocity place strict limits on any theory that possesses a universal rest frame. These limits disprove all theories with "*preferred frames*" that have been examined to date except, perhaps, the one devised by *Will and Nordtvedt* (see the reviews by Ni and by Wesson)].

Another solar-system experiment that might be of interest is the "*Nordtvedt effect*" associated with departure from 'geodesic motion' (this motion is to be defined later. At this point it will suffice to mention that in general relativity a body moves *along a geodesic*, but in most other theories of gravity it does *not*, i.e., unlike its behavior in general relativity and in Newtonian gravitation the acceleration of a body may, according to these theories, be in a *different direction* than the gradient of the Newtonian potential ϕ). Now, the Nordtvedt effect in a theory other than general relativity may be responsible for such effects as a "*polarization*" (an eccentricity in the orbit of, say, earth-moon, that points toward the sun), which may be detected by lunar laser-ranging (the precision required is of one meter or less).

Precise laser ranging may be variously used to evaluate *relativistic perturbations in the lunar orbit* [the so-called *three-body effects* in the lunar orbit associated with

---

** Whitehead, A. N. *The Principle of Relativity*, Cambridge Univ. Press, Cambr., 1922.
*** Will, C. M. *Astrophys. J., 169*, 141 (1971). See also Wesson's book, *loc. cit.*

the pulls of earth, moon and sun, evaluated over and above any Newtonian three-body interactions. These are *nonlinear* perturbations in the motion of earth and moon about their common center of gravity that *may* by measured with high-precision laser ranging and using a radio beacon on the moon's surface].

III.1.2b Cosmological Confirmations of General Relativity

The most important predictions of general relativity are in cosmology. Here the prediction that the universe *cannot* be *static* came first. Einstein could not believe it! After all, the prevailing "*Western philosophy*" of his time, and of all past times, was manifested by the "*universal maxim*": "*The heavens endure from everlasting to everlasting*". Faced with such a contradiction between traditional beliefs and his theory, Einstein weakened; and, reluctantly, he added an *arbitrary constant* to his equations so as to end up with a *stable* universe, i.e., one that does *not* expand or contract.

And physicists were able to live happily in a stable world for another decade or so. By then much verified information had been accumulated by astronomers enabling them to claim, first with hesitation, but later with confirmation, that *distances* between clusters of galaxies *increase with time* according to "Hubble's linear law" (see Lectures I and VI for details). Hence, the actual universe *is*, after all, *not* static and stable, *but expanding*, as Einstein's theory first predicted.

Einstein then rejected the arbitrary constant (called the "cosmological constant") from his original field equations, labeling it as "the biggest blunder of my life". Indeed, had he not added it for the sake of traditional beliefs he could have claimed the expansion of the universe as the most triumphant prediction of his theory of gravity.

Another important set of astronomical, astrophysical and geophysical observations associated with the consistency of general relativity includes *age estimates of the universe*, e.g., age from expansion (for details see §.3 below), ages of oldest stars, age of solar system, age of the moon, and age of the earth. It is a remarkable *quantitative confirmation* of cosmology that such diverse empirical factors are *consistent* with each other and yield *excellent agreement between theory and observation!* It is of general interest to mention some of these results:

a) *Remarkably good agreement* is obtained between (i) the age of the universe as estimated from the inverse Hubble constant $(H_0^{-1} = (18 \pm 2) \times 10^9$ *years giving an age which varies between 10 and 20 billion years*, depending on the value of "acceleration" on which evidence is not yet conclusive)* ; (ii) the ages of the oldest stars $(\sim 10 \times 10^9$ *years)* as calculated by comparing the theory of stellar evolution with the properties of the observed stars (see also Lectures VII and I); (iii) the period since *nucleosynthesis* of the uranium, thorium, and plutonium atoms found on earth

---

* See III.3 below.

$(\sim 9 \times 10^9$ *years*$)$, as calculated from the measured *relative abundances* of various nucleides; and (iv) the ages $(4.6 \times 10^9$ *years*$)$ of the *oldest lunar rocks* and the *oldest meteorites*. (See also Fig. III.4).

b) There is excellent agreement between theory and observation for the linear (low $-z$) parts of the distance-redshift, magnitude-redshift, and angular diameter-reshift relations (see also Lectures I, VI and VII).

c) There is excellent agreement between general relativistic cosmology and observations of the *cosmic radio background radiation, average density and temperature and mass of the galaxies, as well as measurements of helium and deuterium* that provide us with evidence about thermodynamical processes which took place at very large redshifts, i.e., during the early history of the universe (see also III.3 below).

There are other experiments and observations associated with general relativity, notably the ones devised to detect the possible existence of *gravitational waves*, their polarization properties and possible correlation with some astrophysical phenomena (such as supernova explosions, etc., cf. Lecture I). Yet, evidence accumulated so far by these experiments is *not* entirely conclusive. Hence we shall not consider it any further in this lecture.**

To proceed further we must now acquire minimal information on the mathematical tools employed in general relativity and relativistic cosmology. This is done in the next sections, starting with some essential formulations of blue and redshifts in special relativity. *(A reader unfamiliar with the mathematics involved may, however, proceed as follows:* start from III.2.1, read portions of III.2.6 and III.2.6, then III.2.7.3 and III.3.).

III.2 PRINCIPLES AND FORMULATIONS OF GENERAL RELATIVITY AND RELATIVISTIC COSMOLOGY

III.2.1 Doppler Effect and Time Dilation in Special Relativity

What follows is based throughout on a normalized system whereby the velocity of light is taken as unity, and all space-time variables x^{α} ($\alpha = 1,2,3,0$) have the dimension of length. In this notation α, β, γ, μ, ν etc., run over four values 1,2,3,0 with x^1, x^2, x^3, the Cartesian components of the position vector $\mathbf{x}$ and x^0 the time component. In addition, Einstein's summation convention is used, whereby any index i, j, α, β, or γ which appears twice, once as a subscript and once as a superscript, is

---

**Neither we consider the problems associated with the historical *Mach Principle*, except by remarking that Einstein believed that the cause of inertial forces is the sum of the gravitational forces of all the galaxies in the cosmos. This problem is also linked with the questions of cosmological boundary conditions—see below.

to be summed over the four components unless otherwise stated. Repeated Latin indices are summed only over the spatial coordinates 1, 2, 3.

A changeover from Galilean to special relativity has immediate kinematic implications for material objects moving at velocities less than that of light.*

*The principle of special relativity*** states that laws of nature are unaffected under a particular category of operations known as the *Lorentz transformations*, whereby the velocity of light, and also Maxwell's equations, *are invariant* on transition from one (inertial) system of space-time coordinates x^α to another such system $x^{\,\alpha}$. The distinctive fundamental property of these transformations is that they leave invariant the "*proper time*", $d\tau$, (called also the scalar *line element* of interval $d\tau$), defined by:

$$d\tau^2 \equiv dt^2 - d\mathbf{x}^2 = -\eta_{\alpha\beta} dx^\alpha dx^\beta \tag{3}$$

where the *constant* dimensionless symmetric metric $\eta_{\alpha\beta}$ is defined by

$$\eta_{\alpha\beta} = \begin{cases} +1 \text{ for } \alpha = \beta = 1, 2 \text{ or } 3 \\ -1 \text{ for } \alpha = \beta = 0 \\ 0 \text{ for } \alpha \neq \beta \end{cases} \tag{3a}$$

Denoting by cdt the infinitesimal temporal distance (c being the velocity of light in vacuum), and by dx, dy, dz, the rectangular components of the infinitesimal spatial one, eq. (3) becomes

$$d\tau^2 = -dx^2 - dy^2 - dz^2 + c^2 dt^2 \tag{3b}$$

where $d\tau$ describes the elementary interval between two events infinitely close to each other both in space and in time.

In general, we can write that a light wave front has the speed $|d\mathbf{x}/dt| = c$. We can obtain this expression from eq. (3), or from eq. (3b), by subjecting the propagation of light, or of any electromagnetic signal, to the statement

$$d\tau = 0 \tag{4}$$

---

* The simplest, yet most important of these, is the *time-dilation* effect in a moving clock, of which more in a moment.

** In an alternative formulation the special theory of relativity is based on two postulates. The *first postulate* states that *it is impossible to measure or detect unaccelerated translatory motion of a system through free space or through any ether-like medium assumed to pervade it*. We can thus speak of the *relative* velocity of a pair of systems, but it is meaningless to speak of the *absolute* velocity of a single system; consequently, the general law describing phenomena in free space must be dissociated from the motion of a particular observer or a coordinate system, since otherwise we could ascribe some absolute significance to different velocities.

The *second postulate* states that *the velocity of light in free space is the same for all observers*. This postulate—which combines the empirical finding that the velocity of light is independent of that of its source (the Michelson-Morley experiment) with the theoretical premise of the first postulate—has led to drastic revision of the traditional concepts of time and space.

which, in conjunction with (3) yields

$$\sqrt{\frac{dx^2 + dy^2 + dz^2}{dt^2}} = \frac{d\mathbf{x}}{dt} = c = 1 \tag{5}$$

If we now transform to any other system of inertial coordinates x', y', z', t', corresponding to a new set of axes in uniform motion relative to the original ones, the *form of expression* for the interval $d\tau$ must remain unchanged, as befits a Lorentz transformation. Accordingly, the velocity of light will again be

$$\sqrt{\frac{dx'^2 + dy'^2 + dz'^2}{dt'^2}} = \frac{d\mathbf{x}}{dt} = c = 1 \tag{5a}$$

as required by the second postulate of special relativity.

Four-dimensional geometry has thus provided us with a very useful tool for treating the special-relativistic facts. In addition, it is a "language" which is almost indispensable in general relativity and other theories of gravitation.

A space equipped with the metric (3a) at *all* its points is called a *Minkowski* or a *flat* space. This uncurved space is the nest of the special theory of relativity. We shall later see that the presence of mass, energy, and momentum or gravitation, is incompatible with such a simple and invariant metric.

According to the Michelson-Morley experiment, the velocity of light is the same in all inertial systems. Therefore the "proper time" $d\tau'$ for another inertial system

$$d\tau'^2 = -\eta_{\alpha\beta}\, dx'^\alpha dx'^\beta \tag{3c}$$

should equal $d\tau$, i.e.

$$d\tau^2 = d\tau'^2 \tag{6}$$

Thus the Principle of Special Relativity states that *in inertial systems all physical equations must be invariant under a Lorentz transformation.**

An observer viewing a clock *at rest* records consecutive ticks separated by a space-time interval $dx = 0$, $dt = \Delta t$, *with Δt specified by the manufacturer***. We realize that this observer's proper time interval is given by (cf. eq. 3)

$$d\tau \equiv (dt^2 - d\mathbf{x}^2)^{\frac{1}{2}} = \Delta t \tag{7}$$

A second observer, who sees *the same clock* moving with (constant) velocity u, will observe that the ticks are separated by a time interval dt' and also by a space interval $dx = u\, dt'$, and will use this link to obtain

$$d\tau \equiv (dt'^2 - dx'^2)^{\frac{1}{2}} = (1 - u^2)^{\frac{1}{2}} dt', \tag{8}$$

where $u^2 \equiv |u|^2$.

---

* Maxwell's equations, for instance, are *not* invariant under a *Galilean* transformation.
** i.e., for a clock at rest.

Since *both* observers function in *inertial* coordinate systems, their systems are related by a Lorentz transformation. Thus in accordance with eq. (6) the two observers find $d\tau = d\tau'$. In these circumstances eq. (7) and (8) can be equated to yield the famous result

$$dt' = \Delta t (1 - u^2)^{-\frac{1}{2}} \tag{9}$$

thoroughly verified by all experiments involving rapidly moving particles (e.g. in elementary-particle accelerators)*.

The value dt' in (9) should not be confused with the *apparent time dilation* (or contraction) of the *Doppler effect* with which, together with the red-shift factor, we are concerned in astronomical observations. The *red-shift factor* is defined as

$$z = (\lambda - \lambda_{\Gamma})/\lambda = \Delta\lambda/\lambda \tag{10}$$

where λ is the observed wavelength of a characteristic spectral line (of hydrogen, helium, etc.) emitted by the moving source, and λ_{Γ} that of the same line measured (at rest) in a terrestrial laboratory.

In non-relativistic dynamics, the difference $\lambda - \lambda_{\Gamma}$ is related by the Doppler principle to U_r, the radial component of the velocity of the moving source in the direction from the observer to the latter. For small z, this relationship can be formulated as

$$\frac{\Delta\lambda}{\lambda} = \frac{U_r}{c} \tag{11}$$

Thus if our "clock" is a moving source of radiation of (rest) frequency $\nu_{\Gamma} = 1/\Delta t$, the interval dt', between consecutive wave fronts is given by eq. (9) as above. However, during this time the *distance from the observer to the radiation source* will have *increased* or *decreased* by $U_r dt'$, hence the interval between reception of two consecutive wave fronts *in the terrestrial laboratory* will be

$$dt_o = (1 + U_r) dt' \tag{12}$$

or, substituting eq. (9),

$$dt_o = (1 + U_r) (1 - u^2)^{-\frac{1}{2}} \Delta t \tag{13}$$

Incidentally, $U_r dt'/c$ is the change in wavelength—an increase for dilation and a decrease for contraction.

---

* The layman is usually better convinced by the recent empirical evidence provided by Hafele and Keating [*Science, 177,* 166 (1972)] who used commercial jet flights and modern atomic clocks. They found that the *relativistic dilation of time* is a function of the velocity of the clock with respect to an absolute coordinate system *at rest relative to the distant galaxies.* The clocks which circumnavigated the earth in the *eastward* direction ran *slower* than those at rest on the earth's surface by an average of *79 billionths* of a second, while those moving westward ran faster than those at rest by an average of *273 billionths* of a second.

Consequently, the ratio of the frequency of radiation actually measured by the terrestrial observer, v_o, to that of the radiation source *at rest*, v_Γ, is

$$\frac{v_o}{v_\Gamma} = \frac{\lambda_\Gamma}{\lambda} = (1 + U_r)^{-1}(1 - u^2)^{\frac{1}{2}} = \frac{\Delta t}{dt_o} \tag{14}$$

In respect of the above, *four cases* can be distinguished:

1. If the galactic source is moving *radially away* from the earth, then $|u| = U_r$, $U_r > 0$ and eq. (14) predicts a *red shift*, i.e.,

$$\lambda > \lambda_\Gamma \tag{15}$$

The red shift $(\lambda - \lambda_\Gamma)/\lambda$ tends to infinity as U_r (called in this case the *recession velocity*) approaches unity, i.e., the velocity of light.

2. If the source is moving *parallel* to the earth's surface, then $U_r = 0$ and eq. (14) yields *pure time dilation*,

$$\frac{\lambda_\Gamma}{\lambda} = (1 - u^2)^{\frac{1}{2}} < 1 \tag{16}$$

Here the radiation would still appear *red-shifted* — in spite of the fact that $U_r = 0$! Such pure time dilation effects are important in astronomy, since our Milky Way and many other galaxies and stars rotate.

3. If the radiation source is moving radially *toward the earth*, then $U_r = -|u|$, and eq. (14) yields a *violet shift* represented by the factor

$$(1 + U_r)^{1/2}(1 - U_r)^{-1/2} \tag{17}$$

4. If the source is moving *obliquely* (i.e. at an angle between straight toward earth and a right angle to it), radiation would be either red- or violet-shifted, depending on U_r and u (i.e., a radiation source which moves closer to the earth along a trajectory less than right angels may still appear *red-shifted* to a terrestrial observer).

Eq. (14) thus represents astronomical wavelength shifts as a *special relativistic Doppler effect*. In addition to it, however, allowance must be made for certain cosmological subtleties (e.g. a *strong gravitational field* producing a non-negligible *red shift*), which will be discussed later.

● III.2.2 Metric Tensor and Affine Connection

We now proceed to define the metric tensor $g_{\mu\nu}$.

For this purpose, we first consider a particle moving *freely* under the influence of a (purely) gravitational force. According to the *Principle of Equivalence* (cf. III.1.1), there is at every space-time point, in a gravitational field, a possibility of choosing a freely-falling coordinate system, y^α, in which its equation of motion describes a straight-line trajectory in space-time, namely, *where the laws of nature take the same form as in unaccelerated Cartesian coordinate systems* in the *absence* of gravitation, i.e., where

$$\frac{d^2 y^\alpha}{d\tau^2} = 0 \tag{18}$$

and for which

$$d\tau^2 = -\eta_{\alpha\beta}\, dy^\alpha dy^\beta \tag{19}$$

as in eq. (3). Eq. (18) is, thus, related to a "relativistic force" f^α acting on a particle with coordinates $x^\alpha(\tau)$, i.e.,

$$f^\alpha = m\frac{d^2 x^\alpha}{d\tau^2} \tag{20}$$

We now consider the motion of the particle in *any other coordinate system* x^μ, which may be at *rest, accelerating, rotating, etc.* For such systems the freely-falling coordinates, y^α, can always be found (III.1.1), and they become functions of x^μ; thus, the equation of motion (18) becomes

$$0 = \frac{d}{d\tau}\left(\frac{\partial y^\alpha}{\partial x^\mu}\frac{dx^\mu}{d\tau}\right) = \frac{\partial y^\alpha}{\partial x^\mu}\frac{d^2 x^\mu}{d\tau^2} + \frac{\partial^2 y^\alpha}{\partial x^\mu \partial x^\nu}\frac{dx^\mu}{d\tau}\frac{dx^\nu}{d\tau} \tag{21}$$

or, multiplying by $\partial x^\mu/\partial y^\alpha$,

$$0 = \frac{d^2 x^\lambda}{d\tau^2} + \Gamma^\lambda_{\mu\nu}\frac{dx^\mu}{d\tau}\frac{dx^\nu}{d\tau} \tag{22}$$

where $\Gamma^\lambda_{\mu\nu}$ is the (non-tensor) coefficient of *affine connection*, defined as

$$\Gamma^\lambda_{\mu\nu} = \frac{\partial x^\lambda}{\partial y^\alpha}\frac{\partial^2 y^\alpha}{\partial x^\mu \partial x^\nu} \tag{23}$$

In addition, for an *arbitrary* coordinate system, the proper time (19) becomes

$$d\tau^2 = -\eta_{\alpha\beta}\frac{\partial y^\alpha}{\partial x^\mu}\,dx^\mu\frac{\partial y^\beta}{\partial x^\nu}\,dx^\nu \tag{24}$$

Defining the *metric tensor* now as

$$g_{\mu\nu} \equiv \frac{\partial y^\alpha}{\partial x^\mu}\frac{\partial y^\beta}{\partial x^\nu}\eta_{\alpha\beta} \tag{25}$$

eq. (24) becomes

$$d\tau^2 = -g_{\mu\nu}\, dx^\mu dx^\nu \tag{26}$$

● **III.2.3 Newtonian Limit**

We now consider approximation of $g_{\mu\nu}$ in terms of ϕ for a *weak gravitational field*. In this context we consider a particle moving slowly in a *stationary* field of this type, for which all *time derivatives* of $g_{\mu\nu}$ *vanish.* When the motion is sufficiently slow, $d\mathbf{x}/d\tau$ is negligible against $dt/d\tau$ and the equation of motion (22) reduces to

$$\frac{d^2x^\mu}{d\tau^2} + \Gamma^\mu_{oo}\left(\frac{dt}{d\tau}\right)^2 = 0 \tag{27}$$

with the affine connection given by

$$\Gamma^\mu_{oo} = -\frac{1}{2}g^{\mu\nu}\frac{\partial g_{oo}}{\partial x^\nu} \tag{28}$$

in which g_{oo} is the *time-time component* of the metric tensor.

For the weak field in question the metric tensor may be resolved according to

$$g_{\alpha\beta} = \eta_{\alpha\beta} + h_{\alpha\beta} \tag{29}$$

where the approximations in special relativistic coordinate systems are subject to the restriction $|h_{\alpha\beta}| \ll 1$, so that, to first order in $h_{\alpha\beta}$, the coefficient of affine connection becomes

$$\Gamma^\alpha_{oo} = -\frac{1}{2}\eta^{\alpha\beta}\frac{\partial h_{oo}}{\partial x^\beta} \tag{30}$$

where h_{oo} is the time-time component of $h_{\alpha\beta}$.

The equations of motion thus take the form

$$\frac{d^2\mathbf{x}}{d\tau^2} = \frac{1}{2}\left(\frac{dt}{d\tau}\right)^2 \nabla h_{oo} \tag{31}$$

$$\frac{d^2t}{d\tau^2} = 0 \tag{32}$$

Dividing (31) by $(dt/d\tau)^2$ (and noting, from the solution of (32), that the latter is a constant) we obtain

$$\boxed{\frac{d^2\mathbf{x}}{dt^2} = \frac{1}{2}\nabla h_{oo}} \tag{33}$$

Equating (33) with its Newtonian counterpart (2), we find

$$h_{oo} = -2\phi + \text{constant} \tag{34}$$

As regards the constant, we note from eq. (2a) that ϕ vanishes at infinity—where the gravitational field vanishes and the coordinate system becomes identical with the special relativistic metric (3)—hence

$$\boxed{h_{oo} = -2\phi} \tag{35}$$

and from (29) we finally obtain the time-time component of the metric tensor

$$\boxed{g_{oo} = -(1 + 2\phi)} \tag{36}$$

This is a very important result which can link general relativity to the Newtonian theory of gravitation as will be shown later.

In eq. (2) the units of ϕ are those of velocity squared. By contrast, in eq. (36) ϕ is divided by the square of the velocity of light, i.e., it is a *dimensionless quantity*. Its order of magnitude is 10^{-4} at the surface of a white dwarf star*, 10^{-6} at that of the sun, 10^{-9} at that of the earth, and 10^{-39} at that of a proton. Consequently the changes in $g_{\mu\nu}$ produced by varying gravitational fields are very small in most stars and planets.

● III.2.4 Energy-Momentum Tensor of Ideal Fluid in Special Relativistic Fluid Dynamics

● III.2.4a Preliminary Definitions

To derive Einstein's field equations use must be made of the concept of the *energy-momentum tensor***. The meaning of the latter may best be illustrated with the aid of the macroscopic conservation laws derived previously in Lecture II.

We consider first an ideal *fluid* (see definitions in Lecture II), which obeys the *Euler* and *energy equations*, and which, under proper transformation of the co-ordinates $\tilde{x}^{\beta}$ (i.e. those of a *co-moving observer* who follows the motion of a fluid mass element), is *at rest* at some particular position and time. (The co-moving coordinates are distinguished here by a tilde).

At this space-time the energy-momentum tensor, $\tilde{T}^{\alpha\beta}$, takes the simple form

$$\tilde{T}^{\alpha\beta} = \begin{bmatrix} p & o & o & o \\ o & p & o & o \\ o & o & p & o \\ o & o & o & \rho \end{bmatrix} \tag{37}$$

or, in another notation,

$$\begin{aligned} \tilde{T}^{ij} &= p\delta_{ij} \\ \tilde{T}^{io} &= T^{oi} = 0 \qquad (i,j = 1,2,3) \\ \tilde{T}^{oo} &= \rho \end{aligned} \tag{37a}$$

where $\delta_{ij} = 1$ for $i = j$, $\delta_{ij} = 0$ for $i \neq j$ and where the Latin indices are restricted to $i,j = 1,2,3$ for the spatial coordinates and $(\quad)^{oo}$ refers to the temporal coordinate in *non-relativistic* systems. ρ represents the *material* density as usual, but also includes *the energy* of the *rest mass*** of the fluid in question, in addition to its *internal and kinetic energies*. Accordingly, we call it the *"proper energy density."*

Employing now the special relativistic connections between the comoving coordinates $\tilde{x}^{\beta}$ and a reference frame of coordinates x^{α} *at rest in the laboratory*,

---

* See Lecture I.

** Also called *stress-energy* tensor.

*** Since from special relativity $d(\text{energy}) \equiv d(\text{mass})$ when $c = 1$.

the energy-momentum tensor for an ideal fluid takes the special-relativistic form

$$T^{ij} = p\delta_{ij} + (p + \rho) \frac{V_i V_j}{1 - u^2} \tag{38a}$$

$$T^{io} = (p + \rho) \frac{V_i}{1 - u^2} \tag{38b}$$

$$T^{oo} = \frac{\rho + pu^2}{1 - u^2} \tag{38c}$$

● **III.2.4b Conservation of 4-Momentum**

In this and the next paragraphs we discuss the physical importance of the conservation equation

$$\mathbf{V} \cdot \mathbf{T} = 0, \text{ i.e., } \frac{\partial T^{\alpha\beta}}{\partial x^\beta} \equiv T^{\alpha\beta}{}_{,\beta} = 0 \tag{39}$$

which is the *differential formulation of the law of 4-momentum conservation* (also
called the *equation of motion for stress-energy*). The beautiful simplicity and the
powerful applications of this equation will be demonstrated first for classical con-
tinuum physics where $|V_i| \ll c$ and $p/\rho c^2 \ll 1$. The last requirement means that
the pressure in the fluid is small compared to its density of mass-energy (note that
we use $c = 1$ units). For such a fluid, eq. (38) becomes

$$T^{ij} \cong p\delta_{ij} + \rho V_i V_j \tag{38d}$$

$$T^{io} = T^{oi} \cong \rho V_i \tag{38e}$$

$$T^{oo} \cong \rho \tag{38f}$$

and eq. (39) is re-expressed in terms of its components,

$$\boxed{\frac{\partial T^{oj}}{\partial x^j} \equiv T^{oo}{}_{,o} + T^{oj}{}_{,j} \equiv \frac{\partial \rho}{\partial t} + \text{div}\,(\rho u) = 0} \tag{II.20}$$

("equation of continuity")

and

$$T^{jo}{}_{,o} + T^{ji}{}_{,i} \equiv \frac{\partial(\rho V_j)}{\partial t} + \frac{\partial(\rho V_i V_j)}{\partial x_i} + \frac{\partial p}{\partial x_j} = 0$$

or, equivalently, by combining with the equation of continuity:

$$\rho \frac{Du}{Dt} \equiv \rho \frac{\partial u}{\partial t} + u \cdot \text{grad}\,u = -\,\text{grad}\,p \tag{II-36}$$

("Euler's equation")

Therefore the stress-energy tensor, like its classical counterparts, is linked to *conservation of momentum, mass and energy.*

Consequently, we have realized that eq. (39) is a very compact, yet a very power-ful equation which contains much information about constraints on the dynamics of mass, energy and momentum. However, we must stress here that the present formu-lations of the energy-momentum tensor, i.e., eq. (38), are highly simplified since they do not allow for viscous forces, etc. The definition of the general energy-momentum tensor is postponed to §III.2.5.

As another application of the conservation equation (39) one may consider a simplified case in *electromagnetism.*

For instance, if an electromagnetic field interacts with an electrically charged vibrating rubber block (assuming electrically charged beads are imbedded in it) a 4-momentum transfer takes place. As the block vibrates, its accelerating charges radiate electromagnetic waves; at the same time, incoming electromagnetic waves push on the beads, altering the pattern of vibration of the block, causing 4-momentum transfer between block and electromagnetic field. In similarity to the case discussed in Lecture IV one must remember that only the *total* 4-momentum must be con-served, i.e.,

$$[T^{\alpha\beta}(\text{em field}) + T^{\alpha\beta}(\text{block})]_{,\beta} = 0 \tag{40}$$

which means that the rate of change of the momentum of the block equals the force of the electromagnetic field on its beads. The time component of eq. (40)

$$[T^{o\beta}(\text{em field}) + T^{o\beta}(\text{block})]_{,\beta} = 0 \tag{41}$$

means that the mass-energy of the block increases at the same rate as the electric field *does work* on the charged beads. A similar result holds for momentum.

We now turn back to the definition of $T^{\alpha\beta}$.

Clearly for free empty space $T^{\alpha\beta} = 0$ by definition. To verify that this is a *sym-metric tensor*, we condense eqs. (38a) to (38c) into a single equation

$$T^{\alpha\beta} = \eta^{\alpha\beta}p + (p + \rho)U^{\alpha}U^{\beta} \tag{42}$$

where U^{α} is called the *fluid velocity four-vector* (or fluid four-velocity), for which

$$U \equiv \frac{d\mathbf{x}}{d\tau} = (1 - u^2)^{-\frac{1}{2}}u = uU^o \tag{43}$$

$$U^o \equiv \frac{dt}{d\tau} = (1 - u^2)^{-\frac{1}{2}} \tag{44}$$

and which is normalized so that

$$U_{\alpha}U^{\alpha} = -1 \tag{45}$$

Finally we stress that the symmetry of $T^{\alpha\beta}$ enables one to define a conserved angular momentum analogous to the linear momentum (cf. the discussion following eq. 33 in Lecture II).

● **III.2.4c Conservation of Energy and Momentum**

In special relativity, taking the divergence of $T^{\alpha\beta}$ and equating it to zero, we obtain the conservation of momentum and energy (without external forces). Thus, for an *ideal fluid*,

$$0 = \frac{\partial T^{\alpha\beta}}{\partial x^{\beta}} = \frac{\partial p}{\partial x^{\alpha}} + \frac{\partial}{\partial x^{\beta}} \left[(\rho + p) U^{\alpha} U^{\beta} \right] \tag{46}$$

To realize the similarity with the *Euler equation* (II-36) one can break (46) down into a *three-vector momentum equation and a scalar energy equation*. The former is obtained by setting $\alpha = i$ writing* $U^{i} = V^{i} U^{o}$, and then setting $\alpha = o$ in eq. (46) to obtain

$$\frac{Du}{Dt} \equiv \frac{\partial u}{\partial t} + (u \cdot \nabla)u = - \frac{(1 - u^{2})}{(\rho + p)} \left[\nabla p + u \frac{\partial p}{\partial t} \right] \tag{47}$$

which is the *Euler equation in relativistic fluid dynamics*.

Multiplying (46) by U_{α}, we obtain the *energy equation* for an ideal fluid in special relativity

$$0 = U_{\alpha} \frac{\partial T^{\alpha\beta}}{\partial x^{\beta}} = U^{\beta} \frac{\partial p}{\partial x^{\beta}} - \frac{\partial}{\partial x^{\beta}} \left[(p + \rho) U^{\beta} \right] \tag{48}$$

where use is made of the relation

$$\frac{\partial}{\partial x^{\beta}} (U_{\alpha} U^{\alpha}) = 2 U_{\alpha} \frac{\partial U^{\alpha}}{\partial x^{\beta}} = 0 \tag{49}$$

● **III.2.5 Energy-Momentum Tensor with Dissipation**

Dissipation plays an extremely important role in all real (irreversible) processes which, in turn, cannot be described in terms of ideal fluids (see eqs. (39), (49), (50), (59a), and (59d) in Lecture II for the relevant definitions). Therefore, instead of eq. (42) we must write

$$T^{\alpha\beta} = \eta^{\alpha\beta} p + (p + \rho) U^{\alpha} U^{\beta} + \tau^{\alpha\beta} \tag{50}$$

which, as in the case of an ideal fluid, vanishes in empty free space. We thus assume that *all effects of dissipation show up as contributions to* $\tau^{\alpha\beta}$. Inclusion of dissipative effects in relativistic systems entails certain delicate questions which do not arise in the non-relativistic domain. The problems involved have been treated by Landau and Lifshitz, Eckert, Marle, and Eringen.**

---

* See Lecture II Sections II.3, II.6 for definitions of V^{i}, etc.

** Laundau, L. D. and Lifshitz, E. M., *Fluid Mechanics* (Pergamon Press, London, 1959), Section 127; Eckert, C., *Phys. Rev.* 58, 919 (1940); Marle, C., *Compt. Rend. Acad. Sci.* (Paris) 260, Group 4 (1905), and 262, Series A (1966); Eringen, A. C. in *A Critical Review of Thermodynamics* (Stuart, E. B., Gal-Or, B., and Brainard, A. J., eds.), Mono Book Corp., Baltimore, 1970, pp. 490.

Resorting to the linear approximations of the "*Newtonian fluids*" (see Section II.6 eq. 39 Lecture II) and assuming Fourier heat conduction in co-moving Lorentz frames, we can write here

$$\tau^{ij} = -\mu\left(\frac{\partial U_i}{\partial x^j} + \frac{\partial U_j}{\partial x^i} - \frac{2}{3}\nabla\cdot U\,\delta_{ij}\right) - \kappa\nabla\cdot U\delta_{ij} \qquad \text{(II.39), (51)}$$

$$\tau^{oo} = 0\,;\tau^{io} = -k\left(\frac{\partial T}{\partial x^i} + T\dot{U}_i\right) \qquad (52)$$

where μ, κ and k are the (positive) *shear viscosity*, *bulk viscosity* and *heat conduction coefficients*, respectively. Thus, except for the last term in (52) [see eq. (55) below], eqs. (51) and (52) have the same form as in the non-relativistic theory of real fluids given in Lecture II (eqs (50) and (73)).

According to Weinberg* such a dissipation tensor, which remains Lorentz-invariant in all Lorentz frames, is given by

$$\tau^{\alpha\beta} = -\mu H^{\alpha\gamma}H^{\beta\delta}\,\tau_{\gamma\delta} - k(H^{\alpha\gamma}U^\beta + H^{\beta\gamma}U^\alpha)Q_\gamma - \kappa H^{\alpha\beta}\,\frac{\partial U^\gamma}{\partial x^\gamma} \qquad (53)$$

where $\tau_{\gamma\delta}$ is the *shear tensor*, defined, for fluids with negligible bulk viscosity, by

$$\tau_{\alpha\beta} \equiv \frac{\partial U_\alpha}{\partial x^\beta} + \frac{\partial U_\beta}{\partial x^\alpha} - \frac{2}{3}\eta_{\alpha\beta}\frac{\partial U^\gamma}{\partial x^\gamma}, (\kappa = 0) \qquad (54)$$

Q_α, the *heat-flow vector*, defined by

$$Q_\alpha \equiv \frac{\partial T}{\partial x^\alpha} + T\frac{\partial U_\alpha}{\partial x^\beta}U^\beta \qquad (55)$$

and $H_{\alpha\beta}$ a *projection tensor* (on the hyperplane normal to U^α) defined by

$$H_{\alpha\beta} \equiv \eta_{\alpha\beta} + U_\alpha U_\beta \qquad (56)$$

Consequently, the trace of the energy momentum tensor (50) is given by

$$T^\alpha{}_\alpha = 3p - \rho - 3\kappa\,\frac{\partial U^\gamma}{\partial x^\gamma} \qquad (57)$$

Weinberg suggests the following equations for μ, κ and k:

$$\mu = \frac{4}{15}aT^4\tau* \qquad (58)$$

$$\kappa = 4\,aT^4\left[\frac{1}{3} - (\partial p/\partial\rho)_{ci}\right]^2\tau* \qquad (59)$$

$$k = \frac{4}{3}aT^3\tau* \qquad (60)$$

* Weinberg, S., *Gravitation and Cosmology*, Wiley, N.Y., 1972, pp. 56.

where τ^* is a finite mean free time for a material medium with very short mean free times, interacting with radiation quanta. Here a is the Stefan-Boltzmann constant (defined so that the radiation energy density is aT^4), and ρ the energy density of the matter *and* the radiation. In general kT, μ, and κ are comparable in value. When, however, the pressure and thermal energy are dominated by radiation, as in early cosmic times, $(\partial p/\partial \rho)_{ci} \cong 1/3$ and the *bulk viscosity* is small (although it does *not* necessarily *vanish*).

These formulations can be extended to include the contributions of the electromagnetic energy and momentum densities. In general, the energy-momentum tensor is *not* conserved if the medium is subject to forces acting at a distance.

● III.2.6 Einstein's Field Equations in Weak Static Fields

The time-time component of the metric tensor in a weak static field produced by non-relativistic mass density is approximately given by (36), where ϕ is the Newtonian potential as per eq. (2). On the other hand, the time-time component of the energy-momuntum tensor T_{oo} for non-relativistic matter is approximately given by eq. (37) or (38) as $T_{oo} = \rho$. Substituting these results in eq. (2), we have

$$\nabla^2 g_{oo} = -8\pi G T_{oo} \tag{61}$$

as a *"field equation"* for weak static fields generated by non-relativistic matter. (As it stands, eq. (61) is not even Lorentz-invariant). *Note, however, that ∇^2 involves the second space derivative of the time-time component of the metric tensor!*

Eq. (61) is an important link between the Newtonian theory of gravitation and its general relativistic counterpart as represented by Einstein's field equations. There are a number of methods to derive the latter equations. Here we resort to the simplest of them all, namely, suppose from eq. (61) that the general form of the weak-field equations for a general distribution of energy and momentum would be

$$G_{\alpha\beta} = -8\pi G T_{\alpha\beta} \tag{62}$$

where $G_{\alpha\beta}$ contains only *linear combinations of the metric and its first and second derivatives*. It is, however, the Principle of Equivalence which enables us to stipulate that the equations for gravitational fields of arbitrary strength take the form

$$G_{\mu\nu} = -8\pi G T_{\mu\nu} \tag{63}$$

where $G_{\mu\nu}$ *is a tensor which reduces to* $G_{\alpha\beta}$ *for weak fields* and contains *only* terms that are *either linear in the second derivatives of the metric, or quadratic in its first derivatives.**

---

* Eq. (61) suggests that the whole of $G_{\mu\nu}$ *must* have the dimensions of *a second derivative*, hence each term which contains more or less than two derivatives should be multiplied by a constant having the proper dimensions of length. Following Einstein, however, we assume that the gravitational field equations are uniform in scale, i.e., *only terms with two derivatives are allowed!*

Since the energy-momentum tensor $T_{\mu\nu}$ is *symmetric*, then, by (63), $G_{\mu\nu}$ must *also* be a symmetric tensor. Moreover, since $T_{\mu\nu}$ is conserved in the sense of covariant differentiation, so is $G_{\mu\nu}$. For *weak stationary fields* produced by non-relativistic matter the *time-time component* of (63) must reduce to (61), i.e.,

$$G_{oo} \cong \nabla^2 g_{oo} \tag{64}$$

Einstein showed that the most general way of constructing a tensor satisfying these properties of $G_{\mu\nu}$ is by contraction of the curvature tensor, which leads to

$$G_{\mu\nu} = R_{\mu\nu} - \frac{1}{2} g_{\mu\nu} R \tag{65}$$

where $R_{\mu\nu}$ is the *Ricci tensor* and R the *curvature scalar*, $R \equiv g^{\mu\nu} R_{\mu\nu}$. These quantities are constructed *only* in terms of the metric tensor $g_{\mu\nu}$ and its two derivatives. Combining with (63), we finally arrive at *Einstein's field equations*

$$\boxed{R_{\mu\nu} - \frac{1}{2} g_{\mu\nu} R = - 8\pi G T_{\mu\nu}} \tag{66}$$

An additional term, $\lambda g_{\mu\nu}$ was once introduced into (66) by Einstein for cosmological reasons mentioned in §III.1.2b. Its inclusion means relaxation of the condition that $G_{\mu\nu}$ contain only terms that are either linear in the second derivatives or quadratic in the first derivatives of the metric, i.e., the tensor is allowed to contain terms with less than two derivatives of the metric, but no additional terms except $\lambda g_{\mu\nu}$, where λ is the so-called cosmological constant. This term is incompatible with eq. (64), hence its value must be very small, and since the motivation for its inclusion is no longer valid due to the discovery of the expanding universe (see Lecture VI) we do not take it into account.

We now proceed to employ Einstein's field equations in deriving the master arrow of time. But before this step can be taken, the Robertson-Walker metric must be described.

● III.2.7 Robertson-Walker Metric and Derivation of the Master Arrow of Time

Observational evidence, and the general statement of the Cosmological Principle given in Section VI, limit our choice of the co-moving coordinate system, r, θ, and ϕ in the sense that typical supergalaxies (i.e., clusters of galaxies) have constant spatial coordinates r, θ, and ϕ, for which the metric takes the form (see also Lecture VIII);

$$d\tau^2 = dt^2 - R^2(t) \left[\frac{dr^2}{1 - kr^2} + r^2 d\theta^2 + r^2 \sin^2\theta d\phi^2 \right] \tag{67}$$

$R(t)$ being an unknown scale factor which depends *only* on the cosmic time t, and k a constant, which by suitable choice of units for r can be chosen to have the values $+1, 0$, or -1. This is the *Robertson-Walker metric*. It differs from the special relativistic metric for Minkowski space-time only by the presence of the undetermined cosmic *scale factor* $R(t)$ and the constant k. If we consider the system of supergalaxies at a particular moment of cosmic time t_1, then $dt = 0$ and the metric for three-dimensional space at time t becomes

$$d\tau^2 = - R(t)\left[\frac{dr^2}{1 - kr^2} + r^2 d\theta^2 + r^2 sin^2\theta d\phi^2 \right] \qquad (68)$$

At a *later* moment of cosmic time, t_2, we have the same metric except that every interval $d\tau$ would be multiplied by the factor $R(t_2)/R(t_1)$. If this factor is *greater than unity*, intervals *increase with time*, the supergalaxies *recede from each other* (at constant r, θ, and ϕ coordinates!)* and we have an *expanding universe*. Thus, *$R(t)$ represents the change (in terms of cosmic time) in the distance between any two observed supergalaxies*. The spatial coordinates r, θ, ϕ form a co-moving system in the sense *that they are constant for "typical" supergalaxies*. The co-moving coordinate mesh may then be visualized as lines drawn on the surface of a balloon, on which dots represent typical supergalaxies. As the balloon is inflated, *the dots recede from each other but the lines move with them*, so each dot retains its coordinates.

The statement that a supergalaxy has constant comoving coordinates is consistent with the notion that it is in *free fall*. Moreover, the time coordinate t in eq. (67) is not only a convenient *"standard cosmic time"* for all co-moving observers in our universe, but also the *proper time* indicated by a *clock at rest* in any freely-falling galaxy.

It is instructive to think of k as the *curvature* of the three-dimensional space at any cosmic time t (see Fig. IV.6).

We can thus distinguish among three spaces, namely $k = 0$, $k = +1$ and $k = -1$:

(i) $k = 0$: Space is *Euclidean* and *infinite*, i.e., it is "flat" (zero curvature), in particular, the surface area of a sphere of radius r is $4\pi r^2$.

(ii) $k = +1$: Space is *finite* (albeit unbounded) and is said to be *spherical*, since its geometry is analogous (in two dimensions) to that of the *surface* of a sphere in which distances are measured along circles (and a circle is the locus of points at a constant surface distance from a given point on the surface).** The *circumference* of such a circle is *less than 2π times its radius* (as measured along the surface from a given point). The difference between an Euclidean circumference and this one is very small when the radius is *small*. For *larger* radii it increases until the radius of space goes one-quarter way around the sphere and the circumference then reaches a *maximum value, decreases*, approaching *zero* as the radius goes half way round the sphere. Similarly, *the surface area is less than 4π times the radius squared*, it increases to a *maximum* and then shrinks to *zero*. The *volume* of this space is *finite*, and increases

---

* See Fig. IV.5.
** This space is finite but boundless. On this surface *the sum of the included angles in a triangle is always greater than 180 degrees*. Here no two "straight" lines are parallel (and they always intersect). They are called geodesics.

with time according to

$$V = \pi^2 R^3(t) \tag{69}$$

The spatial universe may thus be regarded as a surface of a sphere of radious $R(t)$ four-dimensional Euclidean space. $R(t)$, in this case, may be termed "*radius of the universe.*" (No such interpretation would be possible for a space with $k = 0$ or $k = -1$, but $R(t)$ would still represent the scale of the world geometry and may accordingly be called the cosmic scale factor.

(iii) $k = -1$: In this negatively curved space the geometry of the universe is said to be *hyperbolic*, or "*open*". The surface area of a sphere of radius r is *greater* than $4\pi r^2$, and the volume of the space is *infinite* (except in some specific cases). A saddle-shaped surface represents some of the properties of such a two-dimensional space. Yet, such a shape is misleading since it has a center, whereas the real space defines no preferred position.*

The above results place no restriction on $R(t)$ as a function of time, nor do they impose any constraints on the connection between $R(t)$ and k. The cosmological implications of these spaces are described in 2.7.3.

● **III.2.7.1 Mathematical Derivations**

To proceed further, we must consider now the constraints imposed by Einstein's field equations on the Robertson-Walker metric *for a general isotropic and homogeneous universe*. Accordingly, the metric can be chosen so as to have the form

$$g_{tt} = -1; g_{it} = 0; g_{kj} = R^2(t)\dot{g}_{ij}(x) \tag{70}$$

where i and j run over the three *co-moving spatial coordinates* r, θ and ϕ, and g_{ij} is the metric for the three-dimensional maximally symmetric space;

$$\tilde{g}_{rr} = (1 - kr^2)^{-1}; \tilde{g}_{\theta\theta} = r^2; \tilde{g}_{\phi\phi} = r^2 sin^2\theta; \tilde{g}_{ij} = 0 \text{ for } i \neq j \tag{71}$$

The nonvanishing elements of the coefficient of affine connection (23) for this

---

* A two-dimensional analogue to such a space is the surface of a figure called *pseudosphere* (several of the works of the Dutch artist *M. C. Escher* employ a projection of a pseudosphere onto a plane). However, a complete pseudosphere cannot be constructed in three-dimensional space.

The surface of a pseudosphere is characterized by properties *opposite* to those of a sphere. In particular *the sum of the angles of a triangle is less than 180 degrees*, the *circumference* of a circle increases *more than in proportion to the radius* and the *area* of a circle *more than in proportion to the square of the radius*. Here, through a given point infinitely many lines can be drawn that *are* parallel to *another* line, or geodesic.

(The geometry of such spaces was first investigated in 1826 by *Nikolai I. Lobachevski* (1972–1856). His studies first met with derision but some recognition came during his lifetime. But it was mainly due to G. F. B. Riemann (1826–1866) that the non-Euclidean geometry of Lobachevski, *Johann Bolyai* and *C. F. Gauss* (and other possible non-Euclidean geometries) have become special cases in what is now known as *Riemannian geometry*).

metric become

$$\Gamma^i_j = R\dot{R}\tilde{g}_{ij} \tag{72}$$

$$\Gamma^t_{tj} = (\dot{R}/R)\delta^i{}_j \tag{73}$$

$$\Gamma^i_{jk} = \frac{1}{2}(\tilde{g}^{-1})^{il}\left[\frac{\partial \tilde{g}_{lj}}{\partial x^k} + \frac{\partial \tilde{g}_{tk}}{\partial x^j} - \frac{\partial \tilde{g}_{jk}}{\partial x^l}\right] \equiv \tilde{\Gamma}^i_{jk} \tag{74}$$

with the dot denoting the time derivative of $R(t)$.
The elements of the Ricci tensor (65) then become

$$R_{tt} = 3\ddot{R}/R \tag{75}$$

$$R_{ti} = 0 \tag{76}$$

$$R_{ij} = \tilde{R}_{ij} - (R\ddot{R} + 2\dot{R}^2)\tilde{g}_{ij} \tag{77}$$

where

$$\tilde{R}_{ij} = \frac{\partial \Gamma_{ki}{}^k}{\partial x^j} - \frac{\partial \Gamma_{ij}{}^k}{\partial x^k} + \tilde{\Gamma}^k_{li}\tilde{\Gamma}^l_{kj} - \Gamma^k_{ij}\tilde{\Gamma}^l_{kl} \tag{78}$$

is the spatial Ricci tensor. Following Weinberg,* the Ricci tensor takes the form

$$\tilde{R}_{ij} = 2k\tilde{g}_{ij} \tag{79}$$

for "maximally symmetric space." With eq. (77), (79) yields

$$R_{ij} = -(R\ddot{R} + 2\dot{R}^2 + 2k)\tilde{g}_{ij} \tag{80}$$

Let us now examine the behavior of matter in a Robertson-Walker universe by applying the Cosmological Principle to the tensors that describe the average state of cosmic matter, such as the energy-momentum tensor $T^{\mu\nu}$. We find that the contents of the isotropic universe are, on the average, at rest in the co-moving coordinates r, θ, ϕ (as intuitively expected), and that the energy-momentum tensor of the isotropic universe can then be expressed as

$$T_{\mu\nu} = (\rho + p)U_\mu U_\nu + pg_{\mu\nu} \tag{81}$$

which has the same form as that for a perfect fluid [cf. eq. (42)], and for which the velocity four-vector is

$$U^t \equiv 1 \tag{82}$$

$$U^i \equiv 0 \tag{83}$$

In eq. (81), p and ρ are functions of t alone.
The source term in Einstein's field equations, (66), is then given by

$$S_{\mu\nu} \equiv T_{\mu\nu} - \frac{1}{2}g_{\mu\nu}T^\lambda_\lambda \tag{84}$$

$$= \frac{1}{2}(\rho - p)g_{\mu\nu} + (p + \rho)U_\mu U_\nu \tag{85}$$

* Op. cit., p. 471.

which, with eq. (70), (82), (83), and (85), becomes

$$S_{tt} = \frac{1}{2}(\rho + 3p) \tag{86}$$

$$S_{it} = 0 \tag{87}$$

$$S_{ij} = \frac{1}{2}(\rho - p)R^2\tilde{g}_{ij} \tag{88}$$

Employing eqs. (75), (76), (80), and (86) to (88), the *time-time* component of the field equations yields

$$3\ddot{R} = 4\pi G(\rho + 3p)R \tag{89}$$

and the *space-space* components

$$R\ddot{R} + 2\dot{R}^2 + 2k = 4\pi G(\rho - p)R^2 \tag{90}$$

The space-time components yield $0 = 0$. Eliminating $\ddot{R}$ from (89) and (90), we arrive at the important result

$$\boxed{\dot{R}^2 = \frac{8\pi G}{3}\rho R^2 - k} \tag{91}$$

● **III.2.7.2 Deriving the Master Inequality**

Eq. (91) is employed in the text [cf. eq (12) in Lecture VI] to derive the master *inequality*.
Substituting (91) in (90), we obtain

$$\frac{2\ddot{R}}{R} + \frac{\dot{R}^2}{R^2} = -8\pi Gp - \frac{k}{R^2} \tag{92}$$

which is eq. (13) in Lecture VI.

Neglecting the effect of pressure (when the energy density of the universe is dominated by nonrelativistic matter) and multiplying (92) by $R^2\dot{R}$, we obtain

$$2R\dot{R}\ddot{R} + \dot{R}^3 = -k\dot{R} \tag{93}$$

which integrates to yield

$$R\dot{R}^2 = -kR + \text{constant} \tag{94}$$

Comparison with (91) shows that

$$\boxed{\rho \propto R^{-3}} \tag{95}$$

$$\text{constant} = (8\pi/3)\,G\rho R^3 \tag{96}$$

Eq. (95) is also obtainable from the conservation equation

$$\dot{\rho}R^3 = (d/dt)(R^3[\rho + p]) \tag{97}$$

or, equivalently, from

$$(d/dR)(\rho R^3) = -3pR^2 \tag{98}$$

when $p \ll \rho$.

If the energy density *is dominated by relativistic particles*, such as photons, then

$$p = \rho/3 \tag{99}$$

and eq. (98) gives

$$\rho \propto R^{-4} \tag{100}$$

Knowing ρ as a function of R, we can derive the Friedmann cosmologies by solving eq. (91) for different values of k. The fundamental relationships of cosmology are thus the Einstein equations (91), the equation of state, and a conservation equation [like eq. (98)].

According to eq. (95) or (100), the density ρ must decrease with increasing R at least as fast as R^{-3}. Substituting these functions in eq. (91), we realize that the term containing G vanishes for $R \to \infty$ at least as fast as R^{-1}.

● III.2.7.3 Cosmological Implications

We now distinguish three cosmologies:

(i) $k = 0$: In this case eq. (91) shows that R^2 remains positive-definite so that $R(t)$ increases, albeit more slowly than t. $\dot{R}$ tends to zero as $R(t)$ tends to infinity*, i.e. *the universe goes on expanding forever.* Eq. (91) can be integrated to yield

$$R \propto t^{2/3} \tag{101}$$

which represents the well-known *Einstein-de Sitter cosmological model* shown in Figures III.3 and IV–6.

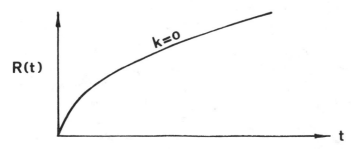

FIG. III.3 **The scale factor of the universe, R(t), for the Einstein—de Sitter cosmological model.** In general relativity this model corresponds to zero pressure and "flat" space; in Newtonian dynamics—to zero total energy (see also Fig. IV–6).

---

* cf. also eqs. (95) and (100).

(ii) $k = +1$: In this case, eq. (91) shows that $\dot{R}^2(t)$ tends to zero, as ρR^2 approaches $(3/8)\pi G$. Since $\ddot{R}$ is negative-definite, $R(t)$ will begin to decrease again, and eventually again reach $R = 0$ at some finite cosmic time in the future. This means that $R(t)$ *will expand to a maximum size*, which corresponds to $\dot{R} = 0$, and then reverse into a contraction, culminating in a singular state at $R(t) = 0$. (See Figure IV-6). This represents an *"oscillating"* universe, although there is no physical justification for regarding it as an endless cyclic process, as some authors claim.

(iii) $k = -1$: In this cosmology $\dot{R}^2(t)$ is always positive-definite and $R(t)$ increases with t for very large cosmic times (see Figure IV-6). In the next section we shall see why this cosmology is the most important one.

III.3 OBSERVATIONS, THE "AGE" OF THE UNIVERSE AND "EQUI-VALENT LOCAL CELLS"

If the known laws of physics are assumed to be correct then all available empirical data indicate that the universe is evolving (i.e., it began with an explosion).

According to the (now abandoned) *steady-state cosmology* the universe *is* expanding but its density is *time invariant* and it is *infinitely old*. To agree with the observed cosmic expansion the steady-state model postulated the continual creation of matter-energy from the void. But the radio background radiation (see Lecture VI) is satisfactorily explained *only* as a relic of an epoch when the universe was very hot and very dense, i.e., *it utterly contradicts the steady-state model!* All other available empirical data tend also to support this conclusion. Hence, the universe must be characterized by a *definite* age.

Consequently, all matter, space and time were, in a sense, "created" at the *initial moment* which was characterized by a state of infinite density. Like a *reversed black hole*, or as a *"white hole"*, the universe started its expansion and everything else has since followed its deterministic dynamics. But the early universe-white-hole was not an object isolated *in* space; it was the entire universe. It is, therefore, meaningless to ask *where* the initial explosion took place since it happened "everywhere".

What "happened" *before* the expansion started? *Spinoza* tells us that *"There was no time or duration before creation"* and *St. Augustine* warns his lay readers that God has created hell for those who ask such questions. Indeed, we simply don't know. And we see no possibility of knowing in the foreseeable future. This "field" belongs to metaphysics. In this matter one can draw a time-invariant borderline between a philosophy of the exact sciences and metaphysics. Thus, if some of my students prefer to think that our universe emerged from a black hole located in *another* universe, or inside other universes or *antiuniverses*, I consider it a matter of personal faith. But faith has not prevented scientists from publishing papers on such topics. Nevertheless I wish to take no part in such speculation.

Thus, we return to our legitimate question: *How old is our universe?* Indeed, today it *is* a legitimate scientific question. Why? First, we note that if the expansion

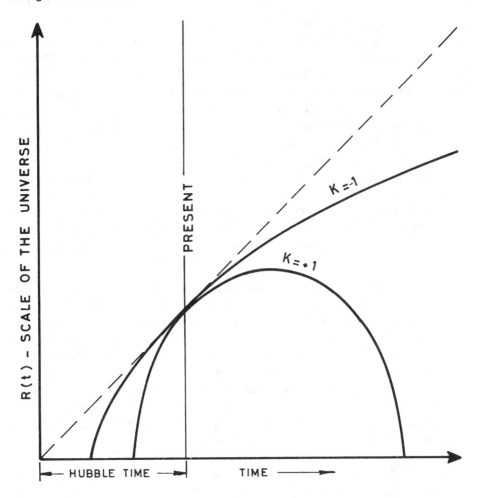

FIG. III.4 **Hubble Time and Age of the Universe**. The $k = -1$ and the $k = +1$ cosmological models shown here describe changes in the scale of the universe with the passage of time. All models must be consistent with the scale and the rate of expansion observed today, so that all their graphs must be tangent at the present moment. If the *rate* of expansion is *unchanging* (broken line), *the age of the universe is the Hubble time. Decelerating* universes are *younger,* and both their history and their destiny depend on the magnitude of the deceleration. With modest deceleration the expansion can continue *indefinitely,* albeit at an ever lower rate. (The "open", $k = -1$ model with positive curvature—see §III.2.7.3 and FIG. IV–6). Greater deceleration implies that the cosmic expansion must stop and then reverse, leading to an eventual collapse (the "closed" model, $k = +1$; see §III.2.7.3 and FIG. IV–6). *Remarkably good agreement* is obtained between (i) the age of the universe as estimated from the *inverse Hubble constant* ($H_o^{-1} = (18 \pm 2) \times 10^9$ years giving an age which varies between 10 and 23 billion years, depending on the value of 'acceleration" on which evidence is as yet inconclusive, (ii) the *ages of the oldest stars* ($\sim 10 \times 10^9$ years) as calculated by comparing the theory of stellar evolution with the properties of the observed stars (see FIGS. I.1 and I.2); (iii) *the period since nucleosynthesis of the uranium, thorium, and plutonium atoms found on earth* ($\sim 9 \times 10^9$ years), as calculated from the measured *relative abundances* of various nucleides; and (iv) the ages (4.6×10^9 years) of the *oldest lunar rocks and the oldest meteorites*. See also §III.1.2.b. [However, errors in the distance modulus may give $H_o^{-1} = (22.5 \pm 1.4) \times 10^9$ years. cf. D. H. Gudehus, *Nature, 275,* 514, 1978].

rate was *not* slowed down at all, the age would be equal to the Hubble time, i.e., about 18 ± 2 billion years (see Fig. III.4 for a graphic demonstration). But since observations show that it *has* slowed down, the age must be somewhat *less!* All three cosmologies discussed in III.2 require a *decrease* in the *rate* of expansion with cosmic time. By combining observational evidence with theory, we can next give a simplified physical explanation of the mathematical equations derived earlier.

At the moment the expansion started the material of the universe was propelled apart by very strong forces. Since then matter has moved under the influence of its own gravitational field (without further propulsion). Since the gravitational force is attractive and its range is infinite, all galaxies continue to interact as they fly apart. This is the *gravitational deceleration*.

Since only a relatively small spherical region of the universe is accessible to most of our astronomical observations, the rest of the universe surrounding it may be regarded as a *hollow shell* which exerts no net *gravitational forces* on masses in its interior*.

Hence, our sphere of astronomical observations can be analysed, in principle, *as if it were isolated and not subject to forces from the outside!* Therefore, our sphere, or *"equivalent cell"* (see §VI.3), *expands by the same rate as the universe as a whole, regardless of the sphere's location or size.*** *To analyze the dynamics of the entire universe we therefore need only to examine the dynamics of a representative sphere within it! This is the first line of reasoning behind the formulation of our Laboratory-Universe Principle of Equivalence* (§VI.3). Moreover, if the sphere chosen is a relatively small one (containing, say, only a few dozen clusters of galaxies), the velocities of the galaxies are much smaller than the speed of light, and *their motion and energy dissipation can be described in terms of Newtonian mechanics.* Our formalism, as in Lecture VI, is therefore based on this conclusion.

Accordingly, a cluster close to the edge of such an "equivalent spherical cell" is affected only by net gravitational forces generated by the matter inside that "cell". But since this matter is distributed homogeneously, the resultant force acting on the cluster attracts it to the center of the "cell"; thereby decelerating its recessional velocity with a gravitational deceleration which depends on the mass inside the selected "cell".

If the "cell" contains a great deal of mass, the clusters must eventually decelerate to a stop and then fall back to the "cell center", i.e., the "equivalent spherical cell" would then begin to collapse and, by the aforementioned principles, so does the entire universe. This is the closed universe ($k = +1$) discussed before.

---

* In "Newtonian" and "Einsteinian" gravitations the mass of a hollow *shell* attracts masses in the interior, but these forces exactly cancel each other. According to the cosmological principle [which is based on observations (see Lecture VI)], the *surrounding matter* is uniformly distributed in all directions, i.e., the universe appears *isotropic* around all observers participating in the expansion everywhere and at all times. So, any observer, including 'Man on Earth', may think that he is "at the center of the universe".

** Above about 10^8 Light-years.

But if there is less than a critical amount of matter the cluster decelerates continuously but will never stop and fall back. Both the "cell" and the universe will keep expanding for ever. These are the "open" and "critical" (ever-expanding) universes ($k = -1$ and $k = 0$, respectively).

The quantity which helps us to determine *the fate of the universe* is, therefore, the *average* ("smeared out") *density*. Estimation of this density may be done through the Hubble time, which is not precisely known.*

A lower limit to the average density is obtained from mass calculations associated only with visible galaxies.** This value could be uncertain by a factor of 3. Even so, observations show that it is *below* the critical value required for a *closed universe!*

It would clearly be of considerable interest to know the actual density and great efforts have been made recently to decide this question. Therefore, much attention has been focused on intergalactic and intercluster spaces. But this is a difficult problem since the "missing" matter may be in the form of individual faint stars, black holes, neutrinos, gravitational waves, ionized hydrogen, other gases, dust and rocks. [An argument against a dominating role played by neutrinos or gravitational waves is that if present in substantial amounts (to affect the observed density), they would have prevented the galaxies and the clusters from ever forming.]

However, the total actual density may also be estimated by extrapolating from present conditions to those prevailing a few minutes after the initial explosion—by *employing the physics of the microwave background radiation—and measuring the relative abundance of the chemical elements.* With this procedure observations provide evidence (not yet conclusive) *that the universe is open and ever-expanding.*

It is a remarkable fact that such diverse factors as the age of individual stars (or planets), the mass of galaxies, the abundance of chemical elements, the microwave background radiation, and the observed rate of expansion of the universe can all be (straightforwardly) interpreted only *in terms of a universe that is infinite in extent (open) and ever-expanding.* This conclusion derives its credibility from the consistency of all available astronomical observations and from the dominating role straightforward interpretations (of each and all pieces of evidence) always play in the history of science.

If, in the future, this conclusion is found to be wrong, a more complicated theory may have to be adopted, perhaps even a modification of the general theory of relativity. One recent modification appears to make a small step beyond Einstein. It is briefly described next.

---

* If $H_o^{-1} = 19 \times 10^9$ years the critical density is 5×10^{-30} gram per cubic centimeter, i.e., about 3 hydrogen atoms per cubic centimeter.

** This density is found by counting the number of galaxies per unit volume of space and multiplying it by the average mass of a galaxy. The mass of a galaxy may be evaluated by observing, say, two galaxies in orbit around each other, estimating their separation and their velocities with respect to each other, and, from these values, determining their combined mass. The procedure for clusters is similar. But the mass calculated in this way must include not only the mass of the galaxies but also the mass of any other matter in the cluster (like intercluster dust, gas or even black holes).

III.4 TIMEKEEPING, ACCELERATED OBSERVERS AND THE PRINCIPLE OF EQUIVALENCE

In 1975 W. H. Cannon and O. G. Jensen, reported an important discovery regarding a *latitude dependence of the proper time rate due to the earth's rotation.** In their paper they relaxed the so-called (and often unstated) *Third Hypothesis of general relativity, according to which the inertial (non-gravitational) acceleration of a clock relative to an inertial frame has no influence on the rate of the clock.* Supported by verified empirical evidence, this relaxation calls for a slight modification of the framework of general relativity. Cannon and Jensen extended the principle of equivalence to include the assumption that *inertial accelerations are capable of a direct influence on clock rates!*

First they note that the length of the world line of an observer in *free fall* in a gravitational field—or equivalently, the length of a timelike geodesic of the curved space-time manifold— is *measured directly* by the timekeeping of a comoving *ideal clock.* They next pose a question concerning observers not traveling timelike geodesics of the space-time manifold, i.e. observers not falling freely in gravitational fields and thus subject to acceleration. In this respect, they note that while the basic hypotheses of relativity and their consequences have been tested and confirmed by a host of experiments, one hypothesis has not. Accordingly, they undertook an empirical test involving almost ideal macroscopic timekeepers; modern atomic clocks. They introduced the notion of *acceleration potentials* $U(r)$ and $A(r)$, which unlike their Newtonian counterparts need not be *single-valued—r—for all observers.* To clarify the physical meaning of these potentials, they reviewed Einstein's reasoning for the effect of gravitational potentials on clock rates. Thus, according to Einstein, the relationship between the time intervals $dt^o{}_1$, measured by an ideal clock in a region of gravitational potential ϕ_1, and the time intervals $dt^o{}_2$ measured in a region ϕ_2, would be

$$\frac{dt_1^o}{dt_2^o} = 1 + \frac{\phi_1 - \phi_2}{c^2} \tag{102}$$

where $\phi \leq 0$ is given by the volume integral [cf. eq. (2a)]:

$$\phi(\mathbf{r}) = -G \int \frac{\rho(R)}{(\mathbf{R} - \mathbf{r})} dv \tag{103}$$

At this point in the development of their theory, Cannon and Jensen pose a question concerning accelerated observers. Since accelerated observers are not considered by the special relativistic postulates, they ask: Does *the same relationship* hold between the length of an *accelerated observer's world line*, $\int d\tau$, and the time-keeping of an *ideal comoving clock?* An affirmative answer is usually given on the

---

* *Science,* **188**, 317 (1975).

strength of the "third hypothesis". This hypothesis, although amenable to empirical confirmation, was never before tested with macroscopic timekeeping. Of the infinity of possible choices, the authors proposed *an extension of the principle of equivalence**, whereby it is assumed that inertial *accelerations are capable of direct influence on clock rates!* Thus, they maintain, nothing compels us to assume that clocks *with identical velocities, but different accelerations*, would be going *at the same rate!* This "compulsive" assumption, built into the theory of relativity, is equivalent to an assertion that *two identical clocks* side by side and *at relative rest* — one in the gravitational field and one outside it—would still run in *synchronism*, while in fact *such synchronism is precluded by the physical meaning of $g_{\mu\nu}$*. On the other hand, while physical realization of this condition is difficult to conceive, the authors maintain and demonstrate that it can be duplicated instantaneously by *inertial accelerations*.

On this basis, they constructed a new *ad hoc* theory, an extension beyond the current general relativity theory, which—at least for the case of *uniform rotation*—incorporates the effects of inertial acclerations on clock rates.

It seems apt to close this section with a quotation from *Albert Einstein***.

> "The theory of relativity is often criticized for giving, without justification, a central theoretical role to the propagation of light, in that it founds the concept of time upon the law of propagation of light. The situation, however, is somewhat as follows. In order to give physical significance to the concept of time processes of some kind are required which enable relations to be established between different places. It is immaterial what kind of processes one chooses for such a definition of time. It is advantageous, however, for a theory, to choose only those processes concerning which we know something certain."

III.5 FROM GENERAL RELATIVITY TO UNIFIED FIELD THEORIES: SOLITONS, INSTANTONS AND GAUGE FIELDS

The analogy between the "creation" and "annihilation" of *solition-antisoliton pairs*, and the corresponding phenomena among quantum-mechanical *particle-antiparticle pairs*, has already been discussed in §II.12. But these soliton-antisoliton pairs are structures that are *localized and confined permanently to a definite region of space.*

The progress of greatest interest to the advance of unified field theories (Introduction) is a recent discovery concerning another kind of soliton, one that is confined to a small region *both in space and in time*. Called *INSTANTON*, this phenomenon is interpreted not as an object, but as an *event*, not as a particle but as a quantum-mechanical *transition* between various states of other particles- a transition equivalent to the quantum-mechanical phenomenon of *tunneling* (which was impossible in classical physics for energy must be conserved at all moments, not merely in the final accounting following the tunnel effect).

* See III.1.1.
** Albert Einstein, *The Meaning of Relativity,* Princeton University Press, 1956, pp. 28

Among the various inconsistencies and difficulties associated with the classical interpretations of quantum theory (Lecture V), one can add the kind of "deficit conservation for short intervals" (i.e., a seeming *violation* of the *conservation law* is required, provided this violation does not last "too long"). In tunneling, according to the classical interpretation, a particle can cross a potential barrier even though it has "*too little energy*" to surmount the barrier.

This difficulty, or rather violation of conservation (cf. §IV.11), is removed by the elegant introduction of instantons as geometrical structures in a field theory which includes the description of quantum-mechanical processes, as well as gravitation and electromagnetism. Considered as solitons in a field theory of four dimensions, instantons are now interpreted as *the evolution of fields in three spatial dimensions and one time dimension*. In fact they appear in a large class of field theories, including the theories most commonly applied to the four fundamental interactions in nature (see below).

Now, we have already noted in the Introduction (Assertion 2), that the *special theory of relativity* exhibits *global symmetry* (for it is limited to observers that are moving with constant relative velocity with respect to each other). Thus global symmetry characterizes 'flat' space-time where the transformation is *the same* for all points in space-time. In contrast, the *general theory of relativity* performs a transformation from *global* to *local* symmetry whenever acceleration or gravitation is present, i.e., *the very transition from global to local symmetry describes the origin of the gravitational field*. Now, field theories with *local* symmetry are called *gauge theories*, i.e., the very transition from global to local symmetry is associated with the introduction of a *gauge field* (see also §III.5.2 below).

We can put this another way: *The introduction of a gauge field has an effect equivalent to curvature of space-time*. Thus, transformation of a frame of reference according to the prescription of a gauge field is analogous to curvature changes that alter the description of solitons and instantons.

It should be noted that there are special configurations of a gauge field where a frame transported around a close loop does return to its *initial* orientation and the rotation of the frame during the displacement is independent of the path taken ("*pure*" *gauge structures*). These structures carry no energy and are in their *vacuum states**. On the other hand, *when the final orientation of the frame depends on the path, the gauge field carries energy* (which increases as the dependence on the path becomes more pronounced).

Another novel result of these revolutionary concepts is the new insight they have given us about the very properties of space and time in the "vacuum state". Indeed, it raises the question '*What is the geometrical structure of the vacuum state?*' As we shall see in Lecture IX, the answer to this question is by no means simple.

---

* cf. IX.2.20.

III.5.1 Spin-2 Gravition and the Introduction of Gauge Fields

At the simplest physical level one may interpret a gravitational plane wave, with wave vectors and helicity ± 2, as consisting of gravitons: quanta with energy-momentum vectors, spin component in the direction of motion, and zero mass (like the photon and the neutrino). Absorption and emission of gravitons (with definite frequencies), is also assumed in quantum interpretations of gravitational radiation. Moreover, gravitons, like photons, may have two independent polarization states. Using proper quantum fields (see below), one may then *create and destroy gravitons* by corresponding *creation and annihilation operators*. One problem encountered here is the fact that such operators cannot be a Lorentz tensors as long as their helicity is limited to the required physical values ± 2.

In general, a true tensor would have helicities 0, ± 1, and ± 2. Hence, to eliminate the unphysical helicities 0 and ± 1, one may introduce a gauge transformation. But this raises a number of difficulties. One of them is related to the fact that *it is impossible to construct a Lorentz invariant quantum theory of particles of mass zero and helicity ± 2 without building some sort of gauge invariance into the theory* (cf. S. Weinberg's works 55 to 61 at the end of the Introduction). *However, the theory of gravitational radiation is gauge-invariant because general relativity is generally covariant*, and general covariance is but the mathematical expression of the *principle of equivalence* (on which the whole of classical general relativity is based).

A few fundamental questions arise at this point.

Since the aforementioned invariance *is* the symmetry on which the general theory of relativity is based, and since supersymmetry transformation moves a particle from one point to another in space-time (see below), a new gauge theory may be constructed in which a gauge field is introduced for each of the symmetries present in the equation (see below). Now, the new gauge fields give rise to new forces (cf. the Introduction). Hence the first question is; *what the nature of the new gauge field should be?* The next three may be stated as follows: If one field is that of spin-2 graviton, what should the other be in order to account for a unified field theory for cosmology and microphysics? How would the quanta acquire mass? What experimental evidence supports or disproves these predictions?

Many names are associated with such (and somewhat different) efforts*. Since I do not intend to review these efforts here, the following will be confined to a few general remarks.

---

* To mention some of them by name: D. Z. Freedman, P. van Nieuwenhuizen, S. Ferrara, S. Deser, Y. A. Golfand, E. P. Likhtman, D. V. Volkov, V. P. Akulov, B. Zumino, J. Wess, J. H. Schwarz, A. Neveu, P. M. Ramond, and, of course, S. Weinberg and A. Salam for uniting the weak and the electromagnetic forces. Other important contributions were made by J. Ward, G. Hooft, M. J. G. Veltman, B. W. Lee, J. Zinn-Justin, P. Higgs, J. Goldstone, J. Strathdee, R. L. Arnowitt, P. Nath, M. Kaku, P. Townsend, M. Gell-Mann, M. T. Grisaru, J. A. M. Vermaseren, J-P. Vigier, D. D. Ivanenko, J. A. Wheeler, and Y. Ne'eman as well as by Yukawa, Finkelstein, Ambarzumian, Schild, Snyder, Koish, Fock, Treder, Möller, Rodicev, Schwinger, Sakharov, Zeldovitz, Brill, Piiz, Vladimirov, Utiyama, Sakurai, Kibble, Brodski, Sokolic, Frolov, Dürr, Mitter, Jamazaki, Naumov, DeWitt, Das, Fradkin and Vasiliev.

C. N. Yang and R. Mills discovered in 1954 that the transition to a local symmetry required the introduction of three gauge fields, each field associated with a massless object having a spin of 1. Later it was shown that the three massless objects can acquire mass through a mechanism called "*spontaneous symmetry breaking*" (which gives mass to some of the gauge objects whereas others—such as the photon and the neutrino—would remain massless). *This property was then used to unify the weak and the electromagnetic forces!* In these unified theories the carriers of the weak force are spin-1 particles with mass called "*intermediate bosons*", while the carriers of the electromagnetic force are the photons. Most important, this unified theory has been recently confirmed by experiments with astonishingly quantitative accuracy (§IX.2.21).

The next problem was to unite particles *with different spins*. Apparently it was solved by the concept supersymmetry (Introduction), which relates particles such as fermions (spin J) and bosons (spin J + 1/2 or J − 1/2). Moreover, it was found that *repeated supersymmetry transformation*, such as from fermion to boson and back, moves the particles in space—a motion that suggests an intimate link with the very properties of space-time (and, therefore, with gravitation).

III.5.2 Supergravity and Spin-3/2 Gravitino

Poincaré invariance is the space-time symmetry underlying *the special theory of relativity*. It means that all laws of physics take the same form in any two coordinate systems, even if they are shifted and rotated and moving with respect to one another (with a constant relative velocity). It is, however, *a global symmetry*, for the special relativistic transformation *is the same* for all points. A *local* Poincaré invariance is obtained when the laws retain *the same form* during independent transformations of *each* point, *as is the case with the general theory of relativity* (which, as shown in this lecture, takes into account acceleration and gravitation). Hence, the very transition *from global to local invariance* (in this case) is equivalent to the introduction of the gravitational field *in the form of a gauge field*. Gravitation is, therefore, the gauge field of the local Poincaré invariance.

Returning now to the concept *supersymmetry*, we note that it is a *global* operation when the angle of rotation is the same at all points in space-time, and *local* when the rotation is different at each point*.

Now, what do we mean by local supersymmetry?

We have already noted that repeated supersymmetry transformation, such as from fermion to boson and back to fermion, moves a particle from one point in

---

* The equations that describe the forces in the atomic nucleus are invariant under arbitrary rotations of an arrow in some imaginary space. If the arrow points "up", say, the nucleon is a proton; if it points "down", it is a neutron; but, most important, the arrow must rotate through the same angle at all points if the nuclear forces are to be invariant ("isotopic symmetry"). Similarly, in supersymmetry, when the arrow points "up", it is a fermion; when it points "down"- a boson. But since the rotation is the same at all points in space-time it is still a global symmetry, just as isotopic symmetry is. The question is therefore: What happens when global supersymmetry changes to local supersymmetry?

space to another. Now, in *local* supersymmetry this displacement of particles is *not* the same at each point in space, i.e., by recalling that *local* Poincaré invariance is obtained when the laws retain the same form during independent transformations of each point (*as is the case with the general theory of relativity*), we may find a deep-rooted connection with the field of gravitation and, in particular, *with the exchange of gravitons*. It is for this reason that local supersymmetry has been given a special name: *SUPERGRAVITY.*

Supergravity theories (there are, at present, more than one) make use of the transition from global to local symmetries by introducing new gauge fields. Now, the most remarkable feature of supergravity theories is that the local supersymmetry on which they are based allows the introduction of only two new gauge fields. These are:

1) *The gauge field of the spin-2 graviton discussed in §III.5.1*
2) *A new gauge field of spin 3/2.*

The spin-3/2 field was chosen because supersymmetry relates only particles with adjacent spin, i.e., $2 \pm 1/2$. Simplicity and other considerations then dictate the choice of the spin-3/2 gauge field.

Now, what is the spin-3/2 gauge field?

To begin with, as a new gauge field, it should introduce a new force and a new particle to be exchanged together with the graviton. A new name was given to this particle: *GRAVITINO.* It is a massless object which is coupled to the other particles only through the feeble force of *microscopic gravitation*. But this requires a word of caution.

III.5.3 Unified Field Theories And The Skeptic

At this point a skeptic may justly ask: Is not gravitation a relevant physical phenomenon only at *large* macroscopic scales? Are not the physicists dealing with unified gauge field theories bitten by the bug of invariance principles, of symmetries and groups? Is the fact that neither the graviton nor the gravitino *have never been observed experimentally*, a good reason to conclude that gravitation and microphysics *cannot* be unified; that Einstein was absolutely wrong in searching for such an illusion? that this search does not belong to physics proper but to metaphysics?

In trying to answer the skeptic's questions we first note that the general theory of relativity is a *scale-free* theory; that there is no inherent scale in it; that whatever applies to bodies of a certain size will apply to bodies much bigger or much smaller. Moreover, the "*weakest*" force of gravitation (in the world of microphysics) becomes the "*strongest*" force when we move our considerations of "*which-is-stronger vs. which-can-be-neglected*" from terrestrial physics to astrophysics of massive stars (Lecture I). Hence, gravitation is highly relevant in both macro and microphysics, and any attempt to prevent the introduction of general relativistic considerations into the world of the so-called elementary particles, if based on the aforementioned claim, is simply false.

On the other hand, the present models employed in quantum theories are somewhere, or somehow, associated with a *characteristic scale* (§IV.13). Moreover, the old quantum theory has raised many questions and inconsistencies (Lecture V), which show that as a theory of the structure of matter it has turned out to be *incomplete*. In spite of this reputation the many users of the old quantum theory believe, even today, that it was Bohr who won that famous battle with Einstein. Even some great admirers of Einstein accept this as a historical fact and, *consequently*, draw some far-reaching assumptions as to the fate of (deterministic) gauge-field theories of the structure of matter (§V.2–§V.6).

Nevertheless, the new supergravity theories of matter encounter some difficult problems. The general theory of relativity can also be deduced from the assumption that the gravitational force *results* from the exchange of massless spin-2 particles and that these particles are described by a quantum field theory. All particles are then considered to be pulled by the gravitons in such a way that they follow the same curved trajectories as those of Einstein's field equations. Hence, *supergravity theories provide corrections to the standard general theory of relativity only at the microscopic scale of quantum physics. For long-range interactions the predictions of general relativity remain as in Einstein's theory.*

What, then, is the nature of the predictions at the quantum level?

The most striking property of the new supergravity theories is their high degree of symmetry. At the quantum level each particle is related to particles with *adjacent* values of spin by the local supersymmetry transformations (supergravity). Thus, a graviton can be transformed into a gravitino and a gravitino into a spin-1 particle. Then, *within each family of particles with the same spin*, relations are formed by a *global* symmetry similar to the isotopic (global) symmetry that relates protons and neutrons. Here the particle may be viewed as equipped with an arrow in an auxiliary space of many dimensions. As the arrow rotates (subject to some topological constrains to be discussed in §IX.2.20), the particle transforms into a graviton, a gravitino, a photon, a quark and so on. Thus, the quanta of all the four forces of nature are derivable from a single unified theory—a unification of physics that has never before been achieved. Yet the unification is not complete nor free from some internal problems (infinities, multiple competing theories, symmetry breaking, etc). While the elimination of infinities in supergravity theories is gradually being accomplished, another problem which, in turn, may be highly significant, remains unsolved. It is connected with the introduction of *the cosmological term* that Einstein enforced in 1917 on his 1916 equations (Introduction). In some extended supergravity theories *it arises in the transition from global to local symmetry.* What is worse, the value of the cosmological term predicted by some of the theories exceeds the upper limit derived from observational evidence. However, before too much doubt lingers in, it should be noted that "symmetry breaking" (which is used to aquire mass to the massless particles of supergravity theories), changes the value predicted by the theories for the cosmological constant. Hence, aside from the fact that experimental evidence for the existence of gravitons, gravitinos, and other particles predicted by the theories, is still not available, and aside from the fact that the present super-gravity theories do not seem to predict all the known particles with spin 1/2 and 1, we face the most intriguing problem of *"symmetry breaking" vs. the cosmological constant.*

In this connection it is of interest to quote the Russian physicist *Dimitri D. Ivanenko*, who already in 1966 speculated that the *expanding space* induces some specific *asymmetry* on the metric of the vacuum (cf. §IX.2.20); that it offers a "direct" cosmological geometrical influence on the *microscopic arrow of time* (cf. §IV.11 on the asymmetry in the decay of kaonic "elementary particles"); that we should keep our persistent efforts to connect local quantum with global cosmological properties *and never treat them separately.* Moreover, according to Ivanenko, the conventional sharp *distinction* between *matter* (ordinary, not gravitational) and *geometry* (space-time) arises from the relative smallness of the transmutations under ordinary conditions of *low energies or temperatures, or small values of curvature**. We shall return to this subject in Lecture VI.

Today we know that of the four interactions, only gravity has resisted incorporation into the unified, renormalizable quantum field theories (§V.6.2.3). In winning the Nobel Prize in physics, *Steven Weinberg* has stated [*Science, 210,* 1212 (1980)] that this may just mean that we have not been clever enough in our mathematical treatment of general relativity. Being skeptic and optimistic alike, he thinks that future developments of the ideas of symmetry, broken symmetries, renormalizability and unification are to be dictated mainly by newer quantum field theories (§V.6, §IX.2.20). He warns, however, against the tendency to draw universal conclusions from the studies of matter only at relatively low temperatures, where symmetries are likely to be spontaneously broken, and where nature does not appear very simple or unified. We shall return to this subject in Lecture V, §6.

---

* D. D. Ivanenko, "*Problems of unifying cosmology with microphysics*", in *Physics, Logic, and History,* W. Yourgrau and A. D. Breck, eds., Plenum Press, New York, 1970. It may be, perhaps, of some relevance to note here that I first read Ivanenko's paper in Sept. 1979. The aforementioned ideas that I found there were very encouraging, for some of them are almost similar to the ones I have worked with since 1965. The 7-page paper of Ivanenko does not, however attempt to give any proof or supporting evidence to his assertions. Yet, they rely heavily on his contributions to the advance of unified field theories—a domain in which I am a new student.

Part II

FROM PHYSICS TO PHILOSOPHICAL CROSSROADS AND BACK

Revolutions are ambiguous things. Their success is generally proportionate to their power of adaptation and to the reabsorption within them of what they rebelled against. A thousand reforms have left the world as corrupt as ever, for each successful reform has founded a new institution, and this institution has bred its new and congenial abuses.

George Santayana

What serious-minded men not engaged
in the professional business of
philosophy most want to know is
what modifications and abandonments
of intellectual inheritance are
required by the newer industrial,
political, and scientific movements . . .

John Dewey

Do you then be reasonable and do not mind
whether the teachers of philosophy are
good or bad, but think only of philosophy
herself. Try to examine her well and
truly; and if she be evil, seek to turn
away all men from her; but if she be
what I believe she is, then follow her
and serve her, and be of good cheer.

Socrates to Crito

THE ARROWS OF TIME

It is inescapable that asymmetric forces must be operative during the synthesis of the first asymmetric natural product. What might these forces be?

I, for my part, think that they are cosmological. The universe is asymmetric and I am persuaded that life, as it is known to us, is a direct result of the asymmetry of the universe or of its indirect consequences. The universe is asymmetric.

Louis Pasteur

The unity of all science consists alone in its method, not in its material.

Karl Pearson

Time is but one thought
taken to be many
Same time for many thoughts
is space
Time and space coming
together is causality
From them arises the World
around us.

An Indian Proverb

Fig. IV.1 **Mind, universe, clock: Which of them generate time?**

Quid est tempus?
Si nemo a me quaerat, scio,
Si quarenti explicare velim, nescio!
St. Augustine

IV.1 TIME AND THE ARROW OF TIME: THE MOST DISTORTED OF ALL IDEAS?

Space and time are the most fundamental ideas of the scientist's picture of the world, and also of the language in which we all describe our thoughts and sensations. We use the concepts of space and time to "relate" ourselves to world phenomena that we and other people observe and to "verify" that what other people report to us as "their observations" *are* "the same phenomena." The agreement obtained by this "procedure" portrays certain *consistencies* and *regularities* which, in turn, enable us to generate *reliable* theories, to plan our lives and even to make successful *"predictions"* about an increasing number of "psychological", "social", "biological" and "physical" phenomena. The success of this "process" may encourage some of us to believe in the self-consistency of our theories and in the existence of one and the same reality.

This "process" reveals another consistency which distinguishes space from time. On a personal basis, one may state it as follows: "I can travel to any direction in space and even return to my starting point. Not so with time." Something fundamental about time, and, to many, something quite mysterious, makes us refer to time with such words as "past", "present", "future", and "irreversible", which have no counterparts in the description of all our experiences with the spatial coordinates.

Consequently, we are led to regard time as "directional", or as oriented with an "arrow" which always points from the "past" to the "future". In physics, and in philosophy, we refer to this fact by such terminology as *"time asymmetry"*, or *"time anisotropy"*. Contrary to the time coordinate, space itself is considered "direction-invariant", "symmetric" or "isotropic".

At this point, two old-new questions come to mind:

(1) Can the origins of the aforementioned "properties" of space and time be attributed to the "ordering properties" of our mind, which, cannot, in principle, be explained by physics as, for instance, in Bergson's doctrine (see Appendix I)?

(2) Can the origins of the aforementioned "properties" be deduced from the

"ordering properties" of our brain *and* from the "ordering laws" of an external physical world?

If we adopt the first possibility we must systematically translate all of physics into a kind of subjective psychology. If we adopt the latter, we are forced to generate a consistant physical theory of "matter", brain and "mind", including a unified theory of the origin of knowledge and life and the origin of all the "building elements" of life, i.e., including the cosmological-geophysical origins of all heavy chemical elements that constitute the "building bricks" of amino acids, DNA, RNA, genes, proteins, living cells, neurons, central nervous systems, brain and "mind".

Before one proceeds to seek answers to such questions, he must develop a consistent terminology associated with certain "biological" and "physical" phenomena on which there is no doubt.

This is precisely the primary aim of this lecture and, indeed, I must use this route before presenting our views in Part III.

It also means that we must try to clear up some historical confusions associated with the most fundamental ideas in science and philosophy. One such confusion, perhaps the most harmful to human understanding of all ideas, *is the conflation of twelve quite distinct ideas huddled under the umbrella of the so-called "arrow of time"* : time asymmetry, time-reserval invariance, irreversibility, causality, retarded causality, determinism, biological clocks, thermodynamic time, cosmic time, general covariance, gravitation and initial and boundary conditions. Let us try now to clear up this confusion.

IV.2 ASYMMETRY-SYMMETRY-SPACE-TIME AND THE UNIFICATION OF THE LAWS OF PHYSICS.

In this lecture we start an analysis of the foundations of the physical sciences in connection with our general assertions (§3 in the Introduction). For that purpose we first employ our second assertion, namely, *that we shall never have any experience or science which we cannot interpret in terms of asymmetry-symmetry-space-time, or their equivalents.*

Secondly, we employ the preliminary conclusions obtained in §2.3.2 and in Table II of the Introduction, namely, *that any unification of the laws of physics should begin with unification of asymmetries* (GROUP II in Table II), *and not with the unification of the mathematics of the differential equations which describe the 'spread' of the four basic interactions in space-time (or its equivalents). I.e., that we should first unify the (apparently) different asymmetric initial-boundary conditions, and, only secondly, proceed with unification of fields through mathematical solutions comparable with observations* (GROUP III in Table II). This means mathematical combinations of *symmetric* GROUP I with *asymmeric* GROUP II in order to obtain *'the solutions'* of GROUP III.

IV.3 METHODOLOGY, AIM AND SCOPE

Since unification of asymmetric initial-boundary conditions is required first, we must, first and foremost, identify the problems which have, so far, *prevented* such a unification. This requirement, as well as our general methodology, dictate the order of priorities in the following sense:

☐ *PREREQUISITE 1*: Identification of a single (but unified) master asymmetry.

☐ *PREREQUISITE 2*: Transformation of all 'independent' asymmetries into a single master asymmetry.

☐ *PREREQUISITE 3*: Resolving the apparent contradiction between *symmetric* 'laws of physics' and *asymmetric* initial-boundary conditions.

Prerequisite 1 will immediately lead us to other, more complicated problems involving the concepts of time-asymmetry, time reversal, reversibility-irreversibility, symmetry breaking vs. 'measurement', 'observation', causality, determinism-indeterminism, 'prediction', probabilities, etc.

The common ground of all these problems is the symmetry-asymmetry of the concept of time. In fact, no fundamental field in science and philosophy of science can avoid running into the problem of time-asymmetry vs. time-symmetry. Yet, as the reader will see, perhaps with surprise, after centuries of trying, occupying the attention of the greatest scholars, it has become clear that the very concept of time is now back in the melting pot.

It is, however, only in PARTS III and IV that the reader will see these topics fused into a single consistent theory. Care should, therefore, be taken to avoid premature conclusions. Hence, I must stress the *preliminary nature* of the concepts to be analyzed in this lecture. Some of these are listed below:

☐ Symmetry and asymmetry (see also Appendix I and Lecture IX).
☐ Anthropomorphic and physical times.
☐ The broken symmetry of time in elementary particles.
☐ The entropic arrow of time.
☐ The (possible) unity of all arrows of time.
☐ Causality and determinism in the theory of relativity.
☐ Cosmological arrows of time and cosmic time.
☐ Time-reversal invariance and reversibility.
☐ Fading memory in classical physics.
☐ Irreversibility of nature and the non-universal meaning of entropy.
☐ Causal violations, 'superspace' and a possible 'failure of time'.
☐ Electromagnetic and microscopic time asymmetries.
☐ Philosophical notions of causality, causation and determinism, including some thought-provoking quotations (Appendices I and V).

Indeed, it may be quite perplexing to realize that the very explanation of such basic aspects of science and everyday experience should remain unclear and paradoxical after consideration by eminent scholars like *Boltzmann, Einstein, Schrödinger, Eddington* and *Milne*. Over the centuries, philosophers such as *Heraclitus, Parmenides, Plato, Aristotle, Maimonides, Descartes, Spinoza, Kant, Bergson,* and *Reichenbach* have attempted to solve the riddles associated with these confusing concepts (q.v. Appendix I, Lecture IX, and Appendix II).

Recent advances in relativistic cosmology, astrophysical thermodynamics, elementary-particle physics and terrestrial timekeeping have added new paradoxical aspects to these concepts. These, in turn, have aroused new interest among contemporary scientists and philosophers.*

Thus, we are now well justified in expecting the emergence of entirely new concepts at the fundamental levels of science and philosophy. But before we plunge into detailed formulations I prefer to introduce a list of the most pressing questions that are to be discussed. This is done to avoid the danger of losing the *over-all perspective* associated with the complex problem at hand. It is also a first attempt in characterizing the actual problems involved. Are these new problems, or old ones in new dress? And, most important, what is their origin? My "answer" to these two questions is presented below in the form of sixteen associated questions, namely:

□ What is the *origin* of time? Is time linked to gravitation and to a cosmic time or is it an (independent) "pre-existing" coordinate?

□ Are relativistic cosmology and astrophysics the proper frameworks for analyzing all time asymmetries that are in evidence? Is the (observed) expanding universe acting as an enduring clock throughout its history?

□ Are the *origins* of thermodynamic and electromagnetic irreversibilities known today? Can they be *linked* with cosmology and gravitation, and *deduced* from general relativity?

□ What are the modern scientific (and associated philosophic) views about causality, causation, determinism, indeterminism and chance?

□ Can irreversibility (or the "established" statistical "law" of increase of entropy, "H-theorem", "mixing", "Markov processes", etc.) be deduced from the mathematics of statistical mechanics, or was it fitted into the theory as a fact-like postulate without waiting for the proof to be completed? Indeed, how is irreversibility *smuggled* into the various analyses of quantum mechanics? And if *no one* can find the proof of irreversibility and time asymmetries in quantum theory, what about its claim to fundamentality?

□ Is quantum statistical mechanics a reliable tool for describing macrocosmic processes among stars, nebulae, and cold space, which, in *contradiction* to statistical-mechanical predictions, do not exhibit a finite (relaxation) time for reaching *thermal equilibrium?*

---

* To mention some of them by name: O. Costa de Beauregard, P. T. Bergmann, H. Bondi, M. Bunge, W. H. Cannon, R. P. Feynmann, M. Gell-Mann, T. Gold, A. Grünbaum, F. Hoyle, O. G. Jensen, T. D. Lee, H. Margenau, C. W. Misner, J. V. Narlikar, Y. Neaman, O. Penrose, K. Popper, S. Weinberg and J. A. Wheeler. Detailed references to some of the works of these authors will be given in Part III.

☐ How do gravitation and irreversibility *"enter"* into our laboratory?

☐ Why do most natural processes operate only in one time direction?

☐ Why should there be any time asymmetry at all?

☐ How is it possible to account for time asymmetry when an examination of the laws of physics (except one) reveals only symmetry?

☐ Why are all time asymmetries in evidence mutually consistent in pointing in the so-called positive directions? Is it a mere coincidence, a matter of definition, or does, perhaps, one of these time asymmetries cause the rest to point in the same direction? If so, which of these is the "master arrow?"

☐ Is there any correlation between the large and small scale motions of the universe? What are the local and astronomical observations which support or disprove such possibilities? Are there any verifiable physical links between the observed mutual recession of the systems of galaxies (the expanding universe), gravitation, the observed irreversible outflow of radiation from all galactic bodies (excluding black holes), and the local, entropic, information, electromagnetic, microscopic and (alleged) statistical arrows of time?

☐ Are the current concepts of entropy and temperature well defined and universally valid? If they are non-universal, or actually misleading, what are the possible alternatives, particularly in strong, non-equilibrium, gravitational systems?

☐ Is entropy an essential element in physics?

☐ Are not thermodynamics, and some aspects of quantum statistical mechanics, redundant fields which in the long run will be fully integrated in a more universal theory?

☐ Is the origin of biological clocks internal or external? Is there a Master Living clock? Do we posses an "innate" ordering of time and space?

Running parallel to all these questions have been the subtle arguments about the status and origin of boundary conditions in cosmology, thermodynamics, statistical mechanics and electrodynamics, and, as we shall see, in perception.

A common ground of all these problems is the concept of time. Thus, by first viewing this concept *within the framework of the physics of time*, one can *remove* many of the *prevailing misconceptions*. Clearly, it would be impossible to cover in detail every historical point of view in a single course of lectures. Instead, emphasis here is on presentation of a choerent discipline supplemented by a critical review of the physical topics involved.

IV.4 CONFUSING CONCEPTS OF TIME AND TIME ASYMMETRIES

It is only in the *technical* procedures of physics that time appears as a primitive concept. The technical-mathematical operations of physics do not explicitly define and

Fig. IV-2 **Simple Spatial Symmetry (see, however, Assertion 2, Introduction, and below).**

explain time but rather specify *operational procedures* for its *measurement* by various observers in units of seconds. To measure time means to define a *"clock"* and its corresponding *"time scale"*.

A clock is judged by *resolution and uniformity*. Resolution pertains to its ability to resolve closely spaced events on its time scale. Today there are stable *atomic oscillators* whose frequencies derive from electromagnetic quanta radiated during "transitions" between excited states of various atoms or molecules. These oscillators can provide temporal resolution of events 10^{-16} seconds apart!

Uniformity of a time scale in physics is a much subtler concept. A time scale is said to be uniform if the *observed* dynamical phenomena of the universe, such as cosmic time or atomic oscillators located at *different* points in *space* with *different inertia* * or *gravitational accelerations*, when measured against that time—*are in accordance with those predicted by a consistent theory.*

Clocks may be "real" *material systems* of which the wristwatch is an example; its spatial "transition" is calibrated in seconds. Or they may be *abstract concepts* by which time is *inferred* from a set of physical *observations* through the intermediary *of a "verifiable theory".* The latter are also known as *"paper clocks",* of which the orbital motion of the planets or the observed recession of the systems of galaxies (the expanding universe), as inferred from Newtonian and /or relativistic theories, are examples.

So far, "to define" only a primitive concept of terrestrial timekeeping, we have had to invoke such perplexing concepts as:

"Measurement", "transition", frequencies of atomic oscillators, electromagnetic quanta, excited states in particle physics, "cosmic time", inertial and gravitational accelerations, orbital motions of the planets, observed recession of the system of galaxies (the expanding universe), Newtonian and/or relativistic theories.

Even a single glance reveals that the whole of present-day physics is required to explain and define time. But, conversely, *time is also a proper physical guide in critical examination of phenomena ranging from the smallest scale of atomic physics to the largest distances of the universe accessible to our observations. It is through time that physics and cosmology become one and the same subject for fundamental studies.*

But the concept of time invokes additional problems associated with *anthropomorphic* notions in science. One of them is to be illustrated next with regard to the *"arrow of time".*

The term *"time's arrow"* was suggested by A.S. Eddington to express the one-way property of time which has no analog in space. Historically, physicists have had a picture of time as an "everflowing stream". Newtonian time was universally uniform; clocks were supposed to agree on the simultaneous rate of time "flow" irrespective of their location or motion. According to Newton, *"Absolute, true and mathematical time, of itself, and from its own nature, flows equally without relation to anything external".*

Einstein brought about the collapse of Newtonian time and a universal "now". Time became part of the four-dimensional space-time. In relativity the "same moment" in two different places is *without physical meaning* (unless we speak of a specific cosmic time, to be defined later).

Newton and pre-relativity physicists made the egocentric mistake of identifying physical time with common human experience. Yet, even today one finds it common practice among philosophers to refer to the passive arrow of time in physics with the

---

* Inertial acceleration is discussed in Lecture III.

implicit meaning of "flow" associated with our own *active* psychological time.

In the absence of an acceptable theory for the mind in science, any analysis within the context of physical space-time must exclude discussions of the muddled concepts of "becoming", "flowing time", "now", or psychological "arrows of time". Present-day physics make no provision whatever for a "flowing time", nor for any "flowing" space coordinates! In relativity the human *"now"* has been replaced with the more modest *"here and now"*, which is not associated with the flow of time but with an instant of time at a single point in space. Moreover, depending on the nature and intensity of the *gravitational field*, the space- and time coordinates vary and may even exchange roles. We shall return to the last specific problem later in our analysis of time in a gravitational collapse or in black holes. At this juncture I wish only to stress two points:

1. The nature of *time* itself is subject to more ambiguity than that encountered in the analysis of *time's arrow*. Hence detailed analysis of the former must be postponed to a later stage.

2. Precise definition of *time's arrow* that would satisfy all scientists and philosophers is not a simple task. We are faced here with many closely-associated terms, such as *time-reversal invariance, irreversibility* of natural processes, *positive* and *negative* time coordinates, direction *of time*, time *asymmetry*, time *anisotropy* and *times's arrow*.

But before we rush to premature conclusions, let me critically examine what two theories of principle,—relativity and thermodynamics—tell us about these concepts. Later, in my discussion on time-reversal invariance, we can arrive at the first important conclusion.

"... . the second law of thermodynamics—holds, I think the supreme position among the laws of Nature".

                                                                                                   Arthur Eddington

IV.5 THE ENTROPIC ARROW OF TIME*

Thermodynamic considerations lie at the foundation of any scientific analysis of natural processes. Since all actual processes in nature evolve along a time coordinate, the thermodynamic conclusions pervade the whole of our experience with time. In fact, much evidence exists today that our own physico-chemical processes as biological organisms are subject to the principles of thermodynamics.**

At the very core of thermodynamics we find the *"second principle"*. It is also called the *"second law"* or *"second postulate"*. The universal significance of this

---

  * The *electrodynamic* arrow of time is explained in §VI.10 where possible *"interactions"* with the *"thermodynamic* arrow" are explained.

  **   A simplified mathematical analysis of the thermodynamics of living organisms is given in my *MDT* (Ref. 4, Intr.).

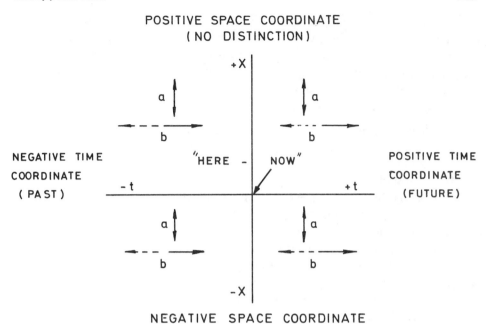

POSITIVE SPACE COORDINATE
(NO DISTINCTION)

Fig. IV-3:

(a) **A simple space reversal;** e.g., transformation from $+x$ to $-x$. This reversal forms a "left-right symmetry". Space reversal has been so deeply ingrained in thought and theory of physics as to be considered a fundamental principle of Nature, both on the macroscopic and microscopic scales of observations. (See, however, discussions on CP and T violations in elementary-particle physics: §IV.11 and §V.6).

(b) Unlike space reversal, macroscopic processes *do not demonstrate time reversal*, i.e., time possesses an asymmetric quality. But why should there be any time asymmetry at all? And, what is the basic nature of the somewhat mysterious time coordinate system in which the basic physical laws are rooted? For details see §VI.3, Lecture VIII, and Figs. IV.4, VI.1 and VI.2.

postulate stems from the fact that it supplies a necessary and far-reaching criterion for temporal asymmetries in nature and to prevalent "driving forces" in all actual processes. It also deals with the question of whether or not the system considered is at equilibrium (a reversible process) or in a non-equilibrium state (an irreversible process). *All* actual processes in nature are irreversible in the sense that in no way can they be *completely* reversed, i.e. that it is impossible, even with the assistance of all agents in nature, to restore *everywhere* the exact *initial* state once the process has taken place.*** This conclusion is based on observations, all of which show that there exists in nature *no macroscopic process entirely free from friction, viscous flow, "heat" and dissipation of energy;* in particular free from *irreversible transforma-*

*** Subjective definitions of the laws of thermodynamics (by a skeptic) run as follows:
    Principle No. 1: You cannot win nature.
    Principle No. 2: You cannot break even with nature.
    Principle No. 3: You cannot get out of the game.

tions of kinetic into internal energy! The source of this apparently limitless kinetic energy is quite a subtle problem (to be discussed later), and the second principle does not supply information about that source. Reversible processes, on the other hand, are merely a conceptual limiting case; no actual macroscopic *process* in nature is exactly reversible. They are, however, of some importance for the theory of time and for applications to states of equilibrium.

The second principle of thermodynamics is a *"fact-like"* empirical law; we speak of its validity insofar as its total purport is *deduced from observations* about which there is no doubt.

Historically, the "second principle" has been stated in various forms, most of which took the form of *"no principles"* or *"impossibility principles"* (e.g. impossibility of various perpetual motions due to energy dissipation or impossibility of an engine which would work in a complete cycle and produce no effect except raising a weight against the gravitational force and cooling of a "heat reservoir"[+]).

Today, the "second principle" is usually stated in terms of entropy, a quantity which is normally assigned the units of energy divided by absolute temperature. Popular descriptions refer to entropy as a yardstick for the "degree of disorder" or "randomness" of the system. (Detailed mathematical expressions for the generation of entropy as function of the space and time coordinates in actual processes involving simultaneous conduction of heat, diffusion, reactions, viscous dissipation, and external forces are given in my MDT).

The "second principle" may be expressed in terms of other physical units, e.g. internal energy, free energy or dissipated kinetic energy. These units are well defined and dispense with the dubious concepts of entropy and temperature. The thermodynamic arrow of time can therefore be well-defined in terms of *entropy-free* physical quantities. Nevertheless, in keeping with historical tradition, I shall first formulate the entropic arrow of time.

For this purpose, we employ the following second-law statement[*]:

"The entropy of any isolated macroscopic system never decreases".

To define the entropic arrow of time, we must merely examine the dynamical behavior of that system at two instants in time, t_1 and t_2, without having *a priori* information about their order or sequence, i.e., without knowing whether

$$t_1 > t_2 \tag{1a}$$

or

$$t_1 < t_2 \tag{1b}$$

[+] For historical account, see, for instance, Max Planck's *"Treatise on Thermodynamics"*, Dover Publications, 3rd ed., N. Y. 1922. For updated reviews see my *CRT* and *MDT*.

[*] This statement should not be confused with some statistical-mechanical *interpretations* according to which the entropy of an isolated system fluctuates and may even decrease for certain time periods. The question of "statistical irreversibility" is a *misconception* in physics and will be set aside until Lecture V.

We next calculate, using the mathematical procedures of thermodynamics, at which instant the entropy S is greater. If, for instance, we find

$$S(t_1) > S(t_2) \tag{2a}$$

we employ the above statement of the second postulate to conclude that

$$t_1 > t_2 \tag{2b}$$

Hence, $S(t_1)$ was evaluated at what we call a *later* instant along the time coordinate. We can then define the general direction of change in time (i.e. from t_2 to t_1) and refer to that direction (or arrow) as pointing toward "future". t_1, according to this specific example, is a "later" instant. In this way thermodynamics show us that our (macroscopic) world is asymmetrical along the time coordinate (see Figures IV-3, and IV-4).

However, this thermodynamic formulation calls for the specific use of certain *a priori* postulates,** namely:

1. A postulate on the "existence" of a physical quantity called entropy, which possesses the following properties: It is an additive quantity, i.e., if a system consists of several parts, the total entropy is equal to the *sum* of the entropies of all parts. The change of entropy dS can, therefore, be split into two additive parts:

d_iS the entropy generation due to irreversible processes inside the system.

d_eS the entropy change (or flow) due to interactions of the macroscopic system with the outside world. We can then write (see also eq. II.76 in Lecture II).

$$dS = d_eS + d_iS \tag{3}$$

2. The statement of the second postulate can then take the simple form

$$d_iS \geq O \tag{4}$$

The internal entropy generation d_iS vanishes when the system attains thermodynamic equilibrium, at which state all irreversible processes stop. For an isolated [+] macroscopic system, d_eS, by definition, equals zero, and eq.s. (3) and (4) reduce to well-known formulation

$$dS \geq O \quad \text{(isolated system)} \tag{5}$$

(which states that the entropy of an isolated system can never decrease, as in the

---

** The reader uninterested in these procedures may proceed directly to §IV.6.

[+] From the very beginning of "modern" science, physicists have kept their sanity and made much technical progress by "isolating" their problems, ignoring unwanted interferences from the rest of the universe. However, the "isolated" system invariably emits or receives neutrino and high-energy photon radiations superimposed on the effects of gravitation (which is always present). Thus the mere use (and assumption of existence) of an "isolated" system is an independent postulate,which must always be questioned and re-examined. No system can ever be completely isolated from the rest of the world, even in principle.

previous example). However, when the system is in complete thermodynamic equilibrium, eqs. (1) and (2) yield

$$S(t_1) = S(t_2) \tag{6}$$

i.e.

$$t_1 = t_2 \tag{6a}$$

If we assume that the very act of "observing" does not upset the equilibrium, the invariable result is that the internal (thermodynamic) time stops! Indeed, no clock, nor any time indicator can exist inside the system at complete equilibrium (since its very existence implies an irreversible process which would destroy the assumed equilibrium). Time's arrow can, however, *reappear* if the system is *reopened* and allowed to *re-interact with the external (irreversible) world*. Therefore we conclude that time's arrow and irreversible processes *cannot* originate *inside* the system. Both are *physically linked* to irreversible events which take place in the *external world*.

It must be recalled that *prior* to its isolation, the material and/or radiation medium enclosed by the system considered *was in physical contact with the irreversible processes taking place in the external world. This is an important conclusion which has been frequently overlooked*. It will be re-examined later from the viewpoints of the electrodynamic, gravitational and cosmological theories. At this juncture let me re-examine it within the framework of the classical and special relativistic theories of space, time and causality.

IV.6 CAUSALITY, CAUSATION AND TIME ASYMMETRIES

The concepts of causality and causation have seized the imagination of many scholars over the centuries, have inspired much thought and controversy, and have played central roles in the development of almost every philosophical school in science. Moreover, continued reevaluations and reexaminations of these concepts have shaped the foundations of modern science as we know it today.

The concepts of causality, time and symmetry can be very powerful tools in science. If used correctly, they can provide a shortcut to important conclusions, and eliminate fruitless theoretical and experimental efforts; if used superficially, they can block exploration of fields which might yield valuable new insights or discoveries.

There is however a wide spectrum of opinions, historically, as to what is ascribable to these concepts. There are the writings, doctrines and interpretations of Plato, Aristotle, Lucretius, Maimonides, Aquinas, Copernicus, Spinoza, Descartes, Newton, Hume, Berkeley, Whitehead, Jeans, Eddington, Einstein, Planck, Heisenberg, Bohr, Neumann, Russell, DeBroglie, Bohm, Popper and many others. To assign credit to each of these variations in a single introductory lecture is impractical and beyond the scope of these lectures. I shall therefore confine my discussion to a few important topics which are of interest in my brief review of these subjects.

* * *

Causality, in general terms, is the relationship between a *"cause"* and its *"effect"*. "Cause" and "effect" are linked with time's arrow: the "cause", according to em-empirical evidence, must *precede* the "effect" *along the same time axis!* The relativity of time has not obliterated this "fact-like" "sequence" or "order" of "events".

The causal doctrine is sometimes compressed into the statement:

The evolution of any self-contained, or isolated system is "determined" by the "initial state" of the system.

In physics we distinguish between *symmetric* and *asymmetric* causality.

Asymmetric causality is the imposition of a single and consistent one-way relation of "cause" and "effect" along the time axis leading to the concepts of time asymmetry, past and future. It should more properly be called *"causation"*. Thus, irreversibility, time's arrow and causation are intimately linked.

Symmetric causality is rooted in the laws of physics. For instance, all our con-servation laws are strictly reversible under time-reversal transformations, i.e. they imply that any alteration in a prescribed state means alteration in the past state *exactly symmetrical* with alteration in its future state; there is no discrimination of "cause" and "effect"; "events" are connected by a *symmetrical causal relation*, which *is the same* viewed from *either side of the time coordinate*. Consequently, by mentally reversing the actual motion of particles, i.e. *by putting them through the same sequence of configurations, but in the reverse order of "cause" and "effect", we have a physical motion which likewise satisfies the conservations laws.* (See also time-reversal invari-ance below.)

The term "causality" should thus be reserved for such symmetric "causal" rela-tions. However, most authors confuse causality with causation and consequently arrive at some misleading conclusions. I shall review some of these misconceptions in their turn.

IV.7 CAUSATION AND DETERMINISM IN RELATIVISTIC THEORIES

Signal velocities exceeding that of light are well known to be ruled out by a combina-tion of the theory of relativity and causality. The possibility of experimentation with, or production of *"tachyons"* (shadow- or ghost-particles moving faster than light) must, therefore, be excluded (but see also below).

Einstein demonstrated that existence of a limiting signal velocity means that "absolute simultaneity" is not valid. This implies an unexpected result, which is of fundamental importance for the problems of causation, causality and time.

In *Newtonian physics* an experimenter could, in principle, *communicate with any system in space*, no matter how distant, in an arbitrarily short (even zero) time interval, i.e. operating at time t_0, he could, in principle, influence any "event" anywhere at all future times, $t \geq t_0$, and conversely, *all "events" occurring anywhere at all past times, $t \leq t_0$, could influence him.*

In *special* relativity, Einstein employed two postulates:

1. The Principle of Covariance, (see Accessory Lecture III, where we also discuss the General Principle of Covariance in General Relativity), which means that the laws of physics should preserve their form under proper coordinate transformation (employing tensor analysis). When properly transformed, the laws are of the same *form* in all frames of reference.

2. In all such systems, the signal velocity in empty space has the same value c.

Since the experimenter (in any such frame of reference) cannot send any signal faster than c, he could, during the time interval $t - t_0$, only reach points at a distance

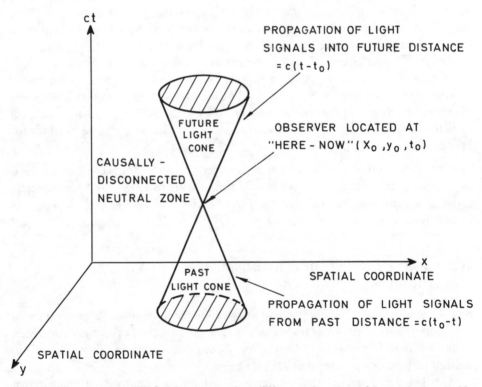

Fig. IV.4 **Causation in special relativity: future and past light cones in an inertial frame of reference.** No signal can reach the neutral zone. *Time's arrow* is directed toward the *future light cone*. Points outside the light cone are said to have *"space-like"* separation from the apex, those inside the cone have *"time-like"* separation, and those on the cone have *"null"* or *"light-like"* separation. The region with "space-like" separation cannot be *causally connected* with an experimenter *"Here-Now"*, i.e., no photon, neutrino or particle of matter can penetrate the causally-disconnected neutral zone since that would require velocities exceeding c. For details see §III.2.1.

$\leq c(t - t_0)$, which lie *inside* the *"future light cone"* shown in Fig. IV-4 (where the third spatial dimension is omitted). In arriving at this conclusion we employed time's arrow and causation, which here mean that an event "Here-Now" (x_0, y_0, t_0) can only cause events in the "future light cone". A photon, neutrino, or any particle of matter cannot travel into the causally disconnected neutral zone (since that would require velocities exceeding c).

On the other hand, time's arrow and causation mean that our experimenter can only be reached by those signals, emitted at an *earlier time t'*, which originate at distances $\leq c(t_0 - t')$. Only the region within or on the forward surface of the "future light cone" can be reached by signals from (causally connected with) its apex (x_0, y_0, t_0), and only signals emitted from within or on the *"past light cone"* can reach the apex.

The entire region outside the "future light cone" *cannot* be causally connected with the observer at the apex. But by a suitable coordinate transformation, any point *within* this region can be made simultaneous with the apex, and thus, in a sense, the entire region outside the light cone constitutes the "present" for the "observer" at the apex. The meaning of the neutral zone (which is causally disconnected from the "observer" at the apex, i.e., can neither be affected by him nor affect him) is a fundamental feature of the theory of relativity. Its significance for our critical evaluation of the role played by "observers" in quantum mechanics will be stressed later.

Other conclusions can be drawn now. For instance, one can modify some formulations in classical physics by setting a maximum rate for the propagation of any signal. (Note that in the heat conduction "law" according to *Fourier*, in the diffusion "law" according to *Fick*, etc., the rate of propagation is wrongly assumed to be infinite; a small correction should therefore be added in principle: this effect may be non-negligible only in astrophysical systems).

In concluding this section, I stress six points:

1. The very *sending of a signal* is an *irreversible process* which points to the "future light cone". Observations show that time, unlike space, is asymmetric for any macroscopic process and for the very process of radiation of quanta. We can never have a sequence of space-relations similar to the sequence of time-relations.

2. The very concept of a signal is based on *macroscopic interference* with a system, as it is this interference which allows *"recording"* or *"recognition"* of physical phenomena as *"signals"*, i.e., transmission of information.*

3. The *"emission"* and *"absorption"* of signals (e.g. photons) is linked with energy *dissipation*, increase in entropy, and (under certain conditions) transmission of information.*

4. In Newtonian mechanics the concept *"prediction"* normally means that the

---

* Critical remarks about the concept of information are given in Lecture IX.

system be isolated from an initial time t_0, for which its state is specified, up to a later time t_1, for which prediction is desired. Open systems can be described in terms of "cause" and "effect", but in "isolated" systems everything is interconnected.

5. In special relativity causally disconnected neutral zones can neither affect "observers" nor be affected by them.

6. Einstein's theory of general relativity imposes similar requirements to those described under (5). Some of these requirements are analyzed next for cosmological considerations regarding the concept of cosmic time.

●IV.8 COSMOLOGICAL ARROWS OF TIME AND COSMIC TIME**

Space-time may be described by a differentiable continuum for which every point, called an event, may be labeled by four independent coordinates x^α, where the indices α or β run over the three spatial coordinates x^1, x^2, x^3 and over the time coordinate x^0. Any index that appears twice, once as a subscript and once as a superscript, is to be summed over the four components 1, 2, 3 and 0, unless otherwise stated (Einstein's summation convention).

The concept of distance, proper time, or interval, in this continuum is expressed through its metrical structure which according to Lecture III, eq. (26), is generally expressed as

$$d\tau^2 = -g_{\alpha\beta}(x)dx^\alpha dx^\beta \tag{6b}$$

The tensor $g(x)$ is called the *metric tensor*, and is a function of the four coordinates x^α alone. The distribution of $g_{\alpha\beta}(x)$ fixes the geometry of the space-time continuum and is determined by the *gravitational field* of the physical system in the neighborhood of the point x^α. The gravitational field is subject to *Einstein's field equations*

$$R_{\alpha\beta} - \frac{1}{2}g_{\alpha\beta}R = -8\pi G T_{\alpha\beta} \tag{6c), (III-66}$$

As explained in Lecture III, the left member of Eq. (III-66) consists of the metric $g_{\alpha\beta}(x)$ and its first and second derivatives only.

Therefore, the left member is a *purely geometrical expression* (the tensor $R_{\alpha\beta}$ and the scalar $g^{\alpha\beta}R_{\alpha\beta} \equiv R$ are formed out of $g_{\alpha\beta}(x)$ and their first and second derivatives). On the right-hand side is the *energy-momentum tensor* (defined by eqs. (37), (38), (42) or (50) in Lecture III). It is determined by the *physical conditions* (mass, energy, pressure and momentum) of the matter and/or radiation at the point of space-time considered. (G is the Newtonian gravitational constant, with appropriate units to connect the different physical quantities).

Thus for any given distribution of energy and momentum, the set of non-linear

---

** Possible links with the electromagnetic and thermodynamic arrows are described in §VI.3 and VI.10. The reader uninterested in the mathematics presented here may proceed directly to Lecture V.

COMOVING COORDINATES

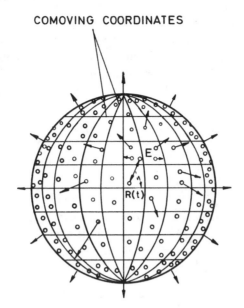

THE DYNAMICS OF THE
BALLOON'S TWO - DIMENSIONAL
SURFACE REPRESENT UNIFORM
EXPANSION OF FOUR - DIMENSIONAL
SPACE - TIME.

NO OBSERVABLE SPACE EXISTS
OUTSIDE OR INSIDE THE EXPANDING
BALLOON'S SURFACE

Fig. IV.5 **The (three-dimensional) inflating balloon analogy for the observed expanding universe** ($k = +1$). As the balloon is inflated (cf. Fig. IV.6), the membrane surface stretches in all directions and every dot marked on it is seen to move away from the others. Yet *co-moving* coordinates painted on the surface of the balloon have *constant values* during isotropic expansion. Any given dot represents a cluster of galaxies. Our local group, marked E, sees itself at the center of the expansion, even though the clusters are in fact always distributed isotropically on the surface of the balloon. The *Robertson-Walker metric* describes the uniform expansion of the four-dimensional space-time (but *not* of a three-dimensional space into "externally available" empty space!). The further the clusters (or "supergalaxies") are from E, the faster they appear to recede. The expansion started when the balloon was at complete deflation (i.e., a point). A two-dimensional analogy would be represented by a widening circle, and a one-dimensional analogy by an extending curved line. $R(t)$ represents the distance between any two supergalaxies. See also Figs. IV-6 and §III.4.

equations (III-66) can be solved (in principle) for the structure of the space-time continuum (including its evolution with the time coordinate). Other theories of gravitation allow similar procedures.

The type of space-time structure described by this general metric is known as *Riemannian* geometry. It is normally very complicated, but under certain distributions of energy and momentum the metric $g_{\alpha\beta}(x)$ reduces to relatively simple expressions.

Perhaps the most important metric employed in modern cosmology is the *Robertson-Walker metric*. Its derivation is given in Lecture III (Section 2.7) and is based on verified observational evidence regarding the expanding universe and its isotropy. Using the spherical *comoving coordinate system*, r, θ, and ϕ, in the sense that typical clusters of galaxies have constant spatial coordinates,* the Robertson-Walker metric takes the form

$$d\tau^2 = dt^2 - R(t)\left\{ \frac{dr^2}{1 - kr^2} + r^2 \, d\theta^2 + r^2 \sin^2\theta \, d\phi^2 \right\} \qquad \text{(6d), (III-67)}$$

where $R(t)$ is the *cosmological scale factor* of the expanding universe which depends only on *cosmic time t* (the change in $R(t)$ may be thought of as representing the change in time of the distance *between any two neighboring clusters of galaxies*). Note that cosmic time possesses some of the properties of Newtonian simultaneity.

An observer at rest in the frame of comoving coordinates observes isotropic expansion; he can therefore label time by the change in $R(t)$, or by the sequence of overall states that the universe, accessible to his observations, passes through on the large scale, as it expands. *This cosmic time is common to all comoving observers, since all experience the same sequence of expansion. Cosmic time and $R(t)$ should therefore represent very fundamental concepts in the theory of time.* In later lectures, I explain their meanings further.

Returning to the Robertson-Walker metric, which for cosmological applications should be used with the field equations (III-66) of general relativity, it should be noted that k can be chosen to have the values of $+1$ (bound, closed, spherical, oscillating universe); -1 (unbound, open, hyperbolic, ever-expanding universe), and 0 (unbound, Euclidean, ever-expanding universe). The standard cosmological models described by these solutions are derived in Lecture III. They are known as the three *Friedmann cosmologies* (see Fig. IV.6).

IV.9 A FEW REMARKS

(i) $R(t)$ represents the *First Cosmological Arrow of Time*. Thus, if, for instance, we photograph a group of supergalaxies at instants t_1 and t_2, their respective separations will be different. Suppose we observe (Fig. IV.5)

$$R(t_1) > R(t_2), \qquad (7)$$

---

* One can visualize the comoving coordinate mesh as lines painted on the surface of an inflating balloon (see Fig. 5), on which dots represent typical supergalaxies (i.e. clusters of galaxies which move together).

then we can only conclude, for $t < t_{max}$, that

$$t_1 > t_2,$$ (8)

or more generally

$$\boxed{\frac{dR(t)}{dt} > 0}$$ (9)

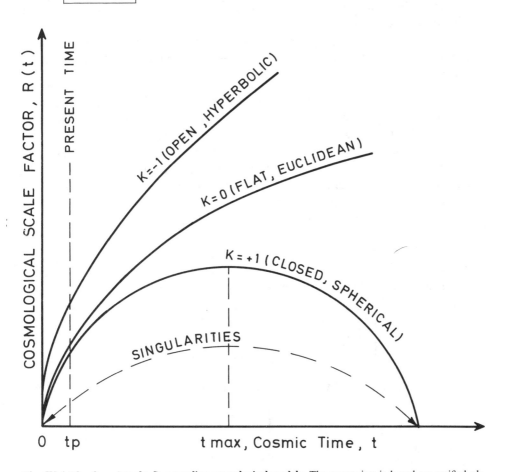

Fig. IV.6 **The three (standard) expanding cosmological models.** The expansion is based on verified observational evidence and on the general theory of relativity (which is a theory of gravitation). The expansion starts at $t = 0$ with $R = 0$. The $k = 0$ and $k = -1$ models expand forever, but the $k = +1$ model would reach a maximum at some future cosmic time, t_{max}, after which it recontracts to another singularity. (For *theoretical analysis* see §III.2.7.3. For *observational evidence* in the late 70's see §III.3. For *empirical evidence* about cosmic time see Fig. III.4 and §III.1.2.b).

Note, however, that *only intercluster space expands*; i.e., only the distances between neighboring clusters of galaxies are becoming, according to all available measurements, larger and larger with time (Fig. VI.1). But, most important, is to note that *the world expansion does not, in any sense, proceed into some hitherto "unoccupied extrauniversal space"*. The reader who thinks so does not, as a matter of fact, have the right geometrical picture in mind (cf. Lecture III and §VI.1).

(ii) This cosmological time asymmetry is independent of any cosmological model (including the steady-state model of Gold, Bondi and Hoyle, which also takes it into account). In fact, the first cosmological arrow of time is *observationally* independent of the general theory of relativity, since it is based on verified astronomical, astrophysical and geophysical observations, e.g. microwave (radio) black-body radiation, Hubble's empirical results, radio source counts, etc. as detailed in Part III. Nevertheless, it is fundamentally important to stress here that it was actually *predicted* by Einstein's field equations of general relativity applied to spaces with maximum symmetries (cf. eqs. III-91 and V-4 as well as §2.4, Introduction).

(iii) The metric tensor $g_{\alpha\beta}(x)$, as well as Einstein's field equations, possess no inherent time asymmetries, i.e. they are invariant under time reversal transformations (see below).

(iv) It is instructive to analyze the properties of the metric in "flat" Euclidean space, which is the domain of the *special theory of relativity*. The transition from Galilean relativity to special relativity had important consequences; the simplest, yet the most important change is the *time dilation* of moving clocks (see Lecture III). The principle of special relativity states that the laws of nature remain invariant under a group of space-time coordinate transformations, called *Lorentz transformations*. The Lorentz transformations leave invariant the "*proper time*" $d\tau$ defined by (see explanation in Lecture III, Section 1)

$$d\tau^2 = -\eta_{\alpha\beta}\,dx^\alpha dx^\beta \tag{10}$$

where the (dimensionless) symmetric metric $\eta_{\alpha\beta}$ is not a field (i.e., it does not depend on x^α as in the general theory of relativity) and is defined by

$$\eta_{\alpha\beta} = \begin{cases} +1 & \text{for} \quad \alpha = \beta = 1, 2 \quad \text{or} \quad 3 \\ -1 & \text{for} \quad \alpha = \beta = 0 \\ 0 & \text{for} \quad \alpha \neq \beta \end{cases} \tag{11}$$

This metric is known as the *Minkowski metric* and may also be written as

$$\eta_{\alpha\beta} = \begin{bmatrix} -1 & & & \\ & -1 & & \\ & & -1 & \\ & & & +1 \end{bmatrix} \tag{12}$$

Here the time coordinate is distinguished from its space counterparts by the positive sign. (Signs can be exchanged between spatial and temporal coordinates, but they cannot *both* have the same sign). If we denote by cdt (where c is the velocity of light in vacuum) the infinitely small temporal distance, and by dx, dy, dz, the rectangular components of an infinitely small spatial distance, eq. (10) takes the form

$$d\tau^2 = -dx^2 - dy^2 - dz^2 + c^2 dt^2 \tag{13}$$

where $d\tau$ is the elementary interval between two events infinitely close both in space and in time.

In general, we can write that a light wave front will have the velocity $|d\mathbf{x}/dt| = c$. We can obtain this expression from eq (10) or (13), by *requiring* that the propagation of light, or of any electromagnetic signal, be subject to the statement

$$d\tau = 0 \tag{14}$$

Thus, eq. (14) and (13) yield

$$\sqrt{\frac{dx^2 + dy^2 + dz^2}{dt^2}} = \frac{d\mathbf{x}}{dt} = c \tag{15}$$

If we now transform to any alternative system of inertial coordinates, x', y', z', t', corresponding to a new set of axes in uniform motion relative to the original ones, the *form of expression* for the interval $d\tau$ must still *be the same* by the nature of the Lorentz transformation. Hence, also in these new coordinates, the velocity of light is given by

$$\sqrt{\frac{dx'^2 + dy'^2 + dz'^2}{dt'^2}} = \frac{d\mathbf{x}'}{dt} = c \tag{15a}$$

as is required by the postulates of special relativity.

The simplest illustration of the metric $\eta_{\alpha\beta}$, for a reader unfamiliar with its mathematics, is given by the following example:

In Euclidean two-dimensional space we employ *Pythagoras' theory* $L^2 = A^2 + B^2$, where L, A and B are respectively the hypotenuse of a *right triangle* and the two legs forming the right angle. If, using the notation of differential calculus, we denote by dx and dy two infinitely small segments at right angles to each other, and by $d\tau$ their hypotenuse, we have

$$d\tau^2 = dx^2 + dy^2 \tag{16}$$

In this form it represents the arc element $d\tau$, in rectangular coordinates x, y; the length of any arc of a curve is expressed by the integral

$$\int d\tau = \int \sqrt{dx^2 + dy^2} \tag{17}$$

Note now that expression (16) is, to some extent, analogous to (13) (which also contains the time coordinate).

(v) A universal "now" cannot exist in a Minkowski space. In relativity the objective physical world simply "is," it does not "happen". Any statement concerning the movement of "now" in time cannot be other than a mere tautology. It is meaningless to discuss motion in time, or of time. In spite of this, modern writers* keep using the concepts "flux" or "flow" of time, often exploiting the analogy of a forward-flowing river. Such misconceptions give the wrong impression that time "passes" or "moves", or that the arrow of time means the direction of movement of the "now", or of the "flow of time".

(vi) From the points of view of both the relativity theory and thermodynamics, it is meaningless to discuss flow of time in an extra (fifth) dimension, or hypertime.** It is, therefore, quite perplexing to find such misleading notions seriously discussed in modern theoretical physics, notably by Nobel Prize winner *E. Wigner*. According to Wigner's interpretation of the foundations of quantum mechanics, the *human mind stands "outside"* the ordinary four-dimensional space-time (as though in an extra dimension), in which irreversibility and all time asymmetries that are in evidence in the quantum measurement theory (a highly controversial subject in itself, see below) are consequences of the *"penetration of human consciousness"* by the outcome of the measurement. Any claim to associate cosmic or thermodynamic times with subjective notions of the working of our minds or with our consciousness cannot be taken seriously, to say the least.

That time should be endowed with such questionable properties is actually due to the confusion of the *objective* physical time asymmetry with the *subjective* illusory forward flow of psychological time. These speculations about subjective time asymmetries are conspicuous in their inability to assert anything new or useful about time; they lack empirical support, and can only serve to transform theoretical physics into psychology. We return to these questions later, in our criticism of quantum-mechanical interpretations.

IV.10 TIME-REVERSAL INVARIANCE AND IRREVERSIBILITY

It is a remarkable fact that our laws of physics (e.g., Newton's equations of motion, the equations of quantum mechanics, relativity, etc.) are symmetrical with respect to time reversal, i.e., they remain invariant under the transformation

$$t \rightarrow -t \tag{18}$$

Thus by *mentally* reversing the actual motion of the planets or of particles (i.e. by putting them through the same sequence of configurations *in reverse order*—like running a movie backward), we have a process which *likewise* satisfies the physical

---

* See for instance H. Reichenbach, *Philosophy of Space and Time*, Dover, N. Y., 1958, p. 138; M. Capek, *The Phisolosophical Impact of Contemporary Physics*, N. Y., 1961, p. 165; A. Grunbaum, *Ann. N.Y. Acad. Sci.* **138**, 374/1967).

** Such misconceptions have been propagated by a number of writers, notably by R. Taylor in *Problems of Space and Time* (Ed. J. J. C. Smart), Macmillan, N. Y. 1964, p. 388; J. J. C. Smart, *Analysis*, **14**, 80 (1954); M. Block, *Analysis* **19**, 54 (1959).

laws. This property is called *"time-reversal invariance"* and reflects the fact that our equations of motion (classical as well as relativistic or quantum-mechanical) of individual particles are symmetric (or impartial) with respect to *choice of sign* of the time coordinate (they are also symmetric with respect to choice of sign of the spatial coordinates). This invariance implies that the individual particles retrace their former paths if all velocities are reversed. The property of time-reversal invariance has been used to derive such theorems as "Onsager's Reciprocal Relations" for macroscopic phenomenological equations of transport phenomena in thermodynamics and statistical physics.***

It is instructive to demonstrate the physical-mathematical consequence of time-reversal invariance in a simple system. Such a system is given, for instance, by a point particle of *constant mass m*, whose motion in a force field **F** is described by a position vector **r**. When **F** does not explicitly depend on the time t (conservative field of force) or on $d\mathbf{r}/dt$ (no velocity-dependent forces), *Newton's second law of motion reads*

$$\mathbf{F}(\mathbf{r}) = m \frac{d^2(\mathbf{r})}{dt^2} \tag{19}$$

With velocity-dependent forces in a gravitational field, the same law reads (see Lecture II).

$$\rho \frac{d\mathbf{u}}{dt} = -\operatorname{div} \mathbf{P} + \rho \mathbf{g} \tag{II.29}$$

where **P** is the *pressure tensor* and **g** the *gravitational acceleration*. Both eqs. (19) and (II.29) are of the form

$$[\text{mass}] \times [\text{acceleration}] = [\text{sum of forces}] \tag{20}$$

The Navier-Stockes equations of motion (eq. II.42) can then be derived from eq. (II.29) by assuming the functional dependence of **P** on velocity gradients. ["Newtonian fluids" are assumed to obey a linear dependence on the velocity gradients (cf. eqs. II.38 to II.40)].

Eqs. (19) and (II.29) may be solved, in principle, subject to a given set of *initial and boundary conditions*. The solution would then allow one to *predict* the trajectories of the particles(s) for all *future times, t*. If $r = s(t)$ is the solution of eq. (19), then

$$\mathbf{F}[s(t)] = m \frac{d^2 s(t)}{dt^2} \tag{21}$$

Now, if we mentally replace the positive time coordinate t by a negative one, $-t$, eq. (21) becomes

---

*** Although Onsager's claim that his results follow from the "principle" of microscopic time reversibility has not been conclusively proved, they are of considerable practical value in empirical studies (see, for instance, Thermodynamics of Living Organisms in my MDT).

$$\mathbf{F}\left[s(-t)\right] = m\,\frac{d^2 s(-t)}{d(-t)^2} = m\,\frac{d^2 s(-t)}{dt^2} \tag{22}$$

Consequently, if $s(t)$ is the solution of the equation of motion, $s(-t)$ is also a valid one and consequently the equations of motion in fluid dynamics are invariant under the time reversal $t \rightarrow -t$. This reversal causes all velocities to change sign according to

$$\mathbf{u}(t) = \frac{d\mathbf{r}}{dt} \quad \text{(for } +t) \tag{23a}$$

$$\mathbf{u}(-t) = \frac{d\mathbf{r}}{d(-t)} = -\frac{d\mathbf{r}}{dt} \quad \text{(for } -t) \tag{23b}$$

Note that $\mathbf{r}$ remains unchanged in the process. Now, one of the most important inequalities in physics reflects the "second law" of thermodynamics, and is given in Lecture II as

$$\boxed{\Phi \geqq 0} \tag{II.49}$$

where Φ is the *dissipation function*. Its physical meaning is simple: an *irreversible* transformation of *kinetic* energy of moving fluid particles into *internal* energy (or "heat"). Its mathematical expression for Newtonian compressible fluids is given by

$$\Phi = \mu\,\frac{\partial V_i}{\partial x_k}\left[\frac{\partial V_i}{\partial x_k} + \frac{\partial V_k}{\partial x_i}\right] - \left[\frac{2}{3}\mu - \kappa\right]\cdot\frac{\partial V_j}{\partial x_j}\cdot\frac{\partial V_j}{\partial x_j} \geqq 0 \tag{II.50}$$

$$(k, i, j = 1, 2, 3)$$

For one-dimensional incompressible flow, (II.50) reduces to

$$\Phi = \mu\left(\frac{\partial \mathbf{u}_x}{\partial x}\right)^2 \geqq 0 \tag{II.50a}$$

Here μ and κ are the shear and bulk (or expansion) viscosities, respectively, and V_i are the velocity components along the space coordinates x_k $(k, i, j = 1, 2, 3)$.

Provided μ and κ remain positive, the mental action of time-reversal leaves this fundamental asymmetry of nature unchanged, i.e.,

$$\boxed{\Phi(t) = \Phi(-t) \geqq 0} \tag{24}$$

At first glance, this simple result should eliminate much confusion from the physics of time asymmetries since, so far, authors in this field have claimed that *time-reversal invariance means reversibility!* This, however, is *not* the case. The irreversible nature of the dissipation function is shown here to be unaffected by the mental action of time reversal; here, irreversibility results from the *positive nature* of the *viscosity coefficients!*

The physical cause behind the positive nature of the viscosities is an interesting subject in itself; it is discussed in more detail in Part III.

The physical meaning of eq. (24) is that any real process in nature is *irreversible* in the sense that in no way can it be completely *reversed*, i.e., that it is impossible, even with the assistance of all agents in nature to restore *everywhere* the exact *initial* state once the process has taken place. But the "initial" state is independent of *choice* of *sign* of the time coordinate! We are, therefore, entitled to replace the *anthropomorphic notion* "initial" by *"final"* if we replace t by $-t$. This, in itself, does not have any physical meaning. The significance of the second "law" of thermodynamics depends on the fact that it supplies a consistent time asymmetry of any evolution in the real world. This time asymmetry does not discriminate between $+t$ or $-t$; *it only requires a consistent, one-way evolution in time.* Once the sign of t is chosen, the second "law" prevents us from reversing it at will. It is in this sense that the time-reversal invariance should be interpreted physically.

By no means can this topic be left at this level; it deserves much deeper analysis and understanding before one can arrive at a universal conclusion. This can be done only later, in Lecture VI.

*   *   *

One must be careful in trying to find physical systems which at first glance do not possess the property of time-reversal invariance. Such an example is given by the motion of a charged particle with mass m and charge z in a magnetic field **B**. The Lorentz force acting on this particle is then

$$\mathbf{F} = m\frac{d^2\mathbf{r}}{dt^2} = z\frac{d\mathbf{r}}{dt}\,\mathbf{xB} \tag{25}$$

Transforming t into $-t$, eq. (25) takes the form

$$\mathbf{F} = m\frac{d^2\mathbf{r}}{d(-t)^2} = m\frac{d^2\mathbf{r}}{dt^2} = -z\frac{d\mathbf{r}}{dt}\,\mathbf{xB} \tag{26}$$

which means that when the particle velocity is reversed it will not move backwards, which would apparently conflict with the time-reversal invariance. But one should note that the magnetic field itself is produced by moving charges. Hence in reversal of these charges the field also changes sign and time-reversal invariance is restored. Other, more complicated examples can be worked out. But the main point in analyzing them is application of time reversal to the *entire system*—including, if necessary, the walls of the laboratory.

To conclude: time-reversal invariance applies to Newtonian dynamics, to Maxwell equations of electrodynamics (§ VI.10) to special and general relativity (Lecture III), to quantum mechanics (Lecture V), and to strong and weak interactions. It was recently claimed, however, that it is violated in certain weak interactions which involve the decay of K-mesons. The physical meaning of these so-called T-violations will be briefly discussed next.

IV.11 MICROSCOPIC TIME ASYMMETRIES IN "ELEMENTARY PARTICLES".

Until recently the postulate of time-reversal invariance, known in particle physics as *T-invariance*, was so deeply ingrained in thought and theory as to be considered a fundamental principle of nature. The impetus towards re-examination of this *symmetry* in microscopic systems was given by the discovery that *left-right symmetry (P)* of the laws of particle physics is violated by the weak interactions. This connection is related to *charge conjugation (C)* which replaces every particle of a system by its antiparticle. In spite of experimental evidence of violation of the latter in ^{60}Co, the two operations apparently leave the laws of motion unchanged, and may be considered jointly as *CP-invariant*. This invariance had been verified by many experiments on β-decay, until the famous experiment by Christenson, Cronin, Fitch and Turlay showed (1964) that the long-lived neutral K-meson can decay into two π-mesons.* The most interesting feature of this is the intimate connection with *time-reversal*. The relevant *CPT-conservation theorem* states that all fundamental physical laws must be simultaneously invariant under *three symmetry transformations*, namely, remain unchanged when the *time* and *space* coordinates are reversed and all particles are replaced by *antiparticles*. The proof of such a statement must necessarily be based on *a priori* assumptions regarding the mathematical nature of the physical laws and on specific hypotheses so general as to rule out a reasonable theory which would violate any of them.** *However, this situation may well be a measure of the limitations of present theoretical physics rather than of the physical universe, and the theorem itself should be subjected to an experimental test. Now, if the CPT-conservation theorem is valid, a CP-violation implies a T-violation (i.e., existence of a microscopic "arrow of time").* In fact, it was established later (1968) that experimental results show *CPT-violation (hence, T-violation)* and *cannot* be explained by assuming T-invariance (Experimental evidence of T-violation is, however, available only in Kaonic systems, but one should not discount it because it is found only in these very rare particles; Kaons seem to have a "special link" to time, as the oscillating time-factor of the wave function is measurable only in their case).

---

* The mass of the K-meson is about half that of the proton, requiring energies of billions of electron-volts to produce it. The neutral K-mesons, K°, are unstable due to their weak interactions. They decay into two to three π-mesons, and also undergo a β-decay (into a π-meson, neutrino, and electron or μ-meson), the 2π-decay mode being dominant by a factor of about 500.

The conclusion that *CP*-invariance is violated here is a straightforward consequence of the theory of K°-meson phenomena, although the connection is by no means obvious [cf. §V.6].

** cf. Assertion 2, §3, *Introduction*, §III.5 and §V.6.

Even if the *CPT*-theorem is abandoned, the *CP*-violation indicates that the fundamental question of *microscopic* irreversibility must be investigated within a theory *more closely unified with the rest of macroscopic science*. The $K°$-meson carries, therefore, a *time asymmetry* which, in principle, could be either *parallel* or *antiparallel* to its *thermodynamic* and *cosmological* counterparts. Moreover, the presence of such a *T*-violating elementary-particle system may be attributed to irreversibility and energy dissipation in *macroscopic systems*. But it is not yet clear whether the properties of Kaons have any relevance to the type of macroscopic asymmetric processes discussed previously [cf. §III.5].

At this juncture it should be stressed that, except for this microscopic arrow, *the laws of physics are all time-symmetric* (exception is the second law of thermodynamics, which, for these reasons, becomes an "outlaw". Whether "law" or "outlaw", we shall show that it is dominated by gravitation and can be deduced from gravitation theories). In other words, the discovery of the microscopic arrow demonstrates, for the first time, a physical system which behaves asymmetrically in time as a result of *an interaction governed by a dynamical physical law which is itself preferentially oriented in time!* This phenomenon may well be a subtle manifestation of an entirely new aspect of the interaction between the micro- and macro worlds, with implications far beyond their immediate evidence (see §IX.2.20, §III.5 and the next section).

But could not *all* observed time asymmetries be attributed to a *master asymmetry* which acts on the 'weak', electromagnetic, and 'strong' interactions? Why, for instance, does an *accelerated electric charge lose radiant energy to outer space?* Why is it never observed to receive *incoming* radiation from outer space, i.e., why we do not observe *advanced* potentials? *Can we work up the details of these interactions, through a unified, but modified Principle of Equivalence between acceleration, gravitation, and dissipation?*

The classical equations of electrodynamics (Maxwell's equations) are fundamentally symmetric under time-reversal, so that every solution of these equations for time t has a counterpart for $-t$ *(§VI.10). These are known as retarded solutions (charge emits energy) and advanced solutions (charge absorbs energy), respectively. Mathematically speaking, both types of solutions are valid! In practice, however, only the retarded solution is referred to, so as to agree with observations! The formal pretext for such an a priori procedure is causality, but to state that the cause must precede the effect is essentially equivalent to an a priori time asymmetry, which is, apparently, equivalent to a priori imposition of electrodynamic irreversibility (§VI.10). However, it has been* demonstrated by Gold* that time-reversal in such systems may only cause the roles of *emitter* and *absorber* to be *interchanged*, and that *these definitions have no significance without an additional assumption regarding the direction of the time asymmetry!*

---

* T. Gold, *"The World Map and the Apparent Flow of Time"*, in *Modern Developments in Thermodynamics*, B. Gal-Or, ed., Wiley, New York, 1974, pp. 63.

Would it be necessary to pursue these thoughts about electrodynamic irreversibility within the frame of the *absorber theory of the universe* with respect to electromagnetic disturbances generated within it? Such an initial work was indeed carried out in 1945–49 by Wheeler and Feynmann.[**]

However, they confined their theory to a *static* universe, so that their arguments are neither complete nor convincing. It is only in the *non-static* (expanding) space that *"perfect absorption"* of radiation becomes feasible.[***] How? I shall return to this fundamental question in Part III, §VI.10 and §VI.3.

IV.12 THE DEATH OF SCALE-BASED PHYSICS?

The dramatic discoveries of modern unified field theories (§III.5) have given the strongest support to the assertion that *physics is fundamentally scale-free*; that there is no characteristic scale in it; that whatever applies to objects of a certain size will apply to objects of any size. Does this mean that scale-based physics is dead? that no characteristic length is involved in quantum gravitation or in more advanced field theories of the structure of matter? Does it mean that *cosmological* conditions are, in principle, as relevant for physics as do the conditions of the *microworld?* Does it unify cosmology with microphysics? And, most intriguing, does it abolish the current separation between *short-range* and *long-range* forces as shown in Table I of the introduction? Should not all fields be treated on the same footing?

Perhaps the best way to deal with these important questions of physics and philosophy, is to clarify the *ultimate aims* of a number of recent efforts in theoretical physics and mathematics, in particular of the following ones:

1) *A conservative synthesis* of the theory of general relativity and old quantum mechanics (without giving up the main tenets of the conservative quantum model). This includes such studies as *quantum geometrodynamics* that are to be discussed in the next section. Their ultimate description of the structure of matter is still based on a minimal characteristic length called *Planck's length*,

$$L = \sqrt{\frac{hf}{c^3}} \cong 10^{-33} \text{ cm},$$

which is the geometrical mean of Compton's wavelength and Einstein's gravitational radius of a particle. Compared to the scales of currently observed subatomic particles (of the order of 10^{-13} cm), there is no feasible way that it would ever be verified or refuted experimentally. Hence, by using *the criterion of scientific testability, refutability and falsifiability* (Assertion 11 in the Introduction), we conclude that this

  [**] J. A. Wheeler and R. P. Feynmann, *Rev. Mod. Phys.* **17**, 157 (1945); **21**, 425 (1949).

  [***] J. V. Narlikar, *"Thermodynamics and Cosmology"*, in *Modern Developments in Thermodynamics*, B. Gal-Or, ed., Wiley, New York, 1974, pp. 53.

irrefutability makes the existence of the Planck's characteristic length a myth. Nevertheless, it is of some interest to discuss it in the next section (and in Lecture V) in the wider context of "determinism vs. indeterminism of physics".

2) *A revolutionary synthesis* of general relativistic and subatomic physics within a grand deterministic theory which would ultimately unify all four interactions and the inner structure of matter. This is not only interesting in itself, but it is also the greatest ambition of physics. Yet, as before, we may never know how in the interiors of subatomic particles the space-time is curved; how they are connected with the outside fields; how they are created and annihilated and moved; and how they develop internal or external asymmetries. As we shall see at the end of the course (§ IX.2.20–21), these problems are linked to the structure of the vacuum; to solitons; to space-time curvature and to symmetry-asymmetry considerations.

IV.13 THE "DUAL" QUANTUM-GEOMETRODYNAMICAL SCHOOL AND "SUPERSPACE"

Historically it is interesting to note that Lotze (1817–1881), considered a pioneer of indeterminism, maintained that both natural events and human acts lie on strings of causal chains, and that such causal chains, once started, have no end in future time. At the same time, he maintained that such causal chains may have *"capricious beginnings"*. This historical viewpoint is in some agreement with a theory that may be called *"Dual Quantum Geometrodynamics"*. It combines the "indeterministic" interpretations of quantum theory* with the "deterministic" general theory of relativity and, by resorting also to the concept of *"superspace"* (see below) ties together the world of the very large (gravitational collapse** and expansible evolution of the world) and the very small (much smaller than what is called today an "elementary" particle).

"*Superspace*" is the term given to a concept which goes back from space-time to "fluctuating" space (Fig. IV.7).*** Einstein gave us deterministic space-time instead of the deterministic "flat" Euclidean space. In "superspace" the indeterministic interpretation of quantum theory is *added* to space as an *ad hoc postulate*, which "causes" space to *fluctuate* from one curved configuration to another. It turns out, however, that these *fluctuations in the shape of the geometry, if valid at all,* are completely *unobservable!* (at least for a long time to come). Even at the scale of dimensions of atoms, nuclei, and electrons, they are *totally* negligible. Only when the scale of consideration is narrowed to the fantastically small size of 10^{-33} cm—the so-called Planck length—these fluctuations *may* have some effect. Space at this size

---

* But see our reservations in Lecture V.

** q.v. Lecture I, Sec. 3.

*** The concept "fluctuating space" is linked to the problem of "vacuum-state fluctuations" that is discussed in §IX.2.20 and in other sections of Lecture IX.

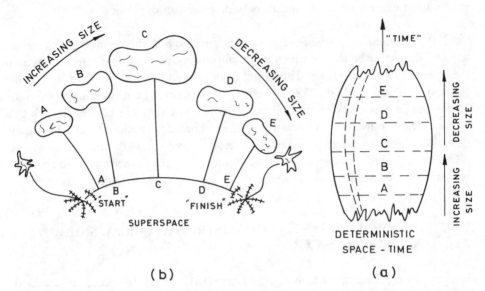

(C)   SPACE

INCREASING SIZE

DECREASING SIZE

"TIME"

DECREASING SIZE

INCREASING SIZE

SUPERSPACE

DETERMINISTIC
SPACE - TIME

(b)   (a)

Fig. IV.7 **Space-time, space, and superspace.** (a) A cut through space-time, such as B, gives a momentary configuration of space. (Space-time, shown here as the outline of an egg, is 2-dimensional. Actually it is a 4-dimensional). Shown is the oscillating cosmological model. (b) and (c): *Superspace*, the arena within which space (c), described as an oscillating cosmological model, undergoes its dynamic evolution from "start" through A, B, C, D, and E to "finish". Superspace is assumed infinite-dimensional, but again shown here as 2-dimensional. Each "point", such as A, stands for an entire 3-dimensional geometry ("potato" A, upper left). At the fantastically small distances of the order of 10^{-33} cm (marked "start" and "finish"), there is no sharp "beginning" or "end", only a succession of *probability waves* characterizing *fluctuations in the geometry* of space which prevent one from using such concepts as the "space-time", "time", "before" and "after", except in the approximations of classical physics. These fluctuations may dominate the final stages of the gravitational collapse of a supermassive star (q.v. §I.3 and Figs. I.4, VIII.2, and IV.6). However, such views cannot be vindicated by experiment. Moreover, they are based on in-deterministic interpretations of quantum physics- interpretations that are now considered false and even misleading (cf. Lecture V).

may be compared to a *"carpet of foam"* in which the millions of tiny "bubbles" have different shapes and are continually "bursting" as new ones are formed. At this fantastically small size space may "appear" like a *"dynamic picture"* in which the geometry is not well-defined; instead, it has a probability for any conceivable shape, a probability dictated by *"probability waves" spread in superspace.*[+]

[+] If the gravitational field is to be quantized along with other fields, opinions differ as to whether such a theory should possess an invariance group or merely be Lorentz invariant. It is not at all clear whether the topological aspects of space-time are to be quantized, or whether a quantum state should resemble an ensemble of (classical) metric fields. For a more fundamental discussion see Lecture V, §III.5, §IV.12 and §IX.2.20.

By contrast, the classical space-time of general relativity remains fully deterministic at all scales. While it is also a dynamic entity which changes according to the distribution of mass, energy, and momentum, it nevertheless remains well-defined, once that distribution has been defined. *It always, at any scale, defines a specific time relationship to the concept of (strict) causality. In other words, causality is strictly obeyed in any formulation of the general theory of relativity.* (The same statement applies, of course, to *special relativity*, restricted to "flat" Euclidean space-time).

IV.13.1 A Possible Failure of Time?

If the postulate of "superspace" is valid, the very concept of time fails to hold at the above scales. There, causality is not obeyed and no longer is there any meaning to such concepts as "before", "after", and "next". With the time sequence gone, *historical continuity* is gone and "initial conditions" cannot be defined. Quantum-mechanical coupling by probability waves "is permitted", but not a defined change with time, nor are there the initial position and the velocity of an object which obey the equations of motion.

During the initial phases of the (currently observed) universal expansion, as well as in the final phases of the *gravitational collapse of a star* [and the tentatively possible eventual collapse of the universe itself),* the phenomenon of indeterminism may dominate. At this stage one evolutionary history may be coupled with other dynamical histories and the usual concepts "before", "after", and causality may lose their all-embracing meaning. At these stages, the relevant physical dimensions reduce to values comparable to the Planck length $(10^{-33}$ cm), and may be considered as forming a "substratum", where the very concept of time fails.

Nevertheless, it should be kept in mind that such a prediction, if valid at all, cannot be observed! Even at small scales (that are many millions of times smaller than today's smallest observable scale)—the fluctuations should have already disappeared—so are the predictions! Already at these scales the "carpet of foam" "looks" completely smooth; and when it curves and changes, time and causality retain their strict deterministic meaning. Time is thus saved and initial conditions can be well-defined. *The concepts "before", "after" and "next" are therefore valid at any observable scale!* Thus, the whole issue of time failure in "superspace" may sound like a new kind of "theology" in which personal taste can be instrumental in forming a final judgment.

The only analogy—and no more than that—that can be mentioned in this context, concerns quantum fluctuations in the electrical field as represented in *"quantum electrodynamics"*. Present throughout empty space, the *electric field fluctuations* are assumed to perturb the motion of every electron in every atom and later the other-

---

* Observations now tend to support an open, ever-expanding universe (see Lecture III for details).

wise expected energy of the electron. And the predictions of quantum electrodynamics in microwave measurements have been confirmed recently with very great precision! (However, the predictions and verification of *quantum chromodynamics* should also be considered (§IX.2.20). Apart from that, no complete and self-consistent quantum-geometrodynamics or quantum theory of gravitation exists at present. Finally, I want to stress here, that if the statistical ("ensemble") interpretations of quantum theory are accepted, the whole approach to this problem must be changed (see Lecture V for details).

IV.14 TACHYONS AND CAUSAL VIOLATIONS

In spite of the central roles played by causality in relativistic theories, a number of physicists have, during the past decade, toyed with the idea of tachyons (shadow or ghost particles moving faster than light),* and even claimed that they might exist or be produced experimentally. Closely connected with these tachyon theories is the question of causal violation which might be simply stated as "an effect preceding a cause on the same time coordinate" (which, in an anthropomorphic analogy, can be illustrated by a tachyon theorist who turns on a "tachyon radiation switch" at t_0, this radiation switch activates a mechanism which kills the tachyon-theorist's ancestor a century before t_0, so that there can be no theorist either to turn on the switch or, for that matter, toy with tachyon theories). It has been claimed that by means of a certain "switching principle" it would be possible to describe processes with positive energy proceeding into the future in a manner violating neither logic nor causality. What has been achieved is a description of tachyons forming part of an isolated four-dimensional space-time continuum.

One can construct such a theory with advanced potentials, but in this case causality is not violated, since in an *isolated* four-dimensional space-time manifold there is *no identification of* one event being the "cause" of another. It is only in an *open* universe, in the sense of doing an experiment and creating a disturbance, that causality becomes a fully meaningful physical concept; there, the disturbance represents a "cause" to an "observer" and he can detect the "effect" of that disturbance.

Such limitations rule out the possibility of designing experiments involving tachyons.

These conclusions remain valid when quantum effects are taken into account, and they remind me the old limerick:

> *There was a young lady called Bright*
> *Who could travel much faster than light*
> *She went out one day*
> *The Einsteinian way*
> *And returned the previous night.*

---

* From the Greek 'tachys' meaning swift.

●IV.15 MACROCAUSALITY AND MICROCAUSALITY IN QUANTUM MECHANICS

Heisenberg devised the S-matrix method* to conform to both quantum-mechanical formulations and the theory of relativity. The method involves no "observables" corresponding to space-time points or to sharply defined space-time regions. Hence, at first glance it might appear that this approach would involve no causality. A closer examination shows, however, that the various formulations of the S-matrix method in quantum mechanics do involve a causality property, which can be related to scattering transition probabilities. These quantities can be measured to high accuracy by the conventional experimental set-ups used today in physics laboratories and may be called *macrocausality*. Macrocausality implies that one can check the causal properties of a quantum-mechanical formulation by examining its physical-region analyticity properties. Thus, if the formulation has a normal analytic structure, it also has the macrocausality property, which means that at least all long-range interactions are carried by physical particles, i.e. the formulation possesses all the general features that it needs to conform to ordinary macroscopic experience in which causality has been verified time and again. Any further causality requirement imposes conditions on the *short-range* structure. Thus macrocausality is equivalent to a set of analytic properties at the physical points (i.e., in finite but very small neighborhoods of these real points), whereas *microcausality* entails analyticity only from these physical points.

In my criticism of the foundations of quantum mechanics (Lecture V) I shall show that initial conditions, causality, causation and imposition of time asymmetries are a logical necessity in any statistical mechanical model (classical or quantal). Their need, however, arises from reasons outside the domain of quantum mechanics.

●IV.16 FADING MEMORY IN CLASSICAL PHYSICS

At the beginning of this lecture, I pointed out that in Newtonian physics the existence of an absolute time and of an absolute simultaneity had been taken for granted. The forces of Newtonian point mechanics are instantaneous *"action-at-a-distance"* forces; no reference is specified to any medium between the particles, or to any *mechanism of transmission* of signals. If the force laws are known and the *"initial conditions"* specified at t_0, the positions of the particles at *future times* are determined as function of t for all $t > t_0$, or at least as long as the system can be considered "isolated". For such a system the number of initial conditions required to fully *determine* its evolution is $6n$ for n mass points (the $3n$ space coordinates and the $3n$ components of the velocities at $t = t_0$). The evolution of the system can be determined in such a way even if the system was open at $t < t_0$ (i.e. prior to its isolation). But the assumption here

---

* See also Lecture V.

is that this evolution is *independent of the history* prior to t_0 (i.e. how it reached the state t_0, whether by having been isolated at $t < t_0$ or by the influence of external phenomena no longer active at t_0). This permits experimental verification of the deterministic predictions of Newtonian mechanics.

In practice, the experimenter assumes that any material he works with has a *"fading memory"*, i.e. that its recollection of long-past experiences fades in time so as to become negligible. Without this assumption, he could never interpret a laboratory experiment, for he cannot know the history of any particular specimen for more than a finite period of time. He may also subject the system, over a short period, to effects that are assumed to *"wipe out" its recollection* of whatever may have happened before (see, for instance, §IX.2.4.1 for shape memory alloy). For instance, a *Newtonian fluid* (Lecture II), characterized by the linear relation

$$\tau = -2\mu\mathbf{D} + (\tfrac{2}{3}\mu - \kappa)\,\mathrm{div}\,\mathbf{u} \tag{27}$$

where

$$\mathbf{D} = \tfrac{1}{2}[\mathrm{grad}\,\mathbf{u} + \widetilde{\mathrm{grad}\,\mathbf{u}}] \tag{27}$$

is in fact assumed to *"forget"* all its past experiences almost immediately and to transfer stresses only according to *"time-smoothed"* velocity gradients.

If we let f^t denote the *history* of the function f up to time t, as the function "cut off" at the present time

$$f^t(t_0) \equiv f(t - t_0) \quad 0 \le t_0 < \infty \tag{29}$$

then a typical expression for strictly causal evolution (i.e. deterministic evolution) is

$$g(t) = \Phi(f^t) \tag{30}$$

where Φ denotes a functional (i.e. a function whose arguments are functions). Eq. (30) states that if the history of f up to time t is known, the value of $g(t)$ is determined. Here f^t may be called the *"cause"* and g the *"effect"*.

Throughout mathematical physics, various special principles of *determinism*, involving the histories of various specific functions, are laid down as *"constitutive relations". All theories of "classical" physics are based on this kind of determinism.* Even in modern theories of thermodynamics or fluid dynamics, a *"cause"* may be the *history of a thermokinetic process,* while its *"effects"* are the *fields* of stress, specific internal energy due to dissipation, and the heating flux (Lecture II).

However, there exist materials with fading memory for which the recollection should be defined in terms of the *complete history.* For such materials the dissipation function Φ, developed in Lecture II may not be exact.

Analysis of shock waves in classical acoustics, for instance, combines two ideas: first, analysis of vibrations of a perfectly elastic continuum (which derives from Newton, Euler, and others), and, second, a thermodynamic theory of changes in pressure, composition, density and temperature. A more realistic approach should take fading memory into account.

In classical physics one can use causality to characterize *open systems*, i.e., systems in which there are *"external" forces*. The external force may be arbitrary, and could thus be characterized as the "cause" of accelerations inside the system. For instance, the external force of *gravity* may be "the cause" of *momentum generation* in a fluid, a cause that must be specified as a function of space and time coordinates.

IV.17 DOUBTS AS TO THE UNIVERSALITY OF ENTROPY

Many fundamental problems have been associated with the concept of entropy. To many minds the universal validity of this concept has failed, and this failure favored the advent of entropy-free thermodynamics.

According to E. T. Jaynes[*] "entropy is a property, not of the physical system, but of the particular experiments you or I choose to perform on it". In relativistic thermodynamics we face the question: Should entropy remain invariant (as is usually postulated) in different frames of reference? (cf. my CRT). A study of the thermodynamics of electrical networks has demonstrated the non-existence of a unique entropy value in a state obtained during an irreversible process.[**] Moreover, the justification for applying the second postulate (expressed in terms of entropy) to characterize spatial systems containing many galaxies or supergalaxies, is the postulate on the *additivity* of entropy (that is, the total entropy is equal to the sum of entropies of all parts of which the observed system is composed). It means that if the second postulate applies *locally*, it must also apply *universally*. An important question is, then, whether entropy remains an additive quantity over large volumes of isotropic space. Up to a point, the amount of information required to describe the contents of a given volume of the *isotropic* universe undoubtedly increases with its volume. However, the content of a volume whose dimensions greatly exceed the scale of the observed local irregularities (i.e., that of galaxies or supergalaxies), is largely predictable (since the universe accessible to our observations is homogeneous and isotropic (see the "Cosmological Principle" in Lecture VI). Since only a finite quantity of information is required to specify (for us) a finite (albeit large) portion of the homogeneous and isotropic world accessible to our observations, then, according to D. Layzer,[***] the entropy per volume approaches zero as the volume increases indefinitely beyond the scale of supergalaxies. If such a conclusion is correct (which is far from certain), the very concept of additive entropy fails in describing very large isotropic systems.

Other problems associated with entropy are methodological. For instance, in most current textbooks on thermodynamics the reader is provided with an arsenal

[*] *Am. J. Physics*, **33**, 392 (1965).
[**] See for instance J. Meixner in *CRT* pp. 37, 47.
[***] In *Relativity Theory and Astrophysics* (Ehlers, J. ed.), *Am. Math. Soc.*, Providence, R. I., 1967, p. 237. But see our doubts as to Layzer's use of the concept "information" in §IX.2.6.

of terms like *"heat bath"*, *"infinite reservoir"*, *"ideal cycle"*, *"quasi-static process"*, *"a system nearly in equilibrium"*, etc.—confusing wordplay which students are trained to use parrot-like in the face of all the logic and mathematics which they have already acquired. They are then presented with the strange "axiom"

$$ds \geq d'q/T \tag{31}$$

and are instructed to believe that a differential can not only be larger than another but can even be larger than something which is not a differential!

*   *   *

Whether or not the concept of entropy fails, one is justified in raising the following objections:

Can entropy-free theories replace classical thermodynamics? In particular, can such theories be useful, when based on sounder foundations than current (entropy-based) thermodynamics?

In trying to answer these questions I next examine a few alternatives to entropy. But even if the alternatives prove logically sounder than entropy, they do *not* necessarily mean that the concept of entropy should be abandoned altogether. Nor do they prove that entropy is an impractical tool for many terrestrial applications.

●IV.18 ENTROPY-FREE THERMODYNAMIC ARROWS OF TIME

From Lecture II we know that, when space and time are considered, the energy equation takes the form

$$\frac{Du}{Dt} = -\frac{1}{\rho}\operatorname{div}\mathbf{J}_q - p\frac{Dv}{Dt} + \frac{1}{\rho}\Phi + b \tag{II.59b}$$

The equivalent of Eg. (II.59b) is the first law of thermodynamics, which can be expressed in the simple form

$$du = d'q - d'w \tag{32}$$

where u, q and w are (specific) internal energy, heat, and work per unit mass, respectively, and where space and time coordinates have been left out. It is a *time-symmetric* law, i.e. it remains invariant under the time-reversal transformation $t \rightarrow -t$ (which does not distinguish directly between the transformation of one energy form to another). It also remains valid for conversion of *internal* into *kinetic* energy, i.e. when

$$\Phi \leq 0 \tag{33}$$

or for the reversal of this process, i.e., when

$$\Phi \geq 0 \tag{34}$$

Furthermore, it cannot distinguish between conversion of *internal* into *electro-magnetic* energy when

$$b \leq 0 \tag{35}$$

or vice versa, when

$$b \geq 0 \tag{36}$$

Consequently, there is no "built-in" time asymmetry, or irreversibility, in any formulation of the first law of thermodynamics.

In modern theories of thermodynamics the essence of the second postulate is reflected in the inequalities

$$\Phi \geq 0 \tag{37}$$

$$b \geq 0 \tag{38}$$

which, in agreement with all macroscopic observations, single out the one-way transformations:

$$\{\text{Kinetic Energy}\} \rightarrow \{\text{Internal Energy}\} \tag{39}$$

$$\{\text{Electromagnetic Energy}\} \rightarrow \{\text{Internal Energy}\} \tag{40}$$

Referring to eqs. II.48, II.49, II.50 and II.59 from Lecture II one obtains:

$$\Phi = -\tau : \text{grad } \mathbf{u} = \rho \frac{Du}{Dt} + \text{div } \mathbf{J}_q + \rho p \frac{Dv}{Dt} - \rho b \geq 0 \tag{41}$$

Equation (41) may be called *Planck's inequality*.
For Newtonian fluids it means that

$$\mu \geq 0 \tag{42}$$

$$3\lambda + 2\mu \geq 0 \quad \text{or} \quad \kappa \geq 0 \tag{43}$$

where

$$\lambda = \kappa - \frac{2}{3}\mu.$$

The last are known as the *classical inequalities of Duhem and Stokes*. Their physical meaning is that *the viscous stress never gives out work in any motion, but may only use up work in transforming it into internal energy*. This includes *radiative stresses*, which may be resolved into an *isotropic pressure of radiation*, P_R, and viscous stresses, τ_R, exactly similar to the material ones τ_m and P_m. The general case can then be defined by

$$\tau = \tau_m + \tau_R \tag{44}$$

$$P = P_m + P_r \tag{45}$$

Inclusion of the radiative stresses aims at generalization of the theory for astrophysical calculations. One must first note that Φ is defined as the *time-rate of dissipation*. To evaluate time's arrow we must focus our attention on the change in the amount of internal energy due to internal dissipation. For that purpose, one must perform at least two measurements at instants t_1 and t_2 (obviously without knowing *a priori* if $t_1 > t_2$ or $t_1 < t_2$).

We recall now that the energy equation, as derived in eq. (II.52), is based on the assumption that the *internal energy is an absolutely continuous additive function*. Thus, if the amount of dissipated energy ϕ, (i.e., the internal energy gain due to internal dissipation in the moving mass element), is observed to be changed, for instance when

$$\phi(t_2) > \phi(t_1), \tag{46}$$

we can use inequality (38) with eq. (41) to conclude that

$$t_2 > t_1 \tag{47}$$

Hence $\phi(t_2)$ was evaluated *at a later instant* and the observer can define the general direction of change in time and state that this direction (or arrow) is pointing towards the future. In this way the dissipation of energy shows (*without* actually applying to the concept of *entropy*) that *actual motion is asymmetrical in time*. Of course, this arrow must vanish when all velocity gradients dissipate.

In a simpler derivation of this arrow we may follow the behavior of an "isolated" macroscopic system into which the medium considered is introduced at a given time. *Heating or cooling* of a solid or a fluid contained in an "isolated" system produces a *macroscopic movement*, i.e., expansion or contraction (cf. eq. (41) when $b = 0$ and $Du/Dt = 0$). This shows that Φ *is the amount by which the increase in ϕ exceeds the heating not produced by the volume $p\, Dv/Dt$, or the radiation work b.*

Introducing the concept of total energy E of the system, the second postulate may be restated as:

> The free energy of any "isolated"
> macroscopic system never increases.

Defining "free energy" as $F = E - U$, where U is the internal energy of the system, another entropy-free arrow of time can be formulated.
Thus if

$$F(t_1) > F(t_2) \tag{48}$$

one concludes that

$$t_1 < t_2 \tag{49}$$

Observations demonstrate that all these arrows point in the same direction.

We can now turn our attention to the inequality $b \geq 0$. Its definition for an unpolarized electromagnetic field is given by eq. (B.59a) and it means an irreversible transformation of electromagnetic energy into internal energy. This leads to a *non-negative resistance tensor* (in a linear approximation) and to the familiar *Joule heat generation*. We distinguish here between *"contact heating"* associated with the *heat flux* $\mathbf{J}_q$ and *"body heating"* associated with b *(generation due to radiation)*, the former being an *absolutely continuous function of surface area* while the latter one of *mass*. The internal dissipation ϕ is therefore influenced by the local heating, be it through the surface heat transfer or the body supply b.

Formulations for b similar to those of eq. (II.59a) may be adopted for polarized systems which, with dielectric relaxation (the magnetic case is analogous) as the single process, leads to the observed positive coefficients characterizing these relaxation phenomena (i.e., in agreement with the familiar Debye equation for dielectric relaxation).

Original painting by Oli Sihvonen, Taos, N. M.

Fig. IV.8 **Thermodynamic interactions between radiation, vacuum and matter? between particles, vacuum and fields? Our perception of the interactions between structures, fields and the vacuum is highly limited (§V.6 and Lecture IX, §IX.2.20).**

THE CRISIS IN QUANTUM PHYSICS

Quantum physics
formulates laws
governing crowds
and not individuals.

Albert Einstein

Everything is foreseeable
But permission is given.

Rabbi Akiba Ben Joseph
(The "Bnei-Brak School")
Killed by Romans between 130–140

Even the categories in which experiences are subsumed,
collected, and ordered vary according to the social position
of the observer.

Karl Mannheim

(Of Einstein's rejection of a probabilistic universe) Many of
us regard this as a tragedy, both for him, as he gropes his
way in loneliness, and for us, who miss our leader and stan-
dard-bearer.

Max Born

V.1 PRELIMINARY REVIEW

This lecture constitutes an attack on what I believe is the Achilles' heel of the more "established" currents in contemporary philosophy of science, in particular those associated, linked, or based on a false philosophy of quantum physics. *My thesis is based mainly on the Einsteinian view of physics and philosophy and, in particular, on Einstein's hope for the ultimate achievement of modern unified field theory.* Indeed, the recent advance of some unified field theories (q.v. the Introduction and Lecture IX), gives the strongest support to the (presently unpopular) deterministic Einsteinian view of physics.

The modest progress achieved so far in the incorporation of probabilistic quantum physics into a deterministic-relativistic physics is *not*, in itself, a proof of the Einsteinian contentions. Yet, it clearly demonstrates the fact that, *today, more than ever, there is an objective need to construct a broad physico-philosophical framework*, which does not, *a priori*, deny such a unification (and may even serve to guide further developments of a unified approach to nature). Consequently, it is imperative to reexamine, with a critical eye, many traditional, or "well-established" interpretations of the conceptual foundations of the quantum theory, especially those that might prevent, or contradict, such a unification (Assertion 5, §3, *Introduction*).

The first group of physico-philosophical concepts which require reexamination includes the *relativistic causality and determinism versus the probabilistic, indeterministic, and "uncertainty" views of some "established" quantum mechanical interpretations.*

Other physico-philosophical concepts that require re-evaluation (and perhaps even modification), include the quantum-mechanical concepts of time, time asymmetries, irreversibility, "information", "memory", initial conditions, "prediction", "retrodiction", "observation" and "observables", as well as the problematic quantum-mechanical views of "measurement".

Some 'well-established' interpretations of quantum theory that have been put into question include the *state vectors* ψ (often called *wave functions*), of which the *time dependence* is given by the time-symmetric (see below), *Schrödinger equation.**

$$ih \frac{\partial \psi}{\partial t} = H\psi \tag{1}$$

Schrödinger first referred to the ψ function as a *"mechanical field scalar"* (see below for some historical notes). But its interpretation "rules" today are entirely different. Each ψ *'predicts'* (see, however, our objections below), a probability distribution over possible 'outcomes' of (successful) measurements of arbitrary observables $\mathbf{A}$. The (theoretically) possible 'outcomes' are the Eigenvalues $\mathbf{A}_n$ of the operators 'corresponding' to these 'observables'. Now, if ψ is expanded in terms of the corresponding eigenfunctions ϕ_n of such an operator, the absolute squares $|\mathbf{C}_n|^2$ of the expansion coefficients are interpreted as 'probabilities for these outcomes'.

One immediately wonders if these state vectors represent something like electromagnetic or gravitational fields, *or are they dependable upon the whims of their user? Can the state vector be measured? To what extent is quantum theory deterministic* (as Schrödinger equation is), and to what extent is it *not*? Can it be assigned to an *individual* elementary object, or is the theory *a priori* limited to *ensembles* of objects? What should be the elements (members) of the ensembles described by quantum theory? Do the state vectors correspond to the method of *'preparing'* a system *before* the measurement takes place?

Quantum theory describes *"pure states"* by the state vectors. But physical states are more often *"mixed states"* than "pure states". (A mixed state may be described by a classical probability distribution over "pure states".)** The 'measurement problem' is then complicated by the fact that the Schrödinger equation does *not* cause transitions of "pure states" into "mixed states". How can this problem be solved if the *initial* state of the entire ensemble (*before* the measurement) was a "pure state"?

Conventional quantum theory claims that *'the result'* of 'successful measurements' is describable as a "mixed state" for the ensemble of micro-objects. But since an *initial* "pure state" of these objects *'in interaction' with the 'measuring apparatus' cannot become a "mixed state" by the Schrödinger equation, we end up in contradictions and various controversies*; e.g., are object and *'measuring apparatus' separable*? How should one draw conclusions about the objects *after* measurement from the state of the ensemble *'interacting with apparatus' and the 'outside world' during the measurement*? Should the theory allow *"reduction of the state vectors" after* the measurement?

---

* This is a *linear* differential equation. Hence, it is subject to the mathematical procedures of *superposition*. H is the *Hamiltonian function*, t is the time co-ordinate, and h is the Planck's constant. H may assume various mathematical expressions, e.g., in the so-called 'interaction picture' it is sometimes transformable to the *Tomonaga-Schwinger equation*, especially when it is used in quantum field theory. See below for the *limitations* associated with the use of this equation in "quantum measurement irreversibility", "pure" and "mixed" states, initial conditions, individual objects, ensembles, interaction with the outside world, time-reversal transformation, etc.

** Note that the entropy of a pure state is zero and that it cannot evolve into a mixed state (see below).

Can one use *"memory* state vectors" to eliminate the need for "reduction of state vectors"? *Is this "reduction" a scientific problem or a question of personal caprice?*

It is a remarkable fact that, with one exception, all known laws of physics are invariant under time reversal transformations, i.e., that they do not, by themselves, provide a direction in time, an irreversible evolution towards the future or the past. Yet, all evidence shows that the world is asymmetric in time, i.e., that it is an evolving, non-static, dissipating ensemble of smaller interacting ensembles, which, in turn, are evolving, non-static, dissipating, and so on.

For many years physicists believed that *Boltzmann's famous 'proof'* of the *"H theorem"* has resolved this difficulty. It was, however, a naive belief.* It can be easily shown that the function H can decrease towards the future as well as toward the past time.** Similar conclusions apply to a series of other "H theorems" that are based on quantum logic. But in quantum theory we must distinguish between two (apparent) time asymmetries; those that arise from *classical* statistical mechanics and those that are inherent to *"quantum measurement theory"* (which is a highly problematic subject in itself).

An unresolved difficulty in quantum theory is *'the introduction'* of time-*asymmetry* into its time-*symmetric* formalism⁺. Elaborate 'procedures', such as *"the elimination of off-diagonal elements of the density matrix"*, have been artificially imposed (together with elaborate, but misleading interpretations–see below), just to make the formalism *'irreversible'*, i.e., just to obtain *entropy increase*, or *H function decrease* with time; just to obtain *"mixed"* states instead of the unrealistic *"pure"* states, etc.

Indeed, I can cite many other 'mathematical', or word-play 'procedures' by which respectable physicists try to conceal the uncomfortable fact that, *to secure agreement with empirical thermodynamics, they are forced to impose asymmetry on the symmetric equations of classical or quantum-statistical mechanics.*⁺⁺ Thus, the object of such

---

* Most textbooks contain the so-called *'proof'* of this theorem using the assumption of 'molecular chaos' *(Stosszahlansats)* which imposes time asymmetry by such 'procedures' as "the positions and velocities of the particles were uncorrelated *before* they collided" (based on the presupposition that "*a future* colision could *not* influence the properties of particles that were about to collide"; see also below and the notes on the *Loschmidt* and *Zermelo* "paradoxes" in Appendix I).

** cf. our discussion of time symmetry in probability theory (§V.5).

⁺ All known physical systems, with the possible exception of those involving K meson (§IV.11), possess Hamiltonian operators H that are invariant under T transformations:

$$T^{-1}HT = H$$

By operating on the Schrödinger equation (1) with T, and by defining a new operator called *antiunitary*, equation (1) remains symmetric. This leaves the physical content of quantum mechanics reversible (unchanged), and is usually summarized in the *'quantum principle of micro-reversibility'*, But Schrödinger equation is not only *time-symmetric* but *deterministic* as well (see below).

⁺⁺ E.g., "correlated vs. uncorrelated outgoing particles"; "neglecting off-diagonal matrix elements for the density matrix"; *"unitarity of the scattering matrix"*; "the matrix element of the interaction between 'in' state—in the infinite past—and the state of free Hamiltonian." " +t in world lines"; "linear relaxation which leads to the *Fokker-Planck* type equation"; "the ergodic hypothesis"; *"Markovian chains"*; "mixing"; *"retarded potentials"*; "*a priori* probability"; *"coarse-graining"*; *"infinitely large* systems";

a quantum-mechanical jargon is often the same: to smuggle irreversibility into reversible equations without declaring the 'contraband'.

V.1.1   The Effect of Gravitation and the Outside World on Quantum Physics

That no physical system can be totally isolated from its environment is a fact accepted by all theorists. For we know that any isolating wall is made of atoms, or a magnetic field, *which is not impervious to external electromagnetic and gravitational effects.* This simple fact carries with it an enormous number of physico-philosophical conclusions that must be examined together. To begin with we make the following *preliminary remarks.*

1) A fundamental assumption in *statistical mechanics* is the existence of a *total Hamiltonian* for the system, i.e., including its *interaction with the walls.* For *rigid (non-expanding-contracting) walls*, a potential energy term may be added to the Hamiltonian operator. This permits the application of *canonical equations* of motion leading to the *Liouville theorem*, and subsequently the *H theorem* and *irreversible evolution in time.* But there are a number of strong objections to this procedure. One of them is that *if the walls are not rigid, but expanding or contracting in some way, then we cannot obtain the Hamiltonian and the canonical methods cannot be valid.* Yet, if the atoms and fields of the expanding-contracting walls are made part of the description, this difficulty may be partially removed. But then the matter and radiation *'outside'* the walls would have to be considered, and so on, till *'the whole universe'* is taken into account. *This is practically impossible; not only in the limited domain of quantum physics, but also in the broader context of general relativistic cosmology* (Lectures VI, VII and IX).

2) Additional objections are as follows.

There is a close relation between the assumed *'emergence of irreversibility'* in the above-mentioned system (by "coupling to the outside world stochastically" through the walls of the container), and the assumed 'emergence of irreversibility' in a *quantum system* due to *"external coupling"* with the *'measuring apparatus'.*

Now, the 'measuring apparatus' is, itself, made up of atoms, which also must be subjected to the laws of quantum physics. If instead of considering only the object system to be described quantum mechanically, with the apparatus considered as *classical* (the *Copenhagen interpretation*), the entire system of object *plus* measuring apparatus is now regarded as one large coupled quantum system, the behavior of which is (in principle) described by an evolution operator for the *whole* system, then this evolution is still *reversible! How, then, has irreversibility entered into the system?*

There have been many unsuccessful attempts to avoid this and similar difficulties (see below), by employing *untestable* interpretations of the foundations of the quantum formalism.

"information-loss in the interaction"; "dominance of *outgoing* waves"; *"A sound physical theory* must entail the *evolution* of the system, therefore. . . ."; "Blind statistical *prediction* is physical, whereas blind *retrodiction is not."*

One famous attempt has been advocated by *Wigner**. It claims that the *origin* of irreversibility and time asymmetry of the quantum measurement process *"is located in the consciousness of the observer"*, i.e., time asymmetry is claimed to be 'introduced' by an *'exterior mind'* (not the brain) of the experimenter who 'looks' at the pointer of the measuring apparatus. This *'psycho-physical parallelism'* has introduced into physics a *subjectivistic interpretation* which, in turn, is frequently associated with subjectivistic-idealistic philosophies. Our objections to this interpretation are enumerated below, in Lecture IX, and in Appendix III.

In another famous interpretation, *Bohm*** proposes a *deterministic* subquantum world of *'hidden variables'* (see below). The effect of measurement is then explained by an equation which *is made* asymmetric *by imposition of asymmetric probabilities*. Our objection to this procedure is explained later in this lecture.

Other unsuccessful attempts have been made by *von Neumann**** and *Everett.*[+] It is, however, the Everett theory which contains some kernels of interest to gravitism. These kernels are not easily distinguishable for, on one hand, we share Belinfante's main objections to the Everett theory,[++] and, on the other hand, *we find Everett's theory of a "universal wave function* ψ*", "memory state vectors", "reduction of state*

---

* Wigner, E.P., *Amer. J. Phys., 31* 6, 1963; *Found. of Phys. 1*, 35, 1970; *Remarks on the Mind-Body Question*, Chap. 98 in I. J. Good's *The Scientist Speculates*, Basic Books, N.Y., 1962; See also Shimony, A., *Amer. J. Phys., 31*,755, 1963; F.J. Belinfante, *Measurements and Time Reversal in Objective Quantum Theory*, Pergamon Press, Oxford, 1975; and P.C.W. Davies, *The Physics of Time Asummetry*, Univ. of Calif. Press, Berkeley, 1974.

** Bohm, D., *Phys. Rev., 85*, 166, 1952; *85*, 180, 1952; *96*, 208, 1954; *Rev. Mod. Phy., 38*, 452, 1966; *Quantum Theory*, Prentice Hall, N.Y., 1951.

*** von Neumann, J., *Mathematical Foundations of Quantum Mechanics*, Princeton Univ. Press, Princeton, 1955.

[+] Everett, H., *Rev. Mod. Phys., 29*, 454, 1957; an expanded version appeared in *The Many-Worlds Interpretation of Quantum Mechanics*, B. S. DeWitt and N. Graham, eds., Princeton Univ. Press, Princeton 1973. See also DeWitt, B. S., *The Everett-Wheeler Interpretation of Quantum Mechanics*, chap. xii of *Batelle Recontres*, DeWitt, C. M. and J. A. Wheeler, eds., W. A. Benjamin, N.Y., 1968; J. A. Wheeler, *Rev. Mod. Phys., 29*, 463, 1957; and B. S. DeWitt, *Phys. Today, 23*, 30, 1970.

Everett's interpretation of quantum theory is essentially that the universe as we know it is just *"one arbitrary member"* of an *"entire ensemble of universes"*, and that at some time in the past, *any* part of the universe has interacted with some other part, and, therefore, it would seem as if we would be forced to reject any talk about a state, of some part of the universe, and that only *"the universe as a whole"* would be in one state or another. But describing the entire universe by a single state vector is *practically impossible*! Moreover, why should we bother to think about other members of "the ensemble of universes" *if we cannot observe them?* These are only two simple objections to Everett's theory. Other objections are to Everett's claim that "with each succeeding observation (or interaction), the observer state 'branches' into a number of different states. Each branch represents a different outcome of the measurement and the corresponding eigenstate for the object-system state. All branches exist simultaneously in the superposition after any sequence of observations."

[++] Belifante, F. J., *Measurements and Time Reversal in Objective Quantum Theory*, Pergamon Press, Oxford, 1975; and *A Survey of Hidden-Variables Theories*, Pergamon Press, Oxford, 1973. Cf.. also. Jammer's book which is mentioned below.

vectors", "branching of the universe", and the origin of 'quantum measurement irrevers-
ibility', as sufficiently instructive to promote further inquiry of the following set of inter-
connected problems:

1) The possibility that *'quantum measurement irreversibility'* is nothing but a myth,
or a hoax, and that the whole issue of asymmetry and irreversibility should be re-
examined in connection with the *asymmetric* effects of *gravitation* as explained by
gravitism (Lecture VI).

2) The possibility that the *'coupling' of a quantum system to the 'outside world'*
is mainly through the deterministic field of gravitation and that an essential property of
the measurement apparatus is its initial condition (Lecture IX).

3) Having understood the nature of the 'quantum measurement irreversibility'
(see below), one should re-examine the following quantum-mechanical 'paradoxes'
and controversies:
   A) Can wave mechanics only predict *future* probabilities, or does it in fact force
us to give equal attention to *'retrodiction probabilities'* for what happened in the *past*?
(see below).
   B) Does quantum physics become *'incomplete'* when its asymmetry in time is removed?**
   C) Was Einstein right in claiming that quantum mechanics is paradoxical and
incomplete?*** *Does it reject causality and determinism?*
   D) Can the *mathematical* formalism of quantum theory *ever* supply us with a
time arrow? (see below).
   E) Since quantum theory is a theory about properties of *'ensembles'*, can its formalism be used to derive *universal-philosophical 'principles'* about *individual* elementary particles and inherent *'uncertainty'* in nature? Can we assume the *SEPARA-*
BILITY of *elements* (members) of the 'ensembles' discussed by quantum physics?

These are subtle issues and we cannot discuss them all in a single lecture. Instead,
we shall focus attention on the *conceptual limitations* of quantum mechanics, beginning with problem E and then proceeding with problems D, C, B, and A.

In this context I shall stress the fact that most of the "well-established" quantum-
mechanical interpretations have gradually been undermined since Einstein first attacked them; others have been exposed as myths, or wishful beliefs, for which no
experimental support was ever found. Indeed, we shall demonstrate that Einstein
was right; Bohr, Heisenberg, von Neuman and others were misled by a philosophy
that would stand neither the test of time, nor that of the theory of time and gravitation.

  ** Cf. Y. Aharonov, P. G. Bergmann, and J. L. Lebowitz, *Phys. Rev. 134, B,* 1410, 1964.
  *** Cf. *The Einstein-Podolsky-Rosen paradox* (originally published in *Phys. Rev. 47,* 777, 1935) as
discussed by numerous authors, e.g., Belinfante, Everett, Wheeler, Davies, Gold, and Jammer. Some of
these authors also discuss the less important issues associated with the so-called *Schrödinger's cat* and *Wig-*
ner's friend paradoxes or rather apparent paradoxes. See also B. d'Espagnat, *Conceptual Foundations of*
Quantum Mechanics, W.A. Benjamin, Menlo Park, Calif. 1971.

V.1.2   The Three Main Schools of Thought*

Over the last decades there have been remarkable developments in thermodynamics and quantum mechanics. Yet many of the basic problems have remained largely unsolved. In the spectrum of opinions expressed by authors who have attempted to solve these problems, one can roughly distinguish three main schools of thought:

1) Traditional *axiomatic* thermodynamics with some refined modifications, which however, cannot explain the origin of irreversibility and time asymmetries.

2) *The statistical school*, which generates *"man-made irreversibility"* or "man-made statistical evolution" by imposing asymmetric conditions on symmetric equations (or concepts) in order to describe the observed behavior of (relatively small) local systems (see below).

3) *The new gravitational school*, which *deduces* the origin of evolution, irreversibility and electromagnetic and thermodynamic time asymmetries from gravitation and the large-scale (nonequilibrium) dynamics of gravitationally-induced processes. The latter, in turm, are intimately related to the *non-static* properties of the (time-symmetrical) *field equations of the gravitational field*. This school includes some new modifications to the physics of time and is part and parcel of gravitism (Introduction).

Because the basic ideas of the first school are by far better covered in the textbooks, and because of the limited length of these lectures, the latter are concerned mainly with the last two schools of thought. We therefore turn now to examine the problems associated with the second school, and, in particular, with false ideas that are traditionally associated with the concepts of *"statistical evolution"*, *"statistical asymmetry"*, etc.

My object in this connection would be to show that, to obtain a consistent mathematical formulation of entropy growth, irreversibility, evolution, and asymmetry in probabilistic theories, one needs *'to break'* the symmetric properties of the statistical-probabilistic equations. So far physicists have done this by means of (*a priori*) imposing on these equations an (often hidden postulate of) asymmetry (which, *in itself*, is equivalent to the very results that the mathematical analysis is aimed "to prove"). *In other words, the (asymmetric) "results" (of classical and quantum-statistical physics) are not really results for they are deliberately enforced on, not deduced from the (symmetric) formalism (§V. 5).*

Such quantum-mechanical tactics are, as I shall explain, unacceptable. Moreover, they lead to a highly misleading philosophy of science, which, in turn, has generated a ludicrously incongruous mode of thinking in education and in general philosophy (see below). *Their advocates all fall into the trap of assuming that which they set out*

---

* Adopted from my article in *Science*, *176*, 11 (1972) and from my reports in *Nature 230*, 12 (1971), *234*, 217 (1971); *Annals, New York Academy of Science*, *196* (A6), 305 (Oct. 4 1972); *Found. Phys. 6*, 407 (1976); *6*, 623 (1976); *Space Sci. Reviews 22*, 119 (1978).

to prove. One cannot get uncertainty for certainty, nor indeterminism from determinism (see below).*

These conclusions lead to somewhat surprising results, which, in turn, are related to the very core of our physical theories and to the role of time, causality, and determinism in philosophy.

My viewpoint, as expressed in these lectures, stems from my long-time search for an explanation of time asymmetries, evolution, irreversibility and thermodynamics within the various frameworks of quantum mechanics; a search which has systematically disclosed that quantum mechanics is *not* a universal theory, but rather should be viewed as a practical "*tool*" in characterizing phenomena on a *limited* physical scale—a "tool" that must be incorporated into a more universal framework.

In spite of their importance in physics and philosophy alike, the fundamental problems of quantum mechanics have rarely, if ever, been evaluated from the combined points of view of the theories of time asymmetries, thermodynamics, evolution, structure, relativity, cosmology and philosophy. The present study is designed to fill this gap.

Contrary to this viewpoint, *most* physicists (and non-physicists) possess today an *unshakable belief* in the ability of quantum mechanics to explain and deduce the origin of thermodynamics, evolution, structure, and time asymmetries from its formalism. This belief stems, perhaps, from the fact that most scientists are today so thoroughly conditioned to the artificial imposition of the quantum-statistical postulates, that they hardly pause to consider their divertive consequences. With the present aversion to physico-philosophical inquiry in physics, an attempt to displace the resulting semi-sacrosanct myth calls for more than a proof of the fallacy involved; for more than the authority of Einstein; perhaps even for more than the full impact of a body of new empirical data in a wide spectrum of interconnected fields of study; it calls for an entirely new approach to the methodology of academe (Volume II).

V.2   EINSTEIN'S OBJECTIONS TO THE 'UNCERTAINTY PRINCIPLE'

Never in the history of physics has there been a theory which scored such spectacular successes in predicting an enormous variety of phenomena (in the domain of microphysics), as quantum mechanics; nor has there been a theory which aroused such unprecedented controversies and dissension in the interpretation of its basic formalism.

---

* §V.5 shows that the hitch sometimes lies in such concepts as 'molecular chaos,' 'a priori' probability, 'prediction', 'superposition', 'ergodicity', 'unitarity', etc. and, sometimes, *in the very act of selecting the mathematics* which 'fit in' with the classical notions of thermodynamics and electromagnetism (Lecture VI). In short, the "results" are smuggled into, not deduced from the mathematics.

Other fallacies associated with "prediction", "observation", "Complementarity", "indeterminism", "Uncertainty Principle", and the so-called "statistical arrows of time" are discussed in §V.4 and §V.5.

Albeit quantum mechanics implies a drastic revision in traditional physical concepts, its *applied* formalism seems to be widely accepted. Thus, it is the *physical meaning* of this formalism that divides past and present communities of scientists and philosophers into embattled camps. The war between the partisans of the various "schools" is waged today unabatedly, and with the discovery of new problematic topics in quantum-mechanical interpretations, its intensity is now at its highest peak. *Einstein, who has started some of these battles almost alone, showed the way to his successors. "Well-established" quantum-mechanical "conclusions" have since been shattered or seriously undermined.*

In the following notes we confine ourselves to preliminary remarks about simple conceptual limitations associated with such "conclusions".

Perhaps the most famous battle fought by Einstein was his objections to the essntially Heisenbergian *"Uncertainty Principle"*. Today most physicists admit that there are essentially two opposing interpretations to this "Principle". These are*:

1) The (Heisenbergian) *nonstatistical interpretation*, according to which it is impossible, *in principle*, to specify precisely the *simultaneous* values of canonically conjugate variables (such as momentum and position, or energy and time) that describe the behavior of a *single (individual)* micro-object. This *non-statistical-single-object* interpretation had been a factor (see however below), in influencing Bohr to claim that nature harbors some kind of fathom *indeterminism* (or 'vagueness'), to which strict causality and determinism 'must yield'. It is an unfortunate fact that this interpretation is advocated today by almost all textbooks on quantum mechanics, without giving any, or at least equal space, to the Einsteinian interpretation which is given below.

2) The (essentially Einsteinian) *statistical interpretation*, according to which strict causality and determinism remain universally valid even in the domain of quantum physics, *because Heisenberg's mathematical uncertainty relation correlates merely the scatter in an ensemble (kollective) of, say, position 'measurements', with that in an ensemble of momentum 'measurements' (see below).*

This interpretaion (originally put forward by *de Broglie*, and later refined by *Margenau*** and others), has recently gained a growing attention in the West as well as in the Soviet Union. Since the acceptance of one interpretation or the other is of paramount importance to *philosophy of science*, as well as to *general philosophy*, we shall return to these and related Einsteinian views in later discussions. At this stage we prefer to stress briefly a few historical processes, beliefs, and controversies that have been, directly and indirectly, associated with these fundamental topics.

* For a full account of these interpretations as well as some newer modifications, see, e.g., Ballentine, L.E., *Rev. Mod. Phys., 42,* 358 (1970); Park, J. L. and E. Margenau, *Inter. J. of Theor. Phys., 1,* 211 (1968) and Appendices A and B in Belinfante's 1975 book.

** Margenau, H., *J. Phil., 64,* (1967); *Found. Phys., 3,* 19 (1973); *Phil. Sci. 30,* 1 (1963); *30,* 138 (1963); *25,* 23 (1958); *The Nature of Physical Reality,* McGraw-Hill, New York, 1950. See also Park, J. L. *The Measurement Act In Quantum Physics,* in *Vistas in Physical Reality,* Laszlo, E. and E.B. Sellon, eds., Plenum, N.Y. 1976.

It is, first of all, a remarkable *socio-economic* fact (and perhaps even a remarkable *socio-historical* issue), that, on one hand, many *unverifiable, unrefutable,* and *non-scientific arguments* have been imposed on these historical controversies, and that, on the other hand, the final adoption of one interpretation rather than the other, has been based on ideology, theology, and a sort of 'voting majority' by which Einstein had, eventually, been 'cast out' (cf. Appendix VI).

Secondly, we note that the non-statistical interpretation advocated by Bohr, Heisenberg, and their supporters, has been based on abstract arguments involving *untestable* (see below), and *unverifiable* claims, whereas the Einsteinian statistical interpretaion has been made as a *straightforward, logical-mathematical consequence of the very formalism of quantum theory* (see below).

Thirdly we note that the highly problematic quantum measurement theory is part and parcel of this controversy–*a controversy in which some interpret the Heisenberg relation rather as a confirmation of causality than its denial.* Indeed, today we know that Einstein understood the physico-philosophical meanings and the important consequences of quantum theory better than Bohr and Heisenberg ever did; that Bohr simply smothered Einstein's cogent analysis in a grotesque form of scientific philosophy based on that fog called *"Complementarity"* (see below); that the majority chorus which rose to condemn Einstein's "naivity", had pushed logical positivism into solipsism until the 'established' philosophy of physics degenerated to a subjectivistic blend of physics and some doubtful claims of psychology (see below).

Fourthly, we note that rarely in the history of science has there been such a famous "principle" *with so little experimental backing.* In fact, no method is available for measuring the position and momentum of an individual electron *simultaneously* and with sufficient precision to evaluate the errors involved.* (In 1959 Ryason** suggested direct examination of the energy-time uncertainty by pulsed-field desorption on a metal tip at extremely low temperatures (0.4°K), applying the technique of field-ion microscopy; the proposed experiment, which Ryason declared to be "technologically feasible" even at that time, *has never been performed.*)

Thus, the (essentially Einsteinian) statistical school, rejecting the nonstatistical interpretation, insists that the latter has never been confirmed experimentally. It also maintains that its own interpretation cannot be rejected without introducing major modifications into the quantum-mechanical formalism.

Jammer, in his comprehensive review***, states that those who argue that the

---

* The concepts "uncertainty principle" and "complementarity principle" (see below) arose from Bohr's and Heisenberg's exaggeration of the physical meaning of the uncertainty relation. *The first step in this exaggeration was elevation of these concepts to the higher rank of "principle"–a term which I avoid– both for the aforementioned reason and for others explained below.*

** Ryason, P. R., *Phys. Rev., 115,* 784 (1959).

*** Jammer, M., *The Philosophy of Quantum Mechanics,* Wiley, N.Y. 1974. This is by far the most comprehensive and updated treatise on this subject, written from a historical perspective.

Heisenberg relation, as applied to a *single* micro-object, has never been confirmed experimentally, certainly have a strong point.

V.3   THE HERESY OF A FEW SKEPTICS

Once announced, Heisenberg's relation, or the "Principle of Indetermination", as it was incorrectly referred to on the first occasion by Ruark,[+] immediately became a favorite topic of discussion in physics and subsequently in other fields of science and philosophy (albeit, its physical meaning and the limitations imposed by it on the formalism, were highly distorted by wishful thinking). Certain writers, taken by surprise, yielded to an old urge to defy the dominant causal rule. Released from the traditional rule, they happily accepted the "natural" rule of "chance", "free will", "uncaused caprice" and indeterminism. Scientists, and non-scientists alike, were thus encouraged to believe that science and the world are really based on "chance", "uncertainty", and "indeterminism". So, they thought, is the case with the human mind: *old-fashioned determinism was to give way to indeterminism and to "anthropocentric subjectivism" of "the free human consciousness". The outcome seemed almost inevitable: most contemporary scientists left old causal doctrines and determinism to join the popular new religion of indeterminism, chance and "free will". Reappraisal of old deterministic doctrines quickly became fashionable among best-seller writers, and, occasionally, theologians.*

*In their wake, an enormous amount of "scientific" literature was produced with the sole aim of spreading indeterminism among students and warning them against the "heresy" of the few skeptics who obstinately adhere to the views of Einstein (and Planck), who consistently retained a deep conviction in rigorous determinism and caus-ality and even claimed that the quantum mechanical interpretations are incomplete representations of nature.***

V.4   MYTHOLOGIZED CONCEPTS OF QUANTUM PHYSICS

Heisenberg's rejection of causality was, at the time of its inception, nothing short of a revolution, especially when set against the background of the well-proven causal relationships of Maxwell's electromagnetic theory, Newtonian mechanics, Newtonian gravitation, and the special and general theories of relativity. Since then, it has become the "*cornerstone of modern physics and philosophy*".

As for Bohr, he first claimed that determinism and causality *fail* in the quantum domain of micro-objects since the *laws of conservation of momentum and energy can have only a statistical validity for individual processes!* He was, however, forced to

[+] Ruark, A. E., *Bull. Amer. Phys. Soc. 2*, 16 (1927); *Phys. Rev. 31*, 311 (1928).

** In his later years, Einstein tried, without success, to derive the quantum representation from relativity rather than have it superimposed on the latter (see §IV.12.13 and Assertion 2 in the Introduction).

change his claim since experiments (notably by Bothe and Geiger) *proved him wrong, in that rigorous conservation holds for individual impacts between photons and electrons. Since this verification of the conservation theory verified also the causal relations (Lecture IV), it could be concluded that the validity of rigorous causality was re-established even for individual micro-processes.*

Consequently, Bohr adopted *"an abstract compromise"* between determinism and indeterminism, short of outright rejection of the causal doctrine. He called it the *"Principle of Complementarity,"* and at one time sought to apply it to such areas as *physiology, psychology, biology,* and *sociology!* (see his *"Atomtheorie und Naturbeschreibung"* (Berlin, 1931), and *"Causality and Complementarity,"* Dialectica, II No. 3–4, [1948], 312–319). He even advanced the claim that *complementarity is potentially valid in all areas of systematic study!*

The Origin of Complementarity

But what is, indeed, the actual origin of complementarity? Meyer-Abich reports* that among German scientists it was remembered that Bohr used to cite *William James* (q.v. Appendix III) and only a few other western philosophers. Moreover, a day before his death (November 18, 1962), he mentioned having read James' book, *"The Principles of Psychology"* (1890) (Dover, N.Y. 1950) before he arrived at his "Complementarity Principle."** In an article in 1929***, he undertook lengthy excursions into psychology in the search for analogies that, in the opinion of both Meyer–Abach and Jammer,+ also refer directly to the chapters on *"The Relations of Minds to Other Things"*, and on *"The Stream of Thought"* in the same book. There (p. 206) *"complementarity in psychology"* is defined in the following words:

> "It must be admitted, therefore, that *in certain persons,* at least, *the total possible consciousness* may be split into parts which coexist but mutually ignore each other, and share the objects of knowledge between them. More remarkable still, they are *complementary.* Give an *object* to one of the consciousnesses, and by that fact you remove *it from the other or others.* Barring a certain common fund of information, like the command of language, etc., what the upper self knows the under self is ignorant, and *vice versa.*"

In the same section James cites experimental evidence which demonstrates that some hysterics could deal with certain sensations *only* in one consciousness or in the other, *but not in both at the same time!*

A few months after the first presentation of complementarity at the International Congress of Physics at Como (September 1927), Høffding (a professor of philosophy at the University of Copenhagen and one of Bohr's mentors), stated that *"Bohr*

* Meyer-Abich, K.J., *Korrespondenz, Individualitat und Komplementaritat,* Wiesbaden, 1965.

** Holton, G., *The Roots of Complementarity,* in *Thematic Origins of Scientific Thought,* Harvard University Press, 1973.

*** Bohr, N., *The Quantum of Action and the Description of Nature, Atomic Theory and the Description of Nature,* Cambridge University Press, N.Y. Macmillan, 1934.

+ Jammer, M., *The Conceptual Development of Quantum Mechanics,* McGraw-Hill, N.Y. 1966.

declares that he has found in my books ideas which have helped the scientists in the 'understanding' of their work, and thereby they have been of real help..."[++]

An admirer, like James, of *G. T. Fechner* (the father of psychophysics), Høffding devoted his first book to psychology. At about the time Bohr took his philosophy course, Høffding used the occasion of the St. Louis meeting of 1904 to visit James in the United States; James, in turn, supplied an appreciative preface for the English translation (1905) of Høffding's *Problems of Philosophy*—a book which Høffding reported later to have originated in his university lectures in 1902. And in the same year of his visit to James, Høffding expressed in his *Moderne Philosophen* his admiration for James' work, to whom the concluding chapter is devoted, with such comments as "James belongs to the most outstanding contemporary thinkers . . . The most important of his writings is *The Principles of Psychology*."[+++]

Additional evidence regarding this origin of complementarity is provided by Meyer-Abich, Jammer, Moore and Klein*. Popper and Bunge** maintain that the *context-dependent character of the quantum representation should be freed from "the ghost called 'consciousness' or 'observer,' or else its name must be changed to 'psychology'."*

Occult entities or qualities are inadmissible to science for they cannot stand the testability criteria (Assertion 11, Introduction). Indeed, Von Neumann and Wigner were wrong in maintaining the intrinsic role played by '*consciousness*' in all physical processes and, hence, in all physical theories.***

V.5 THE FAILURE OF CLASSICAL AND QUANTAL STATISTICAL MECHANICS TO DEDUCE IRREVERSIBILITY AND TIME ASYMMETRIES

Proposition I   Two symmetric probabilities must always be considered, namely, the probability that something *will* happen and the probability that

---

[++] Moore, R., *Niels Bohr: The Man, His Science, and the World They Changed* (Knopf, N.Y. 1966), p. 432. In this book Moore states that "If any doubts existed of Høffding's influence on Bohr's life, it was settled by the placement of his portrait" (next to his father, mother, brother and grandfather, on one wall of Bohr's house in Carlsberg). Indeed, both Høffding and Bohr's father (who was a professor of Physiology at the same university) had, in several ways, shaped much of Bohr's ideas and preoccupations. Høffding was a close friend of the Bohr family.

[+++] Holton, G., *op. dit.*, pp. 144.

* Op. cit.; see also Klein, O., Glimpses of Niels Bohr as Scientist and Thinker in *Niels Bohr: His Life and Work as Seen by His Friends and Colleagues*, Stefan Rozental, ed. (Wiley, N.Y., 1967).

** *Op. cit.*

*** An interpretation of quantum mechanics originally formulated by J. Von Neumann and in recent years notably amplified by E. P. Wigner, states that there are no *eigenstates* unless there is an *interaction-observation* with *"conscious being"*. Otherwise, quantum theorists must regard all states, in macroscopic or microscopic objects, as being distributed superpositions, governed rigorously by the quantum mechanical laws of state functions. Since, according to this interpretation, *"indeterminism"* enters "with

something *did* happen, i.e. statistical *prediction* is precisely symmetric with statistical *retrodiction*! Therefore, *a priori* and *a posteriori* probabilities are symmetrical under time reversal, i.e. *no distinction as to the direction of time can be derived from statistics.* Recourse to statistics (or to any "averaging process over detailed molecular motions," etc.) cannot, *by itself,* result in time asymmetry, time's arrow, or irreversible behavior.

Proposition II   The very choice (or even regular use) of the concept of PREDICTION alone is a *prejudgement which singles out one direction in time.* Consequently, the very choice of, say, an *a priori* probability to characterize the behavior of a physical system, is precisely equivalent to *a priori* imposition of a postulate about time's arrow, and is, therefore, equivalent to the procedure of postulating irreversibility in thermodynamics. The precise point where irreversibility is imposed on quantum mechanical formalisms is frequently unrecognized, even by "professionals." Sometimes it is done deliberately, without due declaration. Such undeclared procedures give a false image to many statistical-mechanical formalisms–an image of fundamentality in which irreversibility is (wrongly) believed to be *deduced* from the formalism *itself.*

Proposition III   The use, or the very *"choice",* of the concepts mentioned in Proposition II, *is precisely equivalent to imposition of the "fact-like" second postulate of thermodynamics.*

Proposition IV   The common concepts, "statistical arrow of time," or "statistical irreversibility", must be abandoned. They possess no physical meaning, and their very use introduces misleading concepts. Nature, therefore, incorporates no "statistical evolution", nor any "statistical arrow of time".

    To conclude: The origin of time asymmetries and irreversibilities in nature cannot be found in any formalism of quantum or classical statistical mechanics (no matter how "sophisticated" the latter). A rigorous statistical-mathematical derivation of irreversible macroscopic equations from reversible microscopic dynamics is impossible.

Proposition V   *Causality and determinism are universal principles of nature. Both are perfectly compatible with classical and quantal statistical mechanics. In no way are they rejected by quantum mechanics.*

Proposition VI   Various formalisms of quantum mechanics are required because of

---

the passage to an eigenstate", *the locus of individual events of Nature is in the consciousness of men. Thus consciousness has immediate relationship in rejecting determinism and in introducing time's arrow to physics!* Wigner's approach is therefore an appeal for an intrinsic role for consciousness in all physical processes and, hence, in all physical theories. It is tantamount to *"hegemonization"* of quantum mechanics.

the dual wave-corpuscular appearance of micro-objects an/or the result of "hidden variables" (Bohm, DeBroglie, etc.) It is not due to an alleged "uncontrollable influence of measurement" (Heisenberg, etc.), nor to a fundamental "indeterminism" governing the behavior of single micro-objects in nature (Richenbach, etc.).

Scholium I. To illustrate, amplify and comment on Propositions I to IV, we may now consider a simple example. Take, for instance, the "statistical theory of coarse-grained density distribution", which we *would like* "to show us" that the entropy of an (isolated) macroscopic system, at time t_2 is greater (under non-equilibrium conditions) than its counterpart at time t_1. The fact that $t_2 > t_1$ is, however, used nowhere in the current "proofs"; hence the entropy can increase both towards the future, and towards the past. Since t_1 has no absolute value (and we must consider consistent evolution of the system from negative to positive time, or vice versa), the system apparently passes through a minimum at t_1, in contradiction with macroscopic observations, the "second law", etc. In these circumstances, textbooks have no option but to claim that "blind statistical *prediction* is physical" whereas "blind retrodiction is not"—an *a priori* postulate *imposing* asymmetric evolution towards positive time.

In other words, what is widely believed to be a *"deduced"* time asymmetry, the "statistical law of increase of entropy", "mixing", "the H-theorem", and so forth, is no more than a deliberately imposed one; an *a priori* opting for asymmetric mathematical evolution towards positive time. The origin of irreversibility in nature is, therefore, not to be found in the "procedures" of statistical mechanics.*

Scholium II. Statistical physics involves other fundamental contradictions (and inconsistencies), summarized by Landau and Lifshitz [13], and more recently by Jammer and others [5, 6, 7, 8, 9, 10, 11, 12, 32, 34, 55, 56, 101, 134;—for these references see Part III]. Most important of these is the observed nonexistence of "statistical cosmological equilibrium". This fact was considered by Landau and Lifshitz as a "serious contradiction" of the statistical formulation of irreversibility and its various attempts to "explain" the "required" approach of a macroscopic system to equilibrium.

Scholium III. At the Cornell Conference on "The Nature of Time" (1963), Rosenfeld

---

* According to Beauregard [CRT], this a priori boundary condition (which is an initial condition), reads "blind retrodiction forbidden," very much as the boundary condition in macroscopic wave theories of electromagnetism reads "advanced waves forbidden." Thus, according to Beauregard, there is a close connection between electrodynamic irreversibility (retarded potentials), the temporal application of Bayes's probability postulate, and the causality concept. The Einstein–Ritz controversy (in which Einstein maintained that the law of wave retardation should follow from the principle of probability increase, while Ritz insisted on deducing the law of entropy increase from the principle of wave retardation) is thus resolved. Even in the remarkable exposition of the foundations of statistical mechanics by O. Penrose [39], the additional postulate is the Markovian one. To ensure that the system is dissipative, the ergodic theory postulates the "mixing" condition [83].

a long-time associate of Bohr, defending Bohr's and contemporary views of quantum statistical mechanics, claimed that *irreversibility and time asymmetries are imposed by the observer, not created by the system!* In short, *the choice of initial conditions* (*a priori probability, initial states, etc.*) *is not a law of Nature, but only a result of our own positions as macroscopic observers!* This view is strongly opposed by our school of thought. The following points are stressed in refuting the above claim:

a) *All laws of Nature are abstractions based on our own "observations."* Aside from the "Second Law", all laws are *strictly invariant* (*and reversible*) *under the time-reversal transformation* $t \rightarrow -t$ (Lecture IV). *Symmetric causality* is involved in all balance and conservation equations in the macroscopic scale (Lecture II), and in all *differential equations* which describe the physical laws (including, for instance, the Schrödinger equation, the Liouville equation, Einstein's gravitation field equations, etc.). Hence, *why should only the "initial" conditions* (*as opposed to "final" conditions*), *or only irreversibility, or causation, be the result of our "consciousness," or of our mind, or of our "observation?"* (But see also Lexture IX).

b) There is no justification for imposing an exclusive "observer" in the case of *initial conditions, but not in other laws of nature* (to say nothing of very distant (*cosmological*) regions of the universe accessible to our "observation").

c) *Cosmological initial conditions are not "created" by anthropocentric effects.* Our "observational" light cone cannot affect most astronomical observations, for these are so universal in scale, time, and distance, that much more than an oversimplified myth is required to account for them.

d) Any successor to the various complementarity interpretations should account not only for the *origin of "local" initial conditions,* but in the same context, for their *"global"* counterparts, such as in *geophysics, astrophysics* and *cosmology* (see below).

*   *   *

Scholium IV. Classification of all laws of nature as *initial or boundary conditions* on the one hand, and *differential equations* on the other, is mathematically *equivalent to integration* (where the two are combined). In fact, both scientists and engineers, in attempting to draw definite conclusions from their premises, must be able to *integrate differential equations* (which, as they stand, are all but *useless!*). Many of the problems encountered by pure mathematicians originated in the need to assist in such integrations. *Why is the integrated equation so important in the exact sciences? And why do we have to distinguish between boundary* (*and initial*) *conditions and differential equations? Indeed, what is the universal physical meaning of such integral-differential equations, and what is the physical origin of the boundary conditions?*

According to the general theory of relativity, free bodies, such as photons or planets, follow the geodesics of 4-dimensional space–time. In the vicinity of matter, space–time becomes curved, so that the bodies pursue curved paths. To estimate these paths, we must calculate the curvature of space–time from point to point, and then plot the geodesics. However, the differential equations of the general theory of relativity do not define this curvature, for the latter depends on the *"distribution"* of energy, matter and momentum considered (Lecture III).

General relativity, or any other advanced theory of gravitation, furnishes a set of mathematical relations which govern the class of possible curvatures throughout space–time. To obtain the exact curvature pattern in a given physical system, we must *integrate* Einstein's equations and then "restrict" the solution so as to conform to the requirements of the situation in question. Determination of the geodesics calls for additional integrations, for these geodesics are themselves controlled by the differential equations.

There are important differences between classical physics and general relativity. In the former case, we must stipulate the *boundary values* of the solution over the *bounding surface!*, although in the general theory of relativity, the boundary conditions *may be furnished directly by the theory!! This means that the general theory, of relativity (and only it!) may tell us about both causality and causation, or, perhaps, about both boundary conditions and differential equations.* These questions are further examined in Parts III and IV (in particular in §IX.2.4), while Table 1 summarizes our preliminary conclusions.

TABLE 1: Differential Equations, Initial and Boundary Conditions, and Physical Laws
(For a broader set and its origin see §IX.2.4.1).

Group I Differential Equations	Group II Initial and Boundary Conditions
1) Symmetric causality, time-reversal invariance.	1) Causation, arrows of time (time asymmetries).
2) Reversible and time-symmetric equations.	2) Irreversible processes, entropic growth.
3) Conservation and balance equations.	3) Second postulate of Thermodynamics.
4) Time-symmetric general relativistic field equations.	4) Dissipation, shear viscosity, stress work, bulk viscosity, etc.
5) Equilibrium Thermodynamics, etc.	5) Non-Equilibrium Thermodynamics, etc.
6) Global, local, and internal symmetries.	6) Symmetry breaking.

Group III

A useful system of complete physical laws results from combination of Group I with Group II; e.g.:

1) Integrated differential equations (e.g., solutions of Maxwell's equations in electrodynamics; §VI.10)

2) Conservation and balance equations which describe actual irreversible processes and are subject to both causality and causation (e.g., the energy equation with asymmetric dissipation, see eq. II.50).

3) Solution of any field equation.

4) Supergravity and the unification of the four basic forces in nature.

V.6 THE EMERGENCE OF QUANTUM CHROMODYNAMICS AND SUPER-SYMMETRY

V.6.1 Spatio-Temporal Approach to Quantum Physics

During the first decades of the development of quantum physics it was often stated that the concepts of time and space are intrinsically *inapplicable* at the quantum level of reality. However, the various arguments most frequently used in support of the *"unreality of quantum-mechanical time"* were eventually shown to be *fallacious* (see also §IV.15). In light of the irrelevancy of the uncertainty relation to the reality of (non-relativistic and relativistic) time, and as a consequence of the great advances of Quantum Electrodynamics (QED) and Quantum Chromodynamics (QCD), we now have sufficient (observational and theoretical) findings to support our 2nd Assertion, namely; that *any universal-unified theory must be based on the concepts of symmetry-asymmetry and space-time, or their equivalents.* Hence, we conclude this lecture by stressing the most important physico-philosophical conclusions that have recently emerged from quantum field theories (see also §IX.2.20).

1) Unlike the old quantum theory, the new relativistic quantum *field* theory describes a particle as an *excitation* of the corresponding *wave field mode*. In fact, the essential feature here is the coupling between different wave fields. This requires a new way of thinking about particles and microphysical processes. Instead of thinking about single subatomic particles, we treat them as forming a field and regard a particle (such as an electron) as an excitation of a mode of this field. As an analogy one may regard the creation of, say, a photon, as plucking of a guitar; as a sudden increment in the excitation of one of the modes of *vibration*. Thus, instead of 'particles' we must think about *'field quanta'*. (E.g., an electron is a quantum of the electron-positron field, a proton a quantum of the proton-antiproton field, and so on (see also under solitons)).

2) By building up diagrams of the type shown in the first Figure of the Introduction, all known microscopic phenomena can be pictorially described in space and time. Such diagrams are called *Feynman diagrams*, after Richard P. Feynman (Nobel physics prize 1965), who first used them as an aid to computation in quantum electrodynamics. This theory, the quantum *field* theory of the electromagnetic interactions of charged particles, is *a most successful gauge theory* (Assert. 2b). It replaces the Maxwell theory of electromagnetism (§VI.10) and is *renormalisable*; i.e., in terms of renormalised expressions all the predictions of the theory are finite (see below). But, most important, QED's predictions for precisely measured quantities, such as the *anomalous magnetic moments* of the electron and muon, and the Lamb shift in hydrogen, *are extremely accurate.* (Agreement with experiment for the magnetic moment, for instance, *is better than one part in 10^9!) This stunning success is one of the greatest triumphs of modern theoretical physics,* and it gives us confidence that the underlying concepts of relativistic quantum field theory are leading to a very promising unified field theory. Moreover, the incredible accuracy of QED demonstrates that *renormalisable gauge theories provide the best tools for unification with*

the other three forces of nature. Indeed, the celebrated *Weinberg-Salam unified field theory* of the electromagnetic and weak interactions (which is also a local-symmetry gauge theory), contains the incredible accuracy of QED.

V.6.2. From Weinberg-Salam Theory to Quantum Chromodynamics

During the last decade we have witnessed a great revolution in our understanding of the weak and electromagnetic interactions. Previously, a phenomenological model of the weak interaction allowed correlation of data concerning nuclear β-decay and the decay of subatomic particles such as muon, pion and the so-called *strange particles* (as well as neutrino-induced reactions and muon capture by nuclei). All these processes were attributed to a self-coupling of a single *'charged current'*.

The new Weinberg-Salam theory is based on the so-called $SU(2) \times U(1)$ gauge[*] group (often called *quantum flavordynamics* (QFD)—see below). It describes and unifies all known weak and electromagnetic phenomena. Moreover, it has been very successful in predicting new phenomena; the now observed neutrino *'neutral current'* scattering cross-sections, and it also treats the *violation of* parity (cf. §IV.11) *symmetries* (in deep-inelastic polarised electron scattering).

Indeed, what is most striking about the new neutrino experiments, is that they reveal an interaction long thought to be absent: *weak transitions with no exchange of electric current!* The spectacular confirmations of these predictions form the basis for the present confidence that there is at last a most promising set of basic concepts that can lead us to a more unified physics. These include the (spatio-temporal) *gauge fields* and the concepts of *symmetry and asymmetry.*

But the subtle symmetry-asymmetry concepts discussed so far are, in fact, more complicated, for they actually involve large cosmological structures (Lecture VI), as well as conservation laws and symmetry principles in subatomic physics (see below).

V.6.2.1 Conservation Laws As Symmetry Principles; And Vice Versa

The fundamental link between symmetry and conservation laws in physics has already been stressed. In light of the spectacular success of the Weinberg-Salam theory we now take a second look at some of its consequences.

The most familiar example of a (local symmetry) *gauge theory* is QED. Here the symmetry group is called $U(1)$ (corresponding to the freedom of changing the phase of any electrically charged field; a symmetry that, in fact, follows from charge conservation).

---

[*] $SU(2)$ means that the group of symmetry transformations consists of 2×2 unitary matrices that are special in that they have determinant unity. The meaning of $U(1)$ symmetry is explained below.

$SU(3)$ is a special unitary transformation in 3 dimensions, most conveniently represented by 3×3 matrices of complex numbers. These matrices must be unitary (i.e., the inverse of such a matrix is the transpose of its complex conjugate).

Now, extending this to a *local U(1) gauge invariance* (Assertion 2b and 2c), requires the introduction of *a single gauge field; the electromagnetic vector potential* (which becomes the *photon field* in the quantum field theory).

A few general remarks must be made at this point.

1) *Newton's mechanics, quantum physics and the special and general theories of relativity all possess the fundamental property that their laws of motion are invariant under symmetry operations such as rotations and translations. This is a basic symmetry law of nature* [which may be appended with other symmetry operations, such as space reflection and time reversal (§IV.10–11)]. But there is more to it. It also means that there is a deep-rooted connection between *geometrical symmetries* of space-time and the *dynamical properties of the so-called MATERIAL objects* (or the quanta that we observe experimentally).

For instance, *the conservation law of energy* follows directly from the *translation symmetry of the time coordinate* (one moment is as good as any other).

Moreover, *the law of conservation of angular momentum* follows from *rotational symmetry*, whereas the *law of linear momentum conservation* follows directly from the *translation symmetry of space* (Lecture II).

2) The underlying basis for these conservation-symmetry principles is *an axiom* which states that *space is strictly symmetric under translations, rotations, etc., i.e., if one part of space is the same as every other, why should one place be the source of a sudden creation of a particle? Or why should a particle start spinning without external propulsion? Why should it rotate, say, clockwise rather than anti-clockwise?* In addition, a rotating object defines an arrow, or a special direction in space. *But if space is symmetric under rotation, no direction can be distinguished or preferred.* Hence, we conclude that *no object can start linear motion, or rotation, or spinning 'spontaneously', or without a force (whose origin is in 'outer space' to the object, or in a structure larger than the object under consideration). This conclusion is a most powerful one, for it leads us immediately to physical conservation laws, on one hand, and to space-time dynamics of larger and larger spaces-structures—up to cosmology—on the other hand* (Lectures VI to IX).

3) Symmetry is not only the basis of conservation laws and 'outer-space' dynamics. It is also the *unifying principle* in physics and philosophy; *a principle without equal!* But it goes hand in hand with *asymmetry* principles; especially those associated with 'time direction' and physical observations. Some of these problems are treated in Lecture IX in connection with human perception of space, time and asymmetry. Here we only introduce a few remarks concerning global, local, isotopic, SU(3), and SUPER symmetries.

V.6.2.2 Global, Exact, Approximate, Isotopic and SU(3) Symmetries

The powerful properties of symmetry allow us to describe objects (or concepts) that appear *unrelated* in *different* theories (or sets of observations). Let us consider a few examples.

1) The special theory of relativity is based on *global symmetry* (Assertion 2 and

Lecture III). It is a *limited* symmetry, and, as such, is not a unifying principle for different theories such as general relativity, QED or the Weinberg-Salam theory. But by examining carefully the last three theories we find *a unifying principle* in terms of *local symmetry* (theories with local symmetries are called gauge theories). Hence, we expect that *any theory unifying the four forces of nature be also based on the local symmetry concept, i.e., to be formulated in terms of gauge fields* (see below and in §IX.2.20).

2) Only systems with negligible "features", "structure", or "coupling" may be characterized with certain *"exact symmetries"*. Thus, *'empty space'* is invariant under continuous translations, rotations and reflections. Here time also possesses symmetry;—the reflection of time reversal. As there is no activity, there may not be a distinction between 'past' and 'future' time directions! However, space is *not* totally empty, nor free from fields and the complex *activities of the vacuum* (Lecture IX). Hence, those purely conceptual symmetries which are *exact* in *'empty space'* are only *approximately* valid, *or not at all,* in, say, *interstellar space.* [Another example is the *'violation'* of reflection symmetry by magnetic charges or by dissipative energy processes (Lecture VI).]

3) A specific symmetry called *ISOTOPIC SYMMETRY* is employed in nuclear physics. It is, however, a *global* symmetry, and, hence, *limited* in its use. It only allows us to establish a relation between particles *with the same spin angular momentum,* such as the proton and the neutron. Both have a spin of $1/2$ and the assumption of isotopic symmetry allows us to regard them as two alternative states of a single undifferentiated object, *the nucleon* (which may be imagined as a particle with an arrow in some *imaginary* space; if the arrow points, say, *'up'* we call the particle a *proton;* if it points *'down',* a *neutron*).

Then, to express the charge independence of the strong interaction, we may say that it is *invariant* under rotations in *isotopic* spin space.

Now, if a system is invariant under rotations in ordinary space, it means that it does not matter *in which direction* the system is oriented, the forces are just the same. In the same sense isotopic symmetry means that the strong force does not distinguish the 'up' orientation (proton) from 'down' (neutron). However, in the same sense that a magnetic field *destroys the symmetry of ordinary space,* the electromagnetic force *destroys the isotopic spin space symmetry.*

4) It is interesting to note that isotopic symmetry, as other symmetry concepts, establishes a conservation law; *the isotopic spin conservation law (which is a much more powerful law than mere charge conservation* that is embodied in it). It predicts that, as far as hadronic processes are concerned, isotopic spin will be conserved. This leads to a certain (but limited) simplification in ordering the variety of subatomic phenomena. However, the weak interactions, which violate time-reversal and parity symmetries, also violate isotopic symmetry! *Hence these symmetries are not universal nor exact!*

All this means that if both the weak and the electromagnetic interactions (as well as the so-called strangeness-discriminating part of the strong interaction), are "switched off" all the particles in a baryon octet (which forms an *hexagonal* figure in a plot of isotopic spin projection against strangeness) will have *the same mass*. This (restricted) type of *'hexagonal symmetry' is called SU(3) symmetry.* Hence, observed differences between these subatomic particles may be regarded as due to SU(3) symmetry violation (or *'symmetry breaking'*). Similar concepts may be devised for other subatomic particles (families of ten- the decuplets, or the singlets).

Yet, it was the use of such limited symmetry groupings that led *Gell-Mann* to predict the properties of the Ω^- meson several years before its discovery. This prediction provides additional evidence of the power of symmetry concepts (even on an approximate basis, or in a limited domain of physics).

V.6.2.3 From SU(3) To Renormalizable Gauge Theories

The main problem of isotopic and SU(3) symmetries is their restricted nature. Being global symmetries they cannot, *as such,* be incorporated into a unified gauge theory (which requires local symmetry). Moreover, isotopic symmetry relates only particles that have *the same* spin! *It was, therefore, necessary to transform SU(3) symmetry from global to local gauge symmetry* (by finding a more universal symmetry concept that would also unite particles having *different* spins). This hope was realized recently by the discovery of *supersymmetry*. What is even more remarkable, it allows unification of the *gravitational field* with the other three fields; not a complete unification, but a very significant one. So important is this discovery that one must also consider some of its historical milestones.

The problem of extending isotopic symmetry from a global symmetry to a local one was solved already in 1954 by *C. N. Yang* and *R. Mills*. Their work showed that the number of required compensating gauge fields equals the number of group generators and that in the case of non-abelian groups [SU(3), etc.] *the gauge fields interact among themselves!* (This transition requires three gauge fields, each associated with a massless particle having a spin of 1—but see below).

These gauge field theories lay dormant until *Faddeev* and *Popov* found a procedure for *quantising non-abelian gauge theories*. It transformed them from *classical* field theories to *quantum field* theories (in which the interacting fields could be identified with quanta.) Another problem was resolved by *Gerard 't Hooft, Benjamin W. Lee, Martin J. G. Veltman* and by *J. Zinn-Justin* in the early 1970's. They showed that the unified theories of the weak and electromagnetic interactions can be *renormalized.** *This put the non-abelian gauge theories on as strong a footing as QED, and made them the best tools for the unification of the laws of physics.*

---

* Renormalization is applicable only in a class of quantum field theories where the *infinities* can be compensated for by correction of such parameters as the mass of the electron. The observed electron mass is the sum of the *'bare mass'* and the *'self-energy'* resulting from the interaction of the electron with its own field. However, the calculations lead to *infinite* self-energy. Hence, renormalization assigns a *negatively infinite value to the 'bare mass'*, thereby *canceling* the two infinities and yielding the observed *finite* mass of the electron. This sounds like a trick, but without it the stunning predictive success of QED and QCD is impossible. Moreover, the experimental confirmation of the Weinberg-Salam unified field theory, which is also a renormalizable theory, is giving further confidence in renormalization (see also §V.6.3.1).

V.6.2.4 Quark Confinement and Asymptotic Freedom In Gauge Theories

Are hadrons composed of quarks? So far only indirect evidence tends to support this idea and the idea of *permanently confined quarks.* In fact, many of the properties of hadrons seem to suggest that the quarks inside them behave, at high energies, as if they are rather loosely bound (i.e., as if they are bound by a force which *increases with distance,* but, when they are close together, the force is small and the quarks are loosely bound). Hence, the quarks may be trapped for ever. But if hadrons are made of quarks, *what happens when hadron-antihadron pairs are created?* Would it be possible to detect them during the brief phase of their creation, before they get bundled up into hadrons? *But, most intriguing, what is the nature of the strong force that binds the quarks together?*

Part of the answer was given in 1973 when *Politzer, Gross,* and *Wilczek* uncovered a remarkable property of non-abelian gauge field theories. What they found is essentially as follows:

1) *Because of the self-interactions among gauge fields, these theories predict that the strength of the interactions mediated by non-abelian gauge fields becomes vanishingly small at asymptotically high energies—or, equivalently, at very small distances!*

2) *In the asymptotic high energy regime these gauge field theories behave almost like free, non-interacting field theories; a novel feature with extraordinary consequences! For this was precisely the property observed in deep-inelastic scattering experiments: at high energies hadrons behave as if composed of loosely bound, free quarks!*

3) Non-abelian gauge field theories are *the only* field theories which exhibit *asymptotic freedom,* i.e., the strong interactions among quarks *must* be described by non-abelian gauge field theories, *and by no other field theory!*

4) Gradually, it became clear, that not only the asymptotic freedom of non-abelian gauge field theories fits successfully with the quark model for hadrons, but, in addition, *it fits with the requirement for three quantum numbers called color.** It were these early discoveries that opened the way for the emergence of QCD.

V.6.2.5 QCD and the Search For Higher Symmetry Principles

QCD is characterized by the following main features and assumptions:

1) Each quark flavor (u, d, s, c, b and so on) comes in *three colors* (see the foot-

---

* *Quarks are fermions.* Now, a particle such as Ω^- contains 3 identical strange quarks (sss) (see Table IV and Fig. 1b in the Introduction). However, *the Pauli exclusion principle* forbids 3 such identical particles to occupy the same quantum state. To avoid this, *Oscar Greenberg* proposed that each quark would be in three different varieties, arbitrary called colors (no connection with ordinary color). Arbitrarily he called them *RED, GREEN* and *BLUE;* so that *RED + GREEN + BLUE = WHITE* (colorless) and the Pauli principle is satisfied (for the quarks are no longer identical). Hence, 'assembled hadrons' are always 'colorless'! Thus, antiquarks must carry 'anticolor' and mesons contain one colored quark and a corresponding antiquark of the appropriate color. *However, the gluons must also be colored!* So, when gluons are emitted or absorbed, the quarks change color; in fact they do *not* remain of one color because of the continual exchange of virtual gluons.

note in the previous section). But *'assembled hadrons'* are always *colorless* (cf. Assertion 2e, Introduction).

2) The three colors of quarks transform as a fundamental triplet of the symmetry group $SU(3)_c$—where the subscript c stands for color. This symmetry is viewed as an *exact symmetry*.

3) The $SU(3)_c$ is a *local gauge symmetry* (i.e., it is expressible in *gauge-field terminology*).

4) The transformation from *global* $SU(3)$ symmetry to *local* $SU(3)_c$-gauge symmetry requires the introduction of *eight vector gauge fields*.

5) The eight gauge fields are associated with *eight colored gluons,* so that when gluons are emitted or absorbed, *the quarks change color.* (I.e., quarks do not remain of one color because of the continual exchange of virtual gluons).

6) *The strong force arises from the exchange of virtual gluons.* (In all quantum field theories the forces between particles arise from the exchange of *virtual* quanta. Note, however, that the basic difference between *classical* field theories, and *quantum* field theories, is that *only in the latter* can particles be *created and destroyed*—but see also our discussion of *SOLITONS* and solitary waves). *Hence, gluons are to QCD what photons are to QED.* (There is, nevertheless, an important *distinction* between gluons and photons; the former carry color, and therefore, *strongly interact among themselves,* while the latter are electrically neutral, and, therefore, cannot directly interact with one another. This distinction results from *the basic interacting properties of the non-abelian gauge fields that cause effective coupling between these fields to decrease at short distances*—§V.6.2.4).

7) *Weak and electromagnetic interactions are color blind* (while color interactions are flavor blind).* Moreover, the *leptons* carry *no color, so they are immune to the strong force.* (Leptons and quarks are considered here as 'elementary' particles).

8) QCD incorporates all the earlier successful predictions of the quark model. It also predicts all the observed strong interactions symmetries, flavor conservation, parity, and time-reversal invariance (and no additional, unobserved symmetries).

9) *If the Weinberg-Salam theory is appended to QCD it does not affect the predictions, except by small, calculable values,* nor does QCD disturb weak and electromagnetic phenomenology! A most promising feature of both theories is the fact that they are based on local-symmetry gauge fields—the same type of fields in which the general theory of relativity is expressible. Therefore, to unite all four forces of nature, one must search for a higher level of symmetry.*

10) QCD *testability* becomes possible in a number of ways. For instance, to compute QCD predictions for lepton-hadron and hadron-hadron scattering processes using perturbation theory. Here one observed effect that perturbative QCD accounts for is the *jet structure* of final state hadrons produced in *electron-positron annihilation at high energies* (the tendency of hadrons to be produced with very little transverse momentum in streams of narrow jet-like cones—a rigorously testing

---

* This does not contradict the requirement that the weak interactions can change the flavor of quarks (e.g., when a pion decays into muon and a neutrino, which are both leptons). Note also that confinement may not be essential to QCD (q.v. Ref. 65 in the Introduction).

procedure which has brought QCD an immediate success with recent high-energy experiments). Eventually one should be able to test rigorously the viability of QCD by evaluating hadron masses, the coupling between mesons and baryons, strong interaction cross-sections, and essentially all strong interaction processes.

11) The next step is likely to be in the direction of finding a *higher symmetry principle* that would bring *QCD* and *QFD* (the Weinberg-Salam theory is sometimes called *Quantum Flavor Dynamics*) closer to each other and eventually to construct a grand unified theory of strong, weak and electromagnetic interactions.

Such efforts are being done now by various investigators. Some of the simplest approaches involve only one gauge coupling constant, whereby all three interaction strengths become equal at very short distances, and, therefore, only at ultra-high energies (around 10^{10} Gev). In these types of models leptons and quarks belong to the same group representation. Hence leptons may be considered similar to quarks; yet not confined as quarks are, for they do not carry color.

An intriguing feature of some of these models is that the baryon number (Lecture II) may *not* be conserved. (Hence, a baryon like the proton may decay, but with a fantastically large half-life instability.)

12) In spite of the optimism surrounding the outstanding success of QCD, too many problems still confront it. Some are related to *the very nature and origin of color*. (Indeed, some make theorists change their own color when asked to explain QCD.) Because quarks, gluons and colored bound states have never been observed, theorists must explain why color is confined. Some believe that the *vacuum state* surrounding a hadron is a strong coupling regime in which the quarks and gluons are bouncing around in a sort of an *"elastic bag"*, like molecules in a ballon *("the MIT bag model")*. Other theorists believe that the color gluon fields form a flux tube (or *"string"*) between color sources *(the "string model")*. Then, if the colored flux is conserved the "potential energy" between sources rises with the separation, thereby implying "confinement of color".

To make matters worse, evidence seems to be mounting for the proliferation of quarks and gluons; a proliferation which makes nonsense the claims that they are "elementary particles". *Should we search for smaller subunits out of which to build the quarks or the hadrons and leptons? Does this uncovering of smaller and smaller structures ever end?* These questions may bring us back to *Salam's strong-gravity* elementary particles (Assertion 2d) or to the problems of "superspace" (§IV.13).

V.6.3 From Quantum Field Theories To Super-Symmetry/Super-Gravity

The search for higher symmetry levels of reality is always marked by the collapse of lower symmetry concepts. It is the unabated war against established physico-philosophical habits that promotes unification of the laws of physics and philosophy.

At stake is not only the unification of the four basic forces of nature, but the very basis of all our concepts; *from higher levels of conservation laws, to a better understanding of the subatomic world; from a more universal view of the macrocosmic world to everyday concepts of structure and perception.*

But there is more to it. The search for higher levels of symmetry principles is strongly linked to another search; the one that seeks higher, or more universal *asymmetries. This paralleled search is also marked by the collapse of lower, restricted asymmetry concepts, and by the continual battle against obsolete views ranging from subatomic to macrocosmic structures, and from initial-boundary conditions in mathematics, thermodynamics and cosmology to asymmetric concepts of time, space, gravity and perception* (§IX.2.20).

Indeed, the concept *SUPERSYMMETRY* is not dissociated from the search for the *MASTER ASYMMETRY* and the *MASTER ARROW OF TIME*. Hence we may view supersymmetry concepts as tentative, or at least as strongly coupled to the super*asymmetry* concepts to be discussed in the remaining part of this course.

But, what is the physico-philosophical meaning of supersymmetry? How is it linked to QCD, to the Weinberg-Salam theory and to the general theory of relativity? Part of the answer was given in Assertion 2b and in §V.6.2.5. It remains to be determined, however, whether these links are testable.

V.6.3.1 On The Limits Of Super-Gravity Unified Field Theories

Subject to the reservations stated above, super-gravity theories are also characterized by the following features:

1) They are based on *local symmetry,* as do QCD and QFD.

2) Using *local gauge invariance* they describe general relativity in the same language of the quantum gauge field theories described in §V.6.1 to §V.6.2.5.

3) As with QED, QFD and QCD, the *infinities* associated with these theories become relatively harmless by proper *renormalization* procedures. (In fact, the problematic infinities in Einstein's theory have turned out to be *finite* by super-gravity renormalization summing procedures.)

4) In the simplest supergravity theory, local supersymmetry becomes possible by the introduction of two gauge fields; the gauge field of the spin-2 *graviton*, and the spin-3/2 gauge field of another virtual 'particle' called *gravitino* (§III.5).

5) To accommodate the quarks, leptons and gluons of QCD, as well as the photons and the three virtual bosons of QFD, and to describe the present zoo of subatomic particles, local supersymmetry must be extended. E.g., it must include particles with spin of less than 3/2. *Indeed, many such extended super-symmetry/super-gravity theories have been devised recently.*

So far, however, this search has failed to produce an extended symmetry principle that would be universal enough to accommodate also all the known spin-1/2 and spin-1 particles. Hence, the search for higher symmetry principles must go on to uncover hitherto undiscovered symmetries and asymmetries. This raises the fascinating prospect that the most powerful symmetries and asymmetries of nature lie buried in the microscopic world of subatomic objects, or are, as yet, concealed from our telescopes. Yet, one of them, perhaps the most universal one, has already been detected by our advanced telescopes. We need only understand *what* these telescopes are revealing. For such an attempt we now turn to a new fascinating subject.

Part III

FROM PHYSICS TO COSMOLOGICAL CROSSROADS AND BACK

Again see you not that even stones are conquered by time, that high towers fall and rocks moulder away, that shrines and idols of gods are worn out with decay, and that the holy divinity cannot prolong the bounds of fate or struggle against the fixed laws of nature? Then see we not the monuments of men, fallen to ruin, ask for themselves as well whether you'd believe that *they* decay with years? See we not basalt rocks tumble down riven away from high mountains and unable to endure and suffer the strong might of finite age? Surely they would never fall suddenly thus riven away, if for infinite time past they had held out against all the batteries of age without a crash.

Again gaze on this, which about and above holds in its embrace all the earth . . .

Lucretius
(99–95 B.C.E., from Book V;
On the Nature of Things)

Every statement in physics has to state relations between observable quantities.

Ernst Mach

LECTURE VI

COSMOLOGY, PHYSICS AND PHILOSOPHY

All the rivers run into the sea
Yet the sea is not full

Ecclesiastes

VI.1 REDUCTION OF THERMODYNAMICS TO GRAVITATION

VI.1.1 Methodology

By now we have completed the preliminary part of the course. Now, in contemplating about the proper methodology to be used during the rest of these lectures, we note that the very nature of preliminary lectures on cosmology is, to some extent, analogous to a piecemeal scanning of terrestrial countryside from a watchtower on a dark night, using *our own* searchlight. In view of the limited sweep of the latter, all it can show, at a given instant, is an elongated ellipse of land, trees and rocks. It is only by the combined use of piecemeal observations, and memory, that the watcher can construct *a preliminary picture* of his surroundings. *But this time the watcher is invited to take a second look, with "starlight" as the only source of "illumination".*

Indeed, it is quite dark outside. Yet the "starlight" ranges from gamma rays, X-rays, and ultraviolet radiation to infrared radiation and radio waves. For the well-equipped observer it also includes polarised radiation, neutrino radiation, spectroscopic "fingerprints" of hydrogen, helium and heavier elements, and it reveals unfamiliar patterns of expanding intercluster space, formation of galaxies, explosions of old stars, clouds, dust, pulsars—in short, a wide variety of events, gigantic in scale and in energy, outrival the most interesting terrestrial view a geocentric observer can ever scan in the wildest countryside. *But, most important, the "starlight" illuminates all objects as a single interconnected whole. Thus, even the "starlight" itself becomes part and parcel of the whole new reality: of the reality exposed by an 'interconnected watchtower' : of the reality which emerges from a careful combination of physics, methodology, astrophysics, astronomy and philosophy.*

Gradually, a careful observer will realize that gravitation is the *prime mover* of all the large-scale phenomena that the "starlight" reveals in the emerging reality of interconnected dynamics of radiation and galactic objects. Gradually, he will come to the conclusion that, indeed, *nothing is so much the beginning of wisdom in science and philosophy as to understand the roles played by gravitation in nature: that gravita-*

tion is the master asymmetry of any unified theory; that it is also the origin of global and local phenomena, ranging from subatomic particles to stars and clusters of galaxies.

However, the relevance of gravitation to science and philosophy is subtle and indirect. It is especially not easy to grasp its indirect roles and relevance to the subject of irreversibility, entropy growth, energy dissipation, time asymmetries and the physics of time. The science that usually groups together all these problems is *thermodynamics*. Hence we shall begin our study of the physics of time and gravitation from the science of thermodynamics. In fact we shall prove that the science of thermodynamics can be reduced to the science of gravitation; that it is the theory of gravitation, whether Newtonian or Einsteinian, which explains the foundations of thermodynamics and the origin of energy dissipation and the arrows of time; both theories enable us *to deduce* the master asymmetry of nature (see below) from their respective formalisms, and from it *to derive* the entire superstructure of the science of thermodynamics. Hence thermodynamics forms an important part of a more general approach called *gravitism*, which unifies some other disciplinary studies in the natural sciences with a unified approach to gravitation and the physics of time asymmetries.

VI.1.2 Dialectical Gravitism: Definition of the First Problems

By employing the aforementioned methodology, in line with the eleven assertions of the Introduction, we can now define the first problems of dialectical gravitism:

■ Traditionally physicists *separate* the causes and the theories of various asymmetries observed in nature, notably, they separate between the thermodynamic, electromagnetic, cosmological, biological, and (more recently) the microscopic arrows of time (Lecture IV). Our first problem is *to unite* the causes (and the theories) of these asymmetries. Hence, our first aim is not limited to a proof that this separation is unjustified and misleading. It is concentrated on the effort to prove that *the separate phenomena are no more than different manifestations of one and the same effect.* For that purpose we shall first collect the observational and theoretical arguments which support or disprove our aim.

■ Lecture V has given us the proof that, after almost a century of trying, *no one has found a proof of time asymmetry within the framework of classical or quantum statistical physics.* Hence, our second problem is *to translate* misleading quantum-statistical asymmetries into a unified *master asymmetry* (or *super-asymmetry*).

■ By proving the physico-philosophical gains from the unification of asymmetries, initial-boundary conditions, thermodynamics and gravitation we shall try to demonstrate the possibilities for further unification with other concepts of science and philosophy. This last problem goes beyond the scope of dialectical gravitism and is treated in Lecture IX within the framework of a broader theory.

VI.1.3 Gravitation as Super-Asymmetry

The definition of our first problem leads us immediately to an enormous number of *secondary problems*, a sampling of which follows:

■ Identification of the observational data which point out to, and characterize, the earliest (and the most universal) asymmetry of nature, and, then, to express it in mathematical terms that can be part and parcel of a universal theory of asymmetry. We shall see below that this requirement leads to the physics of the radio background radiation discussed earlier in the Introduction.

■ Identification of the observational data which point out to, and characterize, the aforementioned asymmetry in *later* cosmic times, and in *smaller* scales, *down to the present time and the terrestrial scale.*

■ Identification of the physical processes which reduce the universal asymmetry to small-scale asymmetries, dissipative processes and irreversibilities.

■ Identification of the physical processes which introduce irreversibilities into our laboratory.

■ Unification of the various irreversibilities under the roof of a single super-asymmetry.

■ Reduction of thermodynamics to a theory of related asymmetries.

■ Incorporation of the super-asymmetry into a theory expressible only in terms of *ASYMMETRY-SPACE-TIME.*

Each of these "subproblems" is subdivided again into *specific* (disciplinary) problems, a sampling of which follows:

■ Formulation of simple Newtonian analog (or a *Gedanken-experiment*) to test the hypothesis of gravitation-asymmetry equivalence or cosmology-local physics equivalence (see below).

■ Following that hypothesis (or *gedanken-experiment*), and in agreement with available observational evidence (in local physics and in modern astronomy), to present the new mathematical formalism that derives the second law of thermodynamics from Newtonian gravitation, or from the theory of general relativity.

■ To prove that the new formalism has *both* theoretical and observational origins.

■ To derive *electromagnetic irreversibility* from the *same* origin of the *thermodynamic irreversibility*, as well as to suggest a plausible common cause for the recently discovered *microscopic irreversibility* in elementary particles.

We now turn to the observational evidence concerning the earliest asymmetry in nature.

VI.2   THE EARLIEST AND MOST UNIVERSAL ASYMMETRY: OBSERVATIONAL EVIDENCE

The evidence of most importance to gravitism refers to the discovery of the 2.7°K background radiation, the distribution of radiation densities in intercluster, galactic, stellar and solar mediums, and the red shift vs. distance relation known as Hubble law. Perhaps the most important of them all is the background radiation whose spectrum is characteristic of a state of thermal equilibrium, achieved through repeated absorption and re-emission. This radiation peaks at the microwave wavelength of about 1 mm and is a relic of the early times when the entire universe was dense, hot and opaque. It has since cooled off and shifted towards longer wavelengths, but retained its thermal spectrum (Introduction).

Einstein's field equations require that the cosmological scale factor, $R(t)$, should have been extremely small at some finite time in the past.* At this early epoch, matter and radiation were presumably in thermal equilibrium, with extremely high temperatures. Thus, prior to formation of local, self-gravitating aggregations of matter (galaxies and stars), the universe had been uniform, with radiation contributing overwhelmingly to the pressure pattern (in equilibrium with uniformly scattered nuclei)**; as the universe expanded, density, temperature and pressure decreased!

As the radiation we are now observing passed through regions with large red shifts relative to us, it would have interacted strongly with free electrons through the Thomson effect. With decreasing temperature, the particles decelerated until, at the key stage of about $4000°-3000°$K, the free electrons became so slow, that they could be captured by nuclei and retained in bound orbits. In this state they could no longer scatter photons, and opacity dropped sharply, breaking the thermal contact between matter and radiation (whereby the universe became transparent).

Whatever radiation existed at that time, has since been enormously red-shifted (by a factor of approximately 1500 since the universe became transparent). But it still fills the space around the earth. As the background temperature today is only $2.7°$K, we conclude that the universe has expanded by a red-shift factor of 1500 since scattering ceased (wavelength is inversely proportional to temperature).*** Hence, this black-body radiation provides us with information of unparalleled value as to the *thermodynamic history* of the universe, at least all the way back to the era when it became transparent. In it we have an *historical pointer*, a cosmological arrow of time. One may call it the second cosmological time asymmetry.

For a mathematical expression of the second cosmological arrow of time, we denote the moment, in past cosmic time,**** at which opacity dropped, by t_1. Then the proper energy density of the leftover photons, in the frequency interval (at the present time t_p) v to $v + dv$, is given by:

---

* For the physical meaning of $R(t)$, see Lectures IV and III.
** For *at least* the first 100,000 years (beginning about 10 seconds after its emergence from the primeval state), the universe must have contained at least 10^7 times more photons than nuclei; a similar conclusion is arrived at by thermodynamic considerations. (In such a gas in equilibrium, nuclei and photons contribute to the total pressure in proportion to their relative abundances).
*** The most distant galaxies known today have a red-shift factor of about one-half.
**** For the definition of cosmic time, see Lectures IV and III.

$$\rho(v)\,dv \cong \frac{8\pi h\, v^3 dv}{[\exp(hv/kT(t_p)) - 1]} \tag{1}$$

where

$$\frac{T(t_p)}{T(t_1)} = \frac{R(t_1)}{R(t_p)}, \tag{2}$$

and h is Planck's constant; k—the Boltzmann's constant; $R(t_1)$—the linear cosmological scale factor at t_1, $R(t_p)$—its value at the present time; $T(t_1)$ and $T(t_p)$—the corresponding temperatures. Equation (2) is therefore a useful mathematical expression of the *measurable* second cosmological arrow of time.

An analogous expression of the second arrow may be based on the decreasing density with time [for instance, by Eqs. (95) or (100) in Lecture III].

Some minor modifications of Eqs. (1) and (2) exist, depending on the opacity of the medium. When no assumptions are made as to the thermal history preceding the drop in opacity, the radiation background should have roughly a blackbody spectrum, with a temperature that tells us the value of $R(t_1)/R(t_p)$. The change in the scale factor $R(t)$ [see also Eq. (III.67) in Lecture III], may be taken to represent the time change in the spacing of a pair of neighboring clusters of galaxies. This enables us to state that Eq. (2) represents the second arrow on a *"local"* basis! (i.e., *without* involving the concept of the *"whole"* universe).

The energy density of the present $2.7°$K radio background radiation is

$$\rho_{\mathrm{R}} = aT^4 = 3.97 \times 10^{-13} \mathrm{erg/cm}^3 = 4.4 \times 10^{-34} \mathrm{gr/cm}^3. \tag{3}$$

The measure of the total density of matter and radiation, even in our own corner of the universe, is much cruder than that of the background radiation. Luminous matter is known today to have a "smeared-out" density of

$$\rho_{\mathrm{M}} \cong 2 \times 10^{-31}\ \mathrm{gr/cm}^3 \tag{4}$$

Assuming that most of the material is in the form of hydrogen, we thus have for the present particle density

$$\rho_{\mathrm{H}} \cong 10^{-6}\ \mathrm{particles/cm}^3 \cong 1\ \mathrm{particle/m}^3 \tag{5}$$

If these values are closer to the actual total density, then the universe would be "open" and the kinetic energy of its expansion would greatly exceed its gravitational energy (Eq. 101 in Lecture III).

The presently observed density is much lower than the one consistent with a "closed" universe. If there exists a substantial amount (i.e., of the order of hundreds of times more than the presently observed density) of still undetected matter, it may be in the form of faint galaxies, neutrinos, gravitational waves, atomic or ionized hydrogen, helium, dust, or black holes (Lectures I and VIII).

It would clearly be of considerable interest to cosmology to know which density is closer to reality, and great observational efforts have recently been directed to this end. However, in the context of our theory, the question is *immaterial*; for our attention focuses only on the fact that $R(t)$ *increases with time*, and that the universe accessible to our observations is *isotropic* on very large scales.

VI.2.1 Which Space Expands and Which Does Not?

The radius of the earth's orbit does not change with $R(t)$, nor does the size of our galaxy. What increases is the spacing of the clusters of galaxies. We have thus a system of non-expanding galactic objects, embedded in an expanding intergalactic space. Thus, in considering the question of scale, immediate distinction is called for between the following spaces:

Space
- ■ "1" Expanding space between clusters of galaxies (intercluster space), character-ized by the lowest radiation energy density, whose value decreases with the expansion (this is demonstrated by the second cosmological arrow of time).
- ■ "2" A virtually non-expanding intergalactic space within a cluster of galaxies (intracluster space).
- ■ "3" Non-expanding interstellar space within a given galaxy*. This space may include the cores of active galaxies (Lecture I).
- ■ "4" Stellar atmospheres (including exploding or imploding stars, protostars, supernovae, white and red dwarfs, blue giants, etc.—for classification see Lecture I)).
- ■ "5" Stellar interiors (including exploding or imploding stars, protostars, black holes, neutron stars, etc.), characterized by high densities, temperatures and pressures.
- ■ "6" Planetary atmospheres.
- ■ "7" Planetary interiors.

■ At a point inside our galaxy (space "3"), but far away from its core, the other galaxies contribute only about 1% of the background "starlight" density. At a point well between the clusters (space "1"), the nearest galaxies are already too remote for a significant contribution, i.e., there exist gradients of "starlight" density; higher density values are in spaces "3", "4", and "6", and the lowest in space "1".

■ Starlight density in intergalactic space is of the order of 10^{-2}ev/cm^3, while that of interstellar space in our galaxy (which is highly typical) is a hundred times higher (at about 1 ev/cm^3). In space "3" the energy density of the cosmic-ray flux is also about 1 ev/cm^3, and comparable to that of the interstellar magnetic field, or to the (turbulent) kinetic-energy density of the interstellar gases.

---

* We may resort to the terms "intergalactic space", or "outer space", which dispense with the distinc-tion between the first three spaces.

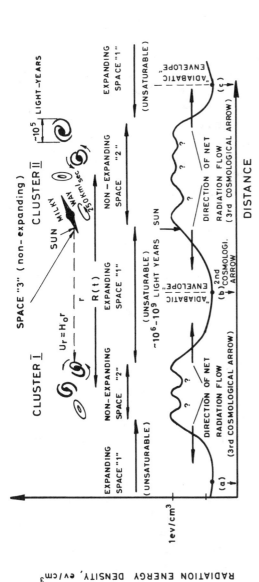

Fig. VI.1 **Schematic presentation of radiation energy density vs distance in expanding intercluster space.** Only space '1' expands (see Section VI.2.1 for definitions) as the result of which the radiation density at points (a), (b), and (c) decreases with time (2nd cosmological arrow of time). The fact that the temperatures at (a), (b), and (c) decrease with time under expansion causes net flow of radiation energy from galaxies to space '1'. Had the latter been static, the radiation density at (a), (b), and (c) would have increased with time, causing the radiation fluxes to decrease and eventually to attain a global thermodynamic equilibrium. Therefore, the expansion renders space '1' radiation unsaturability, as a result of which energy fluxes are maintained in spaces '2' and '3' (3rd cosmological arrow of time).

The energy density of the cosmic background radiation is approximately 0.4 eV cm^{-3}, while the average energy density of intergalactic radiation from the galaxies is only approximately 0.0008 eV cm^{-3} (average energy density of the X-ray background is approximately 0.0005 eV cm^{-3}). In space '3' the energy density of cosmic-ray flux is about 1 eV cm^{-3} and comparable to that of the interstellar magnetic field and to the turbulent kinetic energy density of the interstellar gases. These densities are much lower in space '1'.

Clusters range from pairs, through fifteen or twenty members like our local group, up to clusters as the one in Virgo containing several thousand members. Typical galaxies, like the Milky Way, contain about 10^{11} stars. About 10% of the mass of a galaxy may be in the form of interstellar gas and dust. Clusters may contain a significant quantity of intergalactic gas with a likely density in the range 10^{-28} — 10^{-27} g cm^{-3} (as against 10^{-31} g cm^{-3} for the 'smeared out' luminous matter). Radiation energy density in spaces '3' to '7' may be much higher than indicated schematically here. The connection between expansion, the 'unsaturability' of inter-cluster space, 'irreversible' processes in spaces '1' to '7' and the 'origin of time asymmetries' is first explained in section 3. $R(t)$ is defined by Equations (2).

■ Whether the universe is open or closed, expansion of space "1" causes the radiation density at points (a), (b), and (c) in Fig. VI–1 to decrease with time (the second cosmological arrow of time).

■ The fact that the temperatures at (a,), (b), and (c) (and the radiation densities) decrease with expansion of space "1", cause net flow of radiation energy from the galaxies (more specifically from the stars, planets, gas, and interstellar dust), into space "1". Had the latter been static (i.e., non-expanding), the radiation density at (a), (b), and (c) would have *increased* with time, causing the radiation fluxes *to cease* and, eventually, to attain a global thermodynamic equilibrium. Cosmic expansion renders the intercluster space unsaturable with radiation density. The resulting low radiation density generates gradients, fluxes, and irreversible processes throughout the galaxies, and in all stellar atmospheres. These (unidirectional) fluxes from the galactic masses to space "1" represent the *Third* cosmological arrow of time.

These results, and their connection with local thermodynamics, are explained next, using the gravitation-asymmetry Principle of Equivalence.

VI.3   GRAVITATION-ASYMMETRY PRINCIPLE OF EQUIVALENCE

While lecturing on the above topics, I have frequently noted that part of the audience cannot follow the physical conclusions unless they first examine a simplified model based on familiar concepts. For such a model, I employ a *"laboratory-universe analog"*, within the scope of the Newtonian theory. It is, however, only an analog.

■ Visualize a large number of uniformly distributed masses, of about equal size, each with its own, self-generating energy source, located inside a rigid (non-expanding) laboratory room with reflecting (isolated) walls. It is evident from all known observations that, after a finite time (depending on the laboratory size and on the details of the masses), a thermal equilibrium would be established between sources and surroundings (air or vacuum), at which all temperature gradients and all irreversible processes would vanish.

■ By contrast, a laboratory room with flexible reflecting walls, continuously maintained in a state of sufficiently fast expansion, cannot achieve equilibrium (since the radiation density near the emitters decreases with time), the result being one-directional, unrelaxed, net energy fluxes from the emitters to the unsaturable surroundings! This cause-and-effect link is indirect and involves no "action at a distance," so that if expansion ceases, both the fluxes and the gradients persist, albeit only for a finite relaxation time, during which the system approaches equilibrium (the relaxation time increases with room size and with decreasing number of sources).

■ Since the distribution of the sources is isotropic (though random), a static room of *astronomical size* would reach equilibrium in all its parts at about the same finite time, provided all sources have the same maximum temperature. Since energy always flows in opposite directions along a line of sight connecting any two neighboring sources, there exists a point on this line [see points (a), (b) and (c) in Fig. VI–1], at which the net flux vanishes and each source is enclosed in an imaginary "adiabatic envelope" consisting of an infinity of such points. Each such adiabatic "cell" can now be considered as an "expanding room," equivalent to the original one, but of a smaller size, and its evolution away from or towards equilibrium would be also equivalent to that of its original counterpart. Thus, by extrapolating to astronomical scales, each source assumes the dimensions of a cluster of galaxies and a typical adiabatic "cell" represents the statistically averaged "cell" volume per cluster of galaxies (or per supergalaxies). (See also § III.3.).

■ In fact, it is now immaterial whether the universe is bounded (i.e., closed) or infinite (i.e., open), as for both cases the average evolution observed inside a typical "cell" is equivalent to that of the "larger" laboratory and would suffice for the same general conclusion regarding its dynamics, namely, that the observed astronomical laboratory has not been static for a long time† Transforming the shapes of the sources, resolving them into billions of smaller units, rotating the galaxies, introducing clouds, galactic dust, local explosions, cosmic rays, electromagnetic waves, electromagnetic dissipation, black holes*, etc.—all these, in themselves, cannot affect the earlier conclusion, which is also supported by additional and independent observations, even inside and near our own galaxy [48, 110]**. Even recourse to curved space, and distribution of the galaxies according to size and type, leaves the general result unchanged insofar as the universal laboratory accessible to our observations expands and remains isotropic. (A possible eventual contraction and its effects are analyzed in Lecture VII).

■ Some of these complexities may cause local distortions, or even a net flow of energy from a too-small "cell" to another, thereby forcing us to resort to larger "cells" until isotropy has been restored.***

---

† Whether in "reality" the process consists in expansion or in contraction is essentially a matter of the definition of time (see below); what is important here is the negative conclusion.

* Black holes, if verified, may provide only additional sinks to the already unsaturable sink of expanding intergalactic space. They may play a role in generation of irreversibility in the universe by providing an additional sink which remains unsaturated even if expansion reverses into contraction. (See also Lecture VIII on black-hole physics).

** In passing it may be interesting to note that even the rotational motion of stars is subject to electromagnetic "braking", with mechanical energy dissipated into radiant energy in the corona of the star and eventually lost in unsaturable space [93, 95, 97] (see also below). Observed electromagnetic energy emissions by supernova remnants [91–94] also suggest that similar dissipation effects are due to fluid and radiation stresses [52], magnetic fields [53–56], and turbulent viscosities [52, 57].

*** Gold's *"star in a (static) box model"* [10] has only a weak connection with the cosmological arrow and may still lead to symmetric boundary conditions. This restriction is obviated in the present laboratory-universe equivalence which leads to a *supergalaxy in an expanding adiabatic envelope*. Moreover, Gold's reasonings and motivations, as well as those of Bondi, have been to defend the now outdated steady-state model of the universe [cf. e.g., H. Bondi, T. Gold, *Mon. Not. Roy. Astr. Soc.*, *108*, 252 (1948)].

■ Summing up, we arrive at the following (see also Figs. VI–1 to VI–3) :

1) The (observed) expansion causes spaces "1" to "3" to act as an unsaturable thermodynamic sink.

2) The origin of irreversibility in the universe can be traced all the way back in time and out into expanding intercluster space.

3) The first cosmological time asymmetry (according to our school, the master asymmetry) is that associated with the expansion of intercluster space "1". Thus, if, for instance, we photograph two groups of supergalaxies at instant t_1 and t_2, their mutual separations $R_1(t_1)$ and $R_2(t_2)$ would tell us whether $t_1 > t_2$ or $t_2 > t_1$. This arrow of time is independent of any cosmological model (including the steady-state model, which also takes it into account). For time-dependent models, we have already defined the decreasing density (or temperature) of radiation as the second, and the one-directional energy flow into unsaturable space as the third cosmological time asymmetry.

4) The second and third time asymmetries are (independently) created and dominated by the master asymmetry. To the best of our knowledge today, reverse links are physically impossible, i.e., no physical theory based on observations and on present information can be devised in which the observed expansion would be induced (or dominated) by the decreasing density of radiation and matter or by the one-directional flow of energy from hot galactic sources to cold space.

5) The very existence of a universal sink (which remains unsaturated as long as space keeps expanding) is the indirect cause of the observed, unlimited, one-directional energy fluxes from the surfaces of all galactic stars and nebulae. Gravitation and unsaturable space become the indirect causes (see below) of energy-density and temperature gradients originating at the very cores of galactic emitters (where, provided the temperature is sufficiently high, nuclear reactions generate the energy to be transported to outer regions by radiation, convection, conduction, etc.), and extending through the outer layers of the various planets, stars and nebulae and throughout the galaxies themselves.

6) These causal links are independently supported both by macrocosmic observations and by predictions of general relativity, as well as by the theories of nuclear reactions, electrodynamics, stellar structures, and hydrodynamic stability. Therefore, if relativity is unacceptable as the universal framework of thermodynamics, our argument can still be well based on purely non-relativistic observations and on Newtonian gravitation.

7) In subsequent lectures this principle allows explanation of such unorthodox maxims as :

I see structures for there is gravitation ;
I observe stars for the world has been expanding ;

Same expansion and gravitation
coming together
is time.
From it arise life and the
phenomena around us.

●VI.4 CAN INTERCLUSTER SPACE BE SATURATED WITH RADIATION?

In trying to deal with this problem let me refer first to an old problem originally raised by E. *Halley* (1656–1742) [69] and later by *H.W.M. Olbers* (1758–1840) [68], eventually coming to be known as "Olbers' paradox".* Generally speaking, it states that if we take into account the observed luminosities of galactic stars and nebulae (the sun being a typical emitter), the resulting sky-brightness (equilibrium) temperature *in a static* intergalactic space should be around 6000°K (which is about the surface temperature of the sun), yet the observed night-sky is dark.

The simplest mathematical formulation of the "Olbers' paradox" is as follows. Suppose we have a universe with Euclidean geometry, infinitely old and infinite in extent, and with a uniform distribution of radiating objects. In a spherical shell of thickness dr, located at distance r from earth, there are $n \times 4\pi r^2 dr$ such objects, where n is their number per unit volume. The total contribution of radiation from all such objects is then $(n \times 4\pi r^2 dr)(L/4\pi r^2)$, where we first assume that each object has the same luminosity L. The contribution of radiation from the entire universe would then be infinite. However, a finite result is obtained if a finite cross-section A (normal to the line of sight from earth) is assumed for each object, which would then block radiation from all objects lying beyond it, within the same solid angle A/r^2. Total blockage of radiation results when the entire solid angle 4π is covered, i.e., when

$$\int_0^R 4\pi r^2 n \cdot (A/r^2)\, dr = 4\pi \tag{6}$$

The above yields $R = 1/An$ – the distance where total blockage results, and the total radiation B from all galactic sources within this distance amounts to

$$B = \int_0^{R = 1/An} nL\, dr = \frac{L}{A} \tag{7}$$

which is of the order of the surface brightness of each object, instead of that observed for the night sky.

Numerous attempts had been made to resolve this "paradox". For instance, the hypothesis of absorption of the distant radiation en route to earth implies heating

---

* The use of the term "paradox" may not at all be justified as we shall stress below.

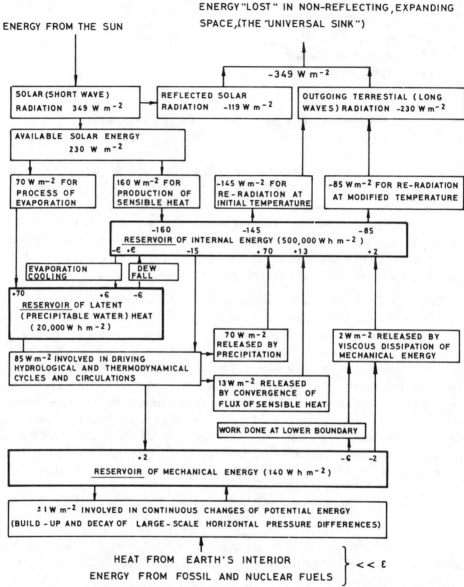

Fig. VI.2 **Intermediate cycles of (gravitationally-induced) solar energy in the Earth's atmosphere prior to its loss in unsaturable space.** In a static, saturable space, solar energy cannot be dissipated, radiation density builds up and equilibrium of the upper atmosphere with the radiation in outer space would cause all input-output energy fluxes to cease. All irreversible processes on Earth will then decay within a finite relaxation time. Gravity and the existence of unsaturable sink in space are the indirect causes of temperature gradients in the Earth's atmosphere. Note that when all energy inputs to the atmosphere cease, and rates of energy expenditure are maintained, the reservoir of internal energy (thermal) will be depleted in about 100 days; the reservoir of latent heat (precipitable water) in about 12 days; and the reservoir of mechanical energy (momentum) in about 3 days. Figure modified from ref [51]. For prebiotic evolution see Lecture IX. What would happen to photosynthesis, food chains and animals if sunlight were temporarily "turned off"? For a recent discovery concerning a past extraterrestrial cause see Ref. 131 (p. 347).

of the absorbing matter, which leaves us with the same problem as before.* On the other hand the red shifts z, systematically observed in the radiation coming from all distant galaxies, signify not only expansion but also modification of the integral

$$B = \int_0^R \frac{nL}{(1 + (H_0/c) r)^2} \, dr \tag{8}$$

made of Hubble's law $z = H_0 r/c$. By this result B converges to a value which may be compared with observations. But no general agreement has been reached yet about the numerical values involved. Disagreement centers around a number of problems:

1) *The observed luminosities are not uniform nor constant with time**.* Hence calculations of the period required for the radiation from all the stars to "saturate" intercluster space vary considerably. The results range from a few hundreds years to 10^{23} years, i.e., from much shorter to much longer than the age of the stars ($\sim 10^{10}$ years). *Consequently the period* (not the tendency!) *required for the saturation of space with radiation remains an open question.*

2) A static universe must reach a *maximum* temperature which equals *the core* (not the surface!) *temperature* T_c of stable, massive, advanced stars in the galaxies. These stars have already exhausted most of their hydrogen, helium, etc. and their core contains the most stable element—iron.[+] Iron, however, cannot reach 10^{10}°K, since it would undergo endothermic decomposition back to helium (which generates energy at lower temperatures, of the order $10^7 - 10^8$°K) [44]. If we now picture that the universe "stops expanding," *its contents would reach thermal equilibrium* after *a finite time* at which all space would be *at the uniform maximum temperature T_c* (and *not* at 6000°K as is currently assumed). This, in turn, would destroy the self-gravitating bodies. Consequently, the expansion generates and preserves structures in the universe.

Naturally, one is tempted to ask what would happen if the universe starts to contract. Apparently, contraction means blue-shifted radiation with negative z values that would cause B to increase beyond the surface brightness of any star. Space temperatures may then exceed T_c, and the radiation would converge from hotter space onto colder galactic masses. Irreversibility would then reappear, with all its specific time anisotropies. Would these irreversibilities and time arrows be pointing in the opposite direction to that we observe today? Would convergence of radiation from space onto matter create advanced potentials instead of the present retarded potentials? These fundamental questions are linked to the very concept of time and its origin—topics reexamined in Lecture VII.

---

*An artificial resolution is obtainable by "chopping up" the universe in space or time or both. Further evaluation of these ideas is given by Bondi [53].

** cf. Fig. I.1.

[+] q.v. Lectures I and VII.

●VI.5   DERIVATION OF THE MASTER ASYMMETRY FROM GRAVITATION THEORIES

In this section I derive the master time-asymmetry from simplified (reversible) Newtonian or general relativistic mechanics. The result is to be combined with the reversible energy equation and produce equivalent inequalities of the "second law", etc. For this purpose I first consider a very large gas cloud which, like the observed universe at large, is isotropic [48]. (See also Lecture III). For the present analysis the exact relation between this cloud and the galaxies need not be specified too closely. The billions of galaxies may be considered either as the particles of the gas, or as localized condensations in an actual intergalactic (atomic or ionized) gas, condensations acting as tracers for the average motion and perhaps for the average density of the gas in their general vicinity. I consider first the Newtonian theory, which is mathematically simpler and leads to many important results essentially identical with those of general relativity. For this case, the theory yields

$$\frac{d^2R}{dt^2} = -\frac{4\pi}{3}\frac{G\rho(t_s)}{R^2}; \qquad \frac{d^2R}{dt^2} \text{ and } \frac{dR}{dt} \neq 0 \tag{9}$$

where R is the time-dependent scale-factor (whose change in time may be taken to represent that of the distance between any two supergalaxies), G the Newtonian constant of gravitation and $\rho(t_s)$ the density of the medium at a standard time t_s with $R(t_s)$ arbitrarily fixed as unity. Conservation of matter then yields

$$\rho(t)R^3 = \rho(t_s) \tag{10}$$

Eq. (9) admits no static solutions (which require the first and second time derivatives of R to vanish), and thereby demonstrates that even the Newtonian theory dictates systematic contraction or expansion on a large scale over which the universe is approximately uniform. Therefore, this otherwise reversible equation incorporates an important time asymmetry, namely compulsory contraction or expansion (see Fig. VI–3). Its integration yields

$$\left(\frac{dR}{dt}\right)^2 = \frac{8\pi}{3}\frac{G\rho(t_s)}{R} - k \tag{11}$$

where k is the constant of integration whose physical meaning is to be given below.

Newtonian dynamics, however, cannot account for many phenomena such as the observed behavior of light in strong gravitational fields, curved space-time, contribution of pressure to gravitation, filling the whole of space with a gas cloud, non-additivity of separate effects, etc. (See Lecture III). For the present analysis Einstein's original field equations reduce to the simple set [eqs. (91) and (92) in Lecture III]:

$$(dR/dt)^2 = (8\pi/3)\,G\rho R^2 - k \tag{12}$$

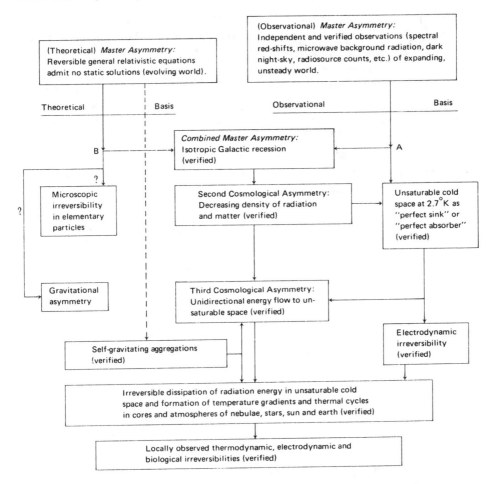

Fig. VI.3 **Preliminary links between concepts in gravitism, showing both theoretical and observational foundations of the theory.** cf. Fig. 1, Introduction, for additional links.

$$(2/R) \, d^2R/dt^2 + 1/R^2 (dR/dt)^2 = -8\pi GP/c^2 - k/R^2 \qquad (13)$$

Referring to Eq. (10), we immediately note that the relativistic equation (12) is identical with the Newtonian equation (11). In other words, despite all the fundamental differences between general relativity and the Newtonian theory, the scale-factor R satisfies the same equation in both theories provided the effect of pressure is neglected—cf. eqs. (35) and (36) in Lecture III. This is the well-known Milne–McCrea theorem. Consequently, the requirement of non-static solutions leads to the same classification of models and evolutions in both theories. In the second case the physical meaning of k is different, i.e. instead of speaking of bound ($k > 0$) and unbound ($k < 0$) clouds, we speak of "closed" and "open" space, or of "spherical" and "hyperbolic" space, or of "oscillating" and "ever-expanding" space*. The case $k = 0$ is then characteristic of Euclidean three-dimensional space. There exist, of course, other important models in relativistic cosmology, in particular those related to the early evolution in which radiation was dominant and its pressure non-negligible. In general, rejecting unrealistic alternatives such as static vacuum fields, negative masses, and single-body models [104], we may conclude that the original field equations of general relativity (which are reversible under time reversal) admit only time-evolving (non-static) solutions! Consequently, the observed expansion of the universe has its origin accounted for within the scope of both Newtonian and relativistic dynamics. *Here, I am concerned with a new thermodynamics based on the derivation and the use of the condition* [cf. Eqs. (9), (11), and (12)].

$$dR/dt \neq 0 \tag{14}$$

which, in accord with observations, means

$$dR/dt > 0 \quad \text{(i.e. expansion)} \tag{15}$$

for past and present epochs. Determination of the actual value of this rate of expansion from any of the cosmological models is irrelevant here and only the inequality sign is of importance, as will be shown later. For that purpose, I rewrite Eq. (15) as

$$\boxed{dv/dt > 0 \quad \text{(expansion)}} \tag{16}$$

where $v = 1/\rho$.

To conclude, both the Newtonian and Einsteinian theories of gravitation predict the master asymmetry or the observed first cosmological arrow of time (see also Fig. VI.3). Thus, sooner or later any of the particles which make up the isotropic cloud will see all others receding at a velocity proportional to distance, and after sufficient time has elapsed, the particles should be moving in accordance with Hubble's law irrespective of any special initial conditions as to their initial motion.** In this manner the field of gravitation generates time asymmetry out of cosmological motion. Gravitation is thus behind the very concept of time asymmetry (or at least behind the master asymmetry); it is also behind the very concepts of time and irreversibility in any process in the universe. How? The first quantitative answer follows.

It should also be stressed that the relative abundances of the chemical elements produced at the beginning of the cosmic expansion (apart from nuclear reactions in the stars—§III.3 and Lecture VII) are in no sense a function of initial chemical composition—in other words, they depend only on what we may call the amount of

---

* For more details see §III.2.7 and §III.2.7.2.
** But see also §VII.3.

matter available, not upon what that matter was originally made of. (Specifically, the hydrogen-to-helium ratio is about 3 to 1 by mass for a very wide range of initial conditions).

●VI.6   IRREVERSIBILITY IN THE NEW GRAVITATIONAL COSMOLOGICAL THERMODYNAMICS

In this section I deduce the *"positive nature"* of the *"second law"* and of "internal dissipation", as well as of the bulk- and shear-viscosities and of the Clausius-Duhem inequality—from time-symmetric conservation equations of energy and mass *in conjunction with the master asymmetry.* To simplify the analysis, I employ a system of *co-moving coordinates* (see Lectures II and III). In the latter case, simple relations are obtained only in the "immediate neighborhood" of a selected point; the principle of energy conservation is assumed to remain valid only in that neighborhood, in agreement with the reversible first law of thermodynamics. *Thus, a local co-moving observer can always select a coordinate system co-moving with the medium: he can then use the same method for measuring heat and entropy, using the criteria of reversibility and irreversibility familiar from classical thermodynamics* [15], [110].* It should be noted that, apart from the need to re-analyze the non-static solutions of the original field equations of general relativity, a relativistic formulation cannot, in itself, create irreversibility out of reversible equations;—it cannot produce a new principle of asymmetry not incorporated in advance in the co-moving classical case. The basic asymmetric equations of our gravitational-cosmological school of thermodynamics can therefore be formulated in the simpler, local, co-moving frame without risk of omission of any fundamental principle of irreversibility.

Since integration over very large sections of the medium is to be avoided here, the following analysis is presented in the form of differential equations. Under these conditions, the familiar (reversible) first law of thermodynamics

$$du = dq' - dw', \tag{17}$$

takes the form**

$$\frac{du}{dt} = \frac{dq}{dt} - p\frac{dv}{dt} + \Phi + b \tag{18}$$

where (see Lecture II for derivation and more details)

$$\frac{dq}{dt} = -\frac{1}{\rho}\operatorname{div}\mathbf{J}_q \tag{19}$$

---

  * For physico-philosophical justifications see also §III.1.1a.
  ** Readers unfamiliar with the details of these mathematical definitions may skip them and proceed with the text from Eq. (36) onwards.

$$\Phi = -\frac{1}{\rho}\,\tau:\operatorname{grad}v \tag{20}$$

$$b = \frac{1}{\rho}\sum_k \mathbf{J}_k \cdot \mathbf{F}_k \quad (k = 1, 2, \ldots, n) \quad (\text{cf. eq. III 59a}) \tag{21}$$

$$\rho = v^{-1} = \sum_{k=1}^{n} \rho_k \tag{22}$$

$$v = \sum_{k=1}^{n} \rho_k v_k/\rho \tag{23}$$

$$\frac{d}{dt} \equiv \frac{\partial}{\partial t} + v \cdot \operatorname{grad} \tag{24}$$

$$\mathbf{J}_k = \rho_k(v_k - v) \quad (k = 1, 2, \ldots, n) \tag{25}$$

$$\tau = \mathbf{T} - p\mathbf{I} \tag{26}$$

$$\mathbf{T} = \mathbf{T}^T \quad (\text{i.e., } T_{ji} = T_{ij}) \tag{27}$$

$$\mathbf{J}_q = \mathbf{q} + \sum_{k=1}^{n} h_k \mathbf{J}_k \tag{28}$$

so far, all these quantities are strictly reversible! Here u is the specific (per unit mass) internal energy, q the "heat" added per unit mass, $\mathbf{q}$ the heat flux due to conduction; p the (scalar) hydrostatic pressure of the total pressure tensor $\mathbf{T}$, as defined in Eq. (26) where τ is the viscous stress tensor and $\mathbf{I}$ the unit matrix with elements $\delta_{\alpha\beta}$ ($\delta_{\alpha\beta} = 1$ if $\alpha = \beta$; $\delta_{\alpha\beta} = 0$ if $\alpha \neq \beta$); ρ the total material density defined by Eq. (22), where v is the specific volume and ρ_k the mass density (per unit volume) of chemical component k in a mixture of n chemical components ($k = 1,2,\ldots,n$); v the center-of-mass (barycentric) velocity vector and v_k the velocity of component k. Φ is defined by Eq. (20) *(the observed fact that $\Phi \geqq 0$ is to be deduced later*)* ; $\mathbf{J}_q$ is the total heat flux defined by Eq. (28) and includes enthalpy transferred by the diffusion fluxes $\mathbf{J}_k$ [defined by Eq. (25)], h_k thus being the specific enthalpy of component k; b is the (scalar) dissipated energy associated with gravitation and radiation. When $\mathbf{F}_k$ reduces to the conservative force of gravitation (per unit mass), we have (see Lecture III):

$$\mathbf{F}_k = -\operatorname{grad}\psi_k \quad (\partial\psi_k/\partial t = 0) \tag{29}$$

with

$$\nabla^2\psi_k = 4\pi G\rho_k \tag{30}$$

---

* The operation symbol in Eq. (20) denotes a double inner product of second-order tensors, e.g., $A:B \equiv tr(AB^T)$, yielding the scalar quantity without any embedded asymmetry!

as an exceedingly close approximation. Here ψ_k is *the Newtonian gravitational potential* (per unit mass of component k). Thus, only in the absence of any type of diffusion currents

$$b = \mathbf{g} \cdot \sum_{k=1}^{n} \mathbf{J}_k = 0 \tag{31}$$

Note also that from the conservation of total mass one obtains

$$\sum_{k=1}^{n} \mathbf{J}_k = 0 \tag{32}$$

For possible interactions with an unpolarized electromagnetic field [see also eq. (81) below]

$$b = \mathbf{i} \cdot (\mathbf{E} + (1/c)\, v \times \mathbf{H}) + \frac{1}{\rho} \sum_{k=1}^{n} \mathbf{J}_k \cdot \mathbf{g} \tag{33}$$

where

$$\mathbf{F}_k = z_k (\mathbf{E} + (1/c)\, v_k \times \mathbf{H}) \tag{34}$$

$$\mathbf{i} = \sum_{k=1}^{n} z_k \mathbf{J}_k \tag{35}$$

z_k being the charge per unit mass of component k, and $\mathbf{E}$ and $\mathbf{H}$ the electric and magnetic field intensities respectively. All available evidence shows that $b \geq 0$, i.e., gravitational and electromagnetic energies are converted irreversibly into internal energy. (This, for instance, leads to a non-negative resistance tensor in the linear approximation and to positive (Joule) heat generation.)**

If we now confine ourselves to fluid media where all torques are moments of forces, then by Cauchy's second law of motion the total pressure tensor $\mathbf{T}$ is symmetric [i.e., eq. (27) is valid for "nonelastic" fluids].

Eq. (18) allows conversion of internal into kinetic energy when

$$\Phi \leq 0 \tag{36}$$

and vice versa when

$$\Phi \geq 0 \tag{37}$$

** Similar (but considerably modified) formulations may be adopted for polarized systems which, with dielectric relaxation (the magnetic case is analogous) as the single process, leads to the observed positive coefficient characterizing the relaxation phenomenon (i.e., in agreement with the Debye equation for dielectric relaxation).

Furthermore the energy equation allows conversion of internal into electromagnetic energy when

$$b \leqq 0 \tag{38}$$

or vice versa when

$$b \geqq 0 \tag{39}$$

My immediate aim is to derive, using the causal links described before, an irreversible theorem based on the reversible energy equation (18) and the verified master asymmetry Eq. [(16)] and express it as equation (37) or (39).

I first consider several simple, well-defined cases, with positive density and total pressure, for a real fluid capable of transferring momentum by viscous mechanisms (i.e., in our theory τ is not neglected). This generalization enables me to derive the local irreversible (dissipative) behavior, unobtainable, for instance, with Tolman's perfect (inviscid) fluid and radiation ([15], pp. 323, 328). I also allow radiative stresses [49] which may be resolved* into an isotropic pressure p_R and viscous stresses τ_R exactly similar to the counterparts τ_G. The general case is then defined by [15,49]

$$\tau = \tau_G + \tau_R \tag{40}$$

$$p = p_G + p_R \tag{41}$$

Since isotropic expansion is the same everywhere (i.e., ρ is a function of time alone, not of spatial coordinates), there exists no net transfer of energy from one local element (or "cell") to another (see VI.3). The whole process is therefore adiabatic, i.e., $\mathbf{J}_q = 0$, $\mathbf{i} = 0$, $b = 0$, and $du/dt = 0$ throughout the medium (as we evidently expect from any simple analysis of local conservation of energy). The energy equation (18) then reduces to

$$-(p_G^a + p_R^a)\frac{dv^a}{dt} - \frac{1}{\rho}(\tau_G^a + \tau_R^a): \operatorname{grad} v = 0 \tag{42}$$

Now, by substituting here the master asymmetry, eq. (16), I immediately obtain the first important physical inequality, namely,

$$\boxed{\Phi^a = -\frac{1}{\rho}(\tau_G^a + \tau_R^a): \operatorname{grad} v > 0} \tag{43}$$

which signifies that the very process of isotropic (i.e., adiabatic) expansion causes the

---

* Radiative stresses may be neglected for most earth-bound thermodynamic calculations. They are included here only for generalization purposes.

mechanical energy $p^a dv^a/dt$ to be converted irreversibly into internal energy (so far as the expansion prevails).* This first inequality is in full agreement with all our macroscopic observations and is due here to the master asymmetry alone! Hereinafter, all mathematical formulations will incorporate this inequality all the way to the "second law" and throughout the entire superstructure of thermodynamics! Furthermore, this result is unchanged when the medium considered is pure fluid (i.e., when τ_R is negligible) or pure radiation (i.e., when τ_G is negligible).

Whenever Φ^a is negligible relative to b, a similar procedure leads to $b \geq 0$, otherwise the general case yields $\Phi^a + b^a \geq 0$. Thus by using the cosmological expansion as a point of departure I finally arrived at the conclusions that $\Phi^a > 0$, $b \geq 0$, or $\Phi^a + b \geq 0$.

These conclusions, as I shall demonstrate later, are equivalent to the second law of thermodynamics! (cf. eq. II.59d in Lecture II and eq. 62 below).

●VI.7   ORIGIN OF DISSIPATION IN NEWTONIAN FLUIDS**

A Newtonian fluid is defined by the linear relation (cf. eq. (40) in Lecture II):

$$\tau = [2\mu\mathbf{D} + (K - \tfrac{2}{3}\mu(\operatorname{tr}\mathbf{D}))]\mathbf{I} \equiv -2\mu\mathbf{D} + (\tfrac{2}{3}\mu - K)\operatorname{div}\mathbf{v} \tag{44}$$

where

$$\mathbf{D} = \tfrac{1}{2}[\operatorname{grad}\mathbf{v} + (\operatorname{grad}\mathbf{v})^T] \tag{45}$$

is the (symmetric) *rate-of-deformation tensor* (also called the *stretching tensor*). Insofar as the positive nature of the bulk (K) and shear (μ) viscosities is not stated, τ may also be considered as "reversible." Substituting eq. (44) in (20), and considering eq. (43) in conjunction with the fact that under isotropic expansion only the radial velocity v_r (and its derivatives) must be taken into account, I obtain

$$\Phi^a = 2\frac{\mu}{\rho}\left[\left(\frac{\partial v_r}{\partial r}\right)^2 + 2\left(\frac{v_r}{r}\right)^2\right] - \left[\frac{2}{3}\frac{\mu}{\rho} - \frac{K}{\rho}\right]\left[\frac{1}{r^2}\frac{\partial}{\partial r}(r^2 v_r)\right]^2 > 0 \tag{46}$$

which must be combined with the continuity equation (assuming no nuclear reactions for this medium)

$$\frac{\partial\rho}{\partial t} + \frac{1}{r^2}\frac{\partial}{\partial r}(r^2 v_r) = 0 \tag{47}$$

---

* If our new definition of time (Lecture VII) is adopted, this result remains unchanged in an isotropic contraction. Under the conventional definition, the opposite irreversible process takes place during contraction.

** Readers unfamiliar with the details of these mathematical definitions may skip them and proceed with the text from Eq. (49) onwards.

where ρ is a function of time alone. Solving eq. (47), we find

$$v_r = -\frac{\rho}{3}\left(\frac{\partial \rho}{\partial t}\right) r \tag{48}$$

$$\boxed{\Phi^a = \frac{1}{\rho^3}\left(\frac{\partial \rho}{\partial t}\right)^2 K > 0} \tag{49}$$

Eq. (49) is a very interesting result, showing that the expansion itself dictates the positive nature of the bulk viscosity! The third inequality of our theory is thus expressed by

$$\boxed{K > 0} \tag{50}$$

The bulk viscosity may therefore be considered as damping resistance to cosmological expansion*. Significantly, the shear viscosity is no longer present in the equation, indicating that if turbulent eddies are absent, damping takes place through the bulk viscosity alone!

When v_r is expressed by the Hubble relation (see Lecture I, §7.3a)

$$v_r = H_0 r \tag{51}$$

the continuity equation becomes

$$\frac{dv^a}{dt} = \frac{1}{\rho}\,\mathrm{div}\,\mathbf{v}^a = \frac{1}{\rho}\frac{1}{r^2}\frac{\partial}{\partial r}(r^2 v_r) = \frac{3H_0}{\rho} \tag{52}$$

Combining Eq. (52) with (49) and (42), I also obtain a relation between the Hubble constant, bulk viscosity, and dissipation; namely,

$$\boxed{K = \frac{\rho p^a (dv^a/dt)}{(1/\rho^2)(d\rho/dt)^2} = \frac{p^a}{3H_0} > 0} \tag{53}$$

Alternatively

$$\Phi^a = 9KH_0^2/\rho = 3p^a H_0/\rho > 0 \tag{54}$$

Taking $H_0^{-1} = 1.77 \times 10^{10}$ years ([47], $T = 2.7°K$ and $\rho = 10^{-30}$ to 10^{-27} g/cm³, we find for ideal interstellar gas a bulk viscosity between about 10^{-5} to 10^{-2} poise. Unfortunately no experimental data on K in space are now available [110].

Referring to the equation of state, eq. (54) shows that the rate of dissipation decreases with decreasing temperature; in particular, as T tends to absolute zero the dissipation also tends to zero.

---

* See also §IX.2.24.

The energy equation may, however, not be satisfied for low-density mono-atomic gases (such as helium, for which $K \to 0$), unless we postulate isotropic turbulent expansion or formation of local, uniformly-distributed, self-gravitating condensations of matter. For both cases, eq. (43) yields

$$\rho \Phi^a = \rho \mu \Phi_v^a > 0 \tag{55}$$

whereby a fourth inequality is obtained, namely

$$\boxed{\mu = \mu_G + \mu_R > 0} \tag{56}$$

and where Φ_v^a is the "dissipation function", whose mathematical characterization is such that

$$\boxed{\Phi_v^a > 0} \tag{57}$$

Similar results are obtained with non-Newtonian fluids. When the contribution of radiative viscosity is negligible, $\mu_G > O$, when radiation is dominant, μ_G is negligible, and $\mu_R > 0$. The radiative viscosity may be estimated from [49,126] :

$$\mu_R = \frac{4aT^4}{15c\rho k} > 0 \tag{58}$$

where k is the coefficient of opacity (i.e., $k > 0$). Thus, for the radiative viscous stress tensor

$$\tau = \frac{4aT^4}{15c\rho k}[2\mathbf{D} - \frac{2}{3}(tr\mathbf{D})\mathbf{I}] \tag{59}$$

as a linear approximation. In thermodynamic studies related to condensation of stars from interstellar material, adiabatic gradients of temperature, stellar stability and internal convection, radiative and molecular viscous stresses are frequently found to be of the same order of magnitude. Radiative stresses are important in treating radiative "braking effects" in stellar rotation [52] as distinct from their counterparts due to magnetic fields [53–56, 91] and to molecular and turbulent viscosities [52,57,110].

●VI.8   TERRESTRIAL THERMODYNAMICS

Turning now to galactic, stellar or terrestrial systems involving non-expanding and non-uniform sub-systems with local volume—(v^L) and velocity (v^L) deviations from the standard* uniform expansion rate dv^a/dt treated before, I apply the first

---

* Including, of course, non-expanding systems in the laboratory which are to be considered as deviations from a uniform expansion rate taken as the standard—for more details on this standard, see Lecture VII.

law of thermodynamics [Eq. (18)] to a local observer, co-moving with the local medium's center of mass velocity. Such an observer can resolve the general mechanical term into $p^a \, dv^a/dt$ and $p^L \, dv^L/dt$ (the last term representing the local deviations from the standard). This last term may be positive, negative or zero, as the local system expands or contracts with respect to the standard (i.e. as it performs positive or negative work). Similarly, the viscous dissipation of energy Φ may be resolved into [86] (cf. eq. 49 for the physical explanation of Φ) :

$$\Phi = \Phi^a + \Phi^L + 6KH_0 \, \text{div} \, v^L$$

$$= \Phi^a + \Phi^L + 2p^a \, \text{div} \, v^L \qquad (60)$$

Eq. (18) can now be rewritten as

$$\delta^{\cdot} = -p^a \frac{dv^a}{dt} + \Phi^a + \Phi^L + 6KH_0 \, \text{div} \, v^L = \frac{du}{dt} + \frac{1}{\rho} \text{div} \, \mathbf{J}_q + p^L \frac{dv^L}{dt} - b \qquad (61)$$

whereby δ is defined.

(Note that for incompressible fluids $\text{div} \, \mathbf{v}^L$ is identically zero. Otherwise, in view of the extremely small value of H_0, the interacting term $6KH_0 \text{div} \, \mathbf{v}^L$ is negligible relative to the locally measured Φ^L.) Now by using Eqs. (61) (42) and (43), we finally arrive at the result

$$\delta = \Phi^L = \frac{du}{dt} + \frac{1}{\rho} \text{div} \, \mathbf{J}_q + p^L \frac{dv^L}{dt} - b \geqq 0 \qquad (62)$$

Here the inequality is based on the fact that for Newtonian fluids we have already shown that $K > 0$, $\mu > 0$ (eqs. 50 and 56). Hence, by eq. 50 of Lecture II we obtain

$$\delta = \mu \frac{\partial v_i}{\partial x_k} \left(\frac{\partial v_i}{\partial x_k} + \frac{\partial v_k}{\partial x_i} - \frac{2}{3} \delta_{ik} \frac{\partial v_i}{\partial x_i} \right) + K \frac{\partial v_i}{\partial x_k} \delta_{ik} \frac{\partial v_i}{\partial x_i} \geqq 0 \qquad (63)$$

The physical meaning of δ is related to our *local* observations which, in all practical thermodynamic and fluid-dynamic considerations, are restricted to the evaluation of Φ^L, $p^L \, (dv^L/dt)$, v^L, etc. As H_o is exceedingly small, only Φ^L is of interest in local studies, while the hidden* effect of Φ^a is of no practical value in our local thermodynamics**. Nevertheless Φ^a has served in this theory to introduce the origin of

---

* Our theory is distinct from various new attempts to restore strict causality in classical physics by reviving earlier proposals called "hidden-variable theories" 89, 90] (but is not in disagreement with them, or with the so-called "Mach principle" associated with non-existence of isolated systems).

** But see Lecture VII.

irreversibility via its induced, cosmological, positive nature. This cosmological link resulted in the positive property of K, μ, μ_G, μ_R, etc.

In Newtonian fluids, with negligible bulk viscosity, δ may be written as Φ_v^L/ρ (where Φ_v^L is the familiar dissipation function) and thus be interpreted as the locally-observed irreversibility.

When $\delta = 0$, it means that only the locally-observed dissipation vanishes (while Φ^a cannot vanish as far as the cosmological expansion prevails). Φ^a may, therefore, be attributed to the embedded irreversibility associated with the observed, one-sided, gravitationally-induced, dynamics of macroscopic space, and also with the unfeasibility of completely isolating a macroscopic system from possible (weak) interactions with high-penetrating neutrinos emitted by galactic sources [48, 82] (see also §I.8).

To understand the physical meaning of Eq. (62), I now examine its compatibility with the "classical" "second law" of thermodynamics and with the "δ-inequality" associated with "internal dissipation" [50].

●VI.9 CONNECTIONS WITH CLASSICAL AND CONTINUUM THERMODYNAMICS

In classical equilibrium thermodynamics, the existence of entropy, s, defined by

$$ds = \frac{(dq)_{\text{reversible}}}{T} \tag{64}$$

is postulated. Here, T is defined by means of the familiar equilibrium function

$$s = s(u, v, c_k, \ldots) \qquad (c_k = \rho_k/\rho) \tag{65}$$

i.e.

$$ds = \left(\frac{\partial s}{\partial u}\right)_x du + \left(\frac{\partial s}{\partial v}\right)_x dv + \sum_{k=1}^{n} \left(\frac{\partial s}{\partial c_k}\right)_x dc_k + \cdots \tag{66}$$

from which the thermodynamic temperature, pressure, and chemical potentials are, respectively, defined as

$$T = \left(\frac{\partial s}{\partial u}\right)_x^{-1} \tag{67}$$

$$p = T\left(\frac{\partial s}{\partial v}\right)_x \tag{68}$$

$$\mu_k = -T\left(\frac{\partial s}{\partial c_k}\right)_x \tag{69}$$

Here the symbol $( )_x$ denotes that all the other quantities are kept constant for each partial differentiation. With these definitions Eq. (66) takes the familiar (Gibbs) form

$$ds = \frac{1}{T} du + \frac{p}{T} dv - \sum_k \mu_k dc_k + \ldots \tag{70}$$

Reverting now to the isotropic processes described previously (in which no net diffusion or chemical reactions take place, i.e., $dc_k = 0$), I obtain

$$\frac{du^a}{dt} = T \frac{ds^a}{dt} - p^a \frac{dv^a}{dt} = 0 \tag{71}$$

Another inequality

$$T \frac{ds^a}{dt} > 0 \tag{72}$$

is thus obtained when the master asymmetry (16) is employed. The same result is obtained if we substitute Eq. (71) in (18) and employ Eq. (43). If, for instance, the medium is a mixture of ideal gas and radiation, for which the equation of state is given by

$$p = \rho RT + AT^B, \tag{73}$$

where A and B are suitable constants [110], we see that *isotropic expansion causes entropy to increase.*

Taking now the time derivative of Eq. (70) for a pure material undergoing a local macroscopic process, we obtain

$$\frac{du}{dt} = T \frac{ds}{dt} - p^L \frac{dv^L}{dt} \tag{74}$$

Substituting the above in Eq. (62) we arrive at the important relation known as *Planck's inequality* [50]

$$\boxed{\delta = T \frac{ds}{dt} + \frac{1}{\rho} \operatorname{div} \mathbf{J}_q - b \geqq 0} \tag{75}$$

which, for an adiabatic process in a gravitational field, reduces to

$$\left(T \frac{ds}{dt} \right)^a \geqq 0 \tag{76}$$

Thus, for cases involving heat transfer

$$\frac{ds}{dt} \geq \frac{(-\operatorname{div} \mathbf{J}_q / \rho)}{T} \tag{77}$$

Eq. (77) is in fact the familiar *"second law"* of classical thermodynamics

$$\boxed{ds \gtreqless \frac{d'Q}{T}}$$

(78)

known also as the Clausius axiom of irreverisbility. Here, however, it was deduced from the δ-inequality, derived in turn from the master inequality, Eq. (16)! This completes the formulation of the foundations of our new theory.

In closing this section, I note also that Eq. (75) is one of the first postulates in continuum (or "rational") thermodynamics (cf. Noll, Truesdell, Miller, and coworkers [130]. Combining it with the Fourier inequality [50], the Clausius–Duhem inequality [50, 87, 88] is obtained. Consequently our new gravitational-cosmological thermodynamics may be viewed as the *master theory* to both classical and continuum thermodynamics.

● **VI.10 ELECTROMAGNETIC IRREVERSIBILITY AND MASTER ASYMMETRY**

No unified theory of thermodynamic and electromagnetic irreversibilities is available to date. In this section we review the main problems involved and the prospect of our school achieving partial unification via a second principle of earth-cosmological links.

Why does an accelerated electric charge lose energy to its surroundings? Why is it never seen to receive incoming radiation?

The classical equations of electrodynamics are fundamentally symmetric under time reversal, so that each solution for time t has a symmetric counterpart for $-t$. These "twins" are known as the retarded (charge-emits-energy) and the advanced (charge-absorbs-energy) solutions, respectively, and mathematically speaking both are valid! In practice, however, only the retarded solution is chosen so as to agree with observations. The formal pretext for such an arbitrary precaution is "causality", but to state that the cause must precede the effect is essentially equivalent to an a priori imposition of electrodynamic irreversibility! Even in quantum electrodynamics, the rates of induced upward and downward transitions must be symmetrical; formally, actual emission is attributed to "spontaneous" downward transitions.

The time symmetry of Maxwell's equations enables us to obtain the familiar retarded solutions

$$\phi_{\text{retarded}}(t, \mathbf{r}) = \int \frac{\hat{\rho}\left(t - \frac{|\mathbf{r} - \mathbf{r}_1|}{c}, \mathbf{r}_1\right)}{|\mathbf{r} - \mathbf{r}_1|} d^3\mathbf{r}_1$$

(79)

$$A_{\text{retarded}}(t, \mathbf{r}) = \int \frac{\mathbf{j}\left(t - \dfrac{|\mathbf{r} - \mathbf{r}_1|}{c}, \mathbf{r}_1\right)}{c|\mathbf{r} - \mathbf{r}_1|} d^3\mathbf{r}_1 \tag{80}$$

The equations themselves, to which the electric charge density $\hat{\rho}$ and the current density $\mathbf{j}$ are subject, read

$$\mathbf{V} \cdot \mathbf{E} = 4\pi\hat{\rho}, \qquad \mathbf{V}\mathbf{X}\mathbf{H} - \frac{1}{c}\frac{\partial \mathbf{E}}{\partial t} = \frac{4\pi}{c}\mathbf{j}$$

$$\mathbf{V} \cdot \mathbf{H} = 0, \qquad \mathbf{V}\mathbf{X}\mathbf{E} + \frac{1}{c}\frac{\partial \mathbf{H}}{\partial t} = 0 \tag{81}$$

(E and H being the electric and magnetic fields respectively) and are supplemented by the Lorentz-force equation

$$\mathbf{F} = Z\left[\mathbf{E} + \frac{\mathbf{V}\mathbf{X}\mathbf{H}}{c}\right] \tag{82}$$

Usually E and H are written in terms of the potentials ϕ and A, namely

$$\mathbf{E} = \frac{-1}{c}\frac{\partial \mathbf{A}}{\partial t} - \mathbf{V}\phi; \qquad \mathbf{H} = \mathbf{V}\mathbf{X}\mathbf{A}; \qquad \mathbf{V} \cdot \mathbf{A} + \frac{1}{c}\frac{\partial \phi}{\partial t} = 0 \tag{83}$$

Set (81) can then be rewritten as

$$\Box\phi = 4\pi\hat{\rho}, \qquad \Box\mathbf{A} = \frac{4\pi\mathbf{j}}{c} \quad \left\{\Box \equiv \frac{1}{c^2}\frac{\partial^2}{\partial t^2} - \mathbf{V}^2\right\} \tag{84}$$

Given $\hat{\rho}$ and $\mathbf{j}$, we can estimate ϕ and A (hence E and H) from equations (79) and (80).

Now, reversing the sign of t, we obtain the *advanced* solutions for eq. (84), namely

$$\phi_{\text{advanced}}(t, \mathbf{r}) = \int \frac{\hat{\rho}\left(t + \dfrac{|\mathbf{r} - \mathbf{r}_1|}{c}, \mathbf{r}_1\right)}{|\mathbf{r} - \mathbf{r}_1|} d^3\mathbf{r}_1 \tag{85}$$

$$A_{\text{advanced}}(t, \mathbf{r}) = \int \frac{\mathbf{j}\left(t + \dfrac{|\mathbf{r} - \mathbf{r}_1|}{c}, \mathbf{r}_1\right)}{c|\mathbf{r} - \mathbf{r}_1|} d^3\mathbf{r}_1 \tag{86}$$

In eqs. (79, 80) ϕ and A at time t are due to disturbances arising earlier from the source, while in eqs. (85,86) they are due to source disturbances at later times. In the former case the solution represents loss of energy from the source to space, while in the latter case incoming radiation is gained from space. In both cases **E** and **H** fall off as r^{-1} for large r, causing the flux of radiation to fall off as r^{-2}. (This result was actually used in the previous formulation of Olbers' paradox.)

A more general solution for eq. (84) is obtainable by combining (79,80) and (85, 86) in the form

$$\phi = \alpha \phi_{\text{retarded}} + (1 - \alpha)\,\phi_{\text{advanced}} \tag{87}$$

$$\mathbf{A} = \alpha \mathbf{A}_{\text{retarded}} + (1 - \alpha)\,\mathbf{A}_{\text{advanced}} \tag{88}$$

where $0 \leq \alpha \leq 1$, and in which both outgoing and incoming radiation is accounted for. In agreement with observations, we impose the condition $\alpha = 1$ with a view to purely outgoing radiation. The reason for this choice goes beyond pure electrodynamics and can now be best illustrated within the framework of the new gravitational-cosmological thermodynamics.

Arguments linking electromagnetic irreversibility to the universe at large were put forward by Wheeler and Feynman [77]. The advantage of their approach is that the value ϕ unity for α is no longer assumed, but derives from the "reaction of the universe" (with the phases of disturbance and of absorption of radiation produced by the intergalactic medium taken into account). This result, however, was actually achieved by imposing asymmetrical initial conditions on the system, of the type leading to the thermodynamic arrow of time and to causality. These questions were examined by Hogarth [25], by Hoyle and Narlikar [4], and by Narlikar alone [16], who pointed out that unlike the static universe analyzed by Wheeler and Feynmann, the actual expanding universe is not time-symmetric. Using such arguments, they also tried to demonstrate which cosmological model agrees with the observed retarded solutions—a topic which is outside the scope of this lecture.

According to Wheeler and Feynmann's "absorber theory" [77–79], symmetry under time reversal is confined to interactions between pairs of particles, while in an asymmetric large-scale ensemble of charged particles, such as the universe, the interactions are themselves asymmetric and make for local time asymmetry. To create the necessary asymmetry, the universe must act as a perfect absorber of electromagnetic disturbances generated within it. How can it do so? Wheeler and Feynman restricted their theory to a static universe, so that their arguments are neither complete nor convincing; it is only in a non-static, unsaturable expanding space that perfect absorption becomes feasible, thereby giving rise to electrodynamic irreversibility! Here again the master time asymmetry is in the center of the picture and can be directly combined (in a variety of mathematical formulations [16] with the reversible electromagnetic equations so as to deduce the observed electrodynamic irreversibility. In a future development the same result may be reached from the change in average distance (i.e., divergence of universal lines according to the observed "red-shifted" master asymmetry) or even from the theory of geometrodynamics or general relativity, which admits no static solutions. Another possibility is via the causal link, which indicates that the master asymmetry makes interstellar space expand towards very low density of radiation and particles [5-7,45,62,63,70-75,91-99,110]. In schematic form:

Master asymmetry → Unsaturable space (particles close to ground state) → Space

acting as "perfect absorber" → Electrodynamic irreversibility.

Accordingly, a wave emitted in a galactic source is absorbed by space particles, which in turn undergo upward transitions in a static universe. Yet at the prevailing background temperature (2.7°K) the particles are maintained close to their ground state by the expansion of intergalactic space. Thus, downward transitions are highly improbable and incoming radiation is perfectly absorbed.

However, closely associated with this chain of causal links is a more fundamental question related to the very nature of the interaction between thermodynamics (of matter) and electromagnetic radiation, for which there are three possible variants:

Master asymmetry → Thermodynamic asymmetry → Electromagnetic asymmetry (a)

Master asymmetry → Electromagnetic asymmetry → Thermodynamic asymmetry (b)

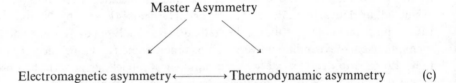

Master Asymmetry

Electromagnetic asymmetry ⟷ Thermodynamic asymmetry   (c)

In mechanism (c) the electromagnetic and thermodynamic asymmetries include embedded photon and material asymmetries which are in effect only observational manifestations of one and the same basic phenomenon of irreversibility, differentiated only by our epistemology. The other versions may also be equally valid. Mechanism (a) attributes the electrodynamic irreversibility to the thermodynamic onesidedness of matter in intergalactic space. (In a completely "transparent" universe, free of aggregations or of non-uniformity due to gravitation, not only does the electrodynamic asymmetry vanish but the "red-shifted" asymmetry is unobservable; in fact, gravitation is behind our ability to observe red shifts and irreversibilities through the motion of self-gravitating sources, etc.). On the other hand, it may be the radiation itself which has a "built-in" asymmetry and generates the observed thermodynamic irreversibility whenever interactions between the radiation and matter take place [mechanism (b)].

If a macroscopic system could be made entirely of superconducting matter, would that enable us to establish proper asymmetric links between thermodynamics and radiation? We do not yet know. To date, we have no means of ascertaining which of the three interactions is the proper one, and we have to assume that they are all equally feasible. On the other hand, these three arrows, and all others now in evidence, are consistent in indicating the same time direction. Is it a coincidence, or perhaps evidence that one of them is dominant and the origin of the others? From what we know now, the answer is affirmative only for the master asymmetry; there is no observational or theoretical evidence that radiation can by itself reverse the causal links, thereby "triggering" the observed expansion. The master asymmetry should, therefore, be the point of departure in laying new foundations of a unified theory of thermodynamic and radiation asymmetries.

COSMOLOGICAL ORIGIN OF TIME
AND EVOLUTION

The ideal aim before the mind of the physicist is to understand the external world of reality.

Max Plank

But those watchful guardians, sun and moon, traversing with their light all round the great revolving sphere of heaven taught men that the seasons of the year came round and that the system was carried on after a fixed plan and a fixed order.

Lucretius

VII.1 TIME: THE ALL-EMBRACING CONCEPT

Are the present concepts of time adequate? Do all evolutions have a common denominator? In trying to answer these questions I put forward a new primary time coordinate whose use resolves a number of standing paradoxes associated with current concepts of time. The origin of such time and its unidirectional coupling with local (geocentric) physics is to be stressed and formulated.

* * *

A striking paradox in the evolution of scientific methodology may best be illustrated by the evolution of the theory of time. As is often done when dealing with an ill understood physical phenomenon, attempts are made to remedy the situation with the aid of additional variables, introduced directly or indirectly; when, however, it becomes necessary to reduce the number of variables and unify apparently disparate phenomena, *these additional variables generate even greater psychological difficulties than those originally encountered.*

In the course of this development, the number of redundant features about the concept of time increased, and in fact they became basic tenets of the tradition which now dominates the thinking of most scientists (a growing number of whom believe now that the very postulation of an *independent* time variable is a fallacy, or at least a redundancy.*)

Many examples can be cited. For instance, most recently Ellis [8] argued that the flow of an elementary particle's time is linked with the expanding universe.

---

* This idea may be traced back to Plato and to Parmenides before him [8]. According to Plato, time and the universe are inseparable; time, unlike space, is not regarded as a "pre-existing" framework into which the universe has been fitted, but is itself a "product" of the universe, being an essential feature of its rational structure [1,23,24,31,33,25].

His analysis unifies gravity and electromagnetism by rejecting orthodox definitions of time. In 1969 Misner [18], linked time to the expanding universe by resorting to an absolute Ω-time, defined as

$$\Omega = - \ln [V^{1/3}] \tag{1}$$

where V is the (dimensionless) volume of space at a given cosmic epoch. According to him *the only enduring measure of evolution of the universe, throughout its history, is its own expansion!* Accordingly, epochs can also be labeled by temperature, which is roughly the same as the red shift z, namely,

$$T = (3°K)(1 + z) \tag{2}$$

Absolute Ω-time (originally suggested by Milne in 1948 [18, 35]) proved to be a most suitable primary standard in labeling Einstein's equations in theoretical studies of relativistic models. From the thermodynamic point of view it is a *unidirectional time* which is *irreversible as far as the universe keeps expanding. Irreversibility is thus ingrained in its very definition.*

VII.2   COSMOLOGICAL ORIGIN OF TIME

Recently, in our studies about the origin of irreversibility, life, and time asymmetries in nature [1,5,6], another primary time has been suggested. While the Milne-Misner Ω-time is a nonlinear function of $V^{1/3}$, our own primary standard of time *varies in proportion to the linear scale factor of the Robertson-Walker metric* (see below). From the thermodynamic point of view it is likewise a unidirectional time, "irreversible" as far as the universe keeps expanding. Furthermore, it not only enables a number of paradoxes (associated, for instance, with the dissipation of energy in a contracting universe) to be solved, but is also consistent with our astrophysical thermodynamics and with the interpretations of both Newtonian and relativistic mechanics. In what follows, I derive this time concept from the formulations of our thermodynamics, and subsequently deduce it from Newtonian and relativistic interpretations.

First I consider the master (time) inequality $(dv/dt) > 0$, where t is the classical periodic time, $v = 1/\rho$, and ρ the denisty of the "smeared out" cosmic mass-energy (eq. VI.–16). This inequality constitutes the basis of our new theory of irreversibility. [It was derived in Lecture VI from both Newtonian and relativistic theories of gravitation]. In agreement with astronomical observations and with the current definition of time, we interpret it as an expansion. But what would its physical interpretation be in a (possible but unlikely*) future phase of contraction? The answer to this question is asymmetric evolution as before, but in the opposite direction (i.e.

---

* See Lecture III.

advanced potentials and convergence of radiation from hotter space onto colder galactic masses; cf. also the discussion of Olbers's paradox in Lecture VI). A more critical examination reveals that with current definitions of time such pulsating (i.e. essentially periodic) evolution — which involves constant amplitudes and violates the "second law" (except for such unrealistic circumstances as ideal fluids [15]) — is not nearly as satisfactory as one more in line with the concept of ever-increasing entropy, i.e. with the pattern of diminishing amplitudes. For this to be feasible, however, the current definition of time must be re-examined. But this is only part of the problem.

Quite clearly, *cosmological and astrophysical processes are not to be measured in earth years.* The earth's motion changes or disappears as its past history is examined; atomic clocks, available only below $10^{9 \circ}$K, have no fundamental advantage over it, and even muon lifetime becomes irrelevant beyond $10^{10 \circ}$K (there is, of course, much recent evidence in support of earlier hotter epochs [18,45,62,63,110]). *Therefore, in any discussion of cosmic, astrophysical and gravitational evolutions, only one primary standard of time is available — the changing spatial dimensions of the hot medium, i.e. the change in v* [defined by Eq. (3)]. For this reason, and for others given below, I find it both logical and useful to examine first the definition

$$t'' = v^{1/3} = (1/\rho)^{1/3} \tag{3}$$

where ρ is the density of a "smeared out" cosmic mass-energy. *This time** is unidirectional as far as space keeps expanding.* Of course, this definition may be significant only at the early states of evolution, where anisotropic rates of change in v may have some effect. The observed (red-shifted) master asymmetry of present and past epochs is, however, isotropic, and for them it is more reasonable to have:

$$t' = r_v \tag{4}$$

where r_v is the radius of v. *This cosmic time admits no reversible transformations during any phase of world evolution, i.e., the unrealistic time-reversal transformation $t' \rightarrow -t'$ is ruled out by this definition.*

Both the t''-and t'-standards suggest that *the whole evolving cosmic system is the most universal and durable "clock", and the entire spectrum of "independent" time asymmetries and irreversibilities may be unified in the very definition of a proper standard. As neither negative volumes nor negative radii are admissible, the act of defining the origin of irreversibilities in nature reduces to defining primary time*.* By this means,

---

** The conventional units of time are, accordingly, to be related to the t-length units by multiplying by the velocity of light. This definition is not yet complete. The general definition of primary time is given later.

   * Both the t'- (or t''-) and Ω-standards can be connected with our local periodic standard of time by means of Hubble constant (see below).

the difficulty referred to earlier with regard to pulsating evolution is obviated, and the master time inequality is converted into a purely spatial one:

$$\frac{dv}{dt} \rightarrow \frac{dv}{dt''} = 3v^{2/3} > 0 \tag{5}$$

where t is the current periodic time standard used in classical and relativistics physics and, therefore, in our master equation (eqs. VI. 16, 43, 49, 50, 62, 75).

It should first be noted that this inequality is independent of any cosmological model—decelerating or accelerating, expanding or contracting! Moreover, under it, internal dissipation and entropic evolution remain always positive (note that dv/dt remains positive even in a contracting phase!)

Still another paradox, resolved by the new definition, is associated with the current ideas of *opposing* directions of irreversibilities in different domains of space-time. According to these [28,59], all irreversible and statistical processes that unwind in the present expanding phase are expected to wind up again in a future contracting phase — in other words, if we postulate complete time symmetry, the direction of time in a closed oscillating model during a contracting phase would be *the reverse* of that of an expanding phase. It has often been argued that if an observer were to survive this *"reversal,"* perhaps by shutting himself inside an *"ideally isolated shelter"* so that his own time, as well as that of his pre-enclosed energy resources, would remain unchanged—then, on emerging from the shelter in the contracting phase, he would collide with opposite local irreversible processes (converging by negative dissipation to lower states of entropy) and find it impossible to coexist with them; in other words, the two phases would be *"uncoupled."* *Under our hypothesis, however, the positive entropic evolution and internal dissipation of the surviving observer would still be in line with that of the contracting phase* [cf. eq. (53) in Lecture VI].

A few conclusions which immediately follow the use of our new standard are to be examined next. First, with regard to observations of the red-shift parameter z (Lecture III) for electromagnetic waves emitted by distant sources, I write the Robertson-Walker metric as

$$d\tau^2 = dR^2 - R^2 \left[\frac{dr^2}{1 - kr^2} + r^2 d\theta^2 + r^2 \sin^2\theta d\phi^2 \right] [c = 1] \tag{6}$$

where the *new primary standard of time is finally defined by the most general motion*

$$t'_R \equiv R; [R = R(t')] \tag{7}$$

Here k is a constant which, by a suitable choice of units for r, takes on the value $+$ 1,0. or $-$ 1, *and R the cosmological scale factor.* This fundamental metric is applicable in general relativity for spaces with *maximally symmetric subspaces* (i.e. it is based on the *"cosmological principle"* which, in agreement with all astronomic observations (q.v. Lecture VI), states that the observable world is spatially *homogeneous* and *isotropic* over regions large enough to contain many nebulae). According to the cosmological principle, we can always place the observer at the origin $r = 0$ and consider an electromagnetic wave traveling towards him in the $(-r)$ direction, with θ and ϕ fixed. Since $d\tau$ is zero along light-paths (null geodesics), the motion of the wave crest is given by

$$d\tau^2 = 0 = dR^2 - R^2 \frac{dr^2}{1 - kr^2} \tag{8}$$

If a light-wave leaves a typical galaxy, located at r_1, θ_1, ϕ_1, at primary time R_1, it will reach the observer at a time R_o, namely,

$$\int_{R_1}^{R_o} \frac{dR}{R} = \ln \frac{R_0}{R_1} = \int_0^{r_1} \frac{dr}{(1 - kr^2)^{1/2}} =$$

$$= \begin{cases} \sin^{-1} r_1 & (k = +1) \\ r_1 & (k = 0) \\ \sinh^{-1} r_1 & (k = -1) \end{cases} \tag{9}$$

Here the terms with r_1 *are time-invariant*, since typical supergalaxies have constant co-moving coordinates r_1, θ_1, and ϕ_1. If the next wave crest leaves r_1 at primary time $R_1 + \delta R_1$ (where δR_1 is small) it will reach the observer at primary time $R_0 + \delta R_0$, which, again, is given by a relation similar to eq. (80) in Lecture VI, except for the different limits of the integral on the left. Consequently $\delta R_0/R_0 = \delta R_1/R_1$ and $v_1 \delta R_1 = v_0 \delta R_0$, where v is *a newly defined frequency*. Hence $v_0/v_1 = \delta R_1/\delta R_0 = R_1/R_0 = \lambda_1/\lambda_0$, where λ_1 is the wavelength near the place and "time" of emission while λ_0 is the wavelength at $r = 0$ "at the moment" of reception. Thus

$$z = \frac{\lambda_0 - \lambda_1}{\lambda_1} = \frac{R_0}{R_1} - 1 \tag{10}$$

and the radial velocity of the moving source, U_r, is related to z by the Doppler effect, whereby (for small z values)

$$U_r \cong \left[\frac{dR/dt_R}{R} \right]_0 r_1 = \frac{r_1}{R} \cong z(r_1 \ll R_0) \tag{11}$$

the subscript referring to the 'present' epoch, for which $t'_R{}^0 = R_0 = H_0{}^{-1}$ if (10) is now identified with *Hubble's law* $U_r = H_0 r_1$. Contrary to current results, the age of the universe $t'_R{}^0$ is here the same for all cosmological models (and equal to 1.77×10^{10} years in standard units of time). In current theories, since r_1 and $R_0 - R_1$ are small, R is expressed as

$$R = R_0 [1 + H_0(t'_R - t'_R{}^0) - \tfrac{1}{2} q_0 H_0{}^2 (t'_R - t'_R{}^0)^2 + \dots] \tag{12}$$

where

$$q_0 = -\frac{(d^2 R_0/d^2 t'_R{}^0)^2 R}{(dR_0/dt'_R{}^0)^2} \tag{13}$$

is the so-called *deacceleration*. However, as expected intuitively by using our definition of primary time, q_0 becomes zero for *any* cosmological model! These conclusions differ from current results, which give q_0 as non-vanishing, i.e., *the t'_R standard transfers all possible cosmological evolutions into a single frame of reference maintained always in a state of uniform motion. But what other, new results can be obtained from this conclusion, especially in evolution theories and in practical applications of fundamental physical laws?*

VII.3 COSMOLOGICAL INTERPRETATION OF NEWTON'S LAWS OF MOTION

In the world accessible to our observations, the motion of individual bodies *deviates from the uniform expansion motion* by unsystematic values. Therefore, from our point of view, and in agreement with both Newtonian and relativistic mechanics, these *deviations* may be *measured* in terms of *acceleration relative* to *the uniform expansion rate*, while the *product of mass and this acceleration gives us a "unified"* but *"measurable local force"* (see below). In all *local* physics it is, therefore, impractical to account for the gravitational field that keeps the world expanding. These interpretations of *Newton's laws of motion* are in agreement with Eqs. (8,35,46,49,55,68, 71,76) in Lecture VI and with the formulations of both Newtonian and relativistic mechanics.

At first sight, omission of the isotropic "work" and dissipation from the energy equation (55) in Lecture VI ascribed to deviations from isotropic adiabatic expansion, may appear surprising. It is only through the positive property of the viscosity of the medium that its trace is left in the dissipation fuction which describes deviations of motions of the same medium from the standard reference of isotropic world expansion.

Newton used rectilinear motion with constant velocity as standard reference, while general relativity (basing itself on the equivalence of inert and gravitational masses) uses the motion of matter under gravitation and inertia. *From the relativity point of view, we can take the universal expansion as the standard, and ascribe deviations from this standard to local thermodynamic dissipation and to terrestrial evolution of structure and life.* We may therefore distinguish between three descriptions of nature:

1) *The unstructured standard "model" of the universe;* in which an average, "smeared out", (unstructured) uniform and non-aggregated medium expands and "increases dissipation" (or "disorder"). This fact is mathematically expressed by equation (VI.43).

2) *The structured reality,* in which deviations from the standard motion are taken into account according to the aforementioned interpretations of the laws of motions, acceleration, structure and order. These facts are mathematically expressed

by equation (VI.62). It is this approach that must be adopted when we deal with such questions as the origin and evolution of life (§IX.2.3).

3) *The subjectivistic viewpoint of reality*, in which human subjectivistic views on order and disorder get the highest priorities and gravitationally-induced structures and order are dismissed as *"subjectively meaningless"*. This viewpoint is the one taught now at our universities according to current textbooks. Mathematically speaking it is expressed by equations such as (VI.78).

Nevertheless, each of these three modes of description is endowed with merit. Taken together they give a pluralistic view about reality as a whole.

It is mode 2 that has been forgotten by scientists and philosophers alike. Hence we continue our efforts to examine it further. This is done in the sections and lectures which follow.

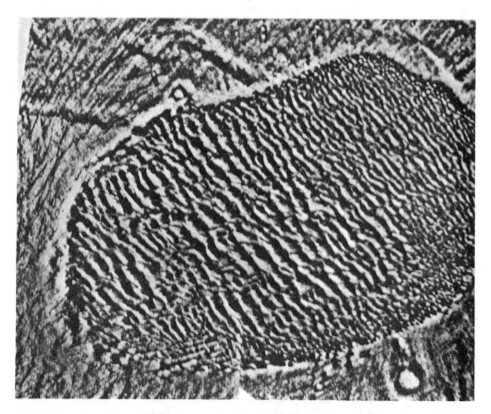

Fig. VII.1: **Gravitationally-induced dune structures on Mars.** The dune field shown is about 130 by 65 km and consists of subparallel ridges 1 to 2 km apart. These structures closely resemble transverse terrestrial dunes. Picture taken by Mariner 9. (From the National Aeronautics and Space Administration).

VII.4   GRAVITATIONAL ORIGIN OF STRUCTURE AND EVOLUTION

Inquiries into the origin and formative stages of all gravitationally-induced processes is one of the central aims of our theory. Hence we stress the need to search for the "common ground" of all evolutionary processes that are normally treated separately within such disciplines as astrophysics, geophysics, geochemistry, geology, paleontology and other related sciences.

This is not an easy task, nor is it free from some traditional trappings. Hence, in selecting, adapting, and synthesizing certain "established conclusions", which have originated from separated points of view, one must be careful and highly skeptical.

Three general assertions appear to be useful here:

1) Since Charles Darwin many scientists have used the findings of paleontology to elucidate evolutionary mechanisms. Among the interesting contributions of the neo-Darwinism of the 1930's and 1940's was G.G. Simpson's synthesis of evolutionary theory and fossil records. Recently, acceptance of continental drift, improvements in dating and stratigraphy (q.v. §I.1), increasing study of contemporary species to interpret ancient life, and advances in microevolution and ecology have stimulated a renewed intellectual revival in macroevolution and paleontology in terms of the changing environment of our planet.

But this revival has encountered a number of difficulties, *mainly those of scale, structure, and causality in all gravitationally-induced processes.* For instance, relating extinctions of Mesozoic tetrapods to gravitationally-induced regressions of the shallow seas that covered large areas of the continents, or placing fossil beds in their proper geographical context as gravitationally-induced basins that narrowly sample the diversity of evolutionary processes, require the development of a *"methodology of interconnectedness"* between *paleontology, geology, gravitation theory, ecology, biology and evolutionary theories. Such a methodology is not available at the present time. Nevertheless we have already witnessed the emergence of some interdisciplinary studies, e.g., the ecologist who seeks to explain patterns of community organization in terms of a changing external environment, or the evolutionary biologist who seeks to interpret the genetic differences between species in terms of interactions with external physical processes.* Indeed, the patterns we observe in evolutionary processes are the sum of many processes and interactions with the environment. Such processes are so complex, that, to the *uncritical* scientist, they appear *randomly* assembled. This appearance is only temporal for it is due *to our own (present) ignorance of nature.* Indeed it is difficult to predict whether the time for the development of a new methodology of interconnectedness has already come, but the attraction associated with such a synthesis may now be strong enough for a modest beginning.

2) *Due to the interconnectedness of nature, and due to the subtle nature of gravitation, no simple definition of gravitationally-induced processes can be universally valid (nor useful).* Hence, one may resort to examples rather than to definitions. A few of these are mentioned below.

i) All matter on the surface of the earth and in the uppermost parts of the litho-sphere participates in a slow, gravitationally-induced migration that causes changes in the structure and chemical composition of rocks. New rocks with new properties are produced in this process. Here the *exogenic* (the product of outside forces) cycle consists of the weathering of rocks, *transportation and sorting* of products, and *re-deposition* of material (usually in new surroundings).

These processes are in many respects similar to a gigantic *chemical rock analysis* involving *chemical separations on a large scale* (§I.1). *Because of its role as a separating and concentrating agent for many chemical compounds, the gravitationally-induced exogenic cycle is of great importance in our theory.*

(ii) For many chemical elements the gravitationally-induced differentiation is the most powerful concentrating and enriching agent. For instance, silicon and zir-conium are highly concentrated in sandstones; aluminium, potassium and boron in shales; calcium, magnesium and carbon in limestones; iron, manganese and barium in oxidates; and sodium, chlorine, magnesium and sulphur in evaporites. Iron, manganese and cobalt are often precipitated as a result of exogenic oxidation re-actions. Along with the hydrogen ion concentration, the differences in the degree of oxidation or reduction cause many enrichment processes and the separation even of chemically closely related elements.

(iii) Geochemical sorption of ions by colloidal particles governs the properties of many gravitationally-induced sediments and is of high importance in the concentra-tion of many compounds in clays.

(iv) The gravitationally-induced *migration of matter in the lithosphere* is charac-terized by its *selectivity*; i.e., under given conditions certain chemical compounds migrate more readily than others. (The elements with high atomic numbers are relatively readily mobilized and, therefore, become enriched in granites and in low-temperature assemblages — see also §I.1.)

(v) Stars and planets are structured by gravitation into concentric layers (Lecture I). On earth, for instance, we normally distinguish among four atmospheric layers, the lithosphere, the asthenosphere, the mantle and the core.* (We shall return to this subject in Lecture IX.)

---

* The lowermost layer, the *troposphere,* extends from sea level to approximately 11 km altitude at middle latitudes. It is the convection region and most atmospheric phenomena take place there. The next layer is the *stratosphere* (up to 95–100 km altitude). In its upper reaches exists a high-temperature region, sometimes called the *ozonsphere,* caused by a relatively high concentration of ozone (which is biologically important because it absorbs ultraviolet radiation of the sun). Above the stratosphere is the so-called *ionosphere* that extends to about 600–700 km altitude. The outermost layer of the atmosphere is called the *exosphere.* (The structure of the earth's environment beyond that is highly asymmetric. There the solar wind forms a shock wave with the terrestrial electromagnetic field. The shock wave bounds one side of the asymmetric region called the *magnetosphere.* Its other side [away from the sun] is not bounded and is extended deep into space.)

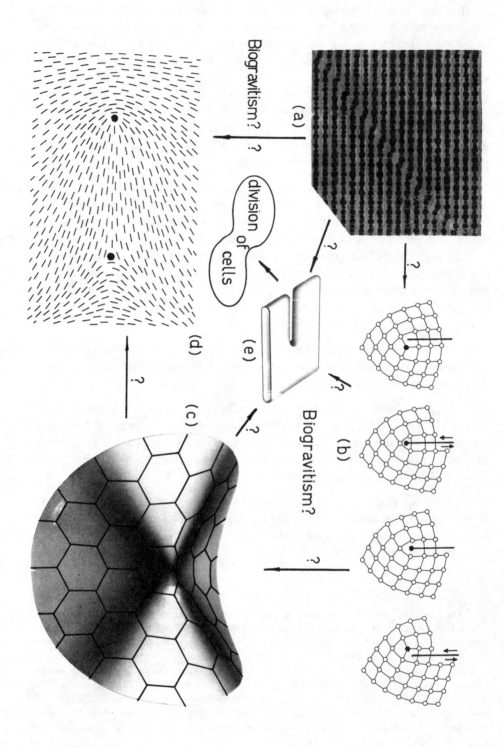

Biogravitism?   ?   (a)

?

?

division of cells

(d)   (e)

?

Biogravitism?   ?

(c)

?

(b)

?

Fig. VII.2: Biogravitism and Cell Division?

(a) is an aerial photograph of an *orchard* in the Imperial Valley of California. It was made shortly after *an earthquake* in 1940. *Note the displacement which disrupted the symmetric pattern of the trees. On a much smaller scale* the gravitational field introduces *imperfections, dislocations, disclinations* and *asymmetry in crystal structures* (cf. §I.1.8). In fact we know today that in a crystallizing suspension of *organic spheres* in water the earth's gravitational field produces an elastic deformation that is readily observed through its asymmetric effect on the evolving crystal structure. Moreover, there is evidence that these very forces play a most important role *in the evolution of crystallized virus systems*. The questions we pose here are therefore as follows:

Do we know the *origin* of the forces which cause cell division in crystallized structures? By what mechanism can a closed structured body be cut in half without breaking the surface? Are the forces of gravity responsible for the *origin* of the cell-division mechanism?

(b) The division of living cells with a structured surface can only proceed by migration and diagonal shifting of disclinations. Here the introduction of (gravitationally-induced) forces [the opposing arrows shown in (b)] causes glide, shears the lattice, shifts the disclination point to a triangle space between the atoms, and then shears the lattice again (whereby the disclination point completes one step in its migration). Thus, if no rearrangement of the cell surface is allowed other than local "tearing and fusing" (as shown in (e)), then this process is *the only way (structured-surface) cells can divide.*

(c) is a simple demonstration of *("negative") disclination* which can form saddlelike objects like the one shown here or much more complicated structures. Thus, *by inserting* a 60-degree sector to a thin structure (a "flat" sheet) one obtains the form shown in (c). If we cut a slit along some radius of the flat disk, and then rotate the cut edges around the center of the disk so that a large gap is opened and then fill the gap with 120-, 180-, 270-, 360-degrees sectors we obtain interesting topological structures. If, instead, we *remove* sectors *("positive disclination")* we obtain various types of cones, capped cylinders, etc. Hence (gravitationally-induced?) disclinations may lead to the construction of closed surfaces like those of *protein coats of rod-shaped viruses*, the *flagella of bacteria* and the *microtubules* [cf., e.g., F. Harris, in *Surface and Defect Properties of Solids*, Vol. 13, The Chem. Soc., London, 1974; and in the *Philosophical Magazine: A Journal of Theoretical, Experimental and Applied Physics, 32*, 37 (1975)]. [Note that in (c) the basic symmetry element is hexagonal, like that of a benzene ring. Indeed layers of fused benzene rings are important structural elements in such "natural" materials as tars, cokes, pitches and graphite (q.v. §I.1 for the structure of graphite and its origin)]. *To conclude: the organized means for a living cell to split into daughter cells without spilling any of the contents is by disclinations.*

(d) This figure represents patterns formed by *stripped animals* (e.g., a *zebra*), by *human hair and by the hair of some other mammals*. A more familiar example is the pattern of ridges on the *fingers of man and the other primates ("fingerprints")*. These structures may also be obtained by (gravitationally-induced) disclinations (e.g., "plus" or "minus" 180- or 360 degrees disclinations in liquid crystals). The resulting patterns are known to the dermatologist as "loops", "triradic" and "whorls". *Due to gravity* the hair can only grow at oblique angle to the skin. (Therefore it has only onefold axes of rotational symmetry perpendicular to the skin). The only allowed disclinations in such a system are therefore those with rotations of "minus" or "plus" 360-degrees (i.e., insertion or a removal of a 360-degree sector).

To conclude: We postulate that *gravity* structures and orders all stars, planets and ecosystems; that it introduces time, irreversibility, asymmetries, dislocations and disclinations into crystals (§I.1); that through it, living cells can divide, form order, generate more structure and enhance life. But through it they also decay, fall, die and precipitate in the ground; sometimes only to reform new structures such as fossils and fossil fuels in sediments.

3) Current astrophysical calculations [41,44,70–75,110,112–119] demonstrate that in all evolving stars, nuclear and gravitational energies are transformed into outward-radiating photons and neutrinos (see Figures VII–3 and VII–4). The stable star subsists on its nuclear resources throughout a series of nuclear-fuel burning stages, each characterized by a different temperature. During these stages a normal stable star adjusts its size, whereby the rate of energy release from a particular nuclear reaction just suffices to offset the radiant energy loss [44] into unsaturable space (plus the extra energy needed for raising the pressure to balance the stronger gravitational force [110]). After exhaustion of the particular fuel involved, a new gravitational contraction sets in (a process converting gravitational into thermal energy, which is also emitted and "lost" in cold space). In the interiors of highly evolved stars the temperature may reach 10^{10} K, sufficient to convert matter, through a chain of exothermic nuclear reactions, into the most stable element—iron [41,44]. Iron, however, may undergo endothermic decomposition back to helium, and this may

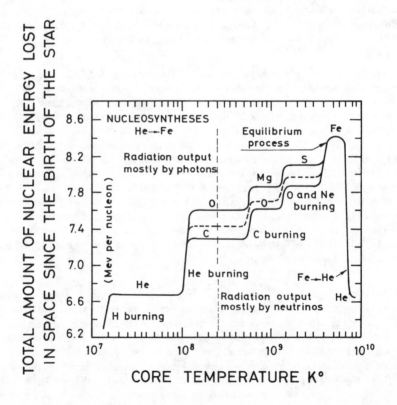

Fig. VII.3. **The overall energy generated by a star** (of about one solar mass), which is transformed into radiation "lost" in unsaturable space as photons and neutrinos. The origin of the radiated energy is identified by "Gravit" for energy release by gravitational contraction, and by chemical symbol of the nuclear fuel, for the radiation losses. Dashed lines represent individual contributions of these two modes, while the solid line is their sum (the total luminosity of the star is in erg/g/sec). The region where luminosity switches from mostly photons to mostly neutrinos is indicated. For more information, see Fig. VII-4.

cause instability, explosion, or implosion in the cores of massive stars in advanced stages of evolution [70]. Numerical calculations have shown [70–75,110,112, 113–119] that this process may result in the reaction of the type associated with the rarely-observed supernova explosions, etc. Normally the energy generated by these processes must be transported by radiation, conduction, convection, and mechanical means (such as hypersonic jet streams, expansions, explosions, or pulsations) from the interior to the surface, where it is dissipated in the form of radiation into the universal sink. Temperature gradients between interiors and outer layers (approximating to about 10^7 K over a radius of 10^8 km in a typical star [43], are believed to be produced primarily by the condition of hydrostatic equilibrium [112] in a gravitating

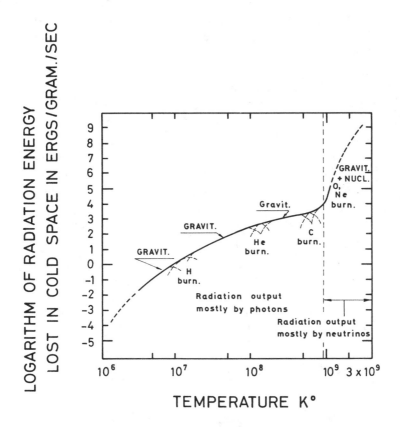

Fig. VII.4: **Total amount of radiant energy "lost" in unsaturable, non-reflecting space from nuclear burning stages (per nucleon)**. During the burning stages, the rate of energy release in the core of the star transforms into an equivalent rate of radiation loss in unsaturable space. Gravitational contraction (represented by the horizontal segments) is resumed after exhaustion of the particular fuel involved. After the He-burning stage, the core contains C^{17} and O^{16} in fractional abundances depending upon the initial mass. The C and O curves are, then, lower and upper limits respectively, as they represent cases in which the core is pure C or pure O. To the left of the vertical dashed line, most of the energy is lost in space as photons; to the right, as neutrinos. The dashed line is a typical case. The same kind of statement applies to the other burning stages.

mass subject to another important thermodynamic theorem, namely the condition of extremum of the energy [110]. This condition dictates, inter alia, the outward energy flow from core to surface (and thence to unsaturable space). Here, again, I note that our thermodynamics and the theory of gravitation are strongly coupled.

● **VII.5   GRAVITATION AND THE OUTFLOW OF ENERGY INTO UNSATURABLE SPACE**

The outermost layers of the stars are so rarefied that radiation passes easily through them outside into unsaturable space. In some of them convection occurs beneath the radiative zone. The situation in which both radiation and convection occur poses difficult computation problems in astrophysics. Convection is normally assumed to occur if the temperature gradient in the radiative zone exceeds that for convective regions, and thus supplements radiation in the transfer of energy from the interior to outer space. Although the *outermost layers* of all stars have radiation as the transfer mechansim (permitting the equations of radiative equilibrium to be applied there as a first approximation), the convection in the lower layers of some atmospheres is so efficient that "*convective equilibrium*" may be assumed for estimating temperature gradients. *In the sun, for instance, convection is in evidence in the form of granulation and cells seen on photographs, similar to the Bénard Cells in convective fluids in laboratory experiments. Each granule is the top of a column of ascending material, while the narrow, darker spaces separating the granules are regions where the cooler gases sink back into the core.*

The differential equations frequently employed for both the radiative and convective domains are [112]:

$$\frac{dM_r}{dr} = 4\pi r^2 \rho \tag{14}$$

$$\frac{dp}{dr} = -\frac{GM_r}{r^2} \tag{15}$$

$$\frac{dL_r}{dr} = 4\pi r^2 \rho E \tag{16}$$

$$\frac{dT}{dr} = \frac{-3}{16\pi ac} \cdot \frac{\kappa}{T^3} \frac{L_r}{r^2} \quad \text{(radiative)} \tag{17}$$

$$\frac{dT}{dr} = 0.4\frac{T}{p}\frac{dp}{dr} = \frac{-0.4T}{P} \cdot \frac{GM_r}{r^2}\rho \quad \text{(convective)} \tag{18}$$

where r is the distance from the stellar center, M_r the mass contained in a sphere of radius r, G the gravitational constant, L_r the net amount of all energy transferred per second (or net luminosity) at r, E the energy-generation rate (often called the "energy-generation coefficient," which is a function of ρ, T, and the chemical composition), κ the Rosseland mean absorption coefficient, a the radiation pressure constant ($a = 4\sigma/c = 7.56471 \times 10^{-15}$ erg-cm^{-3} K^{-4}, σ being the Stefan-Boltzmann constant).

The density ρ may be eliminated by means of an equation of state. The perfect gas law, which is valid both in the radiative and in the convective cases, becomes

$$\rho = \frac{\mu H}{\kappa} \frac{p}{T} \tag{19}$$

where H is the mass in grams of 1 atomic mass unit (in practice, H may be taken as the mass of the hydrogen atom), μ the mean molecular weight, κ the Boltzmann's constant.

In an adiabatic convective domain, we can use the gas law

$$P = K' \rho^{\gamma} \tag{20}$$

which is equivalent to

$$P = K^* T^{\gamma/\gamma - 1} \tag{21}$$

where $\gamma = 5/3$ if ionization is complete, and K', K^* are constants.

Following the procedures summarized by Novotny [112], and adopting Kramer's law for the estimation of κ, we can estimate the temperature gradients in a variety of cases. For instance, in a radiative model with boundary conditions

$$p = 0,\ T = 0 \text{ at } r = R' \text{ and } M_r = M \tag{22}$$

the thermal gradient near the surface is given by

$$\frac{dT}{dr} = \frac{GH}{4.25\kappa} M \frac{\mu}{r^2} \tag{23}$$

$$T = \frac{GH}{4.25\kappa} M\mu \left(\frac{1}{r} - \frac{1}{R'}\right) \tag{24}$$

Consequently, T decreases from center to surface. In the convective case, with the above zero boundary conditions, the result is similar, namely

$$T = \frac{0.4GH}{\kappa} M\mu \left(\frac{1}{r} - \frac{1}{R'}\right) \tag{25}$$

In these cases the nuclear energy generation is zero and L_r, the net outward energy flux, remains constant. Near the center, however, it changes

$$L_r = \tfrac{4}{3} \pi \rho_c E_c r^2 \tag{26}$$

$$T = T_c - \frac{1}{8ac} \frac{\kappa_c \rho_c^2 t_c}{T_c^3} r^2 \qquad \text{(radiative)} \tag{27}$$

$$T = T_c - \frac{4}{15}\pi G \rho_c^2 \frac{T_c}{\rho_c} r^2 \quad \text{(convective)} \tag{28}$$

where the subscript c refers to conditions at the center (which must be specified). Such starting functions, both at the center and at the surface, are required since it is often not possible to integrate (numerically) over the entire model in one direction, and the solutions must be compatible at an intermediate point. Various distributions for T, ρ, L_r, p and composition have thus been calculated and reported in the literature (for a review, see [112]).

Except in a few unusual circumstances, to be mentioned below, the temperature in all stars decreases outwards. From the thermodynamic point of view it is of interest to follow the variation of L_r with r. For instance, in evolution models of approximately one solar mass, L_r rises steeply to a maximum value at about 20 percent of the total mass. However, in the region exterior to a mass fraction of about 40 percent, L_r decreases slightly because energy is being absorbed by the expanding envelope (the radius increases throughout the evolution-cf. p. 90).

VII.6   STELLAR EVOLUTION

Stellar evolution is manifested by the fact that the star is shining. Two requirements must be observed for a star to shine, namely, some generating process must take place within it as a source of the outflowing energy, and continual emission from the surface requires the surrounding space to act as an unsaturable thermodynamic sink—which is only possible in an expanding universe. In other words, stellar evolution is an impossibility in a non-expanding universe! The very fact that we observe shining stars in a dark sky means that the universe is expanding!

The thermodynamics of stellar evolution covers the entire course of events in which an interstellar cloud concentrates until it is held together gravitationally to become a protostar (pre-main-sequence) and later develops through the stages of main-sequence*, red giant, white dwarf, neutron star, nova, supernova, black hole, etc. It shows, in agreement with all astronomical observations to date, that *all later evolution stages of a star, as well as its ultimate fate, are predetermined by its initial mass* (when it starts to burn hydrogen)**. Actually, there is a limiting mass whose value is about $0.1M_0$, depending on composition. Objects with smaller masses never develop sufficiently high central temperatures and densities for normal hydrogen-burning to start, they contract until the core degenerates and eventually become *black dwarfs* whose temperature is not much higher than the intergalactic level.

---

* A star of one solar mass, for instance, remains within the main-sequence evolution for about 10^{10} earth years—the major fraction of its lifetime (see §I.2 for more details).

** See Fig. I.2.

VII.7   TERRESTRIAL EVOLUTION

To illustrate the origin of local thermodynamics in our atmosphere, let us visualize sudden equilibration of the atmosphere with the radiation density in outer space (see Fig. VI-2). At such equilibrium, all input-output energy flow to earth ceases, causing all irreversible processes on earth to decay within a finite relaxation time. In these circumstances, with rates of energy expenditure maintained constant, the reservoir of internal (thermal) energy would be depleted in about 100 days, that of latent heat (precipitable water) in about 12 days, and that of mechanical energy (momentum) in about 3 days [51]. Thus all intermediate thermodynamic cycles and atmospheric circulation of solar energy in our pseudo-steady-state atmosphere, and in other planets, are coupled (through finite, albeit large relaxation times) to the mechanism whereby solar energy is eventually lost to space by reradiation. *How then, one may ask, do we account for equilibrium states in smaller local systems on earth? In the laboratory* on earth, the time constant for such intragalactic transfer may be estimated by *disconnecting* a non-expanding system from its external driving forces of irreversibility, thereby causing the latter to persist for a while *but eventually to decay.* The resulting equilibrium may be extended by imposition of time-invariant boundary conditions *via interaction with external systems which supply the required energy* but (being part of the rest of the earth and unsaturable expanding space) are themselves *not at equilibrium.* Up to now, however, practical geocentric thermodynamics has misinterpreted the irreversible evolution, inferring that its origin lies inside the locally-observed system. Considerations of the overall heat balance of the earth's pseudo-steady-state atmosphere show that artificial energy sources in the atmosphere or on the ground (e.g., fossil or nuclear fuels) either cause additional outward flow (via intermediate cycles and radiation) or, especially if their strength increases gradually with time, transform the current state into a new one. Consequently, we must radically reject any geocentric-traditional notions that the earth's atmosphere, or its oceans, are "heat reservoirs" which absorb any amount of heat and remain at constant temperature. *Neither the prime source of energy nor its ultimate sink are located on the earth.*

The cause of atmospheric circulation in planets is always the same — the long-wave reradiation to unsaturable space and the differential heating by short-wave radiation which, through the action of gravity on the inhomogeneous density field, produces buoyancy forces [109]. The intensities of atmospheric motions and the temperature distributions are determined by such factors as: radiation-energy balance of the atmosphere and its mass and heat capacity, the size of the planet and the angular velocity of its rotation, the angle between the axis of rotation and the ecliptic plane, the absorptive properties of different atmospheric gases, and of course, gravity! On Mars, the bulk of the incident solar radiation is radiated back to space at the same time and place where it is received. The "thermal efficiency" of the Martian atmosphere is thus an order of magnitude less compared with the earth's. Recent calculations based on data obtained from Pioneer 10 indicate that

Jupiter's interior heat is evenly radiated to space from the entire planet at a rate two-three times greater than the heat it takes in from the sun. Jupiter's outer magnetic field reaches a minimum distance of 2.1 million mi. beyond the planet, and sometimes as far as 6.5 million mi. The rapid rotation of the magnetosphere with the planet drives ionized particles outwards. Pioneer 10 data determined also that the moons of Jupiter decrease in density as they lie in sequence away from the planet. This is a result of Jupiter's "heat effect" upon the moons, boiling away the less dense, more volatile materials. In this sense, Jupiter acts as a miniature solar system, since all the planets away from the sun are progressively less dense. Jupiter's temperature distribution was also calculated from Pioneer 10 data. The results: eight miles above its clouds the temperature is -229 F, compared with about -184 F at the topmost cloud layer, 3600 F at a depth of 600 mi, 20,000 F at 15,000 mi, and 54,000 F at its core.

VII.8   SOME OPEN QUESTIONS

■ Thermodynamicists are often surprised to learn that the internal temperature, composition and density of most observed main-sequence stars can now be calculated with greater confidence than the temperature, density and composition of the earth's interior. To a large extent, this is due to the fact that the earth is highly *atypical* in the universe in which about 99% of all matter is composed of *only two elements*: hydrogen and helium. This confidence applies also to the nuclear reactions (in particular the H → He transition) taking place in the cores of many galactic masses. These points are stressed here, because our theory is based, in part, on verified information about the *direction* of net energy flows, and on average temperature *gradients*. The fact that a reverse transfer, namely

[Net radiation from space] → [Net radiation flux to surface of galactic masses, etc.]

has never been observed, is the most striking evidence for what we call the third cosmological asymmetry.

■ Many questions remain unanswered, most of them philosophical in nature. For instance, can one *isolate* a system and analyze its time behavior in disregard of the complexity of the rest of the universe? It is not always simple to argue against such a possibility. The mere statement that the notion (and assumption of existence) of an isolated system is an independent postulate (always to be questioned) cannot, of itself, resolve the difficulty. It is the overwhelmingly observational evidence about the nature of high-penetrating radiation (neutrinos, etc) that must be stressed first. Here, Mach's principle may serve as a guideline, though not as a starting point for a solid new foundation [6].

■ Even if the feasibility of complete isolation of a physical system is accepted, the causal links advocated by our school remain valid. To illustrate this point, it must be recalled that *prior* to its isolation, the system's *medium* formed *an integral part* of the *one-sided* dynamics of the (irreversible) universe (which, in turn, draws its asymmetry from interaction with the master asymmetry via long relaxation times). *Hence, the medium's embedded asymmetry is enclosed with it! This asymmetry continues for a while to evolve inside the "isolated" system, but must eventually decay* (because it is now cut off from its driving (external) asymmetries and gradients). This cutoff is effected through the "isolating" walls, whose agency causes decay of the (*previously enclosed*) dissipation; *the latter being thus wrongly identified by traditional thermodynamicists as an embedded ("independent") thermodynamic arrow of time originated by the process of observation or generated inside the system. This arrow, however, is not an independent one, and cannot reappear unless reconnected with the external driving forces for a fresh irreversible effect.* To see what creates irreversibility, we must first abandon our geocentric approach. It is the "large" system which dominates the evolution of smaller systems, not vice versa!

■ Creation and transfer of irreversibility (from large to small systems) involves electromagnetic asymmetry, thermal gradients, etc., but most important, it involves gravitation! It is gravitation which generated the self-gravitating stars and nebulae. It is gravitation which causes them to contract, warm up and convert gravitational into radiant energy lost in space. It is the theory of gravitation behind the *unsaturability* of intercluster space. Gravitation is also the hidden* agent *behind the very possibility of isolating walls being erected!* In the laboratory, the latter are used for *cutting off* the medium's causal links with *external* (irreversible) driving forces (e.g., electromagnetic waves and various gradients prevailing in the earth's atmosphere). In theory, the gravitational field can also be linked with the condition of an *extremum* of energy, outward thermal flow, etc.

■ Formulation of a more unified theory of thermodynamic and electromagnetic irreversibilities may originate in a coupled theory of Einstein's general relativity and quantum chromodynamics (cf. Assertion 2 in the *Introduction*). Such an ultimate theory is, perhaps, the most challenging field of future research. In the meantime, however, the 'second law' can be derived from the resulting asymmetry of the Newtonian theory (or gneral relativity)—i.e., from the master time asymmetry (coupled with local formulation of the reversible energy equation). This result is independently based on empirical observations. Our approach has, therefore, both theoretical and phenomenological origins, whereas classical and statistical mechanics (as stressed before) have only a phenomenological basis.

VII.8.1 Microscopic T-Violation and the Master Asymmetry: A Possible Connection?

■ T-violation, i.e., existence of a *microscopic* arrow of time**, indicates that the fundamental question of microscopic irreversibility must be investigated within a theory

---

* Unrelated to the "hidden variable theories" of [89,90].

** See §IV.11.

more closely unified with the rest of *macroscopic physics. Irreversibility on the present smallest scale may not be entirely dissociated from that observed on larger scales!*

■ Such an asymmetry may be imprinted on the weak electromagnetic, or gravitational interactions, and may be traced back to the master asymmetry (§VI.3 and §VI.5). Even if this possibility is rejected, we must still avoid an a priori ruling as to the smallest scale for the "penetration" of irreversibility. Isolation of the problem and its restriction to the domain of microscopic physics would merely frustrate the search for a unified origin of asymmetries in nature. The observed phenomena may well be subtle manifestations of an entirely new aspect of the interaction of the micro- and macro-universes, especially through the effect of intercluster expansion on the *number* of neutrinos per unit galactic mass (§IX.20–22).

■ *The Weinberg-Salam theory* (§V.6) has simultaneously accommodated parity violation for weak interaction and parity conservation for electromagnetic interaction by employing spontaneously broken [SU(2) xU(1)] symmetry. It is also conceivable that there is an additional SU(2) gauge interaction whose effects may explain CP violation. [The name SU(2) means that the group of symmetry transformations consists of 2×2 unitarity matrices that are special in that they have determinant unity.] But the observed violations of parity (and other) symmetries cannot be brought about by spontaneous symmetry breaking [which in the Weinberg-Salam theory of the *electro-weak interaction* gives mass to all the vector bosons, except the photon, and is also an *exact local gauge symmetry*]. But these subjects lead us to QCD, supergravity, high gravitational interactions in the early universe and in black holes as well as to the physics of the vacuum, subjects to be treated, among other things, in the next two lectures.

BLACK HOLES AND THE UNIFICATION
OF ASYMMETRIES

VIII-1   INTRODUCTION

Of all the conceptions of cosmology and astrophysics, perhaps the most intriguing is the *black hole* (or "frozen star," as it is sometimes called): a hole in space with a definite edge, over which anything can fall in — but nothing can escape, because of a gravitational field so strong that even radiation is irreversibly trapped and held by it; a thermodynamic sink which drastically curves space and twists time.

When a star of, say, 20 times the mass of the sun has consumed nuclear fuel through its internal reactions for several tens of millions of years, its energy supply runs out. Without it the star can no longer maintain the enormous thermal pressures that normally counterbalance the inward pull of its gravity, the latter prevails, and the star begins to collapse. The outcome of this collapse is governed by two critical masses: the *Chandrashekhar limit,* below which the star may become a *white dwarf,* and the *Landau-Volkoff limit,* below which it may become a *neutron star*.* Above these critical masses, the star may collapse completely to a black hole or shed off its excess mass. In these circumstances, it shrinks to such a small size that the escape velocity from its surface *exceeds* that of light, with the result that rays emitted by it are never able to reach an outside observer.** Our best hope for detecting a black hole would therefore be to find a *binary system* in which the other member is an ordinary visible star (see Lecture I for details).

Existence of black holes was suggested in the 1930's on purely general relativistic grounds, chiefly through the work of J.R. Oppenheimer and his collaborators. However, it has remained in the realm of speculation until a few years ago, when the joint efforts of X-ray, radio and optical astronomers succeeded in providing (a partial) supporting evidence. (Interest in black holes was also associated with the discovery of quasi-stellar objects (QSO's, or quasars) with their starlike optical images (often containing powerful compact radio sources) and with *red shifts* ranging

---

* See Lecture I §3.

** How small may a black hole be? Hypothetically, if the earth became a black hole, its circumference would be only 5.58 cm!

from 0.131 to nearly 3 (See Lectures I and III). However, we do not yet know what QSO are).***

Detailed new observations in the light, radio and X-ray wavelength ranges indicate that the binary source *Cygnus* X–1 (see Table I.1) may include a black hole orbiting around a massive blue supergiant star; other binary X-ray sources listed there may also include black holes.

VIII-2   OBSERVATIONAL EVIDENCE

Two of the eight X-ray binary sources in Table I.1, Cent. X-3 and Her. X-1, clearly do not harbor a black hole. We can be certain of this, because their X-rays arrive in precisely timed periodic pulses: 4.84239 sec. for the first and 1.23782 sec. for the second. Nothing associated with a black hole can, as far as we know today, give rise to such regular behavior. [It can be attributed, however to a *rotating neutron star* (Fig. I.5). A black hole is not likely to produce *oriented beams*, since no off-axis structure such as a magnetic field can survive in it]. Observations of other binary X-ray sources in the table* revealed that each of them comprises a *supergiant* with a *periodically varying Doppler shift*, while there is no sign of spectral lines from the secondary star. In all four cases, the visible spectrum shows lines emitted by gas flowing towards the invisible companion. The X-rays from 3U 1700–37, 3U 0900–40 and SMC X-1 are also *eclipsed* each time the visible star passes between the earth and the invisible companion (Fig. I.7). Accordingly, the latter is almost certainly the *source* of the X-rays, which are most likely generated as the gas is heated by falling into it just as they do when the gas falls into the strong gravitational field of a neutron star (Fig. I.5); at least, astronomers have not yet been able to provide any other quantitative explanation for this origin of X-rays.

*** The QSO's belong to a category of little-understood celestial bodies discovered in recent years, which include the *Seyfert galaxies* (giant elliptical galaxies with very powerful compact radio sources, X-ray sources, and galactic nuclei that appear in some cases to be exploding). General relativistic effects may play an important role in the explanation of their nature. If they are relatively close to us (as some astronomers think) but moving at relativistic velocities, then some unknown sources of energy must be found to account for transformation of matter into kinetic energy and radiation with nearly 100% efficiency. If they are at cosmological distances (as others are inclined to assume), then their *apparent* optimal luminosity indicates an *absolute* luminosity much higher than that of the largest known galaxies, so again a powerful unknown source of energy is to be accounted for. In this context, their strong gravitational fields seem to offer such a source. For this reason the discovery of the QSO's red shifts may be mainly gravitational in origin. Thus they may be so highly compressed that their structure would have to be understood in terms of general relativity rather than of Newtonian mechanics. [The interested reader may find detailed discussions of these questions in the *"Sixth Texas Symposium on Relativistic Astrophysics,"* *Annals of N.Y. Acad. Sci. 224* (1973) and in more recent publications in this field.] Other models have been proposed to explain QSO phenomena. However, no agreement has yet been reached about the nature of these objects.

* Cygnus X–1; 2U 1700–37; 3U 0900–40; SMC X–1 (3U 0115–37).

Only three types of objects have such sufficiently strong gravitational fields: white dwarfs, neutron stars and black holes, and *to distinguish among these three possibilities astronomers have to estimate the mass of the invisible companion.* This is done on the basis of the Doppler shift and spectrum of the *visible* star. The results are:

$$3U\ 1700\text{–}37: \quad M \cong 2.5\ M_\odot \qquad SMC\ X\text{–}1 \quad: \quad M \cong 2\ M_\odot$$

$$3U\ 0900\text{–}40: \quad M \cong 3\ \ M_\odot \qquad Cygnus\ X\text{–}1: \quad M \geq 8\ M_\odot$$

These figures suggest that perhaps one of the X-ray binaries include a black hole. The estimates, however, are fairly rough, and individual astronomers may differ by a factor of two or more in their estimated masses; actually, any astronomer who wishes to play the Devil's advocate may reduce the estimates still further by offering unusual interpretations of the empirical data which alternative contrived explanations and models (with the black hole replaced by a neutron star or a white dwarf) can still be made to fit. It is only through additional empirical evidence, and by working back and forth between theory and fact, that the question of the existence of black holes will be settled in the future. Be that as it may, the black hole can no longer be dismissed as a wild speculation. In fact, some of the best minds are now occupied in studying its properties, whose interpretation has supplied us with a new powerful tool in understanding the nature of space, time, and irreversibility in strong gravitational fields. To follow some of these explanations, we next examine the metric properties of space-time in the neighborhood of a spherical black hole.

VIII-3   SCHWARZSCHILD SOLUTION AND BLACK HOLES

When collapse sets in to form a black hole, we may suppose that the pressure supporting the *surface* of the star is *zero* and that this surface undergoes *free fall* towards the center. If the space outside the black hole is empty — that is, if all the material in the star has been "swallowed" by the black hole — and if the star is rotating, then the final stationary state of the hole must be one of the family of solutions of the Einstein field equations found by *R. Kerr.* However, the Kerr solution (or *Kerr metric*) is not the only stationary space-time containing a black hole: there is also a class of axisymmetric solutions describing a central black hole surrounded by *orbiting disks* (somewhat similar to the rings of Saturn)* which may be the remnants of a rotating star or some other kind of accretionary matter, as shown in Figs. I.9 and I.10.

In this lecture, however, I confine myself to the simplest theoretical case, namely, a *non-rotating spherical* black hole with mass M, located in empty space. For empty space, Einstein's field equations, (III-66) reduce to

---

* In which case, incidentally, the black hole may be visible.

$$R_{\mu\nu} = 0 \tag{1}$$

where $R_{\mu\nu}$ is the Ricci tensor. The most general *metric tensor for a static isotropic gravitational field* is given (in spherical coordinates) by

$$d\tau^2 = B(r)\,dt^2 - A(r)\,dr^2 - r^2(d\theta^2 + \sin^2\theta\,d\psi^2) \tag{2}$$

with functions $A(r)$ and $B(r)$ to be determined by solving the field equations under the proper boundary conditions. The components of the Ricci tensor for this metric are given by¶

$$R_{rr} = \frac{B''(r)}{2B(r)} - \frac{1}{4}\left(\frac{B'(r)}{B(r)}\right)\left(\frac{A'(r)}{A(r)} + \frac{B'(r)}{B(r)}\right) - \frac{1}{r}\left(\frac{A'(r)}{A(r)}\right) \tag{3}$$

$$R_{\theta\theta} = -1 + \frac{r}{2A(r)}\left(-\frac{A'(r)}{A(r)} + \frac{B'(r)}{B(r)}\right) + \frac{1}{A(r)} \tag{4}$$

$$R_{\psi\psi} = \sin^2\theta R_{\theta\theta} \tag{5}$$

$$R_{tt} = \frac{B''(r)}{2A(r)} + \frac{1}{4}\left(\frac{B'(r)}{A(r)}\right)\left(\frac{A'(r)}{B(r)} + \frac{B'(r)}{B(r)}\right) - \frac{1}{r}\frac{B'(r)}{A(r)} \tag{6}$$

$$R_{\mu\nu} = 0 \quad \text{for} \quad \mu \neq \nu \tag{7}$$

With $A(r)$ and $B(r)$ subject to the boundary condition that for $r \to \infty$ (i.e. far away from the gravitational field generated by the black hole) the metric tensor (2) must approach the *Minkowski metric* (eq. III, ...), which in spherical coordinates reads

$$d\tau^2 = dt^2 - dr^2 - r^2\,d\theta^2 - r^2\sin^2\theta\,d\psi^2 \tag{8}$$

we obtain

$$\lim_{r \to \infty} A(r) = \lim B(r) = 1 \tag{9}$$

At the Newtonian limit (see Lecture III)

$$g_{tt} \equiv -B = -(1 + 2\phi) \tag{10}$$

where ϕ is the Newtonian potential given by

$$\phi = -\frac{MG}{r} \tag{11}$$

¶ The primes denote differentiation with respect to r.

r being the distance from the center of the black hole. Applying (10) for $r \to \infty$, the solution becomes

$$B(r) = 1 - \frac{2MG}{r} \tag{12}$$

$$A(r) = 1 \Big/ \left(1 - \frac{2MG}{r} \right) \tag{13}$$

Substituting (12) and (13) in (2), the final result reads*

$$d\tau^2 = \left(1 - \frac{2MG}{c^2 r} \right) dt^2 - \frac{dr^2}{c^2 \left(1 - \frac{2MG}{c^2 r} \right)} \tag{14}$$

where c is the velocity of light in vacuum (in our units $c = 1$) and we do not consider variations in θ and ψ due to spherical symmetry, $d\tau$ is the element of *proper time* (*measured by a clock falling with the surface of the star*), and t *the time as measured by a remote outside observer*. An interesting conclusion is obtained when we set

$$r = r_g = \frac{2MG}{c^2} = 2m \tag{15}$$

The resulting "singularity**" in t and r indicates that when the surface of the star gets down to a radius as per (15) — the so-called *Schwarzschild* or *gravitational radius* — the light cone tips inwards, so that *no further radiation can escape*. In other words, the gravitational field becomes so strong that the radiation is trapped; hence the aptly descriptive term "*black hole.*" The edge of the black hole, known as the *event horizon*, is defined by the wave front that just fails to escape (see Figs. VIII-2 and 3).

The physical meaning of this result becomes clearer when we consider the time rate of fall as it appears to a *remote observer* or, by contrast, to a *local observer* stationed on the falling surface of the star. From Lecture III we know that a photon obeys the equation

$$d\tau = 0 \tag{16}$$

in which case eq. (14) reduces to

$$dt_{\text{photon}} = \frac{dr}{1 - \frac{2MG}{c^2 r}} = \frac{dr}{1 - r_g/r} \tag{17}$$

---

* This solution was found by K. Schwarzschild in 1916.

** What appears to be a singularity in one coordinate system may have a different meaning in another. Indeed such is the case with the *Kruskal coordinates* [*Phys. Rev., 119*, 1743 (1960)].

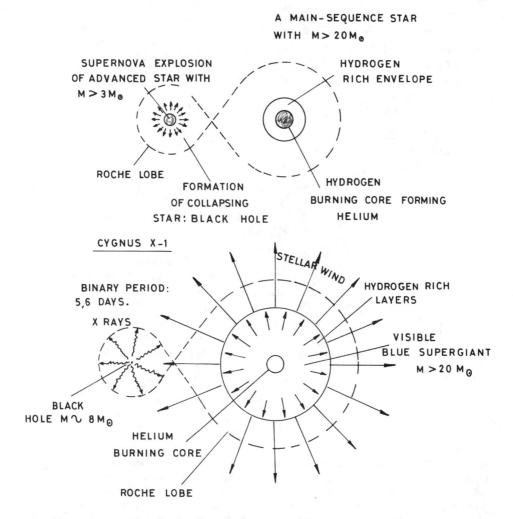

A MAIN-SEQUENCE STAR
WITH M > 20M$_\odot$

SUPERNOVA EXPLOSION
OF ADVANCED STAR WITH
M > 3M$_\odot$

HYDROGEN
RICH ENVELOPE

ROCHE LOBE
FORMATION
OF COLLAPSING
STAR: BLACK HOLE

HYDROGEN
BURNING CORE FORMING
HELIUM

CYGNUS X-1

BINARY PERIOD:
5,6 DAYS.

STELLAR WIND

HYDROGEN RICH
LAYERS

X RAYS

VISIBLE
BLUE SUPERGIANT
M > 20 M$_\odot$

BLACK
HOLE M ∿ 8M$_\odot$

HELIUM
BURNING CORE

ROCHE LOBE

FIG. VIII-1: **X-ray binary source Cygnus X-1 as a black hole in orbit around visible blue supergiant star.**
(above) The Binary system about 6 million years ago at the moment of supernova explosion which led
to the formation of a black hole remnant and prolongation of the binary period. As a result of the explosion the binary system may have been accelerated to a speed of a few dozen km/sec. ("runaway"
binary system). Most such systems remain gravitationally bound after the explosion. (below) The binary
X-ray source Cygnus X-1. The main-sequence normal star has evolved into a blue supergiant star in whose
core helium is fused chiefly into carbon. (X-rays are produced according to the mechanism described in
Figs. I.5, I.6 and I.9. The physical meaning of Roche lobe is shown in Fig—. I.9). See also §I.3, I.4, I.5,
I.8, I.9 and Table I.1).

Recalling that t in eq. (17) is the time as measured by a remote outside observer,
we note that as r approaches r_g — the Schwarzschild radius — dt tends to infinity.
Consequently, *the collapse to the Schwarzschild radius appears to a remote outside
observer to take an infinite time and further collapse to $r = 0$ is therefore unobservable*

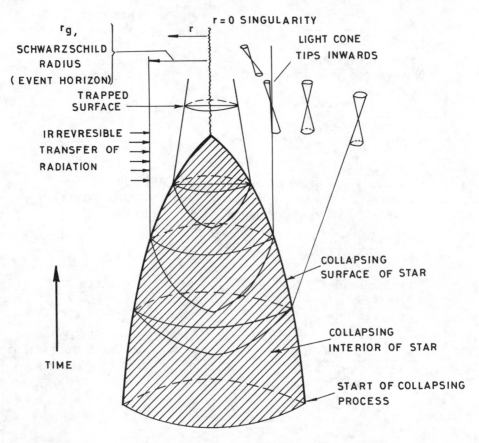

Fig. VIII-2: **The formation of black hole and event horizon in a spherically symmetric collapse of a non-rotating star.** The *Schwarzschild radius* is equal to $2MG/C^2$. The event horizon acts as a one-way membrane; radiation and mass can pass through into the black hole, but nothing can escape to the outside. The gravitational field of a black hole is so strong that even radiation is caught and held inside. In general relativity pressure is also a source of gravitation. Thus, inside a completely collapsing star the 'gravitating effect of pressure' overwhelms its own supportive effect (see §III.1.1. (iii) and eq. III.92). Black holes display their own *distinctive kind of irreversibility*. The event horizon area can never decrease and this promoted a number of physicists to argue that, therefore, it plays a role in black-hole physics very similar to the entropy in thermodynamics. But there is no evidence to support this speculation. For more details see Figs. I.4 and VIII-3 to 4, and Table I.1.

from the outside. (The same result is obtainable by proper integration along the r coordinate).

From a *thermodynamic point of view* the collapse process appears to a remote outside observer as slowing down and approaching an *"equilibrium state"* at which time *t* stops. For such an observer a single, isolated, nonrotating, spherical black hole represents a frozen star *devoid of gravitational irreversibility*. Entirely devoid?

To answer this question we must analyze the time-change of the luminosity of such a star (see below).

What, however, about a *local* observer who is co-moving with the infalling matter and experiencing no gravitational field? Clearly, such an observer is entitled to resort to Newtonian mechanics, and the time required to reach r_g as measured by him is *finite*. Again, we can derive this result from eq. (14). The clock attached to this observer would show the proper time τ, which by eq. (14) is subject to

$$d\tau = \left[\left(1 - \frac{r_g}{r} \right) dt^2 - \frac{dr^2}{c^2 (1 - r_g/r)} \right]^{1/2} \tag{18}$$

or

$$\tau_{(r_1 \to r)} = \int_{r_1}^{r} \left[\frac{(1 - r_g/r)}{(dr/dt)^2} - \frac{1}{c^2 (1 - r_g/r)} \right]^{1/2} dr \tag{19}$$

where dr/dt is the velocity of the falling particles according to the *remote* observer. Eq. (19) converges to a finite limit for all kinds of falling particles: for instance, for parabolic velocity function ($dr/dt = 0$ at infinity), eq. (19) yields

$$\tau_{(r_1 \to r)} = \text{Constant}(r_1 - r)^{3/2} \tag{20}$$

where r_1 is the radial position of the particle at zero time. This result coincides with the Newtonian formula obtained by integrating the expression

$$\frac{dr}{dt} = \left(\frac{2MG}{r} \right)^{\frac{1}{2}} \tag{21}$$

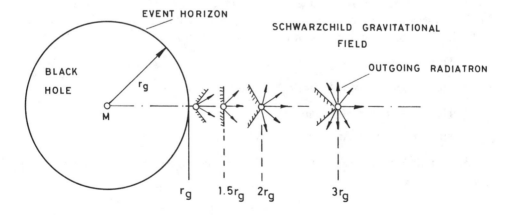

Fig. VIII-3 **Gravitational capture of radiation:** rays emitted from each point into the interior of the shaded conical cavity are captured gravitationally. See also Fig. VIII.2.

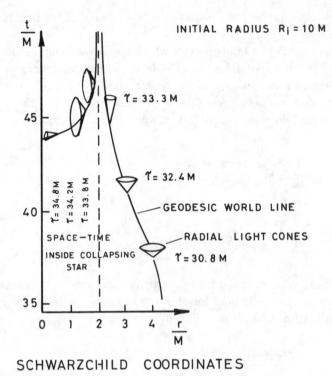

INITIAL RADIUS $R_i = 10$ M

$\Upsilon = 33.3$ M

$\Upsilon = 32.4$ M

GEODESIC WORLD LINE

RADIAL LIGHT CONES

$\Upsilon = 30.8$ M

$\Upsilon = 34.8$ M $\Upsilon = 34.2$ M $\Upsilon = 33.8$ M

SPACE–TIME
INSIDE COLLAPSING
STAR

SCHWARZCHILD   COORDINATES

Fig. VIII-4. **Gravitational collapse to a black hole in Schwarzschild coordinates.** Event horizon is at $r/m = 2$. Note the different orientation of the light cones (see also Fig. VIII.2).

These results can be used to calculate the change in luminosity of a freely collapsing surface as seen by the remote observer. In general, this luminosity is reduced by the following factors, as compared with that of a fixed surface:

1. The *gravitational red shift*: $(1 - r_g/r)$.

2. The *deceleration* of light in the gravitational field: $(1 - r_g/r)$.

3. The *relativistic Doppler shift*: $(1 - v^2/c^2)/(1 + v/c)^2$.

4. The *relativistic Doppler decrease in photon number*: $(1 - v^2/c^2)/(1 + v/c)^2$.

Accordingly the intensity of radiation as measured by the *remote observer* is given by

$$I_\infty = \text{Constant} \; \left(1 - \frac{r_g}{r}\right)^2 \left[\frac{1 - \dfrac{v^2}{c^2}}{[1 + v/c]^2}\right]^2 \tag{22}$$

Here v is just the rate of fall dr'/dt' as seen by the local observer *at rest at radial distance* r, and is obtainable from eq. (14), namely

$$v = \frac{dr'}{dt'} = \frac{dr}{dt} \cdot \frac{1}{1 - r_g/r} = \left(1 - \frac{1 - r_g/r}{1 - r_g/r_0} \right)^{1/2} c \tag{23}$$

where dr/dt is the rate of fall as seen by the remote observer and r_0 the radial distance at which the collapse sets in. For $r \to r_g$, $(1 - v^2/c^2)$ reduces to $(1 - r_g/r)$, and $(1 + v/c)$ equals 2. Eq. (22) then becomes

$$I_\infty = \text{Constant} \left(1 - \frac{r_g}{r} \right)^4 \tag{24*}$$

The luminosity of a collapsing star thus decreases very rapidly as r approaches r_g and tends (irreversibly) to zero when $r = r_g$. However, even though it fades out of sight, the star still has a gravitational field, and measurement of this field at great distances can be used to determine its energy, momentum, and angular momentum.

From the thermodynamic point of view, a black hole behaves like an *infinite, unsaturated sink* for mass and energy; an unlimited amount of energy (or "work") can be pumped into it *without raising its temperature*! Such a system is analogous to a black body at zero absolute temperature with an infinite thermal capacity — indeed, it appears to *violate the "fact-like" second law of thermodynamics*. This creates a serious conflict between general relativity and empirical thermodynamics.

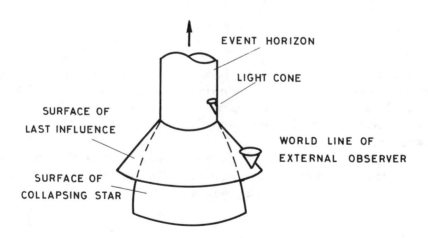

Fig. VIII-5. **The formation of event horizon in a collapsing star.**

* The above analysis applies for a *single source* of radiation falling towards r_g. The fact that absolute luminosity is a *surface* concept introduces certain complications, but the final result is not essentially different.

VIII-4   BLACK-HOLE MECHANICS AND ENTROPY

Black-hole physics gave rise to a number of speculative hypotheses for which *no observational evidence has been found at the time of writing*. One of these hypotheses refers to *"white holes"* — *the time-reversed dynamic solutions* for a collapsing star — whose possible existence is not ruled out by the general theory of relativity, and which may possibly be exemplified in the very *expansion of the universe* away from a singularity (the Friedmann cosmologies).* The point of interest here is that a *black hole CANNOT turn into a white one, while a white hole CAN turn into a black one*; in cosmological models of the conservative type, this conclusion, in itself, ensures a general kind of *"white-to-black-hole" irreversibility* (or *time asymmetry*)!

Black holes add their own distinctive irreversibility to that entailed by expanding intercluster space: no energy or matter can escape over the event horizon, although anything can move in. Black holes, if exist, thus provide an *additional* sink to that provided by unsaturable intercluster space,** but the latter effect *cannot* be explained in terms of the former. It is only Einstein's theory of gravitation which explains both.

A highly loose *analogy* between quantities associated with black-hole physics on the one hand, and the concepts of *entropy* and *temperature* in classical thermodynamics on the other, has been suggested by Penrose, by Christodoulou and Ruffine, by Bekenstein, and by Bardeen, Carter and Hawking.* The above quantities concern the general case of black holes surrounded by rings of matter; the laws of black-hole mechanics may then parallel, and in some ways may transcend, those of classical thermodynamics. The analog of the second law is then claimed to be seen in the fact that the area, A, of the event horizon around a black hole *never decreases with time* — *i.e.*

$$\delta A \geq 0, \tag{25}$$

and when two black holes coalesce, the area of the event horizon around the resulting hole must increase (see also below). There are also analogs of the first, third and zeroth laws. Bekenstein even proposed modified formulations aimed at avoiding "violation of the second law" when entropy flows into a black hole. These hypotheses, however, are strictly ad hoc.

No such formulation can have physical meaning without allowance for the principal source of entropy growth — the expanding intercluster space. Yet it may be intellectually stimulating to play with some black-hole theorems (see below).

---

* See Lecture III.
** See Lecture VI.
*** R. Penrose, *Rivista del Nuovo Cimento*, Ser 1, *1*, 252 (1969); D. Christodoulou and R. Ruffini, *Phys. Rev. D. 4*, 3552 (1971); J. M. Bardeen, B. Carter, and S. W. Hawking, *Comm. Math. Phys. 31*, 161 (1973); S. W. Hawking, *Annal. N.Y. Acad. Sci. 224*, 268 (1973); J. D. Bekenstein, Ph.D. Thesis, Princeton University, 1972. See also ref. 125.

VIII-5   CAN BLACK HOLES "EVAPORATE"?

In recent years theorists have begun to play with some speculative quantum mechanical processes that may occur near a black hole. If valid, these processes may cause creation of elementary particles near a black hole, and, eventually, lead to "evaporation" of black holes. Indeed, quantum field theory considerations show that all species of particles may be created in the strong gravitational field *outside* a rotating black hole (e.g., electron-positron pairs, neutrino-antineutrino pairs, and so on). Thus, a distant observer *may* be seeing a flux of particles which *emerge* from the black hole. As a result of this process, *the black hole may lose rotational energy and eventually spin down to a non-rotating state. However, unless the black hole mass is extremely small, the rate of energy loss is negligible, even over a time scale of billions of years.* Hence, *if the process is valid at all, it cannot* be important for black holes which had originated from gravitational collapse of advanced stars.

Nevertheless, there is one class of black holes in which this "evaporation" may be interesting. It concerns the speculative possibility of the formation of black holes during the early stages of the cosmological expansion.

VIII-6   PRIMORDIAL BLACK HOLES?

Since the density of matter-radiation during the early stages of the cosmological expansion was extremely high, it was possible that *inhomogeneities* in the density could have resulted in the formation of black holes. I.e., if the density was fluctuating in some region, then *rather than expand with the rest of the universe, gravitational collapse of the matter-radiation in this region might have occured.* Such gravitational collapses could form black holes of any mass, *depending only on the scale of the gravitational instability.* Thus, black holes with mass *much less than that of the sun* could have been produced in this manner.

But, according to this view, the temperature of the particles 'evaporating' from black holes is *inversely* proportional to the mass of the black hole. Moreover, the flux of energy carried by these particles is roughly proportional to T^4 (i.e., proportional to $1/M^4$) times the area of the event horizon of the black hole (which is proportional to M^2). *Therefore, as the black hole loses mass, the rate of radiation energy loss from the black hole increases, i.e., as it gets smaller it gets "hotter", till, eventually, it ends with a burst of highly energetic particles from the vicinity of the black hole.* However, no such 'burst' has yet been observed and confirmed!

VIII-7   BACK TO THE MELTING POT OF UNIFICATION?

In the mathematical analysis of possible particle creation vs. black holes 'evaporation', the particles are assumed to obey the *indeterministic* interpretations of quantum physics, while the gravitational field, which is causing the particle creation, is governed by the *deterministic* nature of the general theory of relativity. This contradiction becomes especially pressing when we deal with the final stages of black hole 'evaporation' (when the dimension of the black hole goes below the Planck's scale), i.e., when the *Schwarzchild radius* r_g is less than the *Planck's scale*, viz.,

$$r_g \leq 10^{-33} cm. \tag{26}$$

One can only speculate about what happens beyond these final stages. Would flat space-time prevade the previous site of black hole 'evaporation'? Does indeterminism dominate these final stages? Some of these questions have already been discussed in §IV.13. But, most important, they bring us back to the various issues of unification of the laws of physics (Introduction and §IV.12), in particular to the unification of quantum physics with general relativity and to the issues of asymmetric initial-boundary conditions (Table II, Introduction). They also pose new questions about *the validity of some conservation equations*, in particular the ones associated with possible violations of the laws of conservation of *leptons* and *baryons* (II.12) during 'evaporation' of black holes. (The porocess of black hole formation and 'evaporation' apparently violates these laws. For instance, if a black hole 'evaporates' completely, one will have *begun* a process *with large total baryon number* and would *end up* with *net baryon number zero*.)

One may expect that the ultimate aim of such theoretical studies is, first, to evaluate *the range of validity of the presently known laws* (and interactions—including the validity of conservation equations in quantum physics); secondly, to find the new laws of physics which govern the *observed* phenomena *outside* this range; and finally, *to unify* these laws within the framework of a single consistent theory. The discovery of such new laws is generally accompanied by major breakthroughs in our understanding of nature, and by gradual unification of laws, interactions, symmetries and asymmetries.

It is *the unification of asymmetries* which is gradually emerging as the most important *prerequisite to general unification*; and it is in this sense that we conclude this lecture by stressing the following results:

■ Black holes, if present in the universe, represent another example of gravitational asymmetry [in *adding* localized space unsaturability to the *existing unsaturability* of expanding intercluster space].

■ However, the concept black hole *does not*, by itself, introduce any new law into physics. Hence, *claims to new generalized thermodynamics based on black hole physics have no foundation*. The localized *asymmetry* associated with black holes *is caused and dominated by the master (gravitational) asymmetry of nature (namely, that the universe expands). It is through this master asymmetry that primordial black holes may be formed; it is through it that galactic, stellar and planetary asymmetries are formed* (including gravitational collapse of old massive stars that may lead to the formation of black holes). *I.e., was inercluster space static, stellar evolution would have become an impossibility (due to the relatively short time required for intercluster space to become saturated with radiation). Hence, the very expansion of space renders the possibility of black holes. Black holes cannot evolve in a static universe, i.e., in a uniform universe in which absorption of radiation from evolving stars is an impossibility.*

■ *It is only Einstein's theory of gravitation which can explain both black holes and the origin of the master asymmetry.* Hence, gravitation is *the cause and the origin* of asymmetry in nature. From it alone all the asymmetries of nature emerge. *There is, therefore, no need to render black holes any special status in thermodynamics, in the physics of time asymmetries, or in any study about the origin of entropy growth, irreversibility, and asymmetries in nature.* We can lodge all asymmetries under the same roof of gravitation, and, most likely, still leave enough room for the accommodation of the other three known fields in nature.

But there is more to be associated with asymmetry and gravitation, as will be shown in our last lecture.

Putting a frame around does not make it a unified theory (from *Am. Sci., Sigma Xi*).

REFERENCES (To Part III)

1. Gal-Or, B., *Found. Phys.*, 6, 407 (1976); 6, 623 (1976).
2. Gal-Or, B., ed., *Modern Developments in Thermodynamics*, Wiley, N.Y. (1974). Stuart, E. B., Gal-Or, B., and Brainard, A. J., eds., *A Critical Review of Thermodynamics*, Mono Book, Baltimore (1970) (Proceedings of International Symposium, sponsored by NSF, "A Critical Review of the Foundations of Relativistic and Classical Thermodynamics," at Pittsburgh, Pa., April 7–9 (1969).
3. Gold, T., in *Recent Developments in General Relativity*, Pergamon Press, N.Y. (1962), p. 225; in *The Nature of Time* (Gold, T., ed.), Cornell University Press, N.Y. (1967), pp. 1, 128, 229; in Ref. 1, pp. 63.

4. Narlikar, J. V., in *The Nature of Time* (Gold, T., ed.), Cornell University Press, N.Y. (1967), pp. 25, 28, 62; *Pure and Appl. Chem. 22*, 449, 543 (1970); with Hoyle, F., *Nature 222*, 1040 (1969); *Proc. Roy. Soc. A 277*, 1 (1963); *Ann. Phys. 54*, 207 (1969), *62*, 44 (1971).

5. Gal-Or, B., *Science 176*, 11–17 (1972); *Nature 230*, (1971); *234*, 217 (1971).

6. Gal-Or, B., "Entropy, Fallacy, and the Origin of Irreversibility," *Annals, N.Y. Acad. Sci. 196* (A6), pp. 305–325, October 4 (1972) [N.Y. A.S. Award Paper (1971)].

7. Gal-Or, B., "The New Philosophy of Thermodynamics," in *Entropy and Information in Science and Philosophy* (Zeman, J., ed.), Czechoslovak Academy of Sciences, Elsevier (1974). *Space Sci. Review*; In press.

8. Ellis, H. G., *Found. Phys., 4*, 311 (1974).

9. Rosenfeld, L., in *The Nature of Time* (Gold, T., ed.), Cornell University Press, N.Y. (1967), pp. 135, 187, 191, 194, 227, 230, 242.

10. Bergmann, P. G., in *The Nature of Time* (Gold, T., ed.), Cornell University Press, N.Y. (1967), pp. 40, 185, 189, 233, 241.

11. Beauregard, O. Costa de, in Ref. 2, pp. 75.

12. Aharony, A. (with Ne'eman, Y.), *Int. Jour. Theoret. Phys. 3*, 437 (1970); *Lettere al Nuovo Cimento*, Serie I, *4*, 862 (1970) in Ref. 1, pp. 95.

13. Landau, L. D. and Lifshitz, E. M., *Statistical Physics*, Addison-Wesley, Reading, Pa. pp. 13, 29 (1969).

14. Prigogine, I., in *A Critical Review of Thermodynamics* (Stuart, Gal-Or and Brainard, eds.), Mono Book, Baltimore, p. 461 (1967).

15. Tolman, R. C., *Relativity Thermodynamics and Cosmology*, Oxford Press (1933), pp. 221, 301, 323, 328, 395, 421, 440.

16. Narlikar, J. V., in Ref. 1, pp. 53.

17. There are authors who, unaware of the fundamental problem, hopelessly seek (cf. also [39]) the origin of irreversibility in the Hamiltonian and the interaction terms in it [cf., e.g., Hove, L. van, *Physica 21*, 517 (1955); *25*, 269 (1969)]; or in the coarse graining of phase space, which is required to take account of the fact that all measurements are macroscopic [cf., e.g., Landsberg, P. T., *Thermodynamics with Quantum Statistical Illustrations*, Interscience (1961); *Proc. Roy. Soc. A262*, 100 (1961); or in passage to the limit of an infinite number of degrees of freedom [cf., e.g., Balescu, R. in [2] p. 473; *Physica 36*, 433 (1967); *Phys. Lett. 27A*, 249 (1967)]; or in interpretations of the Liouville equation [cf., e.g., Prigogine, I., in [2], p. 1; or in the impossibility of completely isolating a system from the rest of the universe [cf., e.g., Blatt, J. M., *Prog. Theoret. Phys. 22*, 745 (1959)].

18. Misner, C. W., *Phys. Rev. 186*, 1328 (1969).

19. Beauregard, O. Costa de, *Le Second Principe de la Science du Temps*, Editions du Seuil, Paris (1963); *Pure and Appl. Chem. 22*, 540 (1970).

20. Adams, E. N., *Phys. Rev. 120*, 675 (1960).

21. Beauregard, O. Costa de, in *Proceedings of the International Congress for Logic, Methodology and the Philosophy of Science* (Bar-Hillel, Y., ed.), North Holland, 313 (1964).

22. Gal-Or, B., *Science 178*, 1119 (1972).

23. Feigl, H. and Maxwell, G., eds., *Current Issues in the Philosophy of Science*, Holt, Rinehart and Minneapolis (1962).

24. Grunbaum, A., *Philosophical Problems of Space and Time*, Knopf, N.Y. (1963); (Gold, T., ed.), Cornell University Press, N.Y., p. 149 (1967).

25. Hogarth, J. E. in *The Nature of Time* (Gold, T., ed.), Cornell University Press, N.Y., p. 7 (1967).

26. Layzer, D., in *The Nature of Time* (Gold, T., ed.), Cornell University Press, N.Y. (1967).

27. Mehlberg, H., in *Current Issues in the Philosophy of Science* (Feigl and Maxwell, eds.), Holt, Rinehart and Winston, N.Y., p. 105 (1961).

28. Ne'eman, Y. in Ref. 1, pp. 91.

29. Penrose, O. and Percival, I. C., *Proc. Phys. Soc. 79*, Part 3, No. 509, p. 605 (1962).

30. Prigogine, I., in [2], pp. 3, 203, 461, 505.

31. Reichenbach, H., *The Direction of Time*, University of California Press, Berkeley (1956); *The Philosophy of Space and Time*, Dover, N.Y. (1958).

32. Tisza, L., in [2], pp. 107, 206, 510.
33. Watanabe, S., in *The Voices of Time* (Fraser, J. T., ed.), George Braziller, N.Y. 1966, p. 543; *Progr. Theoret. Phys.* (Kyoto) Suppl., Extra No., p. 135 (1965).
34. Wheeler, J. A., in *The Nature of Time* (Gold, T., ed.), Cornell University Press, N.Y. 1967, pp. 90, 233, 235.
35. Whitrow, G. J., *The Natural Philosophy of Time*, Nelson, London, 1961.
36. Loschmidt, J., *Wiener Ber. 73*, 139 (1876); *75*, 67 (1877).
37. Zermelo, E., *Ann. Phys. 57*, (1896); *59*; 793 (1896).
38. Zel'dovich, Y. B., *JETP Lett. 12*, 307 (1970).
39. Penrose, O., *Foundations of Statistical Mechanics*, Pergamon, Oxford 1970.
40. Phipps, T. E., Jr., *Found. Phys. 3*, 435 (1973).
41. Kovetz, A. and Shaviv, G., *Astrophysics and Space Science 6*, 396 (1970); *7*, 416 (1970).
42. Dicke, R. H., *Phys. Today 20*, 55 (1967).
43. Cox, A. N., "Stellar Absorption Coefficients and Opacities, in *Stars and Stellar Systems* (Kuiper, G. P., ed.), Vol. VIII, University of Chicago Press, 1965, pp. 195–263.
44. Reeves, H., "Stellar Energy Sources," Ibid. pp. 113–193.
45. Sciama, D. W., "The Recent Renaissance of Observational Cosmology," in *Relativity and Gravitation* (Kuper, C. G. and Peres, A., eds.), Gordon and Breach, N.Y. 1971, p. 283.
46. Einstein, A., in *Albert Einstein, Philosopher-Scientist* (Paul Arthur Shilpp, ed.), Harper Torchbooks, N.Y. 1959, Vol. II, p. 687.
47. Sandage, A., *Quart. J. Radio-Astr. Soc. 13*, 282 (1972).
48. Weinberg, S., *Gravitation and Cosmology: Principles and Applications of the General Theory of Relativity*, Wiley, N.Y. 1972, pp. 597.
49. Ledoux, P., "Stellar Stability," in *Stars and Stellar Systems* (Kuiper, G. P. ed.), Vol. VIII, University of Chicago Press, 1965, pp. 499–574.
50. Truesdell, C., *Rational Thermodynamics*, McGraw-Hill, N.Y. 1969, pp. 30, 57, 106, 193.
51. Gringorten, I. I. and Kantor, A. J., in *Handbook of Geophysics and Space Environments* (Valley, S. L. ed.), McGraw-Hill, N.Y. 1965.
52. Huang, S. S. and Struve, O., "Stellar Rotation and Atmospheric Turbulence," in Stars and Stellar Systems (Kuiper, G. P. ed.), Vol. VI, University of Chicago Press, 1960, pp. 321–369.
53. Bondi, H., in *The Nature of Time* (Gold, T., ed.), Cornell University Press, N.Y. 1967; *Cosmology*, Cambridge University Press, 1961.
54. Heelan, P., *Quantum Mechanics and Objectivity*, Nijhoff, The Hague 1965, *1*, 95 (1970).
55. Popper, K., in *Quantum Theory and Reality* (Bunge, M., ed.), Springer, N.Y. 1967, p. 7.
56. Bunge, M., Ibid, p. 107.
57. Motz, L., *Ap. J. 112*, 362 (1952); *Astrophysics and Stellar Structure*, Waltham, Mass. 1970.
58. Katz, A., *Principles of Statistical Mechanics*, Freeman, San Francisco 1967.
59. Cocke, W. J., *Phys. Rev. 160* (5), 1165 (1967).
60. Conant, D. R., in *A Critical Review of Thermodynamics* (Stuart, E. B., Gal-Or, B. eds.), Mono Book, Baltimore (1970), p. 507.
61. De Groot, S. R. and Mazur, P., *Non-Equilibrium Thermodynamics*, North Holland 1962.
62. Sciama, D. W., *Modern Cosmology*, Cambridge University Press, N.Y. 1971.
63. Peebles, P. J. E., *Physical Cosmology*, Princeton University Press, Princeton 1971.
64. Sachs, R. G., *Science 176*, 587 (1972).
65. Christenson, J. H., Cronin, J. W., Fitch, V. L., and Turlay, R., *Phys. Rev. Lett. 13*, 138 (1964); *Phys. Rev. B. 140*, 74 (1965).
66. Lee, T. D. and Yang, C. N., *Phys. Rev. 104*, 254 (1956); Wu, C. S., Amber, E., Hayward, R. W., Hoppes, D. D., and Hudson, R. P., *Phys. Rev. 105*, 1413 (1957).
67. Dass, G. V., Prepring *TH. 1373-CERN* (1971).
68. Olbers, H. W. M., *Bodes Jahrbuch*, 110 (1826).
69. Halley, Edmund, *Phil. Trans. Roy. Soc.* (London) *31* (1720).
70. Fowler, W. A. and Hoyle, F. *Ap. J. Suppl. 9*, 201 (1964).
71. Colgate, S. A. and White, R. H., *Ap. J. 143*, 626 (1966).

72. Arnett, W. D., *Canad. J. Phys. 44*, 2553 (1966).
73. Schwartz, R. A., *Ann. Phys. 43*, 42 (1967).
74. Rakavy, G., Shaviv, G. and Zinamon, Z., *Ap. J. 150*, 131 (1967).
75. Clayton, D. D., *Principles of Stellar Evolution and Nucleosynthesis*, McGraw-Hill, N.Y. 1968.
76. Gal-Or, B. et al., *Intern. J. Heat and Mass Transfer 14*, 727 (1971).
77. Wheeler, J. A. and Feynman, R. P., *Rev. Mod. Phys. 17*, 157 (1945).
78. Feynman, R. P., *Rev. Mod. Phys. 20*, 367 (1948).
79. Feynman, R. P. and Hibbs, A. R., *Quantum Mechanics and Path Integrals*, McGraw-Hill, N.Y. 1965.
80. Lemaitre, G., *Ann. Soc. Sci.* (Bruxelles) *47A*, 49 (1927).
81. Friedmann, A., *Zeits f. Physik 10*, 377 (1922).
82. Dudley, H. C., *Lettere al Nuovo Cimento 5*(3), 231 (1972); Phys. Lett., in press (1972); *Nuovo Cimento 4B*, 68 (1971).
83. Arnold, V. I. and Avez, A., *Problemes Ergodiques de la Mechanics*, Benjamin, N.Y. 1968.
84. Farquhar, I., *Ergodic Theory in Statistical Mechanics*, Wiley, N.Y. 1964.
85. Sinai, Y. G., "On the Foundations of the Ergodic Hypothesis for a Dynamical System in Statistical Gechanics," *Soviet Mathematics 4*, 1818 (1963).
86. Ungarish, M., Internal Report, Technion–Israel Inst. of Tech., Aug. 1972.
87. Eringen, A. C., in *A Critical Review of Thermodynamics*, Mono Book (Stuart, E. B., Gal-Or, B., eds.), Baltimore (1970) p. 483.
88. Kestin, J. and Rice, J. R., ibid. p. 282.
89. Bohm, D., *Phys. Rev. 85*, 166, 180 (1952).
90. de Broglie, L., *Founvations of Physics*, Vol. 1, p. 1, 1970.
91. Finzi, A., "Is Dissipation of the Energy of Orbital Motion the Source of the Radiant Energy of Novae?," Technion Prepring Series No. MT-89, Sept. 1971.
92. Cohen, J. M. and Cameron, A. G. W., *Nature 224*, 566 (1969).
93. Ostriker, J. P. and Gunn, J. E., *Ap. J. 157*, 1395 (1969); *160*, 979 (1970).
94. Finzi, A. and Wolf, R. A., *Ap. J.* (Letters) *155*, 107 (1969); *150*, 115 (1967).
95. Kulsrud, R. M., *Ap. J. 163*, 567 (1971).
96. Pacini, F., *Nature 219*, 145 (1968).
97. Goldreich, P. and Julian, W. H., *Ap. J. 157*, 869 (1969).
98. Zel'dovich, Y. B., in *Advances in Astronomy and Astrophysics*, Vol. 3, Academic Press, N.Y. 1965, pp. 241–375.
99. Janossy, L., *Theory of Relativity Based on Physical Reality*, Akademiai Kiado, Budapest, 1971, pp. 17, 49.
100. Einstein, A., Verh. d. Schweizer, *Nat. Ges. 105*, Teil II, pp. 85–93.
101. Margenau, H., *Philosophy of Science 30*, 1 (1963); *30*, 138 (1963); *Phys. Today 7*, 6 (1954).
102. Heisenberg, W., *Physical Principles of Quantum Theory*, University of Chicago Press, 1931.
103. Schulman, L., *Phys. Rev.*, in press; in *Modern Developments in Thermodynamics*, Wiley, N.Y., p. 81 (1974).
104. Witten, L., ed., *Gravitation*, Wiley, N.Y. 1962.
105. Sperber, G., *Found. Phys. 4*, p. 163 (1974).
106. Kyrala, A., Ibid, p. 31.
107. Rothstein, J., Ibid, p. 83.
108. Nordtvedt, K. L., *Science 178*, 1157 (1972).
109. Oboukhov, A. M. and Golitsyn, G. S., *Space Research XI*, Akademie-Verlag, Berlin 1971, p. 121.
110. Zel'dovich, Y. B. and Novikov, I. D., *Relativistic Astrophysics-I*, University of Chicago Press, 1971.
111. Kantor, W., *Found. Phys. 4*, 105 (1974).
112. Novotny, E., *Introduction to Stellar Atmospheres and Interiors*, Oxford University Press 1973.
113. Paczynsky, B. E., *Acta Astron. 20*, 47 (1970); *Astrophys. Lett. 11*, 53 (1972).
114. Barkat, Z., *Ap. J. 163*, 433 (1971).
115. Barkat, Z., Wheeler, J. C., and Buchler, J. R., *Ap. J. 171*, 651 (1972); *Astrophys. Lett. 8*, 21 (1971).
116. Arnett, W. D., *Ap. and Space Sci. 5*, 180 (1969); *Ap. J. 53*, 341 (1968).

117. Wilson, J. R., *Ap. J. 163*, 209 (1971).
118. Tsuruta, S. and Cameron, A. G. W., *Ap. and Space Sci. 7*, 374 (1970); *14*, !79 (1971).
119. Arnett, W. D., Truran, J. W., and Woolsey, S. E., *Ap. J. 165*87 (1971).
120. Hawking, S. W. in *6th Texas Symp. on Relativistic Astrophysics*, Annals, New York Academy of Sciences, 224, 268 (1973).
121. Davies, P. C. W., *The Physics of Time Asymmetry*, University of California Press, Berkeley, 1974.
122. Jackiw, *Rev. Mod. Phys. 49*, 681 (1977); with C. Rebbi, *Phys. Rev. Let. 36*, 1116 (1976); *37*, 122 (1976); *Phys. Rev. D 13*, 3398 (1976);
123. 'tHooft, G., Phys. Rev. Let. *37*, 8 (1976); *Nuclear Phys. B79*, 276 (1974);
124. Narlikar, J. V., *General Relativity and Cosmology*, MacMillan, London, 1979.
125. Hawking, S. W. and W. Israel, eds., *General Relativity*, Cambridge, 1979.
126. M. Rowan-Robinson, *Cosmology*, 2nd ed. Oxford, 1981.
127. G. Bath, ed. *The State of the Universe*, Oxford, 1980.
128. P. T. Landsberg and D. A. Evans, *Mathematical Cosmology: An Introduction*, Oxford, 1979.
129. G. Burbidge and A. Hewitt, eds. *Telescopes for the 1980s*, Palo Alto: Annual Reviews, 1981.
130. J. Vervier and R.V.F. Janssens, *Spinor Symmetry and supersymmetry, Phys. Lett.*, 108B: 1, 1982.
131. L.W. Alvarez, *et al, Science, 208*, 1095, 1980.

Part IV

BEYOND PRESENT KNOWLEDGE

The philosopher is "the spectator of all time and all existence."

Plato

"Science never makes an advance until philosophy authorizes and encourages it to do so."

Thomas Mann

But as for certain truth, no man has known it,
Nor will he know it; neither of the gods,
Nor yet of all the things of which I speak.
And even if by chance he were to utter
The final truth, he would himself not know it;
For all is but a woven web of guesses.

Xenophanes

It is my belief that philosophy must return to cosmology and to a simple theory of knowledge. There is at least one philosophical problem in which all thinking men are interested: the problem of understanding the world in which we live; and thus ourselves (who are part of the world) and our knowledge of it. All science is cosmology, I believe, and for me the interest of philosophy, no less than of science, lies solely in its bold attempt to add to our knowledge of the world, and to the theory of our knowledge of the world.

Sir Karl R. Popper

HAVAYISM—THE SCIENCE OF THE WHOLE

IX.1 THE FUTILE QUEST FOR FINAL ANSWERS

The reader who has had the perseverance to reach the end of Part III should not expect to find final answers to the questions raised there. Even the material presented in this final lecture is essentially a course without an ending; a sightseeing/mind-seeking guide to a land of complexity; a land for which I can only serve as a 'tourist guide' to those travelers who stop to ask my advice as to the most interesting or most promising sites for a future course of research. I can only indicate those knowledge crossroads that, to my mind, an enthusiastic tourist should cross before arriving at the fields of promising future theories.

It should be pointed out from the outset that some of these cross-roads do not lie in the course of the future; to reach them one must occasionally go back in time to the greatest philosophers of the past. For, what I shall try to show in this lecture is, *inter alia*, based on my first general assertion (Introduction, §3) namely, that, *on the one hand, we must bear in mind that the greatest snare for creative thought is uncritical acceptance of traditional assumptions, and, on the other, that new philosophies are not necessarily more advanced than old ones.*

I shall not concern myself with fields that are of interest to a *single* discipline or category of human interest and knowledge. It is my intention, as will be made clear, to enter a new domain of truly multi-disciplinary interests.

This may come as a disappointment to the reader who expected final answers to traditional or *disciplinary* problems. But let us recall that we are only just embarking on a universal quest for knowledge and are still, in Newton's words, like children playing with pebbles on the seashore, while the great ocean of truth rolls, unexplored, beyond our reach. It is better to debate a question without settling it, says the French moralist Joseph Joubert, than to settle a question without debating it. And, we do have ample (cosmic) time to settle universal questions, for, in Sir James Jeans words "our race has not succeeded in solving any large part of its most difficult problems in the first millionth part of its existence. Perhaps life would be a duller affair if it had, for to many it is not knowledge, but the quest for knowledge, that gives the greatest interest to thought — to travel hopefully is better than to arrive".

IX.2 AN EXAMPLE IN HAVAYISM

IX.2.1 Havayism (or Hwehyism)

In this lecture I confine myself to re-examinations of fundamental concepts by employing a methodology that might be called *havayism* — a word derived from the Hebrew *havaya* (or hwehya).

Roughly speaking, *havaya* means *"the Whole"*, the essence of all existence, nature and god, "things-in-themselves", reality, perception of reality, the most fundamental aspects of society, culture and humanity, historical evolution, and the very essence of life and the physical world. *Alternately*, it means "questions-and-answers", philosophizing or dialectical discourse (see also Assertion 4, Introduction).

It should also be noted that, by reversing the order of the symbols in *Havaya* (or *hwehya*), one obtains the tetragrammaton *YHWH*.

The nature of havayism cannot be defined. Hence, it may be best illustrated by an example. This is done in the rest of this lecture.

IX.2.2. The Interconnectedness of Havaya

There are probably only a few thoughts in philosophy and science that are as important as the thinking about the interactions among the parts of the whole, the universal ordering of phenomena and the evolution of life and intelligence.

In this connection we have already seen that the concepts of *causality* and *causation* are inextricably woven into any interpretation of reality and into all phenomena ranging from subatomic objects to the universe as a whole. However, a wide gap exists between *scientific* interpretations of nature and some *subjectivistic* philosophical doctrines that are related to the *subjective origins of knowledge itself*.

A singular feature of all discussions about the origins of life, thought and knowledge is that the moment one starts discussing them he starts a course without ending, for this course leads to a bewildering variety of subjects the most important of which are:

■ The origin and evolution of the universe, galaxies, stars, planets and the earth.

■ The origin and evolution of heavy chemical elements, molecules, amino acids, DNA, RNA, proteins, cells, organisms, bioaccelerometers, gravity receptors, etc.

■ The structure and functioning of neurons, axons, central nervous systems and the brain. This includes the theory of information, control theory, cybernetics, thermodynamics of life, biophysics, biochemistry of the genetic code, heredity, the philosophy of physics and the philosophy of biology.

■ The philosophy of mind, the philosophy of perception, epistemology, theory of knowledge, physiology, neurophysiology, physiological psychology, psychology, etc.

■ The philosophies of society, state, politics, law, religion, and history; ethics, metaphysics, etc.

■ The history of philosophical doctrines associated with the aforementioned topics.

If anything about this list is certain, it is that one cannot isolate one field from another; nor neglect the connections between its internal components. Take, for example, one of these fields: "philosophy of mind". It deals with such a bewildering variety of mutually connected occurrences as:

Physical memory, biological memory, storage and transmission of information, the essence of information, stimulation, sensations, sense perceptions, images, "innate" perception of space and time, thinking, reasoning, intelligence, learning, etc.

Choice, will, intentionalities, motivation, action, purposeful behavior, believing, personal judgement, the unconscious state, etc.

Obviously the range of topics associated with these lists goes far beyond the capability of contemporary science, beyond what is intended in the philosophical discourses of these lectures or even beyond any series of university courses. Thus, it should be pointed out at the very start that here, as in tourism, driving once across a country does not mean that one has really "seen" that country; nor does it mean that one can then "understand" its social structure. Flying an airplane would not help either. Nevertheless, as most students would probably agree, such introductory tours may be far better than no tour at all.

Indeed, what I most hope for is that these lectures would at least serve as a catalyst for the reader to take up further readings and, later, perhaps, to return to this lecture to compare his own findings with ideas presented herein.

From here on one can employ our principle of *"Universe-Laboratory Equivalence"* (Lecture VI) to describe the dominant role played in every single process in nature by mutually connected, gravitationally-induced phenomena, histories, etc. But for this lecture the most important features are our forthcoming conclusions about the origins of structures and prebiotic-biotic evolutions. These are described next.

IX.2.3 The "Impossibility Principle" of Biological Evolution in

Radiation-Saturated Spaces

The problem of the "observer" in the small and that of the universe in the large, cannot be separated. For life to evolve and exist, with its own biological arrow of time, there must exist heavier chemical elements and temperature gradients in the earth's atmosphere, which, in turn would not exist without the "advanced massive stars" (Lecture I) and the receding clusters of galaxies. In other words, was inter-cluster space *saturated with radiation*, all temperature gradients would have *vanished* and all nucleo-syntheses would have *stopped. Supernova explosions* and o'her sources of heavier elements are ruled out under these circumstances. In fact, the earth, people, and all other forms of life on earth are collections of atoms forged in stellar cores. *All of chemistry, beyond hydrogen and helium, and therefore, all of life, has been formed by stellar evolution.* In other words, with the exception of hydrogen, everything in our bodies and brains has been produced in the thermonuclear reactions within stars which later exploded in galactic space. (The earth was formed from the dust meteorites and planetesimals produced by such an earlier explosion of a massive star that had reached its advanced stage (§I.5.2, §I.6 and below).

Studies of interstellar clouds in different regions of the electro-magnetic spectrum have recently yielded much information about interstellar dust which may contain grains of iron, silicon, carbon, nitrogen, oxygen. Additional constituents include the so-called *organic* compounds: CH^+, CN, OH^-, NH_3, H_2O, NH_2, NH, H_2CO (formaldehyde), CO, CH_3OH (methyl alcohol), amines, imines, cyano-acetylene, etc.[*] Studies of formaldehyde and other molecules found in dark galactic clouds have shown that the temperature inside these clouds varies and may be as low as $5°K$ or even less. Grains of dust floating in interstellar space may therefore provide the required cold surfaces where atoms can combine to form the observed molecules or even more complicated ones.

Can such cold syntheses lead to a biotic evolution? Can other gravitationally-induced phenomena on earth contribute the crucial stages required for the evolution of life? What leads to the evolution of life?

The next section presents preliminary attempts to discuss these questions from the point of view of gravitism.

> Very old are the rocks. The pattern of life is not in their veins. When the earth cooled, the great rains came and seas were filled. Slowly the molecules enmeshed in ordered asymmetry. A billion passed, aeons of trial and error. The life message took form, a spiral, a helix, repeating itself end-leslly, swathed in protein, nutured by enzymes, sheltered in membranes, laved by salt water, armored with lime.
>
> **Thomas Hughes Junkes[+]**

[*] cf. Rank, D. M. *et al, Science, 174*, 1083 (1971), and Biermann, L; *Nature, 230*, 156 (1971). Such spectroscopic studies reveal also the presence of hydrogen cyanide, methyl acetylene, diacetylene, cyano-diacetylene and acetaldehyde.

[+] *Molecules and Evolution,* Columbia Univ. Press, New York, 1966.

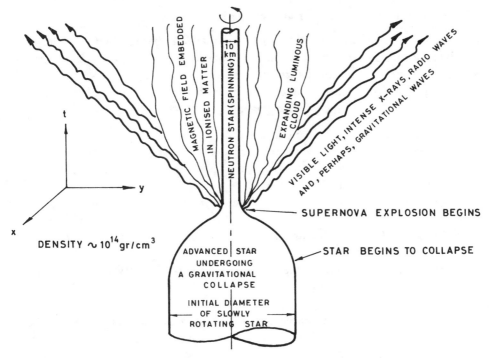

Fig. IX.1. **Gravitational collapse and the dispersal of (gravitationally-forged) heavy elements in space.**
A schematic diagram showing a slowly rotating advanced star (§I.2) undergoing a gravitational collapse
which leads to the observed phenomenon of supernova explosion leaving behind a spinning neutron
star (pulsar) in an expanding luminous cloud that is embedded with strong magnetic fields. Supernova
explosions throw the heavy chemical elements that have been generated in the stellar cores (Fig. VII.2)
(e.g., as dust and rocks) into interstellar space. *It is, therefore, possible that a supernova explosion has
been responsible for the origin of the heavier elements of earth and of the rest of the solar planets (cf.* Figs.
I.3, I.4 and VIII.2 and §I.3).

Type I supernovas form a fairly homogeneous group with relatively little variation between the spectrum of one exploding star and that of the next. They are the only type of supernova found in old elliptical
galaxies (which are known to have few, if any, stars more massive than the sun). They are generally believed
to be members of binary (double-star) systems (§I.4.3) in which matter is first transferred from the companion star till the mass of the star becomes greater than the Chandrasekhar limit (§I.2.3). At that point
the core collapses, releasing energy as a supernova explosion and leaving behind a binary system composed
of an ordinary giant star and an X-ray source (Fig.I.1).

Type II supernovas are believed to involve a star which is much more massive than the sun. This star
can evolve into one of two catastrophies: (i) Ignition of the carbon reaction (Fig. VII.2) could induce
instabilities that would detonate the star as a supernova, leaving behind nothing but an expanding luminous
cloud. (ii) If the carbon reaction induces no instabilities the increasingly high temperatures (Fig. VII.2)
generate high neutrino-radiation fluxes that, leaving the star cool its core and cause it to collapse. In this
case the final burst of neutrinos can carry away to unsaturable space so much energy that it blows off
the entire outer envelope of the star. An explosion of this kind leaves behind an expanding luminous
cloud, in the center of which a pulsar or a black hole (Lecture VIII and §I.5) is formed. Type II supernovas
are found principally in the arms of spiral galaxies which are rich in massive young stars (see §I.6.2).
Their spectra are characterized by strong lines of hydrogen.

About 400 *extragalactic* supernovas have been observed since the first one was detected in 1885.
Since today thousands of galaxies are easily detectable with large telescopes, on the average one supernova
explosion may be detected every few months.

IX.2.3.1 Gravitation, Structure and Evolution: Preliminary Concepts

Here we examine a few primitive concepts that may later become useful in analysing the dynamic interconnectedness of nature, in particular those associated with gravitation as the *prime cause* of structures, irreversibility, time, and geochemical, and biological evolutions. These concepts are based on previous conclusions, namely, that gravitation leads to the following phenomena:

1) Expansion of the universe (Lecture VI).

2) Breakup of the expanding (homogeneous) cosmological gas into individual clouds which evolve from a state of protogalaxies to the present state of stellar galaxies (Lectures VI, VII, VIII, I and III).

3) Generation of all irreversible processes, time asymmetries, time and motion (Lectures VI and VII).

4) Stellar evolution, including the phenomenon of supernova (Lecture I).

5) Geochemical and biological evolutions.

We now turn to examine some of the concepts associated with the last effect.

IX.2.3.2 Gravitation and the Geochemical Evolution of the Atmosphere, Hydrosphere and Biosphere

A. General Concepts

Today there is considerable evidence that the earth and the other solar planets had been formed by gravitational condensation and accumulation from a cloud of dust, meteorites and scattered planetesimals at *low* temperatures.* The details of this evolutionary process were largely developed by *H. C. Urey*, mainly on the basis of (Newtonian) gravitation theory and thermodynamic considerations.** Astrophysical data concerning supernova explosions (§I.3) and the data obtained recently from space explorations (including the origin of old craters on the moon's surface), tend to confirm this theory (see below). Similar supportive evidence comes from the study of meteoroids, which are believed by many to be fragments from the interior of an exploded star (supernova). Therefore, they are used to give evidence of the internal structure of the earth and of the general geochemical character and abundance of the chemical elements. The main tenets of this theory are supported by the following facts:

1) *Gravitation structures the earth in concentric shells, or geospheres, according to their specific gravity.* Combined together, these geospheres roughly resemble the composition of meteoroides.***

* cf. §I.3

** Including the earlier assumptions of J. B. S. Haldane (1927) and A. I. Oparin (1924). [Oparin, A. I. *The Origin of Life*, Macmillan, N.Y., 1938 and J.B.S.Haldane *Rationalists Ann.*, 148 (1929)]

*** cf. the abundance of chemical elements in meteorites—footnote in §I.3 and also below. (All incoming bodies are called *meteoroids*, those that impact the ground are called *meteorites*).

Fig. IX.2 **An Example of Gravitational Structuring, Ordering, and Selectivity in Compositon.**

Saturn's rings obtained with *Pioneer* Saturn spacecraft data. (Computer representation as if viewed from directly over Saturn's north pole with sunlight coming through the rings.)

Gravitation also structures the earth in concentric shells, or geospheres, according to their specific gravity—a fundamental selectivity in prebiotic evolution (see §IX.2.3.2 and Assertion 9 in the Introduction).

After grazing Saturn in 1979, *Pioneer 11* is now heading in the same direction the entire solar system is moving through the galaxy, but faster and ahead of debris from our sun and planets. *It has a good chance to emerge from the solar system into the depths of extra-solar space around 1993* (cf. §I.10 and Fig. VI.1). Voyager I (§I.10) has recently added more information about Jupiter and Saturn (Nov. 1980).

2) Earth is divided by geophysicists into three main spheres—the *core* (mostly composed of iron-nickel), *mantle* (mostly composed of silicates and interstitial iron-nickel that increases in amount with depth), and *crust*. In the crust, the continental areas consist of three continental layers—a deep-seated layer (composed mainly of basalt and granite), a granite layer, and a *sedimentary layer* composed of sediments, sedimentary and metamorphic rocks. The earth's solid exterior crust is called *lithosphere*. Eight elements constitute the bulk of the uppermost lithosphere. These main elements are oxygen (46.42% by weight), silicon (27.6%), aluminum (8.08%), iron (5.08%), calcium (3.61%), sodium (2.83%), potassium (2.58%) and magnesium (2.09%). Thus, the lithosphere is actually an oxysphere.

3) Three additional, outermost geochemical spheres are normally distinguished: (1) the *hydrosphere*, which consists of liquid water, ice and salts; (2) the *atmosphere*, or the gaseous outer envelope of the earth; (3) the *biosphere*, which comprises the

living matter and is in parts of the earth capable of sustaining biological evolution. The biosphere occupies the lower part of the atmosphere and the hydrosphere, and includes a very thin layer of the lithosphere.

According to Urey, the iron core of the earth accumulated slowly by *gravitational differentiation* during the geological history from an almost uniform mixture of accumulated iron and silicates. Thus, the lithosphere was also formed by gravitational differentiation—*a differentiation which is still going on!* Gravitationally-induced *crystallization* (Lecture I) caused the arrangement of the silicate shell according to the *specific gravity* of the crystallizing phases. The gravitational force drives the silicic magmas of *low* specific gravity "upwards" and concentrates the heavy atoms in the *"lower"* levels of the lithosphere. However, *other gravitationally-induced movements*, such as *mountain-building processes, volcanic activity, pneumatolytic and hydrothermal phenomena, and vertical movement of the phases of different specific gravity, gradually caused the present chemical dissimilarity in the upper lithosphere.* Thus, instead of uniform distribution of elements in the various layers of the upper lithosphere a *gravitationally-induced migration* causes chemical inhomogenity* (Lecture I).

According to Urey, the present atmosphere and hydrosphere evolved in a later stage. The early primitive atmosphere was a *reducing atmosphere* containing hydrogen, water vapor, ammonia, methane and some hydrogen sulfide. Gradually the hydrogen escaped from the gravitational field of the earth, and *water was photochemically decomposed into hydrogen and oxygen in the upper levels of the atmosphere.* The oxygen formed was consumed in the gradual oxidation of ammonia to nitrogen and water, and of methane to carbon dioxide and water. Thus, gradually the reducing atmosphere was transformed into an *oxidizing atmosphere.*

The supply of nitrogen, according to this evolutionary theory, may be due to the release of gases from rocks during weathering and volcanic emanations.

The evolution of the hydrosphere was also a gradual process, in part related to *weathering* and *sedimentation* and in part to *water vapor given off by the volcanoes.* In addition, a vast amount of volatile substances, released by volcanic activity, was gradually transported into the evolving oceans. Other substances, partly produced by volcanic activity and partly by rock weathering, have been accumulated constantly in the oceans, lakes, and swamps. However, the geochemical evolution of the earth did not stop with the formation of the atmosphere and hydrosphere, for the earth is changeable chemically, and its evolution still continues.

The first biotic compounds were probably synthesized during the *reducing* stage in the evolution of the atmosphere, most likely in a combination of thermo-

---

* Wind and water erosion and deposition are also important geological processes that lead to inhomogenity and structures. *Both* wind and water movements are gravitationally-induced processes! For empirical data see Fig. VII.3 and for explanations see §VII.7. The crucial effects of lightnings and shock waves are described below.

chemical, photochemical and electrochemical reactions (see below). Plant life probably could not start until there was some *free oxygen* in the atmosphere. Hence, oxygen-based life evolved quite later, probably after the reducing atmosphere had been transformed into an oxydizing atmosphere (see below). The very evolution of life has, in turn, changed the evolutionary patterns of the atmosphere and hydrosphere. (For instance, most or all free oxygen present now in the atmosphere has been gradually produced from water by photosynthesis taking place in chlorophyll-bearing plants).

Another interplay of evolutionary processes is due to the fact that the growth of one plant community alters conditions and creates a new habitat for a second. *Thus tree falls, lava flows, landslides, sedimentation, flooding, and hurricanes all open up new ecological systems.* Here the most important point is the fact that the evolution of life *cannot* be isolated from *gravitational effects* that are generated on the surface of, *inside*, and *outside* earth. The simplest of the later is due to the *moon's gravitational field.* The sea *tides* raised on the rotating earth *by the gravitational attraction of the moon* (and the sun), slowly *reduce the earth's rate of rotation* (about 1.8×10^{-3} seconds per century). But most of the angular momentum *lost* by the earth is transferred, *by gravitational interaction*, to *add* to the moon's orbital angular momentum. *This exchange lifts the moon into an orbit slightly farther away from earth, thereby decelerating its rotation with respect to the stellar background.* Indeed, if the classical works of George Darwin are updated they tend to indicate that the moon was once *much closer* to the earth and was in an orbit more highly inclined to the ecliptic than is its present orbit (see also "Biological Clocks", §IX.2.3.3 below).

B. Time Scale of Biotic Evolution

Before we proceed with the description of more specific processes in the prebiotic evolution of earth it is necessary to form a *time scale* for the various events that lead to the emergence of the *earliest microorganisms.*

Age determination of the earth's crust (by the methods of *radioactive dating and geophysical studies*) suggest that the crust had started to solidify about 4.5 billion years ago.* (Significantly, the moon and various meteorites were also found to have an age of around 4.6 billion years).

On the other hand, the earliest biological organisms now known existed already 3.1 to 3.8 billion years ago. (Confirmations that these are fossils of living organisms can be obtained also from organic molecules, found in dated rocks, which are distinctly associated with living organisms).**

Therefore, the time scale required for the evolution of the earliest living systems is of the order of 1 billion years!

* cf. for instance, Tilton, G. R. and R. H. Steiger, *Science, 150*, 1805 (1965); Barghoorn, E. S. and J. W. Schopt, *ibid, 152*, 758 (1966) [determination of early terrestrial life by radioactive dating of various precambrian rock formations and by searching them for fossils of the earliest organisms.]

** cf. Calvin M. *Chemical Evolution*, Oxford University Press, 1969.

C. The Early Atmosphere

The Earth was accumulated from dust and planetesimals of various sizes *and was cold at its beginning. It warmed up gradually by gravitational contraction which converted gravitational energy into heat* (another important source of heating is radioactice decay).

But the iron can sink into the core and the lighter elements migrate upward only when the earth's interior becomes hot enough (~1000°C) *to melt the external crust.* As the interior becomes hot and molten, entrapped gases undergo through a series of chemical reactions (NH_3 for instance is not stable at elevated temperatures and decomposes. But CH_4 is pyrolyzed to ethane, acetylene and more structured but unstable molecules. It can also be oxidized by water vapor to CO, CO_2 and H_2). These composition changes mean that the composition of gases emerging from the contracting earth have changed gradually with time. In turn they have led to variations in the composition of the atmosphere.

The early atmosphere was most likely composed mainly of nitrogen, ammonia, water vapor, methane, hydrogen and some hydrogen sulphide. Due to gravitational stratification the hydrogen gradually escaped from earth into outerspace and its residence time in the early atmosphere cannot be determined with certainty. It is clear, however, that the early atmosphere was reducing (as was first proposed by Oparin in 1924).

For instance, according to some empirical studies,* the atmosphere did not contain free oxygen *until its age was around 2.5 billion years*; i.e., *when primitive life was already abundant! Consequently life originally evolved in a reducing atmosphere in which the major constituents were methane and nitrogen.*

But both nitrogen and methane are chemically stable at terrestrial atmospheric conditions. Hence, they required strong (localized) *"inputs of energy"* in order to evolve chemically.

D. Energy Sources for the Generation of Biomonomers

Energy sources in the early atmosphere include:
1) Solar Radiation.
2) Cosmic Rays.
3) Radioactive Decay.
4) Volcanic Eruptions and Explosions.
5) Electrochemical heating produced by lightnings.
6) Shock-wave heating produced by meteoroids, lightnings, and volcanic explosions.

---

* cf. Rutten, M. G. *Geological Aspects of the Origin of Life on Earth*, Elsevier, Amsterdam, 1962.

Let us now examine briefly these six sources:

1) The flux of solar radiation 4.5 billion years ago can be estimated from the known gravitationally-induced nucleosynthesis of helium from hydrogen in the sun. Such studies give a value approximately 60% of the current value.** Some initiations of chemical processes in the early atmosphere may then be attributed to the short ultra violet range in the solar spectrum. This energy input gives rise to atmospheric circulation, rain, wind, etc., as well as to ocean surface waves, and cavitation (rising gas bubbles collapse by impact of the waves, etc.).

2) The flux of cosmic rays was probably similar to the current value. It too can cause important chemical variations.

3) Radioactive heating was probably 3 to 4 times greater than at the present because of the decay of radio active elements during 4.5 billion years.***

4) The rate of energy released by volcanic eruptions on the primitive earth cannot be estimated at present. Occasionally, a very large pressure of gasses and water vapors is built up inside the volcano before eruption.

5) Electrochemical heating was probably abundant through the electrical energy released by lightnings.[+] The electric currents produced (10,000 to 100,000 amperes), may raise local temperatures in the atmosphere to 3,000°C and, consequently, cause dissociation, excitation ionization, and radiation. In the lightening path the pressure is also built up and the gas expands outward from the hot core and in a few miliseconds forms a supersonic blast wave, i.e., a thunder shock wave front accross which temperature, pressure and density rise rapidly.

6) Shock waves (formed by volcanic explosions, lightnings, or meteoroids) compress and heat the gas to thousands of degrees, causing dissociation, excitation ionization, and chemical reactions. *But, most important, immediately behind the shock front the temperature drops considerably by recombination, etc.*[++] *This phenomenon helps to preserve complex, unstable, "organic-like" molecules that are formed by the shock front. Hence this fast cooling might be an important stage in prebiotic evolution of long, complex, "organic like" structures.*

Many laboratory experiments indicate that when any of these energy sources is applied to a reduced gas mixture (similar in composition to the early atmosphere), almost the same type of chemical reactions take place, irrespective of the exact composition.*

** Sagan, C. and G. Mullen; *Science 177*, 53 (1972).

*** Bullard, E. in E. P. Kuiper (ed.) *The Earth as a Planet*, University of Chicago Press, Chicago, 1954, pp.110.

[+] A "conservative estimate" puts this value on around 100 lightning strokes per second over the whole earth (Schonland, B. F. J., *The Flight of Thunderbolts*, 2nd ed., Clarendon Press, Oxford, 1964.)

[++] Weihs, D. and B. Gal-Or, *AIAA, II*, 108 (1973). Note also that shock waves induced by meteoroids compress and heat the gas to temperatures of several tens of thousands degrees.

* Sagan C., in S. W. Fox (ed.) *The Origin of Prebiological Systems*, Academic Press, N.Y., 1965; Kenyon, D. H. and Steinman G., *Biochemical Predestination*, McGraw-Hill Book Co., N. Y., 1969; Miller S. L. and L. F. Orgel, *The Origins of Life on Earth*, Prentice-Hall, N. J., 1974.

The products of these reactions include: Hydrogen cyanide (HCN), amines (R-NH$_2$), imines (RCH=NH), alcohols (R-CH$_2$OH), aldehydes (RCHO), organic acids (RCOOH), ethylene (C$_2$H$_4$), acetylene (C$_2$H$_2$) etc. *In turn, these products can react with each other to form more complicated molecules.* For instance, formaldehyde (CH$_2$O) *polymerizes* to form *sugars* (CH$_2$O)$_n$; hydrogen cyanide, ammonia (NH$_2$) and aldehydes can condense to form *amino acids—the building blocks of proteins*: and hydrogen cyanide and ammonia can condense to form *purines* and *pyrimidines—the basic units of DNA, RNA, and ATP.*

Of particular interest is the fact that *shock-wave-induced reactions are thousand to million times more efficient than those produced by radiation or electrical discharge.*** Upon dissolution in the primordial oceans, these products enter into further chemical evolution.

E. Polymerization and Organization

Amino acids can be structured further to *peptides* and *proteins* either thermally or through the participation of stratified clay minerals.*** Radiation promotes formaldehyde structuring into sugars; sugars into *polysaccharides*; and *nucleotide* structuring into *polynucleotides.*$^{++}$

F. Can Polymerization Generate Life?

At this point we reached the end of present knowledge. None has yet produced life in the laboratory, nor explained its generation. The only replication method known

---

** Bar-Nun, A. *et al. Science, 168,* 470 (1970). These authors claim that up to 30 kg/cm^3 of organic material could have been generated by such reactions during the first million years of earth.

*** In 1970 the experiments by Paecht-Horowitz, Berger, and Katchalsky [*Nature, 228,* 636 (1970)], have given evidence of the importance of clays in the polymerization of activated amino acids. According to Cairns-Smith [*J. Theor. Bio. 10,* 53 (1965)], life evolved through natural selection from inorganic crystals, in particular clays. Indeed, the introdiction of nucleic acids and proteins might have provided additional structure to the clays (Cairns-Smith, *The Life Puzzle,* University of Toronto Press, Toronto, 1971). Speculating about the origin of the genetic code, Hartman claims [*Origins of Life, 6,* 423 (1975)], that the interaction with clays requires some ·plausible mechanism which is not a protein but which can distinguish between adapter RNA molecules and amino acids. He claims that one possibility is that of a *lipid bilayer,* for the cell membrane, as a lipid bilayer, can distinguish molecules on the basis of their hydrophobicity.

Now, according to Weiss (p. 75 in Eglinton & Murphy, eds., *Organic Geochemistry,* Springer-Verlag, N. Y., 1969), alkyl amines and alcoholes can interact with clays to form stable lipid bilayers. It may thus be the recognition capability of such a system which is the real reason behind the onion-like evolution of the genetic code. For, as Hartman has stressed, once adaptor and amino acid had been linked, they must remain so even if new amino acids and bases are brought into the coding systems. The origin of the code would then lie in the interaction between clays, alkyl amines, alcoholes, lipid bilayers, and polypeptides under the shearing force of gravity.

++ Fox, S. W.; *Science 132,* 200 (1960); *Nature, 205, 328 (1964); Barker, S. A. et al Nature, 186, 376* (1959); Paecht-Horowitz M. and A. Katchalsky, *Biochim Biophys Acta 140,* 14 (1967); Contreras. G. *et al, Ibid, GI,* 718 (1962).

on earth is through DNA (which can replicate itself and build proteins, etc., according to a chemical code embedded in its structure). *And none has yet been able to produce life by the methods described above.*

Up to this stage science provides evidence for a deterministic and continuous evolution. Everything in this evolution was *predetermined* from the begining of physical time; that is to say up to the polymerization of the aforementioned poly-nucleotides. *But the road from polynucleotides to life may be more complicated than the road from the begining of time to polynucleotides.* Hence, without further evidence we cannot reach any verifiable conclusion. The old fundamental question (and some other ones related to it, such as the unresolved questions about extraterrestrial life) must therefore be left open. However, by sidestepping this question one may, perhaps, proceed by asking a different set of questions:

1) Has biological evolution been *disconnected* from the environment *since the moment of the creation of the first living organism?*

2) Can biological evolution lead us to an understanding of human evolution and, in particular, of human thought and perception?

3) Can cosmological-gravitational evolution end and another, *independent* evolution start?

4) *Do we need to consider time and gravitation in biological evolution?*

I do not believe that definite answers can be given to all these questions. But I do believe that in seeking evidence that might be related to these questions, we can make successive approximations toward better understanding of nature. Let us therefore seek further evidence.

Definition of Life: The continuous adjustment of internal relations to external relations.

Herbert Spencer

IX.2.3.3 Biological Clocks (and "Philosophy of Biology")

The aforementioned problems of evolution and gravitationally-induced processes lead us to a set of bewildering variety of subjects, including those associated with the origin and evolution of various "biological clocks". The latter are described next.

A. Biological Rhythms

Periodic variations in living systems span over a wide spectrum of frequencies from DNA-cell reproduction to wing beats of insects and brain waves. Relatively high frequency of recurrance include such (terrestrial) biological rhythms as heartbeats (20 to 1000 cycles per min.), respiratory rhythms (4 to 250 c/min.), walking and chewing rhythms (60 to 5000 c/min.), brain waves (60 to 4000 c/min.), and wing beats of insects (1200 to 120,000 c/m). Green plants alternate between daytime uses of solar radiation in photosynthesis and nighttime growth processes.

Most plants take a night sleep rhythm, drooping their leaves at night and raising them by daytime. Some, however, prefer to open their blossoms only at night. There are also some animals that prefer the night shift. These include the cockroaches, owls and bats. The day shift includes most "normal" animals, man included. Perhaps the most spectacular *precision* in keeping synchronization with (external) *geophysical* cycles is demonstrated by a unique sea lily. This echinoderm lives in deep sea near Japan and liberates its sex cells annually at about 3 p.m. on the day of one of the moon's quarters. In succeeding years the reproduction time changes progressively into earlier dates in October until about the first of the month whereupon the following year it jumps abruptly to near the end of the month to start the backward progression again. The result is an *18-year cycle! which is essentially the period of regression of the moon's orbital plane!!* Reproductive swarming of mayflies has also been related to moon phase. There is also evidence that in some primates menstruation is linked to the new moon ("menses" means lunar month).

Animal navigation is another interesting aspect of biological periodicity. Navigational guides such as "sun-compass", "moon-compass", or "star-compass" require that animals that appear to possess this kind of ability be able systematically *to alter their orientation at rates to compensate exactly for the earth's rotation relative to each of these celestial references!* There have been many reports that demonstrate some kind of *"celestial navigation"* by such animals as birds, fishes, turtles, spiders, insects and crustaceans. Also known are the *"tidal-timing phenomena"*. Some intertidal organisms, such as green crabs, snails, clams and oysters, become most active when submerged by the rising tide. But before we proceed let me stress again the fact that all the *external* synchronizations mentioned above are *gravitationally-induced!*

> A man, viewed as a behaving system, is quite simple. The apparent complexity of his behavior over time is largely a reflection of the complexity of the environment in which he finds himself.
>
> **Herbert Alexander Simon****

B. Biophysical Links and their Possible Origins

All available evidence now suggests that the evolution of life on earth has developed in the direction of *maximal utilization of (gravitationally-induced) environmental fluctuations,**with specializations related not only to gravitationally-induced *"geophysical timings"* but also to outer space coordinates. Diverse aspects of chemical, physiological and behavioral fluctuations in terrestrial living systems occur at every level from a single cell to the organism as a whole. All periodic biological

---

* For explanation see Assertion 9, p. 53, §3, Introduction
** *The Sciences of the Artificial*, MIT Press, Cambridge, 1969.

processes may be divided into *geophysical correlates* (i.e., those related with terrestrial changes, such as ocean tides, geoelectromagnetic fields, or darker-lighter, cooler-warmer and humid-drier processes), and to those having no such correlates, i.e., *rhythms without any apparent external correlate* (such as heartbeats and respiratory rates). Geophysical correlates have relatively fixed periods which are quite independent upon direct exposure to light and temperature variations. Rhythms without external correlates may however, be altered in response to changes in the immediate environment (yet, when removed from their natural environment and placed in a laboratory under constant environmental conditions they may closely repeat the periodic patterns that they experienced before the isolation).

It is thus concluded that *every living system* contains a kind of *"biological clock"* which involves *specific biological processes that may be "timed" by two kinds of rhythms, one resulting from gravitationally-induced changes* (that take place *outside* the organism), the other being an "innate" or an *internal* process *generated at a very fundamental level of "cellular organization, synchronization and control"* (see below).

Two hypotheses have been proposed to explain the origin and nature of such biological clocks; one involving *"free running", "independent", "innate", "internal thermodynamic processes"*, the other incorporating the effects of *external* rhythmic patterns. Basing first our conclusions on Lectures VI and VII, we tend to support the second hypothesis; namely, that *it is the inflow of information from the external world into the organism which is required to explain the development and the origin of "innate" biological clocks throughout the entire spectrum of biological evolution.*

This claim needs clarification. First it should be stressed again that *internal* rhythms *without* any apparent external correlate have been detected in many living systems. *These rhythms may originate* in the brain (q.v., §IX.2.5.1) and/or in the *genetic code.* Consider, for instance, the well-known *mutation experiments* with drosophila (they have been used for experiments on mutation because of their speed of reproduction and their four large, paired chromosomes), *which demonstrate that at least one gene may be responsible for mutational variations in the circadian clock!* [Konopka, W. and S. Benzer, *Proc. Nat. Acad. Sci., 68*, 2112 (1971)].** Therefore, the questions that we raise here are two:

    i) *What is the natural origin of such mutations?*

    (ii) *What is the origin of the thermodynamic processes which have been responsible for the evolution of DNA, RNA, ADP, cells, and larger living organizations (and for the evolution of synchronization in such systems)?*

** For other references that deal with these problems and/or describe the results of various experiments associated with biological clocks see, for instance, Irwin Bunning, *The Physiological Clock*, The English Univ. Press, London, 1967; R. R. Ward *The Living Clock*, Mentor Book, 1971; J. D. Palmer, *Clocks in Marine Organisms*, Wiley, N.Y., 1974; A. T. Winfree "Resetting Biological Clocks", *Physics Today, 34* March 1975; D. S. Saunders, *Insect Clocks*, Pergamon Press, Oxford, 1976 and references in §IX.2.5.1.

It is in this sense that we employ here the results which emerge from Lectures VI and VII to claim that the answer to the aforementioned questions is the same as that given in response to the questions: What is the origin of thermodynamic processes; of time? of irreversibility? and of all evolution? And it is the rest of the material to be presented in this lecture that will give this claim additional support and, in addition, will demonstrate the fact that *the origins of time, rhythms, biological clocks and life are not only intimately interconnected but may also have a common ground.*

Should we deal with such problems within the framework of a discipline called *"Philosophy of Biology"?* I doubt it. We have already shown the facts that dictate a more universal outlook. Nevertheless, we shall briefly review the background of this emerging discipline before we proceed with our analysis of the problems mentioned above, and, in particular, of their origin and significance.

C. "Philosophy of Biology"—A Side Remark

In recent years we have witnessed the most dramatic increase in human knowledge of fundamental biological processes. This, in turn, has stimulated a demand for a *"philosophy of biology"*; an urge which is also based on an unprecedented interest in evolutionary theory since its development in the 19th Century. Nevertheless, most of the problems proposed by these studies are universal questions now being investigated afresh in the light of new discoveries. Most of the questions remain, however, unanswered. In addition there are newer problems, like the ethics of *"genetic engineering"*, or the questions about *"life technologies"*—including organ transplants and artificial organs. Some of these demonstrate that it is easier, inside the framework of a given discipline, to study life, than to philosophize about it.

Questions about biological evolution are frequently coupled with the problem of *structures and hierarchy in the complexity of the systems investigated.* But disciplinary studies demonstrate, time and again, the impossibility of understanding "biological clocks" (or complex organisms) by studying molecular mechanisms *in isolation*, i.e., *without paying attention to the external origins of life, to hierarchical structures of the whole of nature, to the evolution and complexity of prebiotic organisms, and to the grand temporal order of the non-living universe.* Our preliminary notes on geochemical and biological evolution (§IX.2.3.1) are therefore in line with a general concept according to which the emergence of new characteristics and qualities by living organisms are coupled strongly to evolutionary patterns in the *external* physical universe.

But at this point I want to stress one additional fact: The distinction between living and nonliving, which was widely discussed at the turn of the 20th Century, has lost much of its relevance in light of accumulative evidence against it—in particular—of the now established characteristics of genes and viruses which may be defined as "living" or "nonliving" *without* loss of scientific relevance! A similar "loss of relevance" may be attributed to traditional philosophical doctrines regarding the "definitions of life", e.g., *vitalism* (which holds that there exists in all living things an intrinsic, unmeasurable "life factor" or "vitality"—see also our notes on *Bergsonism* in Appendix I, §B.11).

From *"structures, gravitation and prebiotic evolution"* (§IX.2.3), and from the more specific problems of *"innate" biological clocks"* (§IX.2.3.3-B), we now turn to additional problems that might be *intimately linked* to these topics, namely; *"memory and structures"* (§IX.2.4); *"memory and brain", "memory and symmetry", "brain and innate ordering of time and thought"* (§IX.2.4 to §IX.2.24), and, in particular, *"innate ordering of perception and time asymmetries"* (§IX.2.7 to §IX.2.8).

IX.2.4 From Cosmology to Irreversible Structures and Memory

IX.2.4.1 Asymmetry and Memory: Is there any Connection?

In previous lectures* I showed that we can understand all processes in nature in terms of three distinct groups. In this Lecture we add "human aspects" to the physical concepts discussed before:

Group I: Symmetric conservation laws, time reversal, reversibility, symmetric causality, global and local symmetries, structureless medium, loss of recollection, forgetfullness, meaningless, isotropy, directionless, uniformity, changeless, equilibrium, and loss of information (see below).

Group II: Gravitation, arrows of time, asymmetry, irreversibility, symmetry-breaking, increasing entropy, energy dissipation in expanding intercluster space, boundary and initial conditions, causation, stratification, cosmic and biotic evolutions, natural selection, new structures, recording, memory, information, history, recollection, aggregation, music, speech, language, elementary sounds, codes, syntax, change and non-uniformity (see below).

Group III: It is only by *synthesis (of Group I with Group II) that we obtain all present knowledge about nature.* For instance, the energy equation (II–73) is a classical synthesis of a time-symmetric energy equation with an irreversible term (the term Φ) which represents dissipation of energy due to viscous ("friction") effects.

We now turn to examine the link between *irreversible structures* and *memory* (the link between cosmology and irreversibility has been already explained in Lecture VI. See, in §IX.2.8.1 for the link between irreversibility, language and thought).

Our first notion of physical memory was developed in §IV.16 for materials such as non-Newtonian fluids. Another well-known type of physical memory is the process of *magnetic tape recording.*

---

* See Table II in the Introduction and §VI.3, V.5 and VI.10.

This process offers one of the most accurate and convenient methods of storing, and later reproducing, various types of *"information"* that can be converted into electrical signals.**

Also quite well-known today, is the process of *chemical-biological* recording and reproduction by means of DNA, m-RNA, t-RNA, etc. This process stores the type of information that can be converted into fundamental life processes (e.g., genetic codes, production of proteins, enzymes, etc.).

Less known is the phenomenon of *shape memory*. There is a class of materials that, like patients emerging from psychoanalysis, *"remember their past"*. Members of this class may be drawn from the transition and the precious metals. Shape or structure memory alloys include alloys of nickel and titanium ("Nitinol"); copper and zinc, aluminum or tin; indium and thalium; gold and cadmium; iron and platinum; and some ternary alloys, such as copper-aluminum-nickel.

How does the shape or structure memory alloy work? All of these alloys exist in two "phases", one at high temperatures and another at low temperatures. Internal *(symmetric)* structures characterize the high-temperature phase while the low-temperature phase exhibits a lower degree of internal structure.

The structure memory alloy is first formed into a given "structure", such as a spiral, and may also be annealed at elevated temperatures *"to fix"* the macrostructure. The crystal structure changes as the alloy cools. *Reshaping* becomes possible at the *lower* temperature phase, for example by straightening the spiral. Then, *upon reheating* to a temperature *above* the "phase change" (but well below the annealing temperature), the alloy returns to its initial structure. The uses for such structure memory metals seem to be limited only by one's imagination.

Now, some similar phenomena are observed with macro-organic molecules and polymers.***

** In this process an electrical signal is usually applied to a tape (or a disc) by means of a coil wound around a core of magnetic iron ("head"). The current in the coil produces a magnetomotive force and magnetizes the material particles in the tape. The strength and direction of the final magnetization of a particular point on the tape is determined by the strength and direction of the magnetic field, i.e., energy is *dissipated* in the tape with a "record" containing direction (north or south pole), amplitude, and linear dimension (along the tape) corresponding to the direction (plus or minus), amplitude and time of the original electrical signal.

In reproduction the tape is passed over the same "head" used in the recording, or one similar to it, thereby causing the magnetic flux in the core to change, generating a voltage in the coil. The reproduced signal generally does not exactly resemble the recorded pulse, but does bear a definite "relationship" to it. The "relationship" depends on the quality and details of the recording machine. Various modifications of such a recording principle are now available (e.g., storage (memory) cores in high-speed electronic computers, magnetic tapes, that are used to record television programs, the so-called video tapes, etc.).

*** There is yet considerable controversy over the details of the process. Yet researchers seem to agree that the transformation from high to low temperature phase is what is known as a diffusionless or martensitic transition, as opposed to a more conventional nucleation and growth process. Thus, in the martensitic transition, individual atoms do not diffuse away, but effectively keep the same neighbor atoms as before the transition, and all strains caused by the change in structure are accommodated elastically.

To conclude: Memory, causation, symmetry-asymmetry and gravitation are intimately coupled together. But why gravitation?

It was demonstrated in previous lectures that without gravitation the world would have been "static and uniform" and, therefore, *unobservable*. Without gravitation there would be no aggregations of matter, no production of heavy elements, no gradients, no boundary conditions in thermodynamics and electrodynamics, and, to cap it all, *no "memory", no "observers"* and *no "brain"*; only uniform matter-energy in complete equilibrium. But most important, without gravitation nothing happens, or changes, and even life becomes an impossibility.

"Men are born ignorant, not stupid; they are made stupid by education".

Bertrand Russell

IX.2.5 Brain and "Ordering" of Space and Time

IX.2.5.1 The Brain and Feedback Networks

Now we finally come face to face with the most difficult problem of all; human memory, brain and the subjective "ordering" of space and time (but see also §IX.2.7).

In its most general form human memory implies a capacity to generate *"structure-print"* and recollect "events" that are ordered in *space* and *time* and which are treated as belonging to the *past* life of the individual. From the earliest speculations about memory to most of the latest experimentally-based views, it is assumed that the critical problem is associated with the nature of the "irreversible process" that *"records"* events for later *"reproduction"*, either in their original mode or with the help of signs and symbols which are regarded as equivalent to that mode.

It is generally supposed that when anything happens which affects the behavior of an organism with a *central nervous system*, it leaves behind an *irreversible "structure-print"*, *"trace"* or group of such *"structures-print-traces"*. The experimental psychology of *"remembering"*—and all modern experts claim to base their views upon experimental verification—sets out to explore the conditions for the *persistence* and *length of persistence* of *"structure-print-traces"* and for their *restimulation*.

Consequently, understanding the structure of the central nervous system and of the brain bears special interest to our understanding of subjective ordering of space and time.

The martensitic phase exists in the form of asymmetric platelets or lamellae that have various crystallographic orientations. When the martensitic phase is initially deformed, platelets with prefered orientations relative to the applied stress grow at the expense of others. The key to recovering the high-temperature shape is the *higher symmetric structure*. On the transition to the low-temperature phase, there are many crystallographically equivalent martensite variants to which the transition can proceed. *However, in the reverse transition, there is only one possible place to proceed and that is to the original orientation and shape.*

Today many *physiologists, neurophysiologists* and *psychophysicists* study the structure of the brain and of its associated feedback networks and control systems. *Histologists* try to elucidate internal cell structure, and its birth, growth and death. *Biophysicists* and *biochemists* investigate the complicated physico-chemical processes that form the basis of the human brain and body. *Chemists, physicists* and *thermodynamicists* attempt to build physico-chemical models that would explain the evolution of *autocatalytic chemical reactions* that lead to the *formation of structures, periodic systems and thermodynamic cycles-structures*. Here, cybernetics (see below) is also contributing mathematical models of *"self-directing"* and *"self-correcting"* organic systems. However, the relevance of the evidence that has been accumulated so far by these investigators is mostly indirect. Hence, the few facts and suppositions that are to be mentioned below should be considered tentative rather than conclusive.

First, let me stress that the arguments in favor of mind-body unity that are mentioned below, have been drawn from recent discoveries of detailed *correlation* between *"brain events"* and certain *"mental events"*. Hence, the study of the brain must play a central role in arriving at a better understanding of human perception of space and time.

From evidence that exists today, we know the following:*

1) The birth, development and activity of each living cell is controlled by a great variety of systems. Hence, they *cannot* be studied in *isolation*.

2) The processes involving *exchange of signals* between *individual organs* in the human body and the brain, and the *transmission* of control signals from the *control centers* to the *motor organs* differ substantially. These processes may involve specific molecules that are produced by certain organs (e.g., the thyroid).

3) Transmission of *"information"* and *control commands* is also carried out by signals circulating in the *nervous system*, which, in addition, is responsible for *storage* (memory) and complex *processing* of "information" which ensures the living ability and "purposeful behavior" of the body.

4) The nervous system consists of nerve cells that are generally called *neurons*. The neuron is capable of being excited by *electrochemical processes* and conducts signals in a certain direction. The estimated number of neurons which make up the nervous system of a human individual is on the order of 10^{11}.

5) The neuron consists of a cell body, containing the *nucleus* and the *cytoplasm* (including DNA, t-RNA, m-RNA, enzymes, proteins, etc.), which are enclosed in a *membrane* shell, from which emanate branching extensions, *dendrites* and an axial extension called the *axon*.

---

* By no means should the few facts mentioned below be considered as a review of the subject. Their inclusion is made here for the purpose of comparison with a number of historical notions that are to be discussed below.

6) The terminal branches of the axon are in contact with other cells through the *synaptic contacts*. (There are, on one axon, between 100 to 1000 such synaptic contacts). Other branches, called *collaterals,* emanate from the axon and reach other cells.

7) The *electric potential* of the neuron membrane depends on the *magnitude* and *duration* of the signals fed to the synaptic contacts. When the potential of the membrane reaches a certain *threshold value,* about 40mV, a *neural impulse* occurs — an *electro-chemical wave,* which propagates over the nerve fiber of the axon.

8) After carrying the pulse, the nerve fiber is in a state of complete non-excit-ability for a given period (the so-called *refractory period*), i.e., it does *not* conduct nerve signals whatever the intensity of excitation.

9) Three types of neurons are generally distinguished: *sensory, motor,* and *association neurons.*

10) The central nervous system of human beings is composed of the brain and the spinal cord. Signals are received by this system from sensory elements, called *receptors,* and are transmitted to the motor elements, *effectors,* by means of a network of nerve fibers forming the *peripheral nervous system.*

11) The brain has the following main parts: (a) the *medulla oblongata*; (b) the *cerebellum*; (c) the *midbrain*; (d) the *corpus striatum* and (e) the *cerebrum.* About half of the neurons are in the corpus straitum. The bodies of the nerve cells of the frontal brain are mainly concentrated in its surface layer, the *cerebral cortex,* where, apparently, the main centers responsible for the conscious activity of the brain are located.*

12) Data on the *external world* collected by the sensory organs, transmitted to the receptors and then over the nerve fibers to the peripheral nervous system, are channelled into appropriate *sections* of the central nervous system. Here, the "information" obtained is processed, evaluated, compared with the "stored in-

---

* A number of recent studies [F. K. Stephan and I. Zucker, *Proc. Natl. Acad. Sci.* (U.S.A.) *69*, 1583 (1972); *The Neurosciences: Third Study Program,* F. O. Schmitt and F. C. Worden, Eds. (MIT Press, Cambridge, Mass. 1974) — C. Pittendrigh p. 437 and R. Y. Moore p. 537: N. Ibuka and H. Kawamua, *Brain Res. 96*, 76 (1975)], demonstrate the possibility that the so-called suprachiasmatic region of the rat brain is a *central biological clock* responsible for the generation of several *biological rhythms.* Destruction of this region is associated with a loss of rhythmicity of drinking behavior, locomotive activity, sleep and wakefulness, and adrenal cortical activity, as well as with loss of estrus cyclicity.

But an *additional, independent clock* may be involved *in the generation of rhythms.* This was recently demonstrated by D. T. Kriger and H. Hanser [*Science 197*, 398 (1977)]. They found that suprachiasmatic destruction (by nuclear lesions) do *not* abolish food-shifted circadian adrenal and temperature rhythm-icity. Further studies are therefore required to characterize the central neural mechanisms responsible for the *synchronization and maintenance of the phase-shifted rhythms.* (See also §IX.2.3.3).

Problems associated with asymmetry in neuromotors, structural asymmetries and asymmetrical neural control are discussed in *Lateralization in the Nervous System,* S. Harnad, R. W. Doty, L. Goldstein, J. Jaynes and G. Krauthamer, eds., Academic Press, N.Y., 1977 and in *Evolution and Lateralization of the Brain,* S. J. Dimond and D. A. Blizard, eds., N. Y. Acad. of Sci., N. Y., 1977 (Annals, vol. 299).

formation" (memory) and, *if* necessary, commands are formed which are transmitted over the nerve fibers to the effectors which act on the motor organs (muscles, glands).

13) The activity of the motor organs is monitored by *feedback networks: internal*, which verify the intensity of the action of the motor organs, and *external*, which verify the final effect of the realization of commands.

14) *"Information"* is collected by means of the appropriate receptors and effectors, and commands are sent not only regarding the *interaction* of the body with the *external* world, but also for ensuring the normal functioning of its *internal* organs and systems.

15) Numerous investigations have established a certain role of the nervous system and in particular of the brain in realizing the essence of cognitive phenomena such as *memory* and *thought* (see below) (see also §IX.2.8.1).

16) Study of the *evolution* of the brain in the developmental process of various living systems on earth, lead to the conclusion that it has *gradually developed* as a specialized organ for survival *by adaptation* of the body to the *external* world which is varied and changes with the progress of time.

17) Neurophysiologists have succeeded in mapping the physical sections of the brain, external stimulation of which is *correlated* with the different types of sensory experience. Thus, by direct external stimulation of the brain in the appropriate areas it is possible to produce, for example, visual sensation without the normal stimulus via the eye and the optic nerves.

18) Electrical activity remains confined to discrete pathways that are interconnected with both excitatory and inhibitory consequences. For instance, when a cat's eye is probed by a small spot of light, a specific area on the retina can be found that serves to excite a given brain neuron.

IX.2.5.2 Eye, Brain and Havaya

Due to the finite velocity of light, and the delay in nervous signals reaching the brain, we always 'see' the world in its past states. This remains valid no matter how close or how far away from us an object is. The galaxy Andromeda, for instance, is visible to the *unaided* eye, but we see it 'now' as it 'was' a million years before intelligent mind or consciousness appeared on this tiny planet. This is one difficulty associated with 'seeing'; the other ones involve the human eye, brain and consciousness.

Is it from memory and past experiences—combined with stimulation on the retinas—that we perceive the world of objects in space and time? How is information from the eyes coded into neural signals, into the brain code and into our subjective experience of external objects? Do the senses give us a "direct," "camera-like" picture of "reality?" Do the eye and the brain operate like a photographic or a television camera converging objects merely into pictures or images? Are the chains of electrical impulses—the pattern of neural activity associated with "seeing"—

combined in the brain as an internal picture of the object or are they combined in the brain to form *an internal network-structure* of a "subjective being", which, by means of past experiences, is, in itself, a reality associated with a "verifiable idea" of an external object? And, most important, what do we mean by the concepts "verification" and "reality"?

These problems deserve serious pondering and research, for in seeking their solutions we may advance our knowledge of human perception and our knowledge of what we call "havaya," "things-in-themselves," "reality," "nature," "the world" or, simply, "what lies before us."

To "see" we need only three or four basic entities: (i) *electromagnetic radiation*; (ii) *a non-uniform distribution of matter-energy*; (iii) *eye-brain*; and (iv) *mind-consciousness.* And it is about the independent, distinct, all-controlling or separate existence of item (iv) that many battles have been fought between materialists and idealists, between scientists and theologians, between laymen and scholars, and between various socio-political movements. The battles have not yet been decided, for one reason, at the present time each "movement" can present a logical theory of knowledge which appears to be compatible with "evidence."

In trying to seek solutions to these problems one may first concentrate on some empirical facts on which there is a general agreement. Let us start with the structure of the eye.

(1) A number of individual quanta of light (five to eight) are required to give us the minimal experience of a flash of light.

(2) The individual receptors of the retina are as sensitive as a sophisticated 'man-made' detector can be. Let me stress also that only about 10% of the light quanta reaching the eye stimulate the receptors of the retina, the rest being lost by absorption and scattering within the eye.

(3) In spite of the fact that it is often difficult to conclude whether a visual perception originated from psychological, physiological or external physical origins, a considerable evidence proves that some pure physical properties of light contribute to many well-established phenomena of visual perception—e.g., loss of acuity in dim light which, *until quite recently has been treated as though it were a separate property introduced by the eye.**

(4) *Almost all living creatures, including plants and one-celled animals, respond to light quanta.* In higher organisms there are specially adapted cells to serve as photo receptors. Most primitive eyes respond only to 'light or shadow' or to other well-defined phenomena of radiation. More advanced organisms perceive 'form' and 'color' by means of lens forming images on a mosaic of photon-sensitive receptors and brains sufficiently elaborate to transduce the neural signals from optical

---

* For a summary of various experiments aimed at demonstrating to a fresh student the physical effects of light, the structure of the eye and the brain, see, for instance, Gregory, R. L. *Eye and Brain; the Psychology of Seeing,* World University Library, 2nd. ed., London, 1972.

images on the retinas into perceptions. Here the pinhole pupil of *Nautilus* is an example of the most primitive lens.

(5) *While the human brain is the most complex of all brains, his eyes are not.* Elaborated eyes have generally been developed with simple brains. There is even a kind of eye (of the *Copilia*) whose receptor and its attached lens cylinder *scan* across the focal plane—somewhat similar to the technique of scanning in a television camera.

(6) The human eye performs well over a range of brightness of about 100,000 to 1. Moreover, the iris opens and closes by a control system which can be fully described in terms of a pure servo-mechanism.

(7) The retina is a specialised light-sensitive part of the surface of the brain. It retains typical brain cells and indeed, some of the data processings for perception take place in the eye which, consequently, *is an integral part of the brain!*

(8) *Depth perception* becomes possible with two eyes. As the eyes converge on an external object the angle of convergence is signalled to the brain *as information of distance*—thereby serving as a king of camera-like "range-finder." Therefore, depth perception results from the angle of convergence of the eyes. But to estimate the distances of many objects at the same time the brain must also receive information about which eye is which. This allows the use of *stereoscopic vision,* one of several methods employed by the brain for depth perception. Another one is the so-called convergence-disparity compensation method. Highly developed insects, birds, the tree-living apes and man, demonstrate precise stereo depth perception.

Combining this evidence with the bulk of evidence presented in the previous Lectures one can now draw the conclusion that *all that one 'observes' is a transduced collection of signals that originated in the past by external objects.* Hence, without even plunging into the complicated problems of brain-mind-consciousness one may 'see' some justification for *both* Idealism and Realism or even for *both* Idealism and Materialism in their straightforward meanings. *Each doctrine is compatible with a portion of our knowledge and each is valid within its own context.* But *any* of these doctrines is part and parcel of the *reality of the whole, for any is self-consistent within the pluralistic manifestations of havaya.*

It is the fault of false philosophies, that by insisting on the absolute imposition of their concepts on *all* human ideas and deeds ended up in the *impasse* of human thought. A moment of reflection is sufficient today to show why the concepts of physics do not entirely contradict all the idealistic concepts of "mind", and vice versa. *Life does not escape physics and physics does not escape life; for human life is a matter of duality; of mind and of matter, of objective 'zmann' and of subjective 'moeds,' of time and space, and of symmetry and asymmetry* (see §IX.2.8.1 below and Appendix I).

IX.2.6 The Failure of Cybernetics and Information Theory

Realizing these problems, many scientists have begun to ask whether the central nervous system might normally operate somewhat in the manner of a kind of "super computer" capable of *receiving, storing* and *retrieving information,* of making *references,* and of *"communicating information".*

Actually there are three main possibilities:

(i) To maintain that the more complex of these devices may have *"artificial intelligence"* of a *"brain",* that in the very literal sense can solve problems, *"think"* and demonstrate *"apparent purposeful behavior".*

(ii) To claim that "purposeful behavior" is a unique criterion of the *"mind"* and, therefore, to reject the ideas that such devices meet the requirement of *"mind".*

(iii) To give up the criterion of *"purposeful behavior".*

Defenders of possibility (ii) argue against the possibility of assigning "purposeful behavior" to any physico-chemical device. They may argue, for instance, that the "apparent purposes" of such devices are *built in by their designers* and that it is not up to the devices to *"choose"* their purposes. The emphasis of this line of argument is therefore shifted to *"choice", "free will'* and *"free judgment"* — topics which we have already discussed in regard to causality, etc. (Lectures IV and V). This argument is supported by idealists, Kantianists, etc. (see below).

On the other hand, someone defending the first possibility would point out newly available *empirical evidence* which proves that for a biological organism the basic "purposes" are *also* built in through the *genetic code of heredity,* DNA, etc., and that these codes control the basic biological "drives" for reproduction, growth, energy and safety. Hence, according to this line of argument there is no place for choice, "free will" etc.

The defenders of the second possibility concede that biological organisms have some built-in controls. Yet they maintain that higher organisms have the capacity to *develop* purposive methods and *renew* them while physico-physical devices can only solve problems by the methods designed into them.

In contrast, the defenders of the first possibility hold that today there are physico-chemical devices which, by using *feedback* or *servo-mechanisms,* can analyze the trends and varying "loads", and, accordingly, "adopt" *new programs* that are more successful than earlier ones. In a literal sense such devices might be called *"creative",* for they are capable of "learning" from experience. In fact, most scientists hold the notion that the advancements of science and technology would allow the development of complicated physico-chemical systems that might be acknowledged to have "minds". They point out that various views of this kind have been developed in the field generally known as *cybernetics* or *"theory of information".* However, as I shall explain below, this field has, in fact, failed to become the fundemental ground for explaining these concepts.

Now, cybernetics is commonly regarded as the *"science" of control and communication processes in both animals and machines*. The word was introduced and the field popularized by the mathematician *Norbert Wiener* in a book called *Cybernetics* published in 1948. The founders of this field—*Norbert Wiener, A.Rosenblueth, J. Bigelow, W. B. Cannon, W. S. McCulloch, W. Pitts, W. R. Ashby, C. Shannon and J. von Neumann*—believed they were advancing a generalized theory of such systems as the nerve networks in animals, electronic computing machines, servo systems for the automatic control of machinary and other information processing systems. They also believed that their science overlaps the fields of neurophysiology, computers, information theory and automation and that it also supplies a common ground for other diverse disciplines. However, in their theories, information is normally treated as a statistical quantity; a signal is considered to be a particular choice from a *statistical ensemble* of possible signals. *Hence, they run into the same fundamental difficulties enumerated in Lecture V on quantum mechanics. Moreover, the aspiring discipline of cybernetics did not base itself upon causality, causation, space-time, and time asymmetries, which permit the construction of a universal theory consistent with the rest of modern science*. Instead, it rested on the dubious concept of *"entropy"*, taken from old classical thermodynamics and carrying with it all the questionable and uncertain results that have already been discussed in Lectures IV to VIII. For them, *"God first created information"*; the rest of the world just followed as "reflections" and effects—indeed, a pure idealistic metaphysics. Then they introduced the *subjectivistic* concept of *"disorder"*, or the *"amount of disorder"* in any system. According to this popular approach, *all* "complex organisms" are constantly threatened by an "increase in disorder", with the "end point" of "complete chaos". (The falasy associated with this view was already explained in §VII.3 and in Assertion 8, Introduction.)

Moreover, any discussion of the fundamental meaning of cybernetics must first attempt to elucidate the physico-philosophical meaning of the technical terms used in it, such as *"information"*, *"entropy"*,* *"neg-entropy"*, *"amount of disorder"*, *"noise"*, *"memory"*, *"consciousness"* and even *"life"*. Many papers vainly seeking universal definitions of these concepts have appeared and I do not intend to discuss them here except for the following remarks:

1) Practical adoptions of cybernetic methods, as well as of quantum-mechanical models, need not await the formulation of universal physico-philosophical interpretations.

2) The fact that in the hands of its most enthusiastic proponents, cybernetics has become "an all-embracing science" that deals with the fundamental nature of politics, psychology, economics, pedagogical theory, logic, physiology, biology and even human society as a whole, should not obscure the fact that it is based on con-

---

* The emergence of entropy-free thermodynamics was discussed in Lectures IV and VI.

cepts that are totally opposed to the rest of verifiable science.**

3) A systematic treatment of contemporary cybernetics and information theory is impossible in the framework of a consistent and unified science, i.e., a "disorder-free", "noise-free" or "neg-entropy-free" cybernetics must be developed.

4) The correct physico-philosophical interpretations of the subjective concepts of information, disorder and entropy may be translated to become part and parcel of modern physics and of our entropy-free thermodynamics and dialectical gravitism.***

5) Any physics of time asymmetries or any cosmology based on such subjective concepts as "orderliness", "disorder", "neg-entropy", "information", etc. is nothing but a metaphysically exaggerated mode.

6) Any rational debate on the nature of information as advocated in contemporary cybernetics arrives at a dead end, for it raises such questions as the following:

(a) Does or does not "inorganic" or non-living matter contain "information"?

(b) Does or does not a cybernetic device *constructed by man* contain "information"?

(c) Is "information" describable *only* in terms of "probability"?

(d) *Is all "information" the same?* Should it be classified into different types of human applicabilities, i.e., political, biological, non-living, human, social, etc.?

(e) Does *nature* possess "its own" "information"?

(f) What are the differences between the *"quantity"* and *"value"* of "information"?

(g) Is cybernetics a complete science or an incomplete myth?

The history of science and philosophy testifies to many exuberant appeals to the empty formalism of science as ultimate explanations in areas rather distant from their points of origin and restricted domain of applicability.

It is not accidental that the popularization of Heisenberg's "uncertainty principle" (Lecture V) and of "information theory" and cybernetics was followed by their application in such domains as "human freedom", politics, theology and pragmatic religion. Nevertheless, with the formalism of cybernetics, our vocabulary is "enriched" with models of biological, social, economic and political behavior containing "feedback models". One must consider not only the failures and in-

---

** Even Norbert Wiener contended that *"information is neither matter nor energy" it is just "information"*. W. R. Ashby also warned skeptics by stating that *"Any attempt to treat variety or information as a thing that can exist in another thing, is likely to lead to difficult 'problems' that should never have arisen"*. Thus, information becomes a "ghost", a "soul", a sort of modern "Holy Spirit" that cannot, in principle, be associated or related to other scientific concepts even in the most pluralistic, yet consistent science. There are numerous other allegations that support this pseudo-theology dressed up in the technical terms of cybernetics and information theory.

*** Formation of such a "translation" is to be reported in a sequel volume. A few remarks are given, however, in §VII.3 and in §IX.2.8.1 below.

adequacies of such models, but also their partial success. Every model elucidates a certain attribute of an organism provided, of course, that we do not, at the same time, ignore its potential negative effects on general science and intellectual thought.

Whatever the destiny of cybernetics, there will no doubt be efforts to unite it with the rest of science on the one hand, and with the concepts of modern physico-philosophical thought on the other. But, the latter aim brings us face to face with the Kantian notion of *a priori* information-knowledge on the one hand, and Spinozism and Einsteinism on the other. Thus, sober appraisal of these doctrines is no doubt a prerequisite for any attempt *to revise* cybernetics and information theory. Such an attempt is indeed planned as a sequel to this volume.

IX.2.7 Innate Ordering of Space and Time

The experiments described below are intended to resolve an old dispute about the following questions:

1) Is perceptual "ordering", "structuring" and organization physiologically inborn? In other words, *is it inherent in innate aspects of brain functioning rather than depending on a "synthesizing process" of "learning" through the irreversible interactions with the "external" world?*

2) Is *objective* experience possible and only distinguishable from *subjective* experience to the extent that we apply to the latter the categorical forms of thought? In other words, is objective knowledge made possible by the *ordering which the mind imposes* upon the manifold given in pure sense (Kantianism, etc)?

3) Is knowledge made possible by the mind's contribution to experience? Or, can experience be objective *without* the contribution of the mind?

IX.2.7.1 Use of Cataract Surgery in the Study of Human Perception-Structuring-Memory

Cataract is a disease of the eye in which the normally transparent lens become opaque and clouded. One way of resolving the aforementioned dispute is to deprive people from birth of all *visual* sensory experience and, hence, of *visual perceptual learning.*[+]

In this class of experiments we deal with surgical methods that restore sight to people born *blind* because of cataract.

Following the removal of their cataract, such individuals are initially able to tell when a simple figure is present. However, *at first they cannot discriminate one*

[+] It should be stressed, however, that other organs help us in space perception, e.g., the three semicircular canals in the ear which provide us with built-in Cartesian axes. The role gravity plays in these canals is of particular interest to our theory.

simple shape from another, nor can they readily remember the shape of a just-exposed object. It is only after a long period of *repeated experience* (of the order of several months) that these individuals *can easily distinguish between people's faces or between a square and a triangle!*

This finding gives some support to the assumption that posited neuro-physio-logical mechanisms (e.g., assemblies of brain cells) serve as the medium for *structural changes* that are associated with *perceptual learning and memorizing of shapes* in space.

Additional evidence with laboratory animals reared in impoverished or enriched environments demonstrates also the *requirement of early visual stimulation for perceptual development.* But before we rush into premature conclusion, let me next describe another class of experiments which is, to some extent, even more instructive.

IX.2.7.2 Is Innate Depth Perception of Space Possible?

A somewhat different type of evidence has been accumulated in regard to visual depth perception in laboratory animals and human babies. The so-called *visual cliff* apparatus is one of the techniques whose demonstration depends on apparently *"innate reluctance" of young animals to step off the edge of what seems to be a steep cliff.**

In one of its versions the apparatus consists of a narrow platform on which the subject (e.g., chicks, rats) is placed and two wide platforms on either side of it. Both flanking platforms are equally, but slightly lower than the central one. Thus, emerging onto this platformed system, the infant sees *visual patterns* designed so that one looks much deeper than the other. Typically, but not in all cases, the infant explores the central platform and the flanks, then perceiving the central one as hazardous, finally decides to step down onto the shallow-appearing platform. It is claimed that by this final response, *visually* naive animals *behave as if they are innately capable of pattern and depth perception* (while visual experience serves later to improve it).** Nevertheless, the evidence is not yet complete and the results are still open to different interpretations (see below).

Supportive evidence is claimed to come from a *comparative study of languages.* Though Noam Chomsky claims 4000 languages spoken in the world differ greatly in sounds and symbols, they somewhat resemble each other in syntax to suggest that there is *"a schema of universal grammar"*, or *"innate presettings"* in the human mind itself. Thus, it is argued, that these "presettings", which have their basis in the brain, set the pattern for all experience, fix the rules for the formation of meaningful sen-

---

* R. S. Woodworth and H. Schlosberg, *Experimental Psychology*, Henry Holt and Co., 1954.

** It is evident that *experience alone* can be decisive in depth-perception from light-and shadow cues, i.e., this kind of depth-perception is not innate [cf. e.g., the works of E. H. Hess from the University of Chicago).

tences, and explain why languages are readily translatable into one another (see also *Structuralism* in Appendix III and below).

The aforementioned evidence is not refuting (nor confirming) *old philosophical* ideas about a class of spaces and times that are *inborn* in the human brain; i.e., they form *a priori* perception of space and time as contrasted with that gained from experience. The origin of this *controversial* concept may be traced in historical *times back to Anaxagoras, Plato* and *Aristotle.* But it was mainly through the philosophical works of *Descartes, Locke, Berkeley, Hume, Leibnitz, Kant, Bergson, James, Dewey* and their followers (see below) that these ideas have received wide attention.

IX.2.8 Biological Clocks, Innate Perception and Gravitation

All evidence indicates that behavior deriving directly from *biological heritage* is exhibited by *all* living beings. Here, the branch of science called *ethology* emphasizes such *inborn* behavior.

The available evidence concerning biological clocks was briefly reviewed in §IX.2.3.3, where we stressed the assumption that *the (evolutionary) origin of all temporal behavior in animals can be traced back in time and out to external physical influences.* Here I stress the additional fact that *"innate patterns"* are frequently associated with *simple orientation movement in the field of gravitation,* i.e., as *"up-ward-downward" righting of the biological body in reaction to gravity! Thus, gravity and the geophysical periodicities mentioned above emerge as the prime sources of ordering and simple orientation movement!!*

Animals low on the evolutionary scale characteristically exhibit innate patterns, depend little on *learning*, and have a lesser adaptability to changes in the environment. Consequently, their dependence on *heredity-geophysical-gravitational origins* is high in proportion to animals high on the evolutionary scale. While the latter show some signs of innate patterns, their greater capacity to learn produces a much more flexible mechanism to respond to a variety of other external changes that stimulate them (Assertion 9 in the Introduction).

The evolutionary origin of the so-called *"innate ideas"* may, therefore, be *external* rather than *internal.* But to demonstrate what I mean let me now turn to the problem of thought and asymmetry.

> He gave man speech, and speech created thought,
> Which is the measure of the universe;
> And science struck the thrones of earth and heaven.
>
> **Percy Bysshe Shelley**
> Prometheus Unbound

IX.2.8.1 Asymmetry, Language and Thinking: Preliminary Concepts

In this section I make a preliminary attempt to associate language, word and thought with time asymmetries and external influences. For that purpose I first adopt some of the original views of *Lev Semenovich Vygotsky* (1896-1934). Vygotsky's influence has spread outside the Soviet Union, particularly with the publication of his *Thought and Language* (MIT Press and Wiley, Cambridge and N. Y. 1962, English translation; with introduction by Jerome S. Bruner).

According to Vygotsky, thought and language have different origins—thought in its *prelinguistic* stage is linked to *biological evolution,* while language in its *pre-intellectual* stage is linked to the *social evolution* of the child. In both Vygotsky found a clear link with *external factors*: "Verbal thought is *not* an *innate* natural form of behavior but is determined by a *historical-cultural process* and has specific properties and laws that cannot be found in the natural forms of thought and speech".

There is a moment of discovery when the child sees the link between word and object. From this moment on, thought becomes verbal and speech rational. According to Vygotsky *prelinguistic thought has a similar source to the kind of primitive thought that some animals acquire through biological evolution while speech always has "an external social origin". The origin of both is therefore external!* (but it should be traced back through *all* stages of evolution). *Now, from the moment that we can "associate" objects with words we are capable to express thoughts in an "inner speech". And this "internalization of speech" means that our rational thoughts and words are interconnected, i.e., intelligent thought (not memory or pure imagination) and words are intimately interrelated!* This result is now widely accepted in both Eastern and Western Universities. (However, Vygotsky's theory has not always enjoyed the esteem of official circles in the Soviet Union. In fact this theory was contradicted by Stalin himself[*]).

Vygotsky's coupling of thoughts with words would now be *combined* with our assertion about *the origin of observations.* We have already concluded that observations and all *"aggregated forms"* are coupled together, i.e., that a *uniform* world at *"complete equilibrium"* is *unobservable*: any observation or "picture" without stratified aggregation, structure, "boundary conditions", as well as electro-magnetic relation between *"emitter and absorber" is an impossibility*—(see also the concluding remarks in Assertion No. 8, Introduction).

Therefore, *when observations* [present, *recollections,* or memories of observations], *and thinking* are, at one and the same time, "correlated" in the mind[+] it means that

---

[*] Graham, loc, cit. pp. 368. For references see below.

[+] An almost endless literature exists on this "correlation". Here I mention only a few sources: L. S. Vygotsky's *"Thought and Word"* in *Thought and Language,* MIT Press and Wiley N.Y. 1962; F. Bartlett's *Thinking, an experimental and Social Study,* London, Allen & Unwin, 1958; M. Wertheimer, *Productive Thinking,* Harper, N.Y., 1959; J. P. Guilford's *"Three Faces of Intellect",* The Am. Psych. 14, 469 (1950) J. Dewey's *How We Think* (Boston Heat & Co., 1933); L. Wittgenstein's *Tractatus Logico-Philosophicus unidirection* (asymmetry) *of thinking* [see also G. Humphrey *Directed Thinking* (1948) and *Thinking,* Methuen, London (1951)]. Thinking in the history of philosophy was discussed by many philosophers, notably by Heraclitus, Parmenides, Plato, Aristotle, Maimonides, Descartes, Pascal, Spinoza, Kant, Hegel, Shelling, Husserl, Wertheimer, Wittgenstein, and Dewey (see also §IX.2.10; 2.12; 2.16; 2.18; 2.22).

aggregation and thought and, in addition, *aggregation and words* are associated together and interrelated. (According to Assertion No.2 these correlations and associations must also be related to *space and time* — see also *"Asymmetry and Memory"* in §IX.2.4.1 and *"Brain and ordering of Space and Time"* in §IX.2.5).

Now what is the meaning of this conclusion? According to Assertion 6 (Introduction) *"All laws of nature are abstractions associated with observations"*. These laws are divided into *time-symmetric "laws"* and *time-asymmetric "boundary conditions"*.

Our next problems may therefore be stated as follows: (i) Can we distinguish, in the most general sense, between *symmetric* and *asymmetric* words, sentences, "information", and analytic concepts? (ii) *What is the origin of symmetric and asymmetric concepts in science and philosophy?* (See also §A in Appendix I).

In trying to answer the first question we first stress the fact that stratified *aggregation, boundary conditions and directional structures are related to time asymmetries. Hence, words and thinking must also be associated with time asymmetries!* However, there is a special kind of intellectual thinking, *the mathematical,* which has been partially developed to safeguard the investigator's mind from the usual prejudices and inconsistencies which beset us in analytic thinking. *Mathematics is a peculiar highly refined symbolic language, or way of analytic thought about ideas, including the ideas of symmetry and asymmetry.* For instance, *once a scientist starts working with symmetric equations the consistency of mathematics forces any of his results to remain symmetric, irrespective of his wishes or of his mathematical funambulism* (Lecture V). *Once asymmetry is forced on symmetric equations the same consistency dictates, at any stage of the analysis, asymmetric results. But the very act of forcing the asymmetry on the formalism is linked to previous observations or to mental backrounds. Consequently the ideas of time, asymmetry, stratified aggregation, boundary and initial conditions may be associated with the "order of nature" and/or with "human ordering".* But can *"human ordering"* be separated from *"physical ordering"?* These ideas were first discussed in Appendix I and in Lecture VI. We shall discuss them further below and in §IX.2.10; 2.12; 2.16; 2.18; and 12.22. Here we only point out two conclusions drawn from the contents of lecture V (see also Table V.1 and the preliminary remarks in §IX.2.4.1); namely, that (i) it is only in the *abstract* domains (of aesthetics, mathematics, geometry, physics, etc.) that symmetry, reversibility and symmetrical laws can serve as important tools of *analytic* research; (ii) *it is only by combining* (conceptual, reversible, analytic) *symmetry with* (factual, aggregated, observational) *asymmetry, or by combining reversible equations with initial and boundary conditions, that we can (mathematically) describe the world in agreement with observations.* (See also Assertion 6 in the Introduction).

Our next stage, therefore, is to note that while symbols and analytic concepts may be symmetric, *meaningful words and sentences* (in order, syntax, phoneme, form, sound modulation or other modes), or useful physico-mathematical equations *are basically asymmetric,* i.e., that *asymmetry, aggregation, time and meaningful sentences and words are intimately coupled.* (While some minor reservations are

Fig. IX.3 **Time, Order, and Life are strongly tied up with the gravitational field.** Organells and nuclei of the living cells are "heavier" than the rest of the cell. Such *asymmetry* allows plants to grow *"vertically upward"*. These problems are taken up in the Introduction and in Lecture IX.

What did *van Gogh* hope to convey in this picture? Apparently he did not think about the field of gravitation. Yet, his paintings convey the reality of landscapes and life embedded in a "field" of brush strokes, which, in themselves, in their very shapes, direction and rhythms, convey the presence of flux, structure, direction and a universal field of force which penetrates all things and is at one with land, life, sky and the stars.

justifiable, there is an overwhelming "word of evidence", derived from the languages and the studies of linguistics, cybernetics, information and mathematics that fortifies this contention.) It means that *any meaningful sentence generates "a linguistic arrow of time"*. *Similar conclusions may also apply to rational thought, speech, reading, writing, music and the visual arts*. Another meaning of this result is that *linguistic time-reversal invariance* (§IV.10) or *recorded (e.g., written) space-reversal invariance is impossible*. *Information, meaningful words and intelligent thinking are, therefore, essentially asymmetric processes* (in time); *thinking, writing, and communicating symbols generate time asymmetries, direction, irreversibility, initial and boundary conditions, aggregated prints, syntax, lines, pages, books, etc.*

To conclude: The brain's ability to detect *external* asymmetry, stratified aggregation, boundaries, form complexity (see below) influences *internal* thought and perception. Unfortunately such relationships have so far escaped the attention of students of semantics, mind and history. However, once pointed out in association with a central guiding idea these concepts can unify apparently different fields of human thought.

IX.2.8.2 Evolution and Thinking vs. Complexity and Structure

Is there a fundamental dichotomy between material structuring-asymmetry-aggregation and the direction-aggregation associated with thinking? Or are they but two aspects of one and the same process?

Aggregated thinking and aggregated structures, as previously described, *do not* mean simple aggregation, as it occurs, say, in *a cloud of dust*. They mean *time-asymmetric structures*; a directed heterogeneity in which the aggregated elements are *dynamically* interconnected according to a certain pattern, i.e., they are *"organized asymmetries in time"*. *An increase in complexity* occurs when there are more component elements related by a higher level of organization over time. *Higher organization implies evolution of greater centreity and larger associations of the component elements "closed on themselves."*

By the term "closed on themselves" we do not mean simple *"closed surfaces"* like those formed by gravitation (planets, stars, galaxies), nor the simple interfaces between cells, organs, etc.

First we note that the *inorganic* part of organization and centreity may be associated with *large* systems (e.g., clusters of galaxies, galaxies, stars, planets, atmospheres, eco-systems, etc.) or with *small* ones (e.g., atoms, molecules, crystals, etc.). The *biosocial* part of a given structure, organization, or a centreity may also be associated with *large* systems (e.g., civilizations, nations, states, cities, communities, etc.), or with, relatively, *small* ones (e.g., viruses, cells, clusters of cells, central nervous systems, brains, individuals, etc.). This means that any structure possesses an *innate set* of asymmetric boundary-initial conditions, and in this sense it is distinguishable from other objects and "closed on itself."

Secondly we note that *there is a very fundamental tendency in nature for all things of a given class to clump together, thereby forming units of a new class of higher order, organization and centreity. This fundamental property is an innate tendency of all living and nonliving systems. Both worlds tend to become, on sufficiently large space-time regions, more organized in always more complex structures with more boundary-initial conditions.* *

'Large space-time regions' bring us back to the *subject of evolution*. But the increase in *material* organization, structure and complexity represents only one aspect of evolution. Indeed, evolution is directed both toward higher material organization-structure-complexity and toward higher levels of "reflected thinking".

At this point it should be stressed that the law of increasing organization-structure-complexity-thinking *belongs to the realm* of natural science, for it is derived from *direct observation*; i.e., *it is not an abstract speculation!* What do we mean?

That life originated from inorganic matter is, according to modern science, at least the most likely hypothesis. Moreover, a majority of scientists accepts the hypothesis that *the transition* from inorganic matter to life *happened only once*; that the genesis of life (on earth) belongs to the category of *unique* evolutionary events that, once happened, do continue to evolve into numberless structures, forms, plants and animals. Thus, it is a matter of observational evidence, that, in the living and nonliving worlds, evolution means that higher levels of structuring and complexity have gradually been reached.

What happens when a higher level of structure-complexity is reached? At this stage *entirely novel properties appear along the time coordinate of evolution, i.e., the emergence of new properties occur whenever a certain critical point or threshold is reached!*

According to *Pierre Teilhard de Chardin* all *"levels of reality"*-man, organisms and nonliving matter—are *"genetically related."* [+] Chardin divides the history of our planet into three main evolutionary stages:

1. The period during which the earth crust solidifies by cooling. There is not any life yet.

2. The appearance of life, and the gradual development of the living forms in the *biosphere*.

3. During the last two-million years a new development has occurred. Emerging from the biosphere, "Man" entered the world somewhere in Africa. Rapidly, he has populated the earth and has given it a new appearance. The earth has thus been partially covered by a "second envelope", the "envelope of thinking", what Teilhard

---

* This applies also in the realm of symbols, words, sentences and philosophies.

[+] cf. His book, *The Phenomenon of Man*, Harper Torchbook Edition, New York, 1961, and his essay, *Le Group Zoologique Humain*, Albin Michel, Paris, 1956. See also *Biology, History and Natural Philosophy*, A. D. Breck and W. Yourgrau, eds., Plenum/Rosetta Edition, New York, 1972.

calls the *noosphere*. In short, the three main stages in the history of our world are characterized by "matter", "life", and "thought"—the three *"levels of reality"* that we perceive around us; the three levels which constitute our reality—the *geosphere*, the *biosphere* and the *noosphere*.

These three "levels of reality", according to Teilhard de Chardin, are interconnected by a "genetic" relationship, i.e., the evolutionary process is "genetically imbedded" in matter (and gravitation), for the biosphere proceeds from the gravitationally-induced geosphere, and the noosphere develops from the biosphere.

Teilhard de Chardin's central idea is that nature, as we perceive it, possesses two dimensions or aspects—"the *external*" and "the *internal*". The "exterior aspect" is claimed to be constituted by the "external" dimensions of matter and the relationships among the material elements. The "internal" aspect of reality is the phenomenon of thinking (or "consciousness"), which he claims is *"coextensive"* with matter. Thinking and material structuring are accordingly claimed to be two aspects, or interconnected parts, of one and the same phenomenon. Hence a greater level of (knowledge of) material complexity-structure is always accompanied by a higher degree of thought (or "consciousness"). The "internal" aspect of all things, i.e., our thinking, is, according to Teilhard, a genetic property that must be (deterministically) incorporated in any phenomenological description of the cosmos. The evolutionary law (of increasing complexity—thought) is the core of Teilhard's theory. Being a necessitating law, it allows not only to understand the past but also to estimate future evolutionary trends.

IX.2.9 Spinozism and the External Origins of

"Innate Time Perception"

The evidence about biological clocks, the simple orientation movement that is correlated with the asymmetry of gravity and with geophysical processes, and the gravitational evolution and selection involving the genesis of atoms, molecules, amino acids, nucleic acids, DNA, RNA, enzymes, proteins, etc. — what do they tell us about the possible *gravitational* origins of the so-called *"innate" properties?* Is there any additional, philosophical or empirical support to this unorthodox hypothesis?

In trying to answer this question, I first return to historical rationalism, notably to *Spinoza's philosophy*.

To Spinoza, the habit of observing objects and processes *in isolation, apart from their connection with the vast whole*, is in contradiction with logic and all evidence. Thus in Spinozism *the whole may be different than the sum of its parts*. Indeed, general relativity supports this conclusion (see Lecture III): But only in principle, not in every day experience. Nevertheless, this is a most universal principle and it can be used as a valuable guard against false reductionism and simplistic explanations. In physics we know that the effects, however small, of strong, electromagnetic, weak and gravitational fields, and, in particular, the effect of neutrinos must all be taken into account in any attempt to form universal *conservation laws* and unified theories (see below). Thus, according to this line of rationalism, all physical, and perhaps even all biosocial, and philosophical thinkings must not be studied apart from each other or *apart* from their connection with the vast whole. (See also our assertions in the Introduction).*

Moreover, in the common interpretation of science the habit of studying natural objects and processes in isolation, apart from their connection with the vast whole, leads to misleading reductionism, to the belief that *all complex phenomena can be explained in terms of combinations of simple, "innate" or isolated ideas. Thus, Spinozism is a valuable guard against the devastating reductionist tendencies to understand nature by "building up" from "simple", "innate", or isolated "states" to the vast complicated whole and never in reverse!* But it also guards against the common tendency to emphasize one field, such as physics, at the expense of all other sciences. Rationalism is thus a general methodology which *requires* a two-way street, from simple to the complex whole *and* vice versa, i.e., in the helical, *"feedback"*, evolutionary process of accumulating and comparing rational and empirical knowledge from all fields of science and philosophy (cf. Assertion 10 in the Introduction).

For instance, the origin of life cannot be isolated from geophysical processes and from the vast whole; *nor can it be totally reduced to a set of isolated physicochemical processes!*

In fact, much of the evidence that has been stressed so far in this lecture confirms Spinoza's rational claims for a *definite beginning in cosmic time and for a cosmic evolution.* Looking back over Spinoza's system we now recognize, on the most general level, that all studies must incorporate terms associated with their *origin, evolution, causes* and *histories.* Consequently, examinations should not only be confined to the *"present state"*, but should also include *"direction"*, *origin* and *time rates of change.* We must, therefore, understand the universe in terms of history, evolution, gravitation, time anisotropies, and prediction. Here the roles of time, symmetry-asymmetry, structures and irreversibility must always be emphasized. But we have not left Spinozism with this short note. We shall return to it later in our discussion of Einsteinism.

IX.2.10 From Reason to Kant and Back: A Guide for the Future?

We return now to the discussion of "innate ideas" and in particular to the concept of "innate ordering of space and time" as it appears in a few important doctrines in the history of philosophy.

---

* To demonstrate the practical result of this principle one may, for instance, compare it with *"The First Law of Ecology"*, namely: *Everything Is Connected to Everything Else* (Appendix VI).

Like Descartes, Spinoza, Leibnitz, Locke, Berkely, Hume and other philosophers, *Immanuel Kant* (1724-1804) asked: *How does human knowledge arise?*

It must be pointed out from the outset, however, that for Kant the attempt to answer this question, to study the inherent structure of human memory and mind, belongs to the domain of "transcendental philosophy", for, according to Kant's premises, it is a problem of "transcending sense-experience".

"I call knowledge transcendental which is occupied not so much with objects, as with our *a priori* concepts of objects", he wrote.

The first stage in this mental process, according to Kant, is the *"coordination of sensations"* by applying to them the *"forms of perception"*—space and time! The second stage is the "coordination" of these perceptions by applying to them the "forms of conception"—the *"categories"* of thought.

Thus, according to Kant, the human mind initially employs two simple faculties to classify the material presented to it: *the sense of space and the sense of time.* Therefore, for Kant, they are *a priori to all ordered experience,* which involves and *presupposes* them. Without them "sensations" could never become "perceptions". Thus, space and time are considered *a priori* for Kant and for his adherents because they considered these forms of perception *independent of experience* and *not* given to modification by science!

But in Lecture III we demonstrated that, *according to general relativity, there is no requirement to distinguish, fundamentally, between changes in the geometrical properties of space-time and changes in physical processes, i.e., Einstein and experience itself have modified our concepts of physical space and time. Does this refute Kant's claim of a priori knowledge? The answer to this question is, as we shall see below, quite subtle.*

Our first tendency is to refute Kant's pretensions to *a priori* space and time, for in general relativity, matter acts on space causing it to curve and, in turn, space acts on matter and is "coupled" to it!! *The Kantian isolation of spatial concepts from material concepts must, therefore, be rejected as an impossibility!!* Hence, the Kantian isolation of "*a priori* knowledge", e.g., space and time, from the rest of experience is refuted by modern physics. In general relativity space and time are no longer *the Newtonian "stages" upon which physical processes are displayed.* Thus, *the arrival of relativity brought about the total collapse of (absolute) Newtonian space and time and the rejection of their alleged a priori properties* (Kant's doctrine of the *"thing-in-itself"* is also faced with similar difficulties, which already for Fichte had become a monstrosity). Yet, in spite of its inconsistencies, many Western philosophers still adhere to various versions of Kantianism or neo-Kantianism.

Is this adherence based on religious grounds, on traditional thought, or on pure scientific ignorance? Moreover, *has the development of Western thought been hindered in any sense by Kantianism?*

Never has Western thought been faced with an irrational caricature of the world such as that hurled at its speculative roost by the popular philosophies of Kant,

Schopenhauer, Kierkegaard and Hegel. The route from the philosophies of Descartes, Spinoza and Leibnitz (see below) to Kant's ruling position in the speculative roost of Germany, is the route from theoretical reason without archaic religion to archaic religion without theoretical reason. It is in Kant that reason drastically turned in upon itself.

Actually, the attack began with John Locke (1632-1704). But reason received its greatest blow from Kant's *Critique of Pure Reason*. The development of subsequent irrational movements, such as those of Schopenhauer and Kierkegaard, was then inevitable. Yet, as we shall see below, Kant's theory was an important step in the history of philosophy. To proceed, I shall only introduce a minimal amount of terminology associated with *a priori* knowledge, brief remarks on Kantianism, rationalism, etc., which should not be mistaken for a review of the subject.

First, one must note that Kantianism is also variously known as *"the critical philosophy"*, *"criticism"*, *"transcendentalism"*, or *"transcendental idealism"*. A few additional points must be clarified with respect to *a priori* knowledge.

1) By *a priori* knowledge Kant does not mean knowledge which is "relatively" *a priori*; that is, in relation to this or that experience or to this or that kind of experience. He is thinking of knowledge which is *a priori* in relation to *all* experience. Yet, in a *temporal sense* of the word, he is *not* assuming the "presence" of knowledge in the human mind before it has *begun* to experience anything at all. Kant believes that human knowledge *"begins with experience"* because for the "cognitive faculty", as he puts it, to be "brought into exercise", our senses must be "affected by objects". (Remember the visual cliff experiment!) Thus, given sensations, the "raw material of experience", the mind can set to "work". *So far as these claims are concerned, empirical science appears to support them!!* But, Kant also claims that even if no knowledge is temporally antecedent to experience, it is possible that the "cognitive faculty" supplies *a priori* elements "from within itself!" And it is in this sense, according to Kant, that the *a priori* elements would be *"underived"* from experience.

"Personal judgment" for Kant is, therefore, not a generalization from experience, nor does it stand in need of experimental confirmation before its "truth" can be known! It is on this point that Kantianism diverts most from empirical science. Here, Kant was convinced by Hume's discussion of the principle of causality (see also our remarks in Appendix IV), namely that the "element of necessity" in judgment cannot be justified on purely empiricist lines.

2) We must also keep in mind the fact that in prerelativistic times it was impossible for a philosopher to conceive of any future experience that could change his (essentially Newtonian) concepts of space, time and causality. Thus, it is no wonder that space and time are for Kant *a priori "truths" "before experience"*. To conclude, in Kantianism space and time do not depend on experience, past, present or future. Therefore, they are *"absolute and necessary truths"*. For Kant, it is not merely

"probable", it is "certain", that we shall *never* find a straight line that is not the shortest distance between two points.

Having read Lecture III on Einstein's special and general theory of relativity, as well as Lectures IV to VIII on time anistropies, gravitational fields, curved space-time, modern thermodynamics, etc., one cannot but consider the Kantian dogmata as historical steps toward irrationalism.

Yet Kantianism helps us sharpen modern ideas about space, time, causality, time anisotropies, the origin of knowledge, memory, information and the mind. Philosophy today must be different and more profound, not only because Einstein lived, but also because Kant preceeded Einstein, and because Kant presented problems that must be answered by experience itself.

While the road from Kant to modern thought is strewn with the wrecks of Kantian philosophy lying in the wake of advancing relativity, biophysics, psyco-physics, brain theory, physics, astrophysics, and astronomy, most of Kantian dogmata have not yet collapsed and fallen from their sacramental throne in the Western culture.

Fichte (1762-1814) was the first great critic of Kant. He assumed that conscious-ness, including the representations of physical objects that make up the outer world, is the prodcut of one ultimate cause in the universe. For his thinking was a wholly practical affair, a form of *"action"* (see also Appendix III).

In Fichte's view, the Kantian doctrine of the *"thing-in-itself"* had to be elimi-nated in the interests of idealism.* In his opinion, Kant attempted to have things both ways at once and therefore, ran into hopeless inconsistencies. If the *"thing-in-itself"* is eliminated, it follows that the *"subject"* creates the *"object"* in its entirety and the theory that the subject creates the object becomes a pure metaphysical dogma.**

The history of philosophy testifies that the seeds of what is called German spe-culative idealism, essentially from Fichte to Hegel, were already present in the Kantian system. There is, of course, a difference in atmosphere and interests between Kantian metaphysics and the speculative idealism which followed it in Germany.

The idealistic and subjectivist aspects of Kant's metaphysics were submitted to critical examination by the idealist metaphysicians from Fichte to Hegel. Then, in the middle of the nineteenth century, the cry was raised 'Back to Kant!', and

---

* Kant claimed that within the realm of experience we cannot know "things-in-themselves". We only know things as "phenomena". (See also below).

** Kant's assertions about the subject's construction of experience are based on the individual subject. He introduces the concept of the *"transcendental ego"*, the "I" which is *always "subject"* and *never "ob-ject"*. But if one transforms this into a principle which "creates the object", one can hardly identify it with the "individual ego" without being involved in *solipsism* and a full-blown mythological system, since for one person all other human beings will be "objects", and so they will be "his own creation", and so on, till one is driven in the end to interpret them as the "universal infinite subject" which is pro-ductive both of the "finite subject" and of the "finite object". Indeed, a hopeless doctrine.

Neo Kantians have since tried to develop a "critique" of Kant's *"Critique"* but without falling into what they regarded as the fantastic extravaganzas of the speculative idealists.

Even the Neo-Kantians admitted that Kant had deliberately substituted a new form of metaphysics (of knowledge and of experience) for the older rationalism and empiricism which he had tried to reject.

Kant was brought up on the scholasticized version of Leibnitz's philosophy as expounded by Christian Wolff (1679-1754) and his successors. The *rationalism* of Leibnitz, Spinoza and Descartes (see below) *was* known to him and he called it *"rotten dogmatism"*. He was also exposed to Hume's *empiricist criticism*, which he attempted to refute.

The Kantian *a priori* doctrine can be represented as a development of Leibnitz' theory of virtually *"innate ideas"*, with the difference that the ideas became innate "categorical functions". In addition, Kant employed Hume's ideas about a "subjective contribution" to the formation of certain complex ideas, such as that of causality.

Consequently, Kant's *a priori* doctrine was also influenced by Hume's position on Newtonian physics, which presents a number of synthetic *a priori* propositions. It is here that he went astray from reason and ended up in a sort of irrationalism. Thus, in attempting to combat rationalism and empiricism with his own moral religion, he was driven to irrationalism, adding to it some archaic concepts of "moral freedom".

Kant himself was aware of his difficult position in evolving a consistent theory. He tried in vain to find consistent formulas which would save him from self-contradiction but which at the same time would enable him to retain the concepts of "moral choice" and *a priori* knowledge. All these attempts inevitably led to a transition from theory of knowledge to an irrational metaphysics based on his own beliefs. Indeed, he allowed for belief in the existence of *"noumenal reality"* detached from any sense-experience. But this was inconsistent with his own account of the function of the *"categories"*, for his "categories" have content and meaning *only* in their application to "phenomena".

Hence, it is meaningless, according to Kant's premises, to conceive of "noumenal reality" or of a "supersensuous substrate" as existing. In fact, if reality is itself one of his categories, it is nonsense to talk about noumenal reality at all. Therefore, according to Kant's premises, we are not permitted to employ "moral judgment" as the basis for any kind of metaphysics.

IX.2.11 The Beginning of Time, Cosmology and Causality in Kant's Thought

The failure of Kant's dogmatic system may also be exemplified by his speculations concerning antinomies, thesis-antithesis and the beginning of time.

Many thinkers before Kant, including *Thomas Aquinas* (1225–1274), whose defense of Christianity has been officially declared by the Roman Catholic Church as the basis of theological studies, and Baruch Spinoza (see below) composed philosophical demonstrations about the question of the beginning of time. The latter employed the geometrical system to prove that "There was no Time or Duration before Creation".* The former, on the other hand maintained that it had never been philosophically demonstrated either that the world had a beginning in time or that it had no beginning in time. (Modern astronomy, as we stressed in Lectures III and VI, resolved the dispute by proving that Spinoza was right.)

Kant claimed that any thesis can be proved through the use of synthetic *a priori* propositions. His procedure deliberately leads to a contradiction between two ideas, that is to say to an *antinomy*. He demonstrated a few antinomies. The first one was as follows:

"Thesis": The world has a beginning in time and is also limited in regard to space.

"Antithesis": The world has no beginning and no limits in space, but is infinite in respect to both time and space.**

Then Kant demonstrates his "proofs" for both "thesis" and "antithesis". However, both his "proofs" rest on false and unwarranted assumptions.*** What Kant intended to prove was that one can *avoid* the antinomy *only* by adopting the view of *his* critical metaphysics and by *abandoning the method of rationalism* (though it can hardly be claimed that he does so very clearly). On philosophical grounds, as contrasted with empirical evidence, one must follow neither proposition, accept both, or claim that while one is invalid, the other is valid.

It was only in recent years that accumulative empirical evidence about the (expanding) world accessible to our observations has led to the conclusion of a definite beginning in cosmic time. (See also Lecture III, Lecture VI, Appendix I and the Introduction).

Thus, unlike Kant's methodology, it is Spinoza's assertion about a definite cosmic time that has been finally vindicated by modern science. In fact, as we shall see below, a large part of Spinozism has survived the "test of time" . . . Consequently, most of Kant's celebrated methodology—not his beautifully-written books—should now be committed to the flames of history.⁺

* Baruch Spinoza; *Principles of Cartesian Philosophy* (English Trans. by H. E. Wedeck), The Wisdom Library; Philosophical Library, N.Y., 1961; Ch. 10; pp. 175, also pp. 138 & 73.

** Kant's 2nd edition of the *Critique of Pure Reason*, as edited by the Prussian Academy of Sciences; pp. 454 in the original German edition. See also N. K. Smith's English translation; London; 1933.

*** See, for instance, N. L. Smith; *A Commentary of Kant's 'Critique of Pure Reason';* London; 1930; (2nd edition) pp. 483–506.

⁺ Kant's second antinomy, like the first, rests on false assumptions. It is as follows: *'Thesis': Every composite substance in the world consists of simple parts, and there does not exist anything which is not either itself simple, or composed of simple parts. 'Antithesis': No composite thing in the world consists of simple parts, and there does not anywhere exist any simple thing.* (C. P. R. pp. 462–3). Here the "thesis"

In closing this section, I want to add that the fundamental contradiction between Kant and modern science is frequently clouded by false theological rhetoric. For example, the late C. P. Steinmetz, addressing the Unitarian Church in Schenectady, N. Y., claimed in 1923 that "Modern physics has come to the same conclusion [as in Kant's theory of knowledge] in the relativity theory, that absolute space and absolute time have no existence, but time and space exist only as far as things or events fill them; that is, they are forms of perceptions." One may smile at such wordplay; but one cannot forget that the entire superstructure of Kant's metaphysics was indirectly devised to save Christianity and German idealism.

IX.2.12 The Arrival of the Hegelian System

Returning now to German idealism, one may ask: Who were Kant's main successors?

J. G. *Fichte* (1762-1814) and F. W. J. *Schelling* (1775-1854) cashed in on Kant's obscurity and spun their own webs of metaphysics. But it was left to G. W. F. *Hegel* (1770-1831) to outdo Kant in his writing; indeed, Hegel's works are masterpieces of obscurity shrouded in a weird terminology which, in Hegel's words was *"an attempt to teach philosophy to speak in German"*. What he eventually succeeded in doing was to teach communists to speak hegelianism, but "in reverse". (Karl Marx (1818-1883), assuming that he himself stood on a safe physico-philosophical ground, claimed to have found Hegel standing on his head and put him back on his feet).

Thus, it was through Marx and Engels that German idealism started its reversed transformation into dialectical materialism. It was from Heraclitus that Hegel borrowed the idea that "struggle" is the "law of growth".* Not an intellectual struggle, not even scientific revolution, only the physical struggle that is supposed to be the stress of the world. He believed that a man reaches his full height only through compulsions and suffering. He also maintained that "The history of the world is not the theatre of happiness; periods of happiness are blank pages in it, for they are periods of harmony". For him, history is a dialectical movement, almost a series of violent revolutions. He wrote about *Zeitgeist*, the *Spirit of the Age*, but his enemies called him *"the official philosopher"*, for he allied himself with the Prussian

consists of simple substances, such as the *Leibnizian "monads"* and the antithesis represents the empiricist attack on Leibnizian dogmatic rationalism. But the antithesis represents also the rational standpoint of *Spinozism*. Using the rational method, Spinoza demonstrated that we cannot reach complete knowledge about a given object in itself *for it cannot be isolated from the rest of nature*. (Therefore, complete knowledge about a given object is *unattainable for this would involve the whole of nature*). Here again modern science *sides with Spinoza*. Similar conclusions apply to Kant's assertions about "free causation" and causality (his third antinomy, and, to some extent, his fourth). For the evidence supplied by modern science about these concepts see Lectures IV to VIII. For further readings on Kant see, for instance, W. T. Jones, *Morality and Freedom in the Philosophy of Immanuel Kant*, Oxford, 1940; A. H. Smith, *Kantian Studies*, Oxford, 1947; E. Teale, Kantian Ethics, Oxford, 1951; *A Symposium on Kant*, Tulane Studies in Philosophy, Vol. III, New Orleans, 1954; F. Copleston, *Kant in A History of Philosophy* Vol. 6, Part II, New York, 1960; J. A. Shaffer, *Philosophy of Mind* (1968); G. Ryle, *Concept of Mind* (1949).

* cf. Appendix II.

government. While he devoted himself vigorously to philosophical business, his industry on behalf of Christianity was considerable, e.g., his writings on the life of Jesus in which he attempted to reinterpret the Gospel along Kantian lines and his answer to the question of "how Christianity became the authoritarian religion", namely that the teaching of Jesus was not "authoritarian" at all, but essentially "rationalistic". Above all, he was inspired by a doctrine of the "Holy Spirit", and the religious bias animated the whole of his work. It is no wonder that with all his royal and religious honors he began to think of the "Hegelian system" as the system of the natural laws of the world. So did Germany and all his "forward" and "reverse" followers abroad.

Hegel and Spinoza

Even if almost all of Hegel's doctrines are false, his philosophy still retains important refinements of some previous philosophies. For instance, *Hegel adopted Spinoza's philosophy about the reality of the whole as opposed to the incompleteness of separate things, units and systems. The whole, in all its complexity, is called by Hegel "the Absolute"*. But, he differed from Spinoza in conceiving the whole as a sort of spiritual complex system, not as a simple "substance" or "extension". Hegel, therefore, asserts a disbelief in the reality of time and space as such, for these, according to Hegel, *if taken as completely real, involve separateness and multiplicity!*

This conclusion is partially supported by our doctrine of the origin of time as expounded in Lecture VII from the viewpoint of modern general relativistic cosmology.

In this connection we must also note that when Hegel says "real" he does not mean "the real" that an empiricist has in mind for he asserts that, in a sense, the real is rational, and the rational is real. He therefore claims that *empirical facts must be taken by the empiricist as irrational as long as their apparent character has not been transformed by viewing them as integral aspects of the whole. It is only then, according to Hegel, that empirical evidence becomes rational and, consequently, real. Thus, Hegel asserts that throughout the whole evolution of empirical science there must be an underlying axiom, namely, that nothing can be really universally valid unless it is about Reality as a whole. Spinoza–Hegel underlying axiom has also gained some support from traditional logic*, which assumes that every proposition has a "subject" and a "predicate". Therefore, every fact consists of something having some property. Hence, *"relations" cannot be "real" since they involve two things, not one. Since everything, except the whole, has "relations" to outside things, it follows that nothing universally valid can be concluded about separate objects, and therefore only the whole is real, and any separateness is unreal! Accordingly, the world is not a collection of hard units, whether atoms or "elementary particles, each completely self-subsistent". It follows that the self-subsistence of finite things and finite systems that some modern sciences now advocate are nothing but temporary illusions. To conclude: We can easily recognize that all these axioms are essentially resting on the foundations of Spinozism (see below).*

Hegel and the Theory of Knowledge

How does human knowledge evolve? According to Hegel, knowledge, as a whole, begins with sense-perception in which there is only primitive awareness of a finite object. Then, through skeptical criticism of the senses, it is transformed into a second dialectic movement and becomes highly subjective. Finally, in the third dialectic movement, *knowledge reaches the stage in which subject and object are no longer distinct, for the highest kind of knowledge must be that possessed by "the Absolute", and as "the Absolute" is the "whole" there is nothing outside itself for it to know.*

"The Absolute", says Hegel, *"is a Pure Being"* (thesis); we believe that it just is.* But "Pure Being" without any qualities is "Nothing"; hence, according to Hegel *"The Absolute is Nothing"* (antithesis). Thus, from this thesis and antithesis Hegel moves to the *synthesis*; namely, the union of "Being" and "not-Being" is *"Becoming"*, and consequently, *"The Absolute is Becoming"*. But this conclusion is also not valid for there has to be finite something that "becomes". Thus, according to Hegel, *our knowledge of "Reality" develops by the continual correction of previous errors all of which originate from separateness in which we consider something finite as if it could be the whole thing.*

Hegel asserts further that, in this evolutionary process of human knowledge, each later stage of the dialectic contains all the earlier developments. However, none of these stages are superseded entirely; each is given its proper time and place in the evolution of our knowledge of the "Whole". Consequently, it is impossible to reach the "truth" except by passing through all historical movements of the dialectic. Each of these historical movements may not be entirely false, unless the movement attributes *"absolute truth"* to some *isolated* pieces of information.

Hegel and the Philosophy of History

Hegel's interpretations of history, "freedom" and politics is developed in his *Philosophy of History* and in his *Philosophy of Law*. (They were published together

---

* The various possible meanings of "Being", "Nothingness" and "Becoming" have been discussed by an almost endless array of "philosophers of the self", notably by Jean-Paul Sartre, see his *Being and Nothingness,* Philosophical Library, N.Y., 1953, and *Existential Psychoanalysis,* Philosophical Library, N.Y., 1953. The pursuit of "Being" has led to such contemporary concepts as "essence", "Being-for-Itself", "Being-in-Itself", "Being-for-Others", "In-Itself", "For-Itself", "Divine-Being", etc. While Sartre's books contain the tenets of his own, often criticized philosophy, they nevertheless reveal a highly penetrating analysis of the Hegelian Logic and terminology which are frequently difficult to understand. In this connection it may be easier to say what "Being is not" as in Hegel's words: "It does not matter what the determination or content is which would distinguish being from something else; whatever would give it a content would prevent it from maintaining itself in its purity. It is pure indetermination and emptiness. *Nothing* can be apprehended in it." (Taken from Sartre's *Being and Nothingness,* pp. 17. Here Sartre adds that the Hegelian "Being" is not one 'structure among others', one moment of the object; it is the very condition of all structures and of all moments. pp. 14, see also pp. 15 to 30, 285 to 309, and his *"Key to Special Terminology"*, pp. 769 to 777 in the Washington Square Press edition, 1966).

with his two *Logics* and other works in the *Encyclopedia of the Philosophical Sciences*;
H. Glocker, ed., 2b vol. Stuttgard, 1927-40. A later edition (J. Heffmeister, ed.)
was published in Hamburg in 1952).

As Hobbes and Spinoza before him, Hegel emphasized the *State*. But he glo-
rified it. For him "The State is the embodiment of rational freedom, . . . " Against
Kant he maintained the impossibility to eliminate wars. The state of war has a
positive moral value to Hegel: "War has the higher significance that through it
the moral health of peoples is preserved in their indifference toward the stabilizing
of finite determinations". *Hence, if accepted, Hegel's doctrine of State justifies
every tyranny that one can imagine. In fact, nations, in the Hegelian doctrine, play
the part that classes play in Marxism.*

The principle of historical development, according to the Hegelian doctrine, is
national genius. In fact, he claimed for the State much the same merits that St.
Augustine and his successors had claimed for the Church.

Here Hegelianism and Marxism differ only in terms of the definition of causes.
In Hegelianism war is the prerequisite for positive moral health, dialectic mission,
solution of national conflicts, progress of national genius, etc. In the other doctrine,
all wars have economic causes and became "class struggles" in a dialectic "mission"
of world "revolution and progress". However, in Hegelianism, there is no consistent
philosophical justification for the emphasis on the State as opposed, say, to world
government or other social organizations which Hegel takes as absolutely false.

In his *Philosophy of History* Hegel stresses the roles of Idea, Reason and Spirit.
" . . . The Idea" he says "is, in truth, the leader of peoples and of the world" and
"Spirit . . . is, and has been, the director of the events of the world's history".
"The only thought which philosophy brings with it to the contemplation of history
is the simple conception of Reason; that Reason is the sovereign of the world;
that the history of the world, therefore, presents us with a rational process. This
conviction and intuition is a hypothesis in the domain of history as such. In that of
philosophy it is no hypothesis".

Here the Hegelian meaning of Spirit is the opposite of matter. Then, in describing
the historical development of Spirit, Hegel sees three main phases, namely: The
Orientals, the Greeks and Romans, and finally, who else, but the Germans . . .

Most of Hegel's doctrine is therefore an attempt to unify opposites, e.g., Spirit
and Nature, Ideal and Real, while at the same time it synthesizes contradictory
philosophies of his predecessors, e.g., Spinozism and Kantianism. Small wonder
that conservatives and revolutionaries, believers and atheists have professed to
draw inspiration from Hegel.

IX.2.13 Neo-Hegelianism and Positivism

Hegel's influence has almost vanished for some time after it had first dominated
German universities for some years after his death. Then, in the latter part of the

19th century, it blossomed anew in England and Scotland in the work of such writers as *Benjamin Jowett, T. H. Green, Edward Caird, J. M. E. McTaggart, and Bernard Bosanquet.*

This success was in part responsible for the revival of Hegelianism in Germany, especially under the influence of *Georg Lasson and Wilhelm Dilthey*, and of *Benedetto Croce* in Italy.

During his student days at the universities of Bonn and Berlin, Marx studied history and philosophy and was captivated by the Hegelian system, an influence that always remained one of the most important elements in Marxism. It has been the growing influence of Communism which encouraged many Western thinkers to study Hegel's philosophy of history and politics, as well as his "Logic", because of their influence on Marxism.

In closing this section it should be stressed that Hegel's philosophy of history is partly the effect and partly the cause of the teaching of political science and world history in the German schools of the Pre-World War II era.

One last remark about the connections between Hegel, Spinoza and Kant. *In the following sections I shall draw attention to the somewhat unorthodox thesis according to which the philosophical connection between Hegel and Spinoza is stronger and much more important to the history of philosophy than the philosophical connection between Hegel and Kant.* Both Spinoza and Hegel start their philosophies with an amazingly large number of the same premises. Yet, it is the strong Kantian, idealistic and nationalistic view of Hegel which cause him to end up at some *irrational* conclusions that are *almost diametrically opposed to those of Spinoza.*

But German idealism could not resist the challenges of rationalism and empiricism for a long time. Kantianism, Hegelianism and the like succeeded only in postponing the inevitable confrontation with modern science.

First came the *positivist movement* which, during the last quarter of the nineteenth century exerted a great influence in central and western Europe. At this time Germany was completely under the influence of various versions of *Kantianism*, whose status was almost that of a *state religion*. As a result positivism was largely developed as a critique and rival of Kantianism. *And for these very reasons it was first developed outside Germany, in the universities of Vienna and Prague.* Central European positivism chiefly centered around the Austrian *Ernst Mach*(1838-1916) and in France around *Henri Poincaré* (1854-1912).*

---

* About this time pragmatism appeared in the United States. Both movements interacted later and continued to evolve into newer movements called logical positivism, positivism-pragmatism, linguistic studies, instrumentalism, logical empiricism, etc. (see Appendix III).

Mach, reviving the Humean doctrine, contended that all factual knowledge consists of concepts and elaboration of what is given in the data of immediate experience. In keeping with Comte** he repudiated Kantianism. To the question "What would be left over if all the perceptible qualities were stripped (in thought) away from an observable object?" Mach and his adherents answered, "Precisely nothing".

However, in tracing the origins of Einstein's revolution, we must first mention the rationalist Leibnitz, for he had incisively criticised Newton's concepts about space and time. Thus it was Leibnitz, and to some extent Spinoza (see below), who paved the way for the revolutionary concepts of space and time as exclusively a matter of 'relations' between events. Mach, however, went still further in attacking the arguments of Newton in favor of a *dynamic* space and time. First he carried out a thorough historical and logical analysis of Newtonian mechanics and then demonstrated that it contains no principle that is in any way self-evident to the human mind. He contended that Newtonian mechanics are based on "absolute space" and "absolute time" which *cannot* be defined in terms of *observable* quantities or processes. To eliminate such concepts from science and philosophy, Mach raised the demand (which is now frequenetly called the *"positivistic criterion"*), that *only these propositions should be employed from which statements regarding observable phenomena can be deduced.****

Mach maintained that the inertial and centrifugal forces that arise in connection with accelerated or curvilinear motions had been grasped by Newton as effects of such motions with respect to a *privileged reference medium* imagined as an absolute Cartesian coordinate system graphed upon a real space. Instead, he stressed, *any privileged reference system must be generated by the total mass of the universe, i.e., by all stars, galaxies, etc. It was this universal idea that later served as one of the important starting points for Einstein's theory of gravitation* (Lecture III).

Einstein borrowed from Mach another universal principle that might be called the principle of economy in a theory. By this general assertion Mach demanded simplicity and economy of thought in any physical theory. Thus, according to Mach, *the greatest possible number of observable facts should be organized under the fewest possible principles.* Nevertheless, Einstein was highly critical of Mach's positivistic claims that the general laws of physics are only summaries of experimental results. In his opinion, as well as in his theory and practice, he stressed that *general laws are to be tested by experience, but nevertheless owe their origin to the inventive human mind.*

** Positivism first assumed its distindctive concepts in the philosophy of Auguste Comte (1798–1857) (see Appendix III).

*** For more details see the brilliant and lively description of P. Frank: *Einstein: His Life and Times*, A. A. Knopf, New York, 1972, written in 1947 and 1953. Professor Philipp Frank, one of the leading U. S. scientists-philosophers (born in Vienna in 1884), died in Cambridge, Mass. in 1966.

At this point, at first sight, Einstein seems to side with Kantianism. In fact Einstein liked to read Kant because through him he became acquainted with many of Hume's ideas, and because this reading often aroused emotions and meditations about the world.* He did not share Kant's beliefs and read such works as other people listen to sermons, religious music, or to moral instruction. Nevertheless, the role of mind, as stressed by Kant, as well as by Spinoza (see below), had influenced Einstein's thoughts about the origin of new scientific theories.

It may be instructive to quote here Frank's decription of Einstein's interest in philosophy.**

"Einstein read philosophical works from two points of view, which were sometimes mutually exclusive. He read some authors because he was actually able to learn from them something about the nature of general scientific statements, particularly about their logical connection with the laws through which we express direct observations. These philosophers were chiefly David Hume, Ernst Mach, Henri Poincaré, and, to a certain degree, Immanuel Kant. Kant, however, brings us to the second point of view. Einstein liked to read some philosophers because they made more or less superficial and obscure statements in beautiful language about all sorts of things, statements that often aroused an emotion like beautiful music and gave rise to reveries and meditations on the world."

IX.2.14 Critical Rationalism and Popper's Rejection of Irrationalism

As we move away from Kantianism and Hegelianism toward Spinozism, Einsteinism and dialectical gravitism, I now make a short stop to inquire about the views expressed by one of the greatest thinkers of our time—*Sir Karl Popper*.

With the rise of irrationalism, totalitarianism, and anarchism, Karl Popper (1902-   ) was led to examine the methods of the social sciences, and in particular *why* they are backward.

Popper concluded that it was essential to see how the irrationalist fallacy arose. He was led to a critical evaluation of the intellectual leaders of mankind, particularly Plato and Marx, in order to show why the conflict between rationalism and ir- rationalism *"has become the most important intellectual and perhaps even moral issue of our time . . . "*

Rationalism for Popper "is an attitude of readiness to listen to critical arguments and to learn from experience". He dinstinguishes between two rationalist positions, which he labels *"critical rationalism"* and *"uncritical rationalism"*.*** He insists that "whoever adopts the rationalist attitude does so because *without* reasoning

* See P. Frank's book, *Loc. Cit.*, pp. 51–52 which also stresses the fact that Einstein liked to read Schopenhauer but without, in any way, taking his views seriously. In the same category he also included Nietzsche.

** *Loc. Cit.*, pp. 50.

*** K. R. Popper: *The Open Society and its Enemies* (Princeton University Press, Princeton, 1950) pp. 431.

he has adopted some decision, or belief, or habit, or behavior, which therefore in its turn *must be called irrational.* Whatever it may be, we can describe it as an irrational *faith in reason . . . "*

"The choice", says Popper is therefore "a moral decision . . . For the question whether we adopt some more or less radical form of irrationalism, or whether we adopt that minimum concession to irrationalism which I have termed 'critical rationalism', will deeply affect our whole attitude toward other men, and toward the problems of social life".

Popper's reasoning in regard to 'critical rationalism' is highly illuminating in the humanities, but may be somewhat limited in the physical sciences; for even the very choice of the words 'rational' or 'irrational', 'faith' or 'physics', becomes a 'moral decision'. Nevertheless, the facts presented in Lecture V support Popper's view even in the domain of physics, especially when different ('rational') interpretations of the same formalism (or symbols) are involved. Otherwise a 'moral decision' cannot, by itself, determine rational arguments based on verifiable experience. But my main objections are to Popper's introduction of indeterminism into rationalism. Two of these are mentioned below:

1) The history of rationalism begins with the Eleatics, Pythagoreans and Plato, whose theory of the self-sufficiency of reason became the leitmotif of neo-Platonism. But it is in the philosophies of Descartes, Spinoza and Leibnitz that rationalism reaches the summit of intellectual achievement. Indeed, all first-rank philosophers and scientists—Descartes, Spinoza, Leibnitz, Kant, Locke, Hume, Hegel, Mill, Alexander, Planck, Einstein, as well as many others—were determinists in the sense of admitting the cogency of the arguments of strict causality and determinism. Moreover, in Lecture V we have demonstrated the fact that, contrary to common belief, modern quantum physics does not (and cannot) reject determinism.

2) Rationalism as a classical method, or very broadly, as a theory of philosophy, requires no protection. It is usually associated with attempts to introduce mathematical, experimental, and critical methods into science and philosophy. At the same time it rejects mystic, asystematic, anthropomorphic, geocentric, emotional, and subjective criteria, as well as 'undetermined or uncaused choice' (cf. §3, Introduction and §IX.1).

But at this point, before we enter the intellectual garden of Spinoza, we may use Popper's critique of rationalism as a guide, and perhaps even as a warning against falling into some emotional traps:

> The irrationalist insists that emotions and passions rather than reason are the mainsprings of human action . . . It is my firm conviction that this irrational emphasis upon emotion and passion leads ultimately to what I can only describe as crime . . . The division of mankind into friend and foe is the obvious emotional division; and this division is recognized in the Christian commandment, 'Love thy enemies!'. Even the best Christian who really lives up to this commandment (there are not many, as is shown by the attitude of the average good Christian toward 'materialists' and 'atheists'), even he cannot feel equal love for all men."

Thus, according to Popper, the appeal to emotions "can only tend to divide mankind into different categories"; "into those who belong to our tribe, to our emotional community, and to those who stand outside it; into believers and unbelievers; into compatriots and aliens; into class comrades and class enemies; and into leaders and led . . . "*

IX.2.15 On Spinozism, Einsteinism and Dialectical Gravitism

How Should Spinoza be Studied?

More than any other intellectual activity, philosophy derives its stimulus from unbounded and undisciplinable sources. Nothing is freer than philosophy. Even the philosophy which denies "free will" excersises this freedom. Such a philosophy is to be discussed below.

We have already emphasized that our approach is strictly deterministic, even in the realms of cybernetics, information theory, modern quantum mechanics and microphysics. Hence our theory denies the possibility of "caprice", "chance" and "free will" without an inherent cause. At this point one may be tempted to ask a few fundamental questions:

1) Does the denial of "free will" mean the denial of social and moral responsibility and of the appropriateness of ethics?

2) What do we mean by freedom? And, in particular, what do we mean by freedom of thought, free press, free speech, academic freedom, or "free world"?

3) How can a theory of knowledge and ethics be explained and developed within the framework of a deterministic theory?

4) To what extent can one add to our theory a few selective ideas from existing systems of thought?

It is Spinozism which deals with the first three questions. But I do not intend to review this subject in this volume.+ I shall therefore concentrate on the last question, while making short references to the problem of "free will".

Here again the system of thought that comes first to my mind is Spinozism. After all, Baruch Spinoza is the author of one of the greatest philosophical systems in the history of philosophy, and certain fundamentals of his doctrine do not run counter to our own infrastructure. However, one must approach Spinoza *selectively* and *cautiously*. First he is not to be read but studied; second, even this must begin at a respectful distance from the core of his theory (like the Bible, his writings have often been the subject of false interpretations). The first step in understanding Spinoza is to apply the *historico-critical method*; to determine what he meant by the terminology he used, how he came to say what he said, and why he said it in

---

* K. R. Popper: *Loc. Cit.*, pp. 419.

+ Some of the problems presented by the first two questions are nevertheless discussed briefly in Volume II. Note that Spinozism is also based on Judaism, according to which "one exists where is thought is".

the manner in which he happened to say. As the most penetrating philosopher of all time, his work is to be studied and understood only as a whole.* Moreover, understanding Spinoza is indispensable for a true evaluation of later thinkers such as Kant, Schelling, Hegel, Schopenhauer, Wittgenstein, Whitehead, Einstein and many others.

The second system of thought that comes to my mind is Einstein's views about the world, his physics *and* philosophy. In fact the systems of thought of both Spinoza and Einstein are closely interrelated. *It was Einstein who completed the work which Spinoza started by destroying the mechanistic, anthropocentric and idealistic concepts of an irrational and dogmatic universe.* Einstein clearly realized the philosophical and ideological ties binding him to Spinoza. Spinoza, in turn, merged in his doctrine the best gems of reason he could extract from Greek philosophy, the Talmud and Kabbalah, Maimonides, the Christian Scholastics, Hobbes and Descartes. In the *Tractatus Theologico-Politicus* Spinoza presents a rational and eloquent plea for moral liberty and demonstrates that universal theosophy consists in the practice of a simple but active life, independent of past religious dogmas. Anticipating the methods of modern science, he examines the concepts of space, time, causality, natural laws, determinism, absence of "free will", moral freedom and freedom of thought and speech. In the unfinished *Political Treatise* he develops a universal "theory of government" founded on common consent which has even influenced such systems as the one set out in the Constitution of the United States.

But Spinozism and Einstein's physico-philosophical doctrinces will perhaps not be fully understood even in the twenty-first century. Perhaps by then Einstein's physics and Spinoza's philosophy will have become the hard core of a world philosophy-theosophy for neomodern civilization; a synthesis of those elements in all rational thought which result from a new methodology that enables its followers to establish a grand systematic whole, a pluralistic attempt to comprehend the entire universe. Such a pluralistic system may be grounded in both a mathematical-geometrical ordering of all rational thought and in a physico-philosophical ordering of all observable events presented within the framework of an all-embracing science. Of course it is no compliment to a philosophy, past or present, to say that only by being *"not fully understood yet"* does it become philosophically important. Therefore, one must be prepared to study Spinoza and Einstein, evidently not with the haphazard knowledge that one happens already to possess; not without first updating one's knowledge about past and present philosophies and sciences; and, finally not without first familiarizing oneself with the lives and philosophical backgrounds of these two great thinkers.

Indeed, Spinoza tried to merge his idea of the social world, in which he was outcast and alone, with the universal order of things, making it an almost indistinguishable part of nature. For him *"The greatest good is the knowledge of union which*

---

* Understanding does not, of course, mean "accepting" it.

the mind has with the whole universe". "Freedom", for Spinoza, is a relative notion: human beings are free in proportion as they are able "to think" clearly, "to control" their environment and "to preserve" themselves for the maximum time in an "active" state. A person is therefore "free" insofar as his thinking is "active" rather than "passive" (see also the discussion of "action" in American philosophy).* Had Spinoza been alive today he would have probably defined a man captivated by TV suspense programs as having attained the lowest possible degree of freedom! Is it hard to believe? *Think* about it *twice* before arriving at your conclusion.**

Spinoza anticipated psychoanalysis by emphasizing the need for a complete understanding of the emotions and their causation. He argued that *"liberty"* may be attained and the passive emotions mastered *through understanding of their causes.* And he stated that the *"will"* is a *"mental state"* which is a *"correlate"* of a *"physical state".* The illusion of *"undertermined choice"* arises *from our own ignorance of the preceding causes of thought and action!*

The whole scheme of Spinozism is strictly deterministic. By "cause" Spinoza means roughly what we mean by "explanation". For Spinoza the denial of "free will" means that while there is no causal condition between mind and body, there is a "mental correlate" for every "physical event" and a "physical correlate" for every "mental event". A human being can "know" "external physical objects" only insofar as they affect his own body.

Einstein too, tried to merge his deterministic idea of quantum-mechanical philosophy, in which he was outcast and alone among physicists, with the universal laws of gravitational and electromagnetic fields so as to become an almost indistinguishable part of nature. He did not quite succeed. It is only in recent years that modern physico-philosophical thought has brought about a full understanding of Einstein's difficulties and nearly eliminated them (cf. Assertion 2, Introduction).

IX.2.16 Beyond Descartes

Spinoza studied Descartes with a mind full of Hebrew wisdom and a spirit that revolted against the tyranny of established Scholasticism. Descartes became his teacher and intellectual guide, but he went beyond the philosophy and the methodology of the great Frenchman. The philosophy of Spinoza was first influenced by the Cartesian method but went beyond it to create a whole new system of thought. In this system Spinoza accepted parts of Descartes' physics and also followed Descartes in his attempt to construct a rational psychology. But only in a few instances did Spinoza quote directly from the writings of Descartes. For he did not concur with Descartes' limitation of knowledge and the confines of its procedure. Nor did he accede to Descartes' hidden postulates of Divine Providence.

---

* Appendix III.

** And, to be sure, perhaps try to read Appendices I to VI searching for the alleged "active" role the individual in today's bureaucratic, centralized, and institutionalized society is supposed to play, especially under the control of MCM (Appendix IV).

For example: *"willing" for Spinoza primarily consists in thinking of the course of action to be followed. The denial of "free will" is one aspect of Spinoza's system which has been found highly shocking, as it seems to deny moral responsibility and the justice of punishment. However, anticipating modern criminology, psychoanalysis and psychophysics, Spinoza only pleads for an objective understanding and analysis of causes in undesirable human behavior. He does not deny the sociological and political necessity of punishment.*

Unlike most absolutists, Spinoza stresses the need for *"freedom of thought and speech". In fact, he argues that individuals can be loyal citizens only if they are allowed to speak their minds freely. But he also insists that the state has a right to control certain outward expressions of formal religion.* He was therefore among the first to demand the separation* of church and state. Spinoza is an advocate of democracy as a reflection of reason and considers it *the most stable system.* Moreover, every individual, culture or civilization, according to the philosophy of Spinoza, obeys the *"survival principle"* of nature, i.e., it endeavors to preserve itself and persevere in its existence (§8, Introduction to Volume II).

IX.2.17 From Abuse to the Summit of Philosophical Esteem

How did European intellectuals respond to Spinozism? Spinoza has been reviled or exalted as "monotheist", "pantheist", "atheist", "materialist", etc. In fact, under almost every name in the philosophic vocabulary. But such offhand classifications and definitions are based on hasty readings of isolated passages rather than on understanding of his whole philosophy.

The torrent of abuse which poured forth from both Christian and Jewish theologians served to distort Spinozism in its early days. Today the traces of these distortions still dominate the popular image of Spinoza. Indeed, very few philosophies have been more variously interpreted than Spinozism. Some critics have maintained that Spinoza indulged in generalities without any specific meaning, others blamed Spinoza for filling his doctrine with "irrational emotions".

Let me point out that aside from a work on the philosophy of Descartes it was only the *Tractatus Theologio-Politicus* that was published in Spinoza's lifetime. And it was denounced as "an instrument forged in hell by a renegade Jew and the devil". Nevertheless Spinoza also aroused genuine interest and limited recognition by such intellectuals as Henrich Oldenburg and G. W. Leibnitz. Spinoza met Oldenburg in 1661 when the latter was already a member of the Royal Society in London and was to be appointed one of its two first secretaries in 1663. In 1675 the scientist E. w. von Tschirnhaus visited Spinoza and brought about the resumption of his correspondence with Oldenburg, which had previously been interrupted.

---

* A total separation would, however, run counter to his objections to the possibility of isolation of anything (or of any idea) in nature.

A year later, in 1676, the German philosopher-mathematician *Gottfried Wilhelm Leibnitz* (or Leibniz) (1646–1716) had several conversations with Spinoza at the Hague. Leibnitz read Spinoza's manuscript of the *Ethica* and copied various passages from it, but later did not hesitate to speak of Spinoza in tones of constant depreciation and diminishing esteem. Nevertheless Leibnitz's own accounts of various fundamental concepts in philosophy, such as his account of freedom,* were copied from Spinoza, by whom he was much influenced.

Earlier, in 1671, having heard of Spinoza as an authority, Leibnitz had sent him an optical tract and had subsequently received from Spinoza a copy of the *Tractatus Theologico-Politicus*, which deeply interested him. According to Leibnitz's own account, he "conversed with him often and at great length".

Another instance of recognition came later through the adoption of some of the ideas put forward by Spinoza in his *Tractatus*. It was due to *Hermann Samuel Reimarus* (1694–1768), a respected philosopher and man of letters in Germany. He borrowed heavily from Spinoza in evolving his standpoint of pure *naturalistic deism*. Thus, using Spinoza's rationalism, he denied all mysteries and miracles and proposed the basis of a "universal religion" in which the laws of nature and the equation *"God equals nature" (Deus sive natura**)* were exactly as in Spinozism. Following Spinoza, Reimarus claimed that a "revealed religion" could never obtain universality, as it could *never* be intelligible and credible to *all* people.

What happened in the century following Spinoza's death? In fact, almost nothing. For the next hundred years, Spinoza is described as an arch-heretic with no real importance in his doctrine or scope. It was only in the years of the French and American Revolutions that Spinozism was raised from its position of abuse and obscenity to the summit of philosophic esteem. This was accomplished through the powerful writings of such intellectuals as *Goethe, Heine, Lessing, Mendelssohn, Auerbach, Coleridge, Shelley, George Eliot, Herder, Fichte, Saisset, Hess, Janet, Renan* and many others; most of them not only admired Spinoza but studied him deeply.

The first real recognition came from *Gotthold Ephraim Lessing* (1729–1781), a contemporary of Kant, Mendelssohn and Jacobi, who found in Spinoza a powerful system of thought and the solace he had sought in vain elsewhere, though he never accepted the system as a whole. He used to claim that he knew no other philosophy but Spinoza's.

Johann Gottfried Herder (1744–1803), *Johann Gottlieb Fichte* (1762–1814) and *Friedrich Wilhelm Joseph von Schelling* (1775–1854) were greatly influenced by Spinozism. Even Hegel, who formally did not accept most of the ideas of Spinozism, declared that *"to be a philosopher one must first be a Spinozist"*. Spinoza was greatly

---

* For Leibnitz freedom is a relative notion, and something is free in proportion to its activity; i.e., in proportion to the clarity of its perceptions. However, Leibnitz's account does not give a clear meaning to the freedom of choice he wishes to allow to human beings.

** See Spinoza's *Ethics*.

admired by *Johann Wolfgang Goethe* (1749–1832). He kept rereading the *Ethics* as a precious human document.

Like Hegel, Friedrich Heinrich *Jacobi* (1743–1819) rejected most of Spinoza's conclusions; yet he too concludes that there is no real philosophy other than Spinoza's and that any consistent philosophy logically leads to the determinism found in Spinozism. (When Lessing, whom he met in 1780, told him that he knew no philosophy but Spinoza's, Jacobi took up the study of Spinozism). But the rationalism of it did not attract him and he denounced it in *Über die Lehre des Spinoza, in Briefen an den Herrn Moses Mendelssohn* (1785; 2nd ed. 1789).

Moses Mendelssohn (1729–1786) and other Enlightenment figures attacked Jacobi's assertions as groundless. It was this famous controversy that made the greatest contribution to the spread of Spinozism, attracting the attention of most German intellectuals. Many of Spinoza's ideas were also incorporated in Mendelssohn's *Jerusalem* (1783). *Nevertheless Mendelssohn does not acknowledge the fact that he borrowed heavily from Spinoza.* Spinoza's name is mentioned only once in Mendelssohn's work and even that only as the author of the *Ethics* (but not as the author of the *Tractatus* from which Mendelssohn borrowed the major part of his philosophy). *This omission was probably required, for Spinoza's doctrine was still considered highly dangerous. In fact, in corresponding with Jacobi, Mendelssohn made a great effort to "clear" his close friend Lessing from the dangerous charge of Spinozism.* But these attempts only intensified the interest in Spinozism among such philosophers as Kant, Herder, Fichte, Maimon and later Schelling. For these reasons Mendelssohn was wrongly identified as anti-Spinozist. But the fact was that he admired him and was one of the first philosophers to see Spinozism in its proper historical perspective. In Spinozism he also saw the beginning of the Leibnitz' doctrine of predetermined harmony.

How did Kant Come to Know Spinozism?

Little reliable information is available about Kant's education. At the university he was certainly introduced to the fashionable philosophy of Leibnitz and *Christian Wolf* (1679–1754) and the ideas of *Isaac Newton* (1642–1727). Kant's earliest work, *Thoughts on the True Estimation of Living Forces*, written in 1746, was concerned with a point in Leibnitz' physical theory. In fact, some of his critics claimed that there was nothing in the *Critique* which could not be found in Leibnitz' works. But it was only in Kant's first period (up to around the '70's) that he can be characterized as a critical follower of Leibnitz. Accepting the main concepts of Leibnitz' doctrine, he nevertheless had many doubts about the details. His doubts probably sprang from his great admiration for Newton (to whose writings he is said to have been introduced by his Königsberg teacher, *Martin Knutzen*).

In the second period of his work (which ended around 1790), Kant developed his basic philosophy. But between 1770 and 1781 he published virtually nothing. *Only from 1781 onward did he pour his astonishing series of philosophical works on reason, metaphysics, morals, natural science and judgment, and it was these works that brought him immediate fame in Germany.*

In this connection it is of particular interest to note the following historical facts:

1. *Moses Mendelssohn visited Kant in Königsberg in 1777.* A few days later (August 20, 1777)[+] Kant wrote to M. Hertz, one of Mendelssohn's students, how much he valued the intellectual conversations with Mendelssohn and how much he needed someone like Mendelssohn as an intellectial friend in Königsberg. Kant, of course, kept corresponding with Mendelssohn.

2. When Kant published his *Critique of Pure Reason* in 1781 he sent out only four copies with personal dedications, two of them to M. Mendelssohn and to M. Hertz.

3. There is an amazing similarity in arguments, subjects and methodology between Spinoza and Kant's *Critique.* The systems which Kant criticized most thoroughly were therefore those advocated by Descartes, Spinoza, Leibnitz and Wolf on one hand and those advocated by the British empiricist Hume on the other. It is true, Kant once remarked, that Spinoza's doctrine is not easy to refute.[*]

To conclude: Like any other philosopher, Kant was subject to influence by his contemporaries and by his predecessors. Opinions differ about the degree of influence on Kantianism which should be ascribed to Leibnitz, Wolf, Newton and Hume. Here we have only pointed out the generally unknown influence of Spinoza.

Another great admirer of Spinoza was *Moses Hess* (1812–1875), the rational philosopher of socialism and Zionism. Hess combined Spinozism with Hegelianism to produce a number of the fundamental concepts of modern socialism that were partially adopted by Karl Marx. Hess saw in Spinoza the greatest thinker and the greatest philosopher of all times. The influence of Spinoza's thinking can be seen in all his writings, from his early general socio-political essays, through the first theoretical expression of Zionism— *Rome and Jerusalem* (1862; Eng. transl., 1918)— and down to his later works.[**] Hess also underscores the global and historical importance of Spinoza in unifying theoretical philosophy with experimental science and in creating a rational basis for political life based on social justice, equality and freedom of the individual.

[+] See *Kant's Works*, Academy Publication, Vol. 10, p. 195.

[*] See Kant's *Critique of Judgment*, Translated by J. H. Bernard, London, 1931 (2nd ed.), pp. 304–5.

[**] See *Moses Hess, An Annotated Bibliography*, Burt Franklin, New York, 1951; *The Works of Moses Hess.* An inventory of his signed and anonymous publications, manuscripts and correspondence; E. J. Brill, Leiden, 1958; Edmund Silberner; *Moses Hess.* E. J. Brill, Leiden, 1966; see also the important works of Sir I. Berlin.

It seems fitting to close this section by quoting Dagobert D. Runes from his Preface to Spinoza's *Principles of Cartesian Philosophy*.***

"Spinoza declared when others hinted, and raised his voice to a clamor when others dared little more than whisper. He thus drew the brunt of orthodoxy in attack upon himself and became the anti-Christ of seventeenth century Europe, that grand epoch which marked the end of Scholasticism and the dawn of the Enlightenment. He walked the crimson path of martyrdom, bevenomed by Christians, denounced by his kinsmen and denied even by his early admirers, who, though sparked by his brilliance, fled in fear of its inherent danger.

The flame of Spinozism shone across the lands of Europe, cold and cutting to the bigots, an inspiration to those spiritually yearning in the chains of traditional circumspection."

IX.2.18 Theory of Knowledge

According to Spinoza we are aware of external things only *in relation* to one another and *cannot* know them *"as they are in themselves"*. *We cannot state the "truth" about a given object in itself, for it cannot be isolated from the rest of nature. Consequently we cannot reach complete knowledge about a given object since this would involve the whole of nature.*

This is a grand generalization concerning all knowledge. In fact, all our experience, including the most advanced aspects of bio-physics, physics, astrophysics, cybernetics, astronomy and relativistic cosmology, has not refuted a single word in this universal statement. We can therefore incorporate it with confidence into our own system.

Next we examine a few other universal maxims of Spinoza which demonstrate his use of the Cartesian, mathematical-geometrical ordering of reason.

Lemma III (Ethics, Pt. 2, Prop. XIII).* A body in motion or at rest must be determined to motion or rest by another body, which was also determined to motion or rest by another, and that in its turn by another, and so on *ad inf.*

Demonstration: Bodies (Def. I, Pt. 2) are individual things, which (Lemma I) are distinguished from one another in respect to motion and rest, and therefore (Prop. XXVIII, Pt. 1) each one must necessarily be determined to motion or rest by another individual thing, that is to say (Prop. VI, Pt. 1), by another body which (Ax. I) is also either in motion or at rest. But this body, by the same reasoning, could not be in motion or at rest unless it had been determined to motion or rest by another body, and this again, by the same reasoning, must have been determined by a third, and so on *ad inf.*—Q.E.D.

Corollary: Hence it follows that a body in motion will continue in motion until it be determined to a state of rest by another body, and that a body at rest will continue at rest until it be determined to a state of motion by another body.

*** The Wisdom Library, Philosophical Library, New York, 1961, pp. 1.

* Translated by W. H. White (1883) and revised by A. H. Stirling (1899). Edited by J. Gutmann, Hafner, New York, 1949.

Next I list a few of Spinoza's propositions, but without their corollaries, demonstrations, etc.

Lemma I (Principles of Cartesian Philosophy, Pt. 2).* Where there is extension or space, a substance is necessarily there.**

Proposition II (Principles of Cartesian Philosophy, Pt. 2). The nature of a body or matter consists solely of extension.**

Corollary: Space and body are not really different.**

Proposition VII (Ethics, Pt. 2). The order and connection of ideas is the same as the order and connection of things.

Proposition IX (Ethics, Pt. 2). The first thing which forms the actual being of the human mind is nothing else than the idea of an individual thing actually existing.

Proposition XVI (Ethics, Pt. 2). The idea of every way in which the human body is affected by external bodies must involve the nature of the human body, and at the same time the nature of the external body.

Proposition XXIII (Ethics, Pt. 2). The human mind does not know itself except insofar as it perceives the ideas of the modifications of the body.

Proposition XXVI (Ethics, Pt. 2). The human mind perceives no external body as actually existing unless through the ideas of the modifications of its body.

Proposition XLVIII (Ethics, Pt. 2). In the mind there is no absolute or free will, but the mind is determined to this or that volition by a cause which is also determined by another cause, and this again by another, and so on *ad inf.*

Since the vindication of Einstein's theory of general relativity Spinoza's Propositions have gained much support. First, if space and body are not really different and if the velocity of light in empty space is independent of the motion of the observer (or, respectively, of his "subjective choice" of the inertial system to which it is referred), no absolute meaning can be assigned to matter, space or the conception of subjective time or the simultaneity of events that occur at points separated by a distance in space. It is in consequence of this discovery that space, time, matter, energy, gravitation and acceleration were welded together in Einstein's deterministic theory of gravitation.

IX.2.19 From Spinozism to Einsteinism and Unified Field Theories

From 1920 until his death in 1955, Einstein worked on his unified-field theory. In this theory he hoped to find equations that would allow one to conclude that electric charges curved space as did masses. One aspect that needed answer was the question of opposite curves of space, i.e., should the so-called "positive" and "negative" charges a priori confer opposite curves on space?

---

* Translated by H. E. Wedeck, edited by D. D. Runes, Philosophical Library, New York, 1961.

** Note the agreement with the concepts of general relativity (Lecture III). However, it should be noted that some of these concepts were borrowed from *Descartes*.

Probably the most revolutionary aspect of Einstein's unified field theory was his proposal to reduce quantum theory to a simple consequence of the unified theory of fields. This hypothesis was rejected by most physicists. According to their charges the notion of the quantum action is indispensable.

Vindication of the general theory of relativity was only the beginning of a long process that slowly brings Spinozism and Einsteinism closer to each other. Developments in the second phase of the process took place mainly in three areas: first, *cosmological theories* (these were described in Lectures III and VI); second, *unified field theories*, which have aimed at showing that not only gravitation, but also the weak, the electromagnetic, and the strong nuclear forces are manifestations of the geometry of curved space-time (see below and Assertion 2 in the Introduction).

In the general theory of relativity the usual forces, such as electro-magnetic, elastic, etc., produce *accelerations* by modifying the geometry of space-time, which, in turn, produces a gravitational field, i.e., for every *interaction,* whether it be pure gravitational, electromagnetic, etc., there is a *corresponding* energy-momentum tensor; and the geometry of space-time is *fully determined* by this energy-momentum tensor (see Lecture III). However, general relativity did not originally supply the proper formulations for these tensors, i.e., *how,* for instance, the *electromagnetic field* modifies the geometry of space-time. Early unified theories (unnecessarily) attempted to accomplish this by making the geometry higher dimensional or non-Riemannian.

Did Riemann consider gravitation and electricity? At the age of 28 *G. F. B. Riemann* (1826–1866) produced the mathematical machinery for defining and calculating the curvature of space (see the discussion of the metric in Lecture III). There is, perhaps, nothing more instructive in the history of relativity than the fact that the great Riemann spent his dying days working to find a *unified account of electricity and gravitation!*

Among the various reasons for his failure to make the decisive connection between gravitation and curvature was the idea of the curvature of space rather than the curvature of space-time. Yet, significantly, he was the first person to realize the importance of *multiply-connected topologies,* fundamental concepts that may eventually open the door to modern general relativistic views of electric charge as *"lines of a field trapped in the topology of space"* (see below).

IX.2.20 Beyond Present Physics: Is Matter a Manifestation Of Solitons and Instantons of Space-Time?

IX.2.20a Vacuum States in Gauge Fields

We have now come face to face with the most difficult problems associated with modern physical studies of time, space, symmetry-asymmetry, vacuum states, gauge fields, elementary particles and unification of the four fundamental forces in nature. In comparing these problems with the aforementioned philosophical theories of time and space we need evoke almost all of the physical concepts that have been discussed in previous lectures. Of particular use are the new concepts about the analogy between elementary particles and solitons and instantons of space-time in modern gauge theories (cf. Introduction, Assertion 2, §II.12 and §III.5). But closely related to these problems is the physical characterization of space-time structures which describe "vacuum states". Hence, we first stress that what we call empty space is not what we think intuitively it is, for it is the arena of "violent phenomena" that need clarification. Secondly we note that this area of study is basic to any understanding of the structure of the physical world.

Now, what do we mean by "violent phenomena" in the vacuum state?

According to the modern quantum theory of fields, even empty space is not completely uneventful and "void". This empty space, *"the vacuum"*, is endowed with certain potentialities. Pairs of electrons-positrons, mu mesons, photons, and other elementary particles may be "created" there, even if only to be "annihilated immediately". These virtual physical processes have an observable influence on certain physical phenomena, e.g., on the emission of light by atoms.

One is, therefore, tempted to ask: *How will these virtual processes influence the curvature of space-time? Is there any (additional) field associated with these processes? Does the ever present gravitational field influence these processes? And, most important, what else is there out of which to build an "elementary particle" except geometry itself?*

IX.2.20b Quantum Chromodynamics as a Partial Answer

In trying to answer the aforementioned questions we first make a short stop at the workshop of a modern study called quantum chromodynamics (QCD). But, as before, the reader is warned to suspend his judgement till we reach the end of the "tour".

QCD describes the strong interactions between quarks (cf. §V.6) by employing a (non-Abelian) gauge-field theory (§II.12, §III.5). This, in fact, is a generalization of *quantum electrodynamics (QED)*, in which the quarks are the sources of a *"gluon field"* and their *"charge"* is given by their trivalued "color degree" of freedom.* Thus, in QCD the gauge field carries charge. (An essential property of

---

* Quarks are considered as particles with spin 1/2; they carry fractional charges, are pointlike and not very strongly bound to each other when confronted with a large momentum transfer. Several different types (("flavors")) of quarks may be distinguished, with different masses (Table IV, Introd.).

Protons and neutrons behave "as if" composed of only two types of quarks, the so-called u and d quarks. But more types of quarks are required to explain the numerous mesons and baryons identified experimentally so far.

An important theoretical feature of quarks is the fact that they transform into each other only through weak interactions. Moreover, it is necessary to ascribe an internal degree of freedom to each quark (a trivalued spin called "color", which permits three distinct quantum states. — Assertion 2, Introd.).

a field is that it can carry energy. The most familiar field is the electromagnetic one described by Maxwell's field equations (§VI.10). Gravitation, as we frequently noted in these lectures, is also described by field equations, those of Einstein's general theory of relativity).

So far, the calculable results of QCD seem to agree with experimental findings (such as the weakness of interactions for large momentum transfers). But the most important problem is the fact that quarks have not been yet observed experimentally as free particles.* This fact gave rise to the idea that quarks must be confined, under presently obtainable experimental conditions, to within hadrons. Yet, the idea that quarks are the internal constituents of baryons and mesons (hadrons) is hardly questioned nowadays. Indeed, most physicists now say that hadrons behave *"as if"* they are made of quarks, i.e., that atomic nucleons, and their excited states, behave "as if" they are composed of three quarks and that mesons "as if" made up of quark-antiquark pairs (see Assertion 2, Introduction and §V.6).

Such new understanding of the structure of matter should, by no means, be considered as final and conclusive. Not only that we are still far from having a reliable quantum theory of quarks and quark interactions, we are also far away from any understanding of the fact that more and more quarks are required to explain the properties of "elementary particles". Morevover, QCD has not only demonstrated that what we called "elementary particles" is a totally false physical concept (for they do *not* form a basic starting point for the description of matter), but it has raised new questions, notably: *What else is there out of which to build a quark except geometry itself?*

IX.2.20c The Vacuum as "Pregeometry"?

The question of the geometrical structure of the vacuum state is central to our understanding of the problems raised in this lecture. We shall, therefore, make a second stop at the dock of a newer factory that "creates" solitons, instantons, and the like out of the vacuum state.

This "factory" is run together by mathematicians and physicists, but the mathematicians are usually the first to come out with new "products". So let us see now what they can tell us about their newest findings.

Indeed, during the past few years these mathematicians have produced many important theorems in gauge theories (for which they normally use a different terminology associated with what they call *CONNECTIONS*). The regular working tools here are various theorems about multiply-connected topologies, domains, arrows rotating around closed loops, lines of field trapped in the topology of curved space, etc.

---

* So far it has not been possible to "liberate" a quark experimentally from a hadron and observe it as a free particle. However, in high-energy collisions, jets of mesons have been observed with large momentum exactly in the direction in which quarks should have been ejected (see below).

One of their theorems is based on a procedure of carrying an arrow around a closed loop in a field. It states that if the field has no discontinuities (and does not vanish at any point), then the arrow must rotate an integer number of times during any circuit. The arrow can fail to rotate at all, in which case the integer is zero, or it can turn once, twice and so on, *but it cannot make half a turn.* This theorem forbids, therefore, a loop with a fractional rotation in fields without discontinuities. In physical fields it can then be associated with *a lump of energy that cannot spread out over space,* i.e., *it is confined by a "twist" in the geometry where the net rotation of the field changes by one turn.* The three-dimensional solitons discussed in §II.12 and §III.5 are similar in structure, and their confinement can be explained by such theorems.

Now, what does a soliton tell us about the structure of the vacuum?

The most important conclusion of these studies is the fact that topological solitons can exist only in fields that have *multiple vacuum states.* Now, vacuum states of a field are the *zero-energy* configurations, i.e., the state of a field that has minimum or zero energy is called the vacuum state. Thus, in the vacuum state the field must be constant throughout both space and time, since any variation would give, say, the kinetic or the potential energy a value greater than zero. Hence, the vacuum is the state of a uniform field that has zero intrinsic energy. But because the intrinsic energy is determined *by the magnitude of the field,* a plausible state is one with intrinsic energy at its minimum *when the field itself is zero everywhere.*

Perhaps surprising is the fact that there are a number of field equations for which the intrinsic energy *vanishes* at some *nonzero value of the field. A true vacuum may therefore be described by a space permeated by a uniform field.* Indeed, certain field equations give rise to a multiplicity of vacuum states. But the most interesting feature of these field equations is the fact that *each vacuum state* is at a *different* value of the field itself. *Hence, despite the fact that they all have the same zero energy, they are also distinct.* The next important feature of these fields is the fact that *they give rise to solitons when at least two vacuum states exist, i.e., it is a simple mathematical procedure to create a soliton-antisoliton pair from a state of global vacuum.* According to modern quantum field theory, this corresponds to the *creation* of a *particle and its antiparticle* in the vacuum state. (In the reverse process a soliton and antisoliton collide and annihilate each other, as is observed with actual particles and antiparticles.)

But, most important, the solitons and antisolitons are not enforced *a priori* on the field equations. *They arise naturally from the mathematics involved.* (Outside their topological domain the vacuum can continue indefinitely, not so inside. So is the unequivocal result obtained with gauge fields.)

Now, if all solitons, antisolitons and instantons arise naturally in some of the available gauge theories, one is tempted to ask: *Why not use that gauge theory that would most naturally express them only in terms of space and time? Why not use the gauge theory expressible in terms of the metric? Why not we try first the gauge theory that is compatible with the gravitational-geometrical field?*

IX.2.21 A Unified Gauge Field as the Greatest Ambition of Physics?

Having considered the most plausible aspects of time, space, asymmetry, symmetry, solitons, particle creation in the vacuum state, etc., we now return to our starting point, namely: Would unification of the four fundamental fields be feasible along the deterministic guidelines Einstein envisioned? Is Einstein's seemingly futile dream, of bringing together, under a single set of equations, all of nature's basic forces, more realistic today than in Einstein's lifetime? Were Einstein's colleagues right or wrong in scorning him for advocating such "a far-fetched" or outrightly false view of physics and philosophy? For being preoccupied with a lonely quest that his contemporary physicists have utterly rejected as meaningless (Lecture V)?

Now, about a quarter of a century after his death, the doubts about Einstein's greatest intellectual quest are dwindling as a result of a number of highly exciting experimental and theoretical discoveries. Most of these discoveries are in the domain of subatomic particles, the nuclear "zoo" of quantum theory—a zoo that, at last count, contains some 200 subatomic particles.

It is mainly in the domain of quantum theories that Einstein's philosophy has encountered the greatest opposition (Lecture V). Hence, in bringing these formal lectures to a close, it gives me a great pleasure to report here those extraordinary discoveries in quantum field physics which give the strongest support to the Einsteinian doctrine.

The first exciting development took place in the domain of a unified gauge field theory of weak and electromagnetic interactions.* This early theory unifies the two forces by introducing a common gauge field transmitting these interactions. The new field has several components, one representing the photons as carriers of electromagnetism, others representing the carriers of the weak interactions. Fine, a skeptic may say, two forces are now expressed by a common langauge, so what? Why should I prefer the new over the old formulation, except, of course, for aesthetic reasons?

The clearest response to the skeptic's claim is now evident: Not only the new gauge field theory resolved inherent contradictions in the previous theory of weak interactions, it also *predicted* the existence of neutral currents (i.e., weak interaction processes *without* transfer of charge). Indeed, weak interaction processes without charge transfer have *not* been assumed, nor detected, by the time the new theory was published. It took only a few years after their prediction by the unified theory

---

* Weak interaction processes, it was long assumed, are all associated *with* charge transfer; a radioactive nucleus changes its charge when it decays, and the charge is transferred to the emitted electron (emitted with a neutrino, cf. §II.12a and Assertion 2, §3, Introduction). The gauge field theory referred to is the celebrated Weinberg–Salam unified field theory discussed previously.

that their existence was verified in neutrino-induced reactions. The first evidence came from *CERN* and the most recent one from *SLAC* (where the neutral interaction was also detected between electrons and protons by way of a subatomic asymmetry — the violation of parity discussed in §IV.11). *But, most important, the unified field theory predicts these and other weak interactions with astonishingly quantitative accuracy.* Yet, the existence of the predicted carriers of the weak interactions (the so-called intermediate bosons), has not been experimentally verified so far.

Another significant discovery surfaced recently in the domain of QCD. It was made by the powerful *PETRA* colliding beam accelerator in Hamburg (which accelerates electrons to energies of 15 billion electron volts and sends them head-on into positrons, coming at high speed from the opposite direction). In previous experiments, involving other accelerators operating at *lower* energies, the debris from electron-positrons collisions has consisted of only *two* "jets", or streams, of hadrons. This time *three* jets were detected; two were from a quark and its antimatter equivalent, the antiquark; the third apparently from a *gluon* (cf. Table I in the Introduction). This means that, if gluons (and quarks) are verifiable physical objects, the PETRA experiment has given *the first strong proof of QCD*. Here the strong interactions between the quarks are described by a (non-abelian) gauge-field that carries charge. Hence, it may well be that quarks and electrons are also *composite systems*, and that the proliferation of new types of quarks (evidence indicating the need of a fifth quark has surfaced recently and many more may follow), is nothing but the beginning of a long series of smaller subatomic objects that future experimental efforts will bring forth. The growing evidence about the existence of gluons as carriers of the strong force (which binds together the objects that make up the world of the atomic nucleus) is, therefore, a great step towards the eventual union of all four of nature's basic forces within the framework of a single *deterministic* field theory. Hence, Einstein's deterministic view of small and large-scale physics can no longer be dismissed as an impossibility, nor can his objections to indeterministic interpretations of the original quantum-mechanical theory be ridiculed any more. In fact, we witness now the opposite; as the evidence supporting the validity of deterministic unions of all sciences is growing fast from day to day, the remaining supporters of indeterminism are pushed deeper and deeper into a meaningless corner of the physical sciences. All freedom left to them is now cornered at the extremely small and *unobservable* scale of about 10^{-33} cm. (§IV.15).

Indeed, it is only at this fantastically small scale that indeterminists may still talk about *chance, fundamental uncertainty, probability waves, and "strict free will"*. This is the only unexplored corner left untouched by the sweeping advance of modern physics, a corner where idealists, Kantians, subjectivists and the like can still hold on to the old collapsing concepts of indeterminism and "strict free will". Thus, the chimerical concepts of Kantianism, idealism, subjectivism, etc., may only apply at three levels of "collapse" in nature, namely:

The initial phase of the universal expansion ("white-hole vs. black-hole physics"— cf. Lectures I,III, and VIII).

The end phase of a gravitational collapse which leads to the formation of a black hole (Lectures I and VIII).

Unobservable topological fluctuations of space at the scale of about 10^{-33} *cm.*

But, given the collapse of the so-called uncertainty principle (Lecture V), the advance of soliton-instanton physics, and the recent progress in the deterministic unification of the basic fields of nature, that last hold of chance and indeterminism collapses too. All that is left is Einstein's strict causality down to the smallest twist in the geometry of space-time. There is no doubt today that, following Einstein's philosophy, we will find an unending series of *worlds within worlds* when we continue to penetrate deeper into matter, to *smaller* distances and *higher* energies. Continual construction of ever higher-energy accelerators and colliding beam devices will most likely bring forth many new subatomic structures and deliver positive or negative evidence as to the validity of extant quantum models.

IX.2.22 Pregeometry, Kant, Sakharov, and Skepticism

At this point a skeptic may say, *well, all this "progress" is nothing but a combination of mathematics, philosophy and experimental physics that will reduce everything to space-time-symmetry-asymmetry and would, therefore, only transfer the question to 'what is the metric'? Is pregeometry the fundamental "building material" of nature? Are not we bound to end up with square one when this "progress" is completed? Are not we going back to the Kantian concept of pregeometry?*

These are difficult questions for which no simple answer can be given. All that can be stressed at this stage is that the concept pregeometry should not be confused with the *Kantian "a priori geometry"*. Pregeometry is linked to Einstein's geometro-dynamics, gauge fields, solitons, instantons, and subatomic objects. Contrary to the Kantian concept it should, in principle, be subject to *experimental verification*. Yet, what do we mean by pregeometry?

One possible answer may, perhaps, be associated with some preliminary concepts proposed by the Russian physicist *Sakharov*. In his view, what *elasticity* is to atomic physics, *gravitation* is to elementary physics. The energy of an elastic deformation is nothing but energy put into the "bonds" between atom and atom by deformation. Similarly, according to Sakharov's premises, the energy that it takes "to curve space" is nothing but "perturbation in the vacuum of fields plus particles brought about by that curvature". Thus, for Sakharov, an atom is built out of "elasticity of the metric". But this only transfers the question to *"what is elasticity of the metric"?* Is elasticity the pregeometry of curved geometry? Is it linked to the newly discovered properties of the gluons?

For Sakharov the Newtonian G (Lecture III) is simply the *"elastic constant of the metric".* Some physicists believe that such a view of nature should be explored further. For the moment, however, it remains beyond verifiable knowledge. So are *Salam's 'strong gravity' constant* (eq. 11, Assertion 2d, §3, Introduction), the solitons (§II.12, §III.5), the gluons (§V.6) and the prequarks, preons and pre-preons.

IX.2.23 Prequarks, Preons and Pre-Preons?

In his Nobel Lecture, *Abdus Salam* announced [*Science, 210,* 723 (1980)] that he and J. C. Pati have recently embarked on a new theory which incorporates prequark objects called preons. Such preons, they assume, carry magnetic charges and are bound together by very strong short—range forces, with quarks and leptons as their magnetically composites.

According to Salam, even the gauge fields, themselves, may be composite, and, with the growing multiplicity of quarks and leptons, we must think of quarks (and possibly of leptons) as being composites of some more basic entities (prequarks, or preons, or pre-preons) and of some more basic gauge fields.

Thus, motivated by the ideas of extended super-asymmetry, super-symmetry and super-gravity theories (§V.6), future science may reveal wheels within wheels, fields within fields, geometry within pregeometry, and so on, perhaps as limitless, inter-connected worlds within worlds.

IX.3 THE SKEPTIC OUTLOOK

Finally, to cap these formal lectures, I feel it fit to introduce a few informal remarks.

Once asked to define my viewpoint in three sentences I answered: At heart I am indeterminist; in actual life, a skeptical pragmatist-pluralist; in classical physics, an empiricist-determinist; in quantum physics an Einsteinist; and in the philosophy of science a gravitist-Havayist; that is to say that I have abandoned all classical divisions, and, therefore, no simple answer can be given for philosophy proper. Indeed, I had acquainted myself with all previous schools of thought but could identify with none. Moreover, I maintain that *definitions* that are shot from the hip, *contrary to intent,* often miss their target, or even bounce back.

This is to warn the reader against falling into the habit of shooting out *"concise definitions"* which isolate ideas (Assertion 2), and, therefore, miss their very goal. In fact, I even have doubts if the writing down of these lectures can *"pen down"* the target originally set up for them.

More recently, in the International Congress *"Levels of Reality"* (Florence, 1978), I was asked to state *briefly* my viewpoint on the possible links between gravitation and *non-physical* qualities, like the human spirit and long-standing expectations. To that I answered:

Even though we know that in the end we cannot *"get up"* again; that a day will come when staying *"upright"* becomes an impossibility, we do not *"break-down"*. By the same knowledge that foresees our own *"collapse"* in the journey *"from dust to dust"* we *"withstand"* that which will *"bring us down"*. In the end we know that it will *"tear down"* almost everything that we build; that ultimate *"downfall"*, *"decline"*, and *"defeat"* of every motile creature is the fate of individual life—a life that must terminate in *"de-aggregation"* and gravitational *"precipitation"*. Yet we *"stand up"*. Perhaps not all is *"on the downgrade"*? Thus, in the face of *"gravitational collapse"*, we hope to preserve some *"lasting structure"*, or *"spiritually"*, get *"up-and-away"* from that *"gravitating earth"*. Do these *"upholdings"* lead people to adopt *"spiritual"*, *"idealistic"* or *"heavenly"* ideologies? To construct towers and sky-high cathedrals? To launch programs for the exploration of "outer" space? or to preserve some "lasting" structure in the composition of music, the writing of books and the construction of new theories? I do not know.

It was about fifteen years ago, during my affiliation with the Johns Hopkins University, that I started the work on the foundations of dialectical gravitism and Havayism. And it was this work which gradually led me to become an autodidact of philosophy and of the general theory of relativity, a process which later prompted me to give these lectures as formal university courses.

But following the publication of my recent works, I came to realize that they contain technical concepts, formalism and mathematics that could be clearly understood only by specialists. Hence I decided, that, this time, I would try to write an entirely different book, that, with a minimal amount of technical formalism, would follow the general guidelines mentioned in the Introduction.

Ever since I have been working on this book, which reproduces, if not the letter, then the spirit of my lectures—mainly those delivered as undergraduate and graduate courses at the Technion-Israel Institute of Technology, Johns Hopkins, Pittsburgh and New York (SUNY) Universities. For this reason the book is presented as a series of lectures.

No series of lectures was ever a set-piece; they have remained in a state of flux until now when the final writing and printing has "frozen" them. This "freezing" is very apprehensive to me, but it cannot be helped. My hope is to complete a second volume in this series within the next few years. Meanwhile I hope that at least some of my readers will "thaw out" the printed lines and give them greater dynamical force through their own critical attention.

Indeed, it was the work on the present volume that gave me the opportunity to develop more fully than in my earlier works the general and physico-philosophical principles of this field of study. Arriving at this end, seemed, at times, like reconstructing a vast jigsaw puzzle from scattered and (apparently) unrelated pieces of information. But the eventual emergence of new regularities, which, at the beginning, I had not even suspected to be a part of the overall picture, was often a rewarding surprise.

I can only invite the interested reader to join me in further work on some of these thought-provoking excursions along trails leading to new fields of inquiry.

Verifiable evidence cannot yet be admitted to a number of topics discussed in this last lecture. Therefore, if scientists (and philosophers) are counted as "moderate skeptics", as no doubt they should, they must adopt the practice of "suspense of judgement". By this I do not mean radical empiricism nor the denial of any rational addition to the advancement of scientific thought.

What I mean has already been stated in Assertions 1 and 10 in the Introduction, namely, *that intellect is given us to speculate, to observe, to compare, and to speculate again*. Here, "to speculate", is to make a small step forward by "rational reason" that is based on all "empirical evidence" that has been accumulated till the time the step is taken. Yet, a critic may say here, "reason", even incremental and based on previous evidence, is unreliable. But if reason is unreliable, the philosophical arguments which support this conclusion are themselves unreliable. Here Kant and other thinkers have tried to distinguish between "legitimate" and "illegitimate" uses of reason, or between "true" and "untrue" sentences. But this methodology, carried out to its extreme, can only lead us to a "dead-end street". Therefore, it is the advancement of all sciences, by small, consequent steps of "rational reason" based on "empirical evidence", that demonstrates how verifiable, universal-unified knowledge grows with historical time (as is also evident from our global success in applying it in practice). This, in itself, is a reliable fact which justifies *a limited use* of "rational increments" *to unify* separate fields of science and philosophy.

Indeed, all denials of the possibility of rational additions to knowledge are difficult to substantiate for they imply knowledge of the existence of the very subjects which are claimed to be unknowable (e.g., reason). And I do not refer only to radical subjectivism. I mainly refer to any scientist who formally, or informally, implies such claims in regard to every single step in his teaching and research. He who believes that he advances without reliance on rational principles is self-deluded. For as we have stressed in Assertion 1, scientific theories always advance, stagnate or decline under the domination of a rational philosophy, whether declared or undeclared. Therefore, to undertake a scientific approach to the whole of nature, one must consciously evolve a unified (and universal) philosophy of science and not pretend to avoid it. This methodology does not necessarily lead to a *single* science; it leads to a *pluralistic*-but basically *unified* science. For by employing the same basic assumptions scientists working in different disciplines can, and do use several concepts and methods to describe "the same phenomena". Nothing contradictory is involved in that, as far as it advances our understanding. It is only when the very advancement of a *wider* field of study *is inhibited by previous practices, that a translation of concpets and modification of thought become a necessity.*

There is, however, one particular objection which I believe is essential for all scientific and philosophical studies. Consequently it was stressed in a number of paragraphs—especially in Lecture V. My objection may be best directed against Eddington's subjectivistic conclusion:

> "We have found a strange footprint on the shores of the unknown. We have devised profound theories, one after another, to account for its origin. At last, we have succeeded in reconstructing the creature that made the footprint. and lo! it is our own."

While I take pleasure from Eddington's flowery prose, I cannot agree with his implied conclusion. It is not only *our* "footprint" that counts. There are *other* "footprints". One may find them anywhere in nature; "on the shores of the unknown", on the beaches of Haifa, or in the fields of physics and astronomy. And they do not *exclusively* belong to *man!* Man is not the center of science, nor is he the sole "footprinting maker" in the universe.

Anthropomorphism and subjectivism would not advance us far away. They can only prevent us from opening gates to new horizons, or block the doors to the advancement of science and philosophy. Many historical examples demonstrate this fact and the more recent ones have been described in Lecture V and in the opening remarks to this book (see also Appendix VI and especially Appendix III).

It is perhaps better to conclude on a dual note of "skepticism" and "optimism" as exemplified by Einstein's dictum:

> "The hypotheses with which (a development in modern science) starts, become steadily more abstract and remote from experience. On the other hand it gets nearer to the grand aim of all science, which is to cover the greatest possible number of empirical facts by logical deduction from the smallest possible number of hypotheses or axioms. Meanwhile the train of thought leading from the axioms to the empirical facts of verifiable consequences gets steadily longer and more subtle."

Additional conclusions, or formulations, would, perhaps, be peripheral. Thus, as we bring these formal lectures to a close, we are left with a new theory and its potentials. But whatever the outcome of this theory it is only a slice through our *humanistic* views (which are discussed in Volume II).

VOLUME II

CRITIQUE OF
WESTERN THOUGHT

(six addenda in the form of six lectures)

INTRODUCTION

"Perfection of means and confusion of goals seem—in my opinion—to characterize our age."

Albert Einstein

1. CRISIS AS CONSTRUCTIVE CATALYST

Crisis is the father of new theories and practices. Confronted with nihilism, anomaly or crisis, people assume a different attitude towards existing traditions, and, accordingly, the nature of their thought changes. History testifies that this prerequisite holds true in every field of human endeavor, be it science, philosophy, politics or theology.

Indeed, many of the fundamental concepts of Western philosophy are still in the melting pot. A number of rival camps, each defending its own interpretation, surround the pot. The battles are waged unabatedly, almost daily; their intensity is now at an all-time high. They have created a situation of muddled suspense which may just be the precondition for arriving at valuable new ideas.

The growing willingness in the West to try out new ideas, the manifestation of great discontent and nihilism, and the intensifying debate over fundamentals *are all symptoms of a transition to new phases. Are these phases leading us to social and intellectual revolutions? to a fresh look at nature? or to stagnation followed by decline?*

Today, more than ever, the established traditions of Western philosophy and science are in a state of acute crisis. They lack the momentum to stem the tide of decline in certain areas of thought and practice, or to open gates to new horizons. Indeed, it is not a minor tragedy of our society that, with few exceptions, educational, professional and intellectual standards, especially those associated with *academe* (see below), and *mass communications* (Appendix IV), have not been kept at suitable levels. These declining standards are part and parcel of a wider general crisis which encompasses the whole of Western society. *Are the origins of this crisis known today? Are they associated with certain hidden, stagnating phenomena in science, philosophy and society? Or with a false philosophy? Indeed what are the actual problems involved? And, most important, what are the prospective alternatives?*

Such questions, and related ones, are dealt with below where, *inter alia,* we try to find some *"correlations"* between the present crisis and the objective needs of *"suitable methodologies." But we also examine the present crisis in the light of certain "doomsday theories"* (Vico, Spencer, Spengler, Toynbee, Marxism, etc.) *which variously claim that the West is doomed to extinction and that the Slavic or Sinic civilizations, having entered their proper new phases, will in time replace it.*

It is now commonly felt that *we cannot entirely ignore such "doomsday theories" and that by examining today the origins of our crisis we may be able to generate means to prevent "a-not-impossible tragic outcome." A critical study of the problem may also answer two cardinal questions, namely:*

i) Should contemporary academe spearhead the regeneration of the very philosophy which has given the West its life?

ii) Should such needs prompt us to reexamine, with a critical eye, the very foundations of our current ethics, doctrines and practices, especially if "standbys" or alternatives are required?

There can be no doubt that the present crisis will force Western states to create "new institutions" capable of uniting ideology, philosophy, science, industry and national planning, and applying them to the total ("multi-disciplinary") and global problems that confront us (see below).

By *"new institutions"* I do not mean the kind of *"Institutes of Advanced Studies"* (e.g., the one attached to Princeton University). Nor do I mean any simple combination of the "interdisciplinary" methodologies now employed by the *Brookings, Hudson or Rand Institutes in the U.S.,* or by some national academies. *I mean an old-new kind of study about the whole of our problems and interests* (q.v. Assertion 10, §3, *Introduction,* and below).

Problems in the real world are not separable into disciplines.

While disciplines and departments are an *administrative* convenience (and provide *needed* specialization), a 'methodology of interconnectedness' may grapple problems and interests *which do not fall into separable activities along disciplinary lines* (e.g., those of ethics, ideology, philosophy, policy, priorities, decision-making, and quality vs. quantity of academe, research and defense). Another possible task is to try to achieve *cross-fertilization of ideas demanded by the solution of such problems* (and, thereby, to create new fields of inquiry, planning, and development).

In trying to comprehend the aforementioned questions we may proceed by reexamining the methodology associated with the central triangle of our "collective leadership," namely: *Government, Industry and Academe.* Elsewhere, in *Government, Industry and Academe; In Search for a Policy,* I have tried, in the framework of a "national think tank" composed of senior government officials, directors of major industries, university rectors, and senior editors from the mass communications media, to argue against certain dangers associated with the absence of policy and philosophy, and to show that the absence of national planning, the growing crisis in Western leadership, and the decline in academe, originate, in part, from a gross misunderstanding of the role of philosophy in Western thought and practice, and especially from the neglect of a unifying methodology. While engaged in the systematic analysis of these subjects, I tried to develop possible guidelines for such a methodology. The notes collected for that purpose became the basis of the ideas presented in this Introduction.

2. Our universities: Are they adequate?

The number of people involved in education, research, development, and welfare has increased, since World War II, at a greater pace than at any other period in history. Most of this change resulted from increased financial support by *national governments,* which have thus become the chief *patrons* of most such activities. *By these means, the governments have been able to profoundly influence the evolution of methodology and practice in the public and private sectors of the economy and in the state and private universities.*

The present situation in which even the persons in the private sector largely *depend upon government sources for financial support of their work and in many instances, their livelihoods as well, is an unhealthy one, especially in academe.* The *autonomy* of the classical university, historically and necessarily one of the most independent institutions of Western society, *has therefore been gradually eroded. In this situation it is difficult to expect individuals to be critical of their patron. This leads to a clear conflict of interests. On one hand the needs of society dictate a close cooperation betwen academe and government. On the other hand any nation-state needs a review and, if warranted, a criticism.*

This conflict cannot be resolved by disciplinary academic debates. One way to resolve it is to take it *outside* the domains of both government and academe. Thus, in the absence of a consistent philosophy and exceptional leadership, an independent review of methodology and policy may be best done in a relaxed atmosphere by a kind of an *ad hoc* 'Sanhedrin'; a fluid, sitting-together assembly that is financially independent of both government and the university; a selective group consisted of free scholars, educators, industrialists, economists, news-paper editors, members of governmental cabinet, etc. *In the absence of a proper philosophy they may, perhaps,* provide, for the time being, a kind of *tentative* guide for national policy. Yet, it should be stressed *a priori,* that while such an assembly may, perhaps, avoid political and subjective intricates of various pressure groups that now rule the captivated roosts of government, academe and industry, *it cannot replace the need to develop an interconnected system of thinking about the totality of our problems; a new procedure for science-based inquiry about the global interdependence of any national policy.*

3. A Proper Time To Reassess Priorities?

The difficulty of speculating about priorities and the various needs of our society cannot mask one important fact, namely, that "up-dated" concepts that were useful and valid only four or five years ago, are now *outdated; concepts and methodologies must be continuously relearned and, sometimes, entirely new ideas must be devised to meet the (ever-coming) new needs. Consequently it is safe to say that our society is becoming, above all, "knowledge society" in which proper leadership in interconnected thinking are bound to be the most critical commodities.*

The traditional methodology currently employed in our universities cannot respond to such needs. Therefore, they *cannot* house interconnected thinking, for they would dominate it by *discipline-oriented modes.* On the other hand, experience shows that society *does* require the services of some discipline-oriented modes. Hence, one possible solution is to parallel interconnected modes of thinking to existing (discipline-oriented) university activities. Such methodologies must reject any trace of *a priori* departmentalization, i.e., *although some of these studies may be distributed among the various departments of a traditional university, there is an objective need to regenerate an old tradition that cannot be associated with any specialism because the ideas with which it deals are common to all studies, or not involved in any.* A number of themes, such as the inseparability of natural and social processes, the problems of academe, industry and the commons, and the importance of large-scale natural processes undergoing dynamic and evolutionary change appear today, more than ever, to warrant a *unified approach.* While some of these themes can be recognized in a wide variety of specific problems, no one of them provides a unified core for such a study, nor do they *collectively* encompass the whole field (cf. Spinozism in Lecture IX).

4. The Fluctuation Of Western Science

The first example in history of a large-scale financing of academe by government is Alexander's fundings of Aristotle's activities.

Alexander instructed his hunters, gamekeepers, gardeners and fishermen to furnish Aristotle with all the zoological and botanical material he might desire. *At one time, according to some writers, Aristotle had at his disposal a thousand men, scattered throughout Greece and Asia, collecting for him specimens of the fauna and flora of every land.* In addition, Alexander gave Aristotle, for physical and biological equipment and research, and for corps of aides and secretaries, the sum of 800 talents *(more than 100 million dollars in contemporary purchasing power).* He also sent a costly expedition to explore the sources of the Nile and discover the causes of its periodical overflow. Thus the *Aristotelian Lyceum* was the forerunner of contemporary *"Big Science"* (with its Big Governmental Funding) in the same sense as the *Platonic Academy* was the precurser of *"Little Science"* (with its intellectual, utopian and broad philosophical perspectives and with its *deliberate, active* and *planned* interventions in social, cultural and political life).

Ever since academe fluctuates between Aristotelianism and Platonism. Thus, when funding cannot be stretched further, it becomes a time for reassessing priorities, of asking hard questions about every concept and methodology and about the aim of science, medicine, defense, education, politics, or area of technological development seeking support. This transition, in itself, is a very healthy one for it leads to higher levels and to greater selectivity in any field of human activity (cf. Appendix IV).

But there is more to it. The very notion of *"academic freedom"* is strongly affected by these modes. *Thus, in the Aristotelian mode of "Big Science" one can hardly maintain an independent enterprise. It is easier, perhaps, to regain a (relative) "degree of independence" in the Platonic mode, for it does not involve Big Funding nor the patronage of government.* Yet, it does intervene with governmental, social and political issues—on its own, self-propelled philosophy and along guidelines of its own ideology. For these, and for later tragic events in the history of science, academe has been trying, with varying degrees of success, to keep a critical eye on any trace of governmental (and other) patronages of its affairs and to seek the (time-dependent) optimal course between the *Platonic* and *Aristotelian* modes of thought and practice.

But how can we find a middle way between universities influenced by the State and universities influenced by private wealth? *This problem has never been really solved.* Hence, one may ask: *Is there any other possibility?* Apparently there is. It also had flourished in Greece and was later advocated by such free thinkers as Spinoza. According to it, *ideology and philosophy should not come from any "established university" but from free individuals who teach, review, criticize and create independently of either public or private control* (cf. Lecture IX).

Keeping these remarks in mind we now return to contemporary academe and focus attention on its actual ethics, methods and practices.

5. Reforging a Link with Classical Western Thought?

. . . no science is immune to the infection of politics and the corruption of power . . .

Jacob Bronowski

The rise of bureaucracy and clergy in academe; the absence of a unifying philosophy; the decline of avant-garde and elitism; the lack of intellectual leadership; are they symptomatic to the mounting failures of our ideology and education? or, perhaps, to a kind of growing impotence? Do they indicate that we need a fresh look at our system? Or are we satisfied with the present situation?

Baruch Spinoza

ΠΛΑΤΩΝΟΣ ΤΙΜΑΙΟΣ, Η ΠΕΡΙ ΦΥΣΕΩΣ.
ΤΑ ΤΟΥ ΔΙΑΛΟΓΟΥ, ΠΡΟΣΩΠΑ.

Σωκράτης. Κριτίας. Τίμαιος. Ἑρμοκράτης.

ΣΩ. Εἷς, δύο, τρεῖς· ὁ δὲ δὴ τέταρτος ἡμῖν, ὦ φίλε Τίμαιε, ποῦ τῶν χθὲς μὲν δαιτυμόνων, τανῦν δ' ἑστιατόρων; ΤΙ. Ἀσθένειά τις αὐτῷ συνέπεσεν, ὦ Σώκρατες· οὐ γὰρ ἂν ἑκὼν τῆσδε ἀπελείπετο τῆς συνουσίας. ΣΩ. Οὐκοῦν σὸν τῶνδέ τε ἔργον καὶ τὸ ὑπὲρ τοῦ ἀπόντος ἀναπληροῦν μέρος; ΤΙ. Πάνυ μὲν οὖν, καὶ κατὰ δύναμίν γε οὐδὲν ἐλλείψομεν· οὐδὲ γὰρ ἂν εἴη δίκαιον, τοὺς ὑπὸ σοῦ ξενισθέντας, οἷς ἦν προσῆκον ξενίοις, μὴ οὐ προθύμως σε τοὺς λοιποὺς ἡμῶν ἀνταφεστιᾶν. ΣΩ. Ἆρ' οὖν μέμνησθε, ὅσα ὑμῖν καὶ περὶ ὧν ἐπέταξα εἰπεῖν; ΤΙ. Τὰ μὲν, μεμνήμεθα· ὅσα δὲ μή, σὺ παρὼν ὑπομνήσεις. μᾶλλον δὲ εἰ μή τισοι χαλεπόν, ἐξ ἀρχῆς διὰ βραχέων πάλιν ἐπάνελθε αὐτά, ἵνα βεβαιωθῇ μᾶλλον παρ' ἡμῖν. ΣΩ. Ταῦτ' ἔσται. χθὲς που τῶν ὑπ' ἐμοῦ

ΑΡΙΣΤΟΤΕΛΟΥΣ ΑΝΑΛΥΤΙΚΩΝ ΠΡΟΤΕΡΩΝ ΠΡΩΤΟΝ·
ΠΕΡΙ ΤΩΝ ΤΡΙΩΝ ΣΧΗΜΑΤΩΝ.

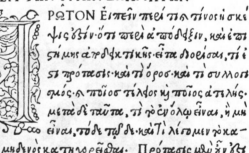

ΠΡΩΤΟΝ Εἰπεῖν περὶ τί καὶ τίνος ἡ σκέψις, ὅτι περὶ ἀπόδειξιν, καὶ ἐπιστήμης ἀποδεικτικῆς· εἶτα διορίσαι, τί ἐστι πρότασις· καὶ τί ὅρος· καὶ τί συλλογισμός· καὶ ποῖος τέλειος καὶ ποῖος ἀτελής· μετὰ δὲ ταῦτα, τί τὸ ἐν ὅλῳ εἶναι, ἢ μὴ εἶναι τόδε τῷδε, καὶ τί λέγομεν τὸ κατὰ παντὸς ἢ μηδενὸς κατηγορεῖσθαι. Πρότασις μὲν οὖν ἐστι λόγος καταφατικὸς ἢ ἀποφατικός, τινὸς κατά τινος· οὗτος δὲ ἢ καθόλου, ἢ ἐν μέρει, ἢ ἀδιόριστος. λέγω δὲ καθόλου μὲν, τὸ παντὶ ἢ μηδενὶ ὑπάρχειν· ἐν μέρει δὲ, τὸ τινὶ ἢ μὴ τινὶ ἢ μὴ παντὶ ὑπάρχειν· ἀδιόριστον δὲ, τὸ ὑπάρχειν ἢ μὴ ὑπάρχειν ἄνευ τοῦ καθόλου ἢ κατὰ μέρος· οἷον τὸ τῶν ἐναντίων εἶναι τὴν αὐτὴν ἐπιστήμην, ἢ τὸ τὴν ἡδονὴν μὴ εἶναι ἀγαθόν. διαφέρει δὲ ἡ ἀποδεικτικὴ πρό-

Excerpts from Plato's and Aristotle's Books

It is our conviction, that it is precisely these questions which should prompt us to *reexamine, with a critical eye, the very foundations of our doctrines and practices; but to reexamine them without traditional or temperamental biases and without prejudice in favor of academe (to which so many of us belong nowadays).* Since academe cannot be separated from the powerful media, the latter must also be reexamined, with a critical eye.

Indeed, it is mainly the mass communications media which today drives the Western trend to reject the classical, the traditional, the intellectual in favor of the "modern," the mindless, the popular. As such, it has dissolved many classical methodologies, and discredited most institutions and values (Appendix IV). *Under these conditions the Western avant-garde can hardly survive and the most hedonistic or destructive temptation becomes accepted overnight. This situation is echoed in education, in politics, in ethics, in philosophy, in the arts. Many young people have thus lost faith in anything intellectual or traditional and, indeed, elbow one another aside as they push their "rights" to an extreme, or as they demand services and "academic-degrees-for-all." Decline of Western thought results, and with it, the whole Western structure is undermined* (Appendix IV). Perhaps the only solution for this crisis is to try to reforge a link with the classical Western thought; perhaps by using a workable methodology of interconnectedness, of feedback, of scholarship.

But whether under Aristotelian or Platonic influence, academe cannot be separated from the influence of political communities. Hence, it is imperative to reexamine this interface at the same time that we look on the links between academe and the media.

The relationship between academic and political communities is one of constant mutual frustration. Actually there is always some kind of tension between government and academic institutions. In public institutions it is the tensions between the patrons of the institutions—government authorities and state "councils"—and the *institutions themselves.*

The *roots* of these tensions may be traced to such factors as different goals, different skills and different ethical and philosophical traditions. Nevertheless, the two camps are closely *interdependent* from the point of view of the "goods" which each possesses and the other needs. Thus government depends on those who have the highest skills, while those who have these skills often have no place to use them, or at least not as effectively, if they are not also linked to the public sector or funded by government agencies.

These tensions have become more intensified in modern times, especially with respect to national goals. In Western society they often focus on two sensitive issues: *"government funding"* of higher education, research and development, and participation of "academia" in *"centralized"* or *"classified"* research and development.

It is about time that Western scientific communities, especially those in the United States, start deal with government in a more realistic fashion, both with regard to the *individual's* piece of research and with regard to *"institutional funding."*

First, with respect to the *pretense of virtue* that some exhibit when it comes to such questions as the conduct of organized or classified research and military recruitment on campuses. A somewhat typical example of behavior in most U.S. universities is provided by the following *"decision-making process":*

In the spring of 1970, during the protest over Cambodia, the faculty and students of *Johns Hopkins University* voted "to bar Defense Department activities" from their campus. The president of the University accepted this recommendation and affirmed it as the University's policy. The next fall, when it became obvious that the Defense Department would require the statement of cooperation previously— attached to *NASA grants,* and that a multimillion dollar contract Johns Hopkins had with the Defense department was in danger of being lost, the University declared its campus *open* to "Defense Department activities."

We have here a clear example of *academic hypocrisy: when it is a matter of money or principle, principle goes out the window! And it does not matter which principle.* (Cf. Appendices III, IV and VI).

Secondly, with respect to *priorities* or *ethical principles.* If a certain national goal gets the highest priority, university policies should consistently reflect it. *Whenever priorities are agreed upon, scientists should be willing to stick to them-not as a pretense for going after the research dollar, or the latest fashion; but as a genuine aim.*

But who is to determine priorities?

The common practice in going after research money is that after scientists have outlined very clearly what their research proposal is, shown its validity in scientific terms, shown the soundness of their approach in pursuing that research, in the great majority of cases they add a strange statement that has almost become a ritualistic incantation. The magic spell which attracts the favor of "grant-givers" goes something like this (cf. also Appendix IV):

"Let me now assure you how helpful my research is going to be in furthering the objectives of your agency."

And it does not matter which agency. And it does not matter at all whether the research has any relevance to the genuine objectives of that agency. This brings us, face to face, with the central issue: Is there any National Policy or Master Plan, especially with respect to government agencies, their decisions and their goals? And if there is none, how are priorities to be evaluated?

Passé is the fruitful, mission-oriented, cooperative effort of the famous *Manhattan Project;* for no longer is there the sense of national urgency and the guiding spirit of leaders like *Einstein* and *Roosevelt*. Is this evidence of the *failure* of "peaceful science" to promote, coordinate and achieve significant mission-oriented projects? or a *prerequisite* for reducing the tension between government, scientists and other professionals? *Indeed, it is well known that the Manhattan Project has (later) created a number of controversies concerning its moral and ethical implications.* The central issues may be condensed into two universally applicable questions, namely:

1. Do scientists and engineers have a responsibility for the welfare of society in their professional role *over and above* their role as responsible citizens? And if so, then what guidelines can practically be developed, without a new Moses, for the ethical exercise of this responsibility?

2. What effects do *decision-making* in regard to the various kinds of scientific research and the funding of research and teaching have on the goals and moral and social responsibilities of scientists and engineers *as individuals,* or as a *community?* In particular, what are the *corporate responsibilities,* toward the rest of society, of organizations of scientists?; be they organizations and societies of scientists in similar fields, or be they the institutions, universities and research institutes harboring these scientists.

These are tough questions. And many Western scientists have been confronted with them. But usually they manage to sidestep the issues involved, most frequently by declaring themslelves members of the club called *"Pacifism,"* or [without a formal declaration], semiofficial members of the club called *"Militarism."* Such attitudes result from the *split "American philosophy"* in the postwar era [Appendix III]. But this pragmatism, and its by-products, are dangerous tools that work well in the short run and fail under prolonged stress. Western academe may thus not be able to avoid this central issue; it must therefore establish *"a code of ethics for a time of crisis."*

When once a nation begins to think, it is impossible to stop it.

Voltaire

6. Militarism, Pacifism and the Einsteinian Ethics

Einstein's portrait received a place of honor alongside those of Gandhi and Albert Schweitzer in the headquarters of a European peace society in the 1920s; two decades later he was regarded as *"the father of the atomic bomb."* After many years of service on behalf of pacificism and the ideals of democratic enterprise, it was intolerable to him to think that the Germans might get a hold of the atomic bomb and he considered it imperative *that the Americans be the first to develop this devastating weapon.* On August 2, 1939, he addressed his fateful letter to President Roosevelt:

> "Some recent work by E. Fermi and L. Szilard which has been communicated to me in manuscript leads me to expect that the element uranium may be turned into a new and important source of energy in the immediate future ... A single bomb of this type ... exploded in a port ... may very well destroy the whole port, together with the surrounding territory ..."

This letter initiated the famous Manhattan Project.

Was Einstein a militarist or a pacifist? Indeed, those who thirst for superficial definitions may have a difficult time here. One must avoid such offshoots of conventional wisdom and replace empty definitions by examinations of the important goals of humanity.

Civilization has one higher task than the achievement of social goals, the pursuit of happiness and the alleviation of the hardships of nature. It must also, as far as possible, ensure its own survival (for the longest possible time).

History has not happened to others. We live it every day, every place. Western civilization and its scientific revolutions are no less vulnerable than the now extinct civilizations of the Aztecs and the Incas, the Sumerians and the Hittities, the early civilization of China, and what is referred to as Graeco-Roman civilization. We have, however, reason to hope that the Vico-Spengler cyclic 'law' may not, after all, be the most universal one. Perhaps we can still buy the time to develop the proper methodology and employ it to improve our long-term chances to stem the tide of decay. It is admitted, however, that no government can advance this aim without the full cooperation of academe. And no government can convince academe without academe convincing itself as to the gravity of this subject.

Indeed, in their heart, scientists all over the world know the fact that what they uncover in their work has, in many instances, *a marked effect on the society around them*, and, yet, the "philosophy" behind their work is still based on the belief (even in a state university) that whatever they uncover is *not* of interest to anyone but themselves, and that they are *not* responsible to anyone but themselves for the kind of research they do, and for the technology that may be extended from that research. Thus, despite the *subjective* impact wielded by scientific research on society, some Western scientists keep proclaiming a *disinterest in consequences* and a *disclaiming of values* external to research itself! What hypocrisy!

Perhaps the optimal course for scientists to take at this time of crisis in philosophy and society is, on the one hand, to try to change the way they themselves evaluate their work vis-á-vis society as a whole, and, on the other hand, to refrain from aloof conduct and uninspired "academic solutions" to the problems facing this society. So far scientists have proceeded too far in the latter direction. And they have ended up with scientism and the superstition of academic freedom. There are many ways to academic freedom; there are many ways of looking at society and at the physical world; many scientific methods of exploring the universe around us. The present Western methodology is only one of them.

No style of thinking will survive which cannot produce a usable product when survival is at stake.

Thomas Favill Gladwin

It is essential for the men of science to take an interest in the administration of their own affairs or else the professional civil servant will step in—and then the Lord help you.

Lord Ernest Rutherford

7. On Academe, Power and Survival

Many Western scientists fall into the trap of assuming the 'total autonomy' of higher education. They fight against state intervention—especially when it involves classified research and development. U.S. universities in particular have led the rebellions against any direct involvement with national defense, classified, or organized research. These trends have generated a serious gap between U.S. academe and those who are preoccupied with long-term planning for survival—a crisis that is bound to influence the future of many democratic states.

Those who support this movement with a kind of fanatical naiveté have ignored one fact of life: that society has one higher task than to pursue its goals, engage in the pursuit of happiness and harmony and strive to eliminate pain, poverty, and the ubiquitous curse of ignorance. It must also, so far as possible, ensure its own survival! And survival is not the sole responsibility of government.

But a dangerous, indeed an infinitely more dangerous trend, is the association of an ever-expanding body of academe with long-term potentials for survival. Some scientists have been quick to argue that an ever-expanding body of university graduates will serve society best. But the fact cannot be eluded: as matters now stand, the welfare of our society depends on priorities in education, research and development, on our furnishing the most favorable environment and support while allowing science to develop according to its most advanced methodology; according to a scientifically-based pan-philosophy which does not ignore, nor neglect any aspect of global interdependence.

It is only with the help of such a philosophy that the relative autonomy of academe can be preserved; it is the only way to guarantee that 'indicative guidance' becomes the limit of government involvement with academe. But if production of graduates qua production of graduates is the primary concern of our academe, then long-term survival takes second place. And so it does.

Only when we get science into greater perspective will our priorities become more consistent with the global interdependence of society. There is much middle ground between a *laissez-faire* national policy for science and complete intervention by the state. On the one hand, we must develop the methodology which best influences the growth of science and technology, and which best ensures the existence of conditions which facilitate intellectual thinking. On the other, we must *encourage, respect, and reward* those scientists and civilian institutions that are, *by their own motives or by central design and purpose,* directing their best participation towards better defense capabilities. *No national policy can ever succeed in directing the best talents to conducting mission-oriented research and development without the genuine and unfeigned involvement of all academe. And no academe deserves its title without total involvement.*

8. Power and Survival in the Political Philosophy of Spinoza.

In closing this lecture it seems fitting to mention a few of the concepts of *Spinozism* regarding survival, ethics and political philosophy. Seldom has a single book enclosed so much wisdom and fathered so much rational morality as Spinoza's *Ethics*. His ethics flow from his universal philosophy (see also Lecture IX), and he builds his systematic ethics not on altruism and the "natural goodness of man," like utopian reformers do, nor on the natural wickedness of humanity, like cynical philosophers, but on what he considers to be an inevitable and justifiable egoism. *A system of morals that teaches a civilization to be weak is worthless in the eyes of this gentle philosopher: "The foundation of virtue is no other than the effort to maintain one's being; and man's happiness consists in the power of so doing"* (Ethics, IV, 18, note).

Survival is a central and fundamental law in Spinozism. Every individual being and every society obeys this "law of nature" and strives to preserve itself and continue in existence. Virtues are forms of ability and power!

According to Spinoza, all political philosophy must stress that mutual need begets mutual aid *"since fear of solitude exists in all men, because no one in solitude is strong enough to defend himself and procure the necessities of life, it follows that men by nature tend towards social organization."* And he adds:

"The formation of society serves not only for defensive purposes, but is also very useful, and, indeed, absolutely necessary, as rendering possible the division of labor.

If men did not render mutual assistance to each other, no one would have either the skill or the time to provide for his own sustenance and preservation: for all men are not equally apt for all work, and no one would be capable of preparing all that he individually stood in need of." For these reasons Spinoza stresses the central role the State and its aims must play in society: ". . . Laws should in every government be so arranged that people should be kept in bounds by the hope of some greatly good, rather than by fear, for then everyone will do his duty willingly."

"The last end of the state is not to dominate men, nor to restrain them by fear; rather it is so to free each man from fear that he may live and act with full security and without injury to himself or his neighbor" (Tractatus Politicus, Ch. 6; Ch. 20).

Spinoza formulated a *political philosophy* which expressed the liberal and democratic hopes of many nations. It became one of the main sources of thought that culminated in Rousseau and the French Revolution. In fact all philosophy after him is permeated with his thought.

9. Why Higher Education Has Failed

In examining what is possible today in science and what is not one may first focus attention on the system of higher education in the United States. Here one of the central questions is this: Does the U.S. spearhead a decline in the West, as some people claim, or is it only temporarily captivated, as I believe (Appnd. III), in the hands of an obsolete "philosophy of action and anti-elitism"? Understanding this problem may be a valuable clue as to why U.S. intellectuals are not really in the position they should.

Holding professorships in a number of American universities and working closely with some of the U.S. most advanced industries, I feel at home with higher education and industry in that great country. And it is mainly upon this experience that I base the following criticism—a criticism that is aimed to challenge American designers of future learning systems to formulate curriculums and educational systems that will *re-connect thinking to practice;* that would make people *more accountable* not only for the capacity to think but for the impact on others, and especially on other nations—of using that thought.

This is especially difficult because the Western system, in particular in the U.S., fosters *separatism of action from thought* (Appendix III). The universities and the press have led many Americans to believe that having the good idea is enough, on the one hand, or, at the other hand, that a mere 'delivery' is enough in the absence of the inspiration of a guiding philosophy (Appendix IV).

The historical and educational roots of this attitude are discussed in Appendix III.

All I want to stress here is *the isolation of American young from the challenging external world and from their own society at large.*

Prior to entering the university, and during their studies, most American young make no personal contribution to society at large, nor to their country; that is to say, they do not encounter any military service, nor a minimal *civil* (national) service to help society in its various difficulties. All that is required is a performance in class—a performance that is mainly oriented *'towards the self'*. Thus, they cannot relate themselves to the *'outside world'*. Instead they pursue their own whims. Moreover, in the absence of a guiding philosophy for thinking and action, in the absence of the teacher's ability to say 'what is best', 18-year-old students just pick up university courses from a curriculum with 500 courses or more. They "design" their own curriculums according to their whims *for their professors lack the courage and the authority of superior guidance; for their professors throw up their hands in despair in response to the anti-system outbursts in the 1960's and 70's.*

The result is that the young are quite isolated from the external world and from hard-core courses of instruction for an important period of their life, simply to learn *apart* from the exigencies of hard practice and the reality of society at large; mainly to learn to put much emphasis upon *"the discovery of the self"*. Indeed, up until relatively recently, they (and their teachers) imagined that the horizons of their country are *unlimited!* They had assumptions of so much excess and surplus that they have norished the idea that "you don't have to try hard; you can make mistakes and still win!"; that without struggle you can be "a good student".

Then, emerging from the universities, they find that all great actions and any real leadership in science, business and government, inevitably fall—or rise—along some philosophical guidelines that connect thinking and action; *the very guidelines that have been abolished from their system of higher education; the missing solutions to most of their problems:* those that point to their own ignorance; the ignorance of a working philosophy; of a unifying philosophy of thought and practice. But as we shall see below, this problem is not unique to the U.S. The whole Western system is now confronted with it! Today, more than ever, there is a sense of deep aversion to compulsive philosophical studies in *all* Western universities!

Who is to be blamed for this lacuna? Let us examine next the role of the 'peers of academe', of contemporary intelligentsia, of the 'ruling elite' in our 'best universities'.

10. The Case For Elitism and Excellence

Thus I saw that most men care for science so far as they get a living by it, and they worship error when it affords them a subsistence.

Johann Wolfgang von Goethe[+]

If a professor thinks what matters most
Is to have gained an academic post
Where he can earn a living, and then
Neglect research, let controversy rest,
He's but a petty tradesman at the best,
Selling retail the work of other men.

Kalidasa (between 200 b.c.e. and 400 a.c.e.)[++]

Today, more than ever, the scattered fragments of Western thought are cut off from consistent philosophical principles. They lack a guiding idea that can inspire the young, revitalize intellectual leadership, or stem the tide of bureaucrats and officialdom. Penetrating academe en masse, the latter replace academic with bureaucratic hierarchy, collect administrative power, and impose their low standards on the young.

By traveling to endless 'organizational meetings', on every mediocre subject that one can image, and by publishing technical jargons that are *(deliberately?)* made nonsensical to other "officeholders", many "jobholders" make their livings as "experts" and "professionals". Lecturing in these meetings they demonstrate professionalism by verbal funambulism; they hide behind organizational titles, and come on big with politicians and ill-informed newsman (Appendix IV)

Intellectual freedom, without fear of unwarranted dismissal, or other sanctions, is *exploited* nowadays to defend (and propagate) *meaningless* research and courses of instruction. This, in turn, causes "jobholders" to multiply in geometric progression, and limits the graduation of bright minds to an arithmetic growth. *A new type of clergy is thus beginning to establish itself, primarily through the very impact of its masses. Will this bring Western academe to a modern version of the cloistral dark ages? or to a revolution followed by a renaissance?* This remains to be seen. For the moment, however, officialese and officiants dominate some of the most vital domains of academe, and endless students, whose ability and motivation are questionable, to say the least, shape the future of "our avant garde".

In contemporary Western academicism there is a pervasive idea, adhered to by many, that given time and effort anyone can master anything of an intellectual nature. Thus the concept of elitism in education has become almost obsolete.

Granted that there is a natural range of *intellectual,* as well as *physical* abilities, why do most Western universities persist in trying to diminish, or even eliminate elitism in education? *Not everyone is equally educable!* To grant academic degrees to all may, eventually, harm society. It is not even clear what mankind is ultimately supposed to gain by it. It is clear, however, what is lost. When we destroy the environment of elitism, we destroy the highest level of creative endeavor, whether it be scientific, artistic or lit-

---

[+] From Eckerman's *Conversations with Goethe;* 15 October, 1825.
[++] From *Scientific Quotations,* A. L. Mackay, Crane, Russak, New York, The Inst. of Phys., 1977. Malavikagnimitra i.17; In *Poems from Sanskrit,* trasl. John Brough, Pengwin, London, 1968.

erary, and we replace it with endless mediocre activities. *Thus we have raised a new type of clergy; a clergy which reaches down to the lowest intellectual level; a clergy which admits elitism only in such areas as "sports and entertainment"; a clergy which does not win the respect of the public.*

It is therefore imperative that we put a halt to the Western obsession with tainting elitism. *We must recognize elitism in education for what it is; the generation of avant-garde and the opportunity for highly creative individuals to pursue ever-higher goals, or to reexamine the very foundations on which their culture rests.*

To begin with, grade inflation should be stopped. Next, the gates leading to higher education should be kept half closed to prevent mass inflation. Finally, government support of state universities should be dissociated from the current practice of allocating financial support in proportion to the number of heads in academic herds.

It is the *philosophical basis* of higher-education that should undergo *a drastic change:* from peak sales of pedigrees to endless breeds of mediocre graduates, to granting authentic degrees to a few who deserve them. *This is an unpopular policy. But it is a timely one.*

But it should be stated at once, that we must reject most past, and present meanings of elitism, i.e., those associated with *hereditary nobility, aristocratic elite,* or *upper class.* Instead of the past *hereditary and plutocratic aristocracies,* instead of the present rule of *bureaucrats, officialdom, or proletarian hierarchy,* we must search for merit in the possession of *virtue,* or *arete;* in academe, in the professions, in the arts, in politics, in the press (Appendix IV).

None in a democratic academe can have tenure with elitism; for academe must incessantly strive to find 'the best', not *apart* from society, not by *ignoring* people's wishes, but as an *interdependent service,* as an integral part of the structure of a dynamic, changing society; a society which fluctuates between the *Platonic* and *Aristotelian* modes of academe, government, and philosophy.

Whereas Plato stressed wisdom, in the transcendent sense of his idealistic philosophy, as the prime basis of virtue, Aristotle took a more traditional view of virtue, though he incorporated the capacity to reason profoundly and thus achieve wisdom along with other kinds of superior performance.

Thus, academe must incessantly strive for an *interconnected elitism,* instead of an isolated aristocracy; for a *humble elitism,* instead of an arrogant one; for *sociable elitism,* instead of a ruling one; for a *conditional elitism,* instead of an hereditary one. But, most important, we should bear in mind that elitism *cannot* be simply defined, as was often done in the past, by defining *selectivity* and *excellence,* or by reciting the traits that have been prevalent in western culture; for in our rapidly changing culture they should have a much broader meaning in the context of globally interconnected mode of thinking (cf. the Introduction and Lecture IX).

11. Territorialism and Internal Politics in Science?

At the beginning of my lectures, students often ask me to *define* the aim, scope and methods of the "philosophy of science". My reply is that there is no "philosophy of science", but only the individual philosophies of certain scientists. Moreover, even the terms "science" and "philosophy" cannot be clearly defined. Major disagreements about the definition of "science" have arisen from premature commitments by a philosopher unfamiliar with modern scientific methods, or, vice versa, by a scientist operating within the narrow disciplinary boundaries of his profession.

In this series of lectures I express a skeptical view of today's methods of academe, "science," "philosophy," and the "philosophy of science." My view is based on a number of considerations*, some of which are as follows:

---

* My skeptical view derives from Assertions 4 and 11 (Introduction; §3), which, *inter alia,* uphold Aristotle's view that dialectical reasoning is a "process of *criticism* wherein lies the path to the principle of all inquires."

1. *A definition often tends to isolate an idea, field of study, or object.* Such isolation stands in glaring *contradicition* to the aim of a *unified* approach which I adopt in these lectures. Indeed, ideas like *"object"* and *"isolation"* are coupled. The concept of "isolation" (and therefore also the concept of a definition) is an *idealization.* In nature, there is no perfect realization of such a requirement as the isolation of a *physical* system. * The same situation is found at every step in education, science, communication and philosophy. Yet we usually accept definitions without any thorough discussion! It is often *the very act of definition* that, *a priori*, limits the accuracy of our description, not vice versa!

2. It is usually a *"jobholder"* who is eager *"to define"* (or protect?) his *"territory"* ** of professional activity. In the scientific community we find "jobholders' and "enthusiasts." *** The "jobholders" carry out duties connected with the jobs entrusted to them. They are highly motivated by a quest for immediate status (or by economic and political considerations). *Consequently, by and large, they tend to defend their "territory" with well-fenced definitions and titles.* The "enthusiasts", on the other hand, are anxious to carry out their own self-appointed tasks, and can hardly do anything else. They accept no "borderlines" or "established territories".

But it is really the "jobholders" who keep the scientific community going with the necessary continuity and smoothness; they are the defenders of tradition and custom, the builders of departmental disciplines and professional committees. It is they who do all the routine work without which everything might degenerate into chaos; yet by and large it is the "enthusiasts" who are the men of science, the inventors, the discoverers. They are the main instruments of change and progress (Appendix II).

A famous enthusiast, *Galilei Galileo,* had, however, some complaints:

> "To those who show with great brevity and clearness the fallacy of propositions commonly held true by the generality, it would be fairly bearable injury to be repaid only by contempt instead of gratitude; but very unpleasant and grievous is a certain other sentiment which is sometimes aroused in some persons who claim to be at least the peers of anybody in the very same studies, persons who are seen to have maintained as true conclusions which another has then by a short and easy discourse unveiled and proclaimed false. I shall not name that sentiment envy, which usually turns into hate and anger angainst those who discovered the fallacies, but I will call it rather annoyance and yearning to be able to maintain hoary old errors rather than allow newly discovered truths to be received. This yearning sometimes induces them to write in contradiction of truths even they themselves admit in their hearts, just so as to keep down the fame of others in the opinion of the numerous and ill-informed common folk."

Later, in modern times, complaints as skeptical as *Max Planck's* have been voiced:

> "A new scientific truth does not triumph by convincing its opponents and making them see the light, but rather because its opponents finally die, and a new generation grows up that is familiar with it."

A. L. Mackay (*Scientific World, 13,* 17 (1969)) puts this another way: "How can we have any new ideas of fresh outlooks when 90 percent of all the scientists who have ever lived have still not died?"

Indeed, even Einstein was vigorously attacked and quite isolated by his opponents (Lecture V).

"In science", says Darwin, "the credit goes to the man who convinces the world, not to the man to whom the idea first occurs." (cf., e.g., Einstein, Bohr and quantum physics - Lecture V).

---

*   cf. Assertion 2 (Introduction, §3), and Lectures VI, VII and IX.

**   cf. Assertion 9 (Introduction, §3).

*** These terms have been borrowed from Sarton's *A History of Science* (Norton & Co., N.Y., 1970) pp. xii–xiii.

But these are extreme cases, so that, in general, one may say that *success in science depends on the emergence of appropriate social structure.* Science 'detached' from social structure can, perhaps, do without "jobholders", but a university community cannot. *Both "jobholders" and "enthusiasts" are equally important,* for *without inertia and tradition, "the enthusiasts" are unable to polish and sharpen their ideas.*

New ideas are never completely independent and original; they hold together and form causal chains, the golden chains that we call progress. New chains may later become embarrassing and dangerous to further progress. Sometimes they become heavy, like iron shackles, and there is no way to escape but to break them. *Such revolutions are part of the history of science, but they are also essential parts of the structure of science.*

Perhaps the best description of everyday crises in academe has been given by *Erwin Schrödinger:*

> Our age is possessed by a strong urge towards the criticism of traditional customs and opinions. A new spirit is arising which is unwilling to accept anything on authority, which does not so much permit as demand independent rational thought on every subject, and which refrains from handling any attack based upon such thought even though it be directed against things which formerly were considered to be as sacrosanct as you please. In my opinion this spirit is the common cause underlying the crisis of every science today. Its results can only be advantageous; no scientific structure falls entirely into ruin; what is worth preserving preserves itself and requires no protection.

3. Many contemporary scientists concede that the product generated by them is *"nothing but a new truth about nature,"* yet, to some extent, their preoccupation is with *"careerism"* connected with the process of publication rather than the quality of the thing generated. *"Territorialism," "invisible colleges"* and nondemocracy exist almost paradoxically in the midst of the traditional democracy and openness of the scientific community.

Hidden variables of subjective influence and (local) *cultural tradition actively shape large portions of scientific and philosophical perceptions.* Consequently the best I can do in trying to answer the students' question that I mentioned above is to quote *Albert Einstein:*

> *The whole of science is nothing more than a refinement of everyday thinking.*

Science and culture in a given civilization are therefore intimately interrelated. There is no scientific subfield, nor even a single scientist totally isolated from contemporary society. Academe is therefore a microcosm of the society in which it functions.

12. The Need of Philosophy in a Time of Crisis

According to *Karl Jaspers* (1883–1969), one of the most respected philosophers in Germany since the end of World War II and one of the leading founders of existentialism, the basis of technology, which is propelling our civilization, is modern science, whose emergence represents a profound turning point in the history of mankind.

In *The Perennial of Philosophy* (Philosophical Library, New York, 1949), Jaspers analyzes our era of radical change, of modern science and technology with its consequences for man's working habits and of the unity of the globe created by modern communications, together with today's nihilism which rejects everything, *including philosophy,* as worthless. Nothing that *mass education* makes people *blind* and *thoughtless,* he poses two central questions:

> 1) Is all this a spiritual revolution, or is it an essentially external process arising from technology and its consequences?
>
> 2) What should philosophy do in the present world situation?

In trying to answer these questions, Jaspers stresses that *"nihilism, as intellectual movement and as historical experience, becomes a transition to a profounder assimilation of historic tradition. From an early time, nihilism has not only been the road to the primal source—nihilism is as old as philosophy—but also the acid in which the gold of truth must be proved."* And he adds:

> To achieve this, philosophy proper must, among other things, reject the idea of progress, which is sound for the sciences and the implements of philosophy. The advocates of this idea falsely believed that what comes later must supplant what comes earlier, as inferior, as merely a step to further progress, as having only historical interest. In this conception the new as such is mistaken for the true. Through the discovery of this novelty, one feels oneself to be at the summit of history. This was the basic attitude of many philosophers of past centuries. Over and over again they believed that they had transcended the whole past by means of something utterly new, and that thereby the time had finally come to inaugurate the true philosophy . . . .

Then, stressing that philosophy is bound to science and works in the medium of all the sciences, he warns against too much scientism:

> Superstitious belief in science must be exposed to the light of day. In our era of restless unbelief, men have snatched at science as a supposedly firm foundation, set their faith in so-called scientific findings, blindly subjected themselves to supposed experts, believed that the world as a whole could be put in order by scientific planning, expected science to provide life aims, which science can never offer—and expected a knowledge of being as a whole, which is beyond the scope of science.

13. Historicism vs. Theoretical Physics

Scientism has also been discussed by *Sir Karl R. Popper* (1902–   ), one of the leading philosophers of this century. He is best known for his studies in the history of philosophy and the philosophy of science and also for his political and ethical philosophy. His book *The Open Society and Its Enemies** has become a contemporary classic in political philosophy. It examines some of the principles of social reconstruction and outlines the main difficulties faced by modern civilizations.

One of Popper's main arguments is that we cannot predict the future because human history is strongly influenced by *"the growth of knowledge"* and because it is logically impossible to predict "the growth of knowledge" itself. Accordingly, he claims that *"we must reject the possibility of a theoretical history; that is to say, of a historical social science that would correspond to theoretical physics."* In particular he stresses that history has no meaning. *"It is we who introduce purpose and meaning into nature and into history."*

Popper, like Jaspers, refutes the idea that "history progresses" or that we are "bound to progress". "For to progress is to move towards some kind of end, towards an end which exists for us as human beings. 'History' cannot do that; only we, the human individuals, can do it." Facts as such, according to Popper, have no meaning; they gain meaning only through our own decisions and interpretations. *Historicism, like scientism, is thus a superstition.* For it tries to persuade us that if we merely follow its laws and fall into step with its progress everything will and must go right, and that, *consequently, no fundamental decision on our part is required;* it tries to shift our own responsibility onto history and science, and thereby onto the play of irresistible forces beyond ourselves.

A critical comparison of Popper's inherent optimism with Plato's and Toynbee's pessimism (Appendix II), does not necessarily lead to a contradiction. *Global historical facts concerning local processes of decay in civilizations or individual attempts at regenerating the underlying cultures are not incompatible with one another. We can make it our fight to regenerate our culture and oppose its antagonists, and, in particular, we can return to the very roots of philosophy and ideology which initially gave Western civilization its life. We must, in Popper's words, learn to do things as well as we can, and to look out for our mistakes.*

---

* Rowtledge & Kegan Paul, London, 1945, *Volume I: Plato, Volume II: Hegel & Marx.*

And when we have dropped empty scientism, of expecting everything from sacred academicism, when we have given up the present nihilism, then, one day, we may succeed in leading our culture into another renaissance.

To recognize our lacuna is, however, not a simple task. For one thing, our lacuna has been sustained by a false methodology, the lack of consistent philosophy, and a set of interconnected global trends. World War II marks a turning point in our civilization, when a cluster of overwhelming issues started to accelerate a decline. Four major issues converge to give powerful impetus to this decline: the growing challenge of the Soviet and Chinese civilizations founded on the Marxist philosophy; the growing challenge posed by limited resources and the Third World; the fate of democracy, unification and ideology in the West; and, what I assert might be the most important issue: *the shift of the center of Western thought from Western Europe to the U.S., a shift which resulted in a drastic modification of Western politics, culture, science and philosophy* (Appendix III).

Today there are many Westerners who are attracted to dialectical materialism and Marxist philosophy. Both constitute a real challenge to the West, for they are employed as tools by governments that now rule more than a billion people. They form an *operative element* in *creating* the Western crisis, for they *export* an *anti-Western philosophy* and consistently use it to provoke a protracted Western crisis that is very wisely, *and passionlessly,* being exploited by the Communists and accepted by various parties around the globe as *"The unquestionable science"* in "explaining *all* facts" and "guiding *all* action." Although there have been drastic alterations in the original Marxist philosophy, there remains a basic core of theory which promotes Communist world strategy by providing it with a flexible guide to action, and, most important, by serving as a catalyst to change the rest of the world.

Why cannot the West develop a consistent, testable, workable philosophy, and employ it as a flexible guide to thought and action, as a catalyst, and as a practical tool in shaping the course of events in reference to its own goals? But, most important, what are these goals?

Our most popular slogans—democracy, human rights, free enterprise, personal comfort and property—by themselves, or within the frameworks of idealism, pragmatism, logical positivism, etc.—*cannot* be effective for long without being part and parcel of a *whole* system; i.e., without being part and parcel of a TOTAL way of thinking; of a *consistent pan-philosophy* that can be employed as the (testable) *compass* for reflection and action.

APPENDIX I

A FEW HISTORICAL REMARKS ON TIME, MIND, AND SYMMETRY

A. Symmetry and Asymmetry at the Dawn of Science

The ideas of *symmetry* (duality, twoness, pairing, etc.) and *asymmetry* (directing, aiming, beginning-end, evolution, attraction, repulsion, polarity, oppositeness, etc.) must have occurred to the human mind early in the history of science, for there are many "obvious" symmetric and asymmetric pairs in nature. People have two eyes, two nostrils, two ears, two hands, two feet, etc. *Yet, the hands are asymmetric;* for eating, drinking, using tools, or fighting imply different tasks for each hand. The two hands demonstrated the right and left (or preferable) aspects of everything. In addition all animal life is dominated by the *"polarity"* of sex, every animal is either male or female, i.e., human and animal life is, at the same time, symmetric *and* asymmetric. *Moreover, every symmetry appears necessarily under an embedded asymmetric aspect;* things are soft or hard, hot or cold, dry or moist, large or small, up or down, gravitated or dispersed.

Consequently, it should not surprise us that primitive concepts betray a kind of *"utopian symmetry"*— like that of the perfect spherical world in Platonism.* Similar findings emerge from prehistoric China. (Tracing symmetry–asymmetry concepts to their origins would very probably take us back to the most remote antiquity.)

Chinese ideology has been dominated by the symmetry–asymmetry concepts of *yang* and *yin,* the male and female, "positive" and "negative" principles of nature. *Yang* is hot, light, energetic, male; it is the sun, rocks and mountains, heaven and goodness. By contrast, *yin* is cold, dark, passive, female; it is the earth, the moon, water, trouble and evil (evidently the early Chinese philosophers were all males, and in our own time, many male chauvinists in the West are in complete accord with this view). But, aside from this *'sexual physics',* the concepts of *yang,* and *yin* were extended by the Chinese to the whole universe. And they ended up with a special kind of sexual cosmology.

Other civilizations generated similar cosmologies, using a different terminology. Indeed, *the principle of symmetry was extensively applied in early religious, philosophical, scientific and artistic activities.* The desire for symmetry (and music) was especially marked in *Pythagoras.* Yet, it had been pursued before Pythagoras' time in the Orphic mysteries and other religious ceremonies. In fact dualism and symmetry are rooted in the deepest recesses of human consciousness (e.g., the *Zoroastrian religion*).

Each language evidences the presence of a *symmetric–asymmetric number base* and numbers (the former was often five [among many American tribes], sometimes twenty [among the Mayas] or sixty [among the old Sumerian-Babylonians], but more often ten or a mixture of decimal and sexagesimal notions.* The symmetric–asymmetric numerical philosophy had far-reaching consequences for science, for it initiated *quantitative* studies of symmetric and asymmetric phenomena in nature. One might even conclude that physicists of all ages, the natural philosophers, have been constantly allured by the hope of finding new symmetric–asymmetric–numerical–material relations that would allow them to penetrate the secrets of nature. Here, Pythagoras was probably the boldest in claiming that *numbers are immanent in things!* He

---

* In this connection see our results in §IX.2.8.1. For sexagesimal notions see next page.

Symbols of *yang* (white, hot, energetic, male) and *yin* (dark, cold, passive, female) in the center and the eight diagrams around.

maintained that numbers are the essence of things, i.e., the better we understand them, their symmetry and associated asymmetry, the better shall we be able to understand nature. These views are the *seeds* of other, fully developed *modern theories* that employ *mathematics and geometry* in the description of the *physical evolution* of the universe (see Lecture III on the foundations of Einstein's general theory of relativity, the discussions on time reversal and the arrows of time in §IV.6 to §IV.12, and the discussion of time asymmetries in words, language, information, and thinking in §IX.2.8.1).

B. Mind and Order

The ideas of symmetry–asymmetry and time have often been associated with *"human ordering"* or with the *"order of nature"*. Indeed, many philosophers have been attracted by the hope of finding differences between these two ordering systems, physical VS. mental ordering, and the possibility for *"separation"* between "body" and "mind". But can "human ordering" be separated from "physical ordering"? Is there any correlation between human time and geophysical time, or between biological and geophysical clocks? These problems have been discussed in Lexture IX. What we have attempted there was to understand time by seeking relationships between apparently different clocks. This has allowed some to be explicable in terms of others, thereby *reducing the number of independent variables* (or independent "time machines"). To supplement this study we present here a few historical ideas associated with the separation of human time from physical ordering, and of "mind" from "body".

1. 甲子	11. 甲戌	21. 甲申	31. 甲午	41. 甲辰	51. 甲寅
2. 乙丑	12. 乙亥	22. 乙酉	32. 乙未	42. 乙巳	52. 乙卯
3. 丙寅	13. 丙子	23. 丙戌	33. 丙申	43. 丙午	53. 丙辰
4. 丁卯	14. 丁丑	24. 丁亥	34. 丁酉	44. 丁未	54. 丁巳
5. 戊辰	15. 戊寅	25. 戊子	35. 戊戌	45. 戊申	55. 戊午
6. 己巳	16. 己卯	26. 己丑	36. 己亥	46. 己酉	56. 己未
7. 庚午	17. 庚辰	27. 庚寅	37. 庚子	47. 庚戌	57. 庚申
8. 辛未	18. 辛巳	28. 辛卯	38. 辛丑	48. 辛亥	58. 辛酉
9. 壬申	19. 壬午	29. 壬辰	39. 壬寅	49. 壬子	59. 壬戌
10. 癸酉	20. 癸未	30. 癸巳	40. 癸卯	50. 癸丑	60. 癸亥

* Cf. the sexagenary cycle shown above. In this cycle the ten symbols of each first column are alike; they are ten celestial stems. [H. A. Giles, Chinese-English dictionary (Shanghai, ed. 2, 1912), Vol. 1, p. 32.]

B.1 The Hebrew 'Et,' 'Moed,' 'Zmann,' 'Schahoot,' 'Meschech' and 'Va-Yehi'

A few distinct concepts of time play a central role in Judaism and in Hebrew philosophy. The difficult problem before us is related to the fact that it is almost impossible to translate the Hebrew words associated with the concept of time without losing much of their essence and original meanings. The English words 'time' and 'duration' have something very abstruse in their nature, while the six Hebrew words associated with these concepts convey a much greater value of distinct meaning. One way out of this difficulty is to turn right to the original. However, it is only within the context of *Havayism* (Lecture IX) that the meaning of such words can be fully understood. Hence, in this lecture, we shall only cite a few examples, starting with the words 'et' and 'zmann.'

> "To *everything* there is *'zmann'*, and *'et'* to every purpose under the heaven . . ." says Kohellet (III, 1 to 8), and to distinguish 'zmann' from 'et' he adds clear cut examples. " 'et' to be born, and 'et' to die, 'et' to plant, and 'et' to pluck up that which is planted". . . ". . . 'Et' to weep and 'et' to laugh". . . ". . . 'et' to love and 'et' to hate, 'et' of war and 'et' of peace."

Everything in the world *may* therefore be associated with an *impersonal* 'zmann', while the concept 'et' is exemplified by *purposeful season, predetermined "era,"* and *subjective* happenings, changes and motions in *human* life. Zmann is, therefore, both objective and subjective, while 'et' is semi-subjective and is tied up with life on earth. Zmann is symmetric, infinite and universal. 'Et' is somewhat fixed, and can take place only *under* the heavens, closer to man and his feelings, mind, and purpose. While Kohellett may have something important in mind to tell us in making the distinction between 'zmann' and 'et', modern Hebrew does not. Nevertheless, even in modern Hebrew, *the word 'et' can never be used with respect to cosmological times*. The last concepts are exclusively reserved for the word 'zmann'. Hence, 'zmann' in modern Hebrew is normally accepted as the master word covering all other concepts of time.

The word *'moed'*, on the other hand, is usually used in relation to "date", *"fixed time," "moment," "event"* or *"appointed* time", but, again, as with 'et', 'moed' is a *subjective* time, associated with human deeds, such as holidays, meetings, birth, or predetermined signals for *human* actions.

The word *'schahoot'* may be variably translated as *"irreversible finite duration",* or as having sufficient time to perform a specific human or nonhuman action. Thus, it may be used on both objective and subjective grounds. But the emphasis here is on the *finiteness of a period of time;* the temporal *limitations* on both sides of an *advancing* time clock.

It implies a *unidirectional advance of a time arrow that cannot be reversed,* i.e., a period which has a definite *beginning* and a definite *end,* irrespective of the rest of the universe.*

The word, *'Meschech'* carries with it some subtle ideas. It implies, continuity, dimension, extension, causality, directionality, duration, movement and motion.

It may be associated with motion *"from source to sink",* such as when we observe the *direction rivers flow away from their sources* (Rashi, Beithza, XXXIX), or with the *creation of existence out of the void* ("Rashbag", "Keter Malchoott"). 'Meschech', in scientific language, may best be associated with the concepts *"arrow of time", "time anisotropy", "time asymmetry",* or *"irreversible zmann."* It is also interesting to note that "Meschech" does *not* distinguish between time and spatial coordinates. Both dimensions are treated on the *same* basis (Eirovin, X).

The last word is *'Va-Yehi,* the first in the Bible associated with time in an *observable* world.* It means genesis, coming-to-be, becoming, and creation.** Its roots are in the word *'Hayoh',*** which, in turn, is associated with such concepts as *'Haya'* (was, past existence), *"Hoveh"* (now, present existence), *'Yiheyeh'* (will be, future existence), *'Hythavoott'* (formation, generation, creation, becoming, appearance, etc.), and *'Havaya'*—the basic concept behind *Havayism* (Lecture IX).´

It is quite surprising to see that, over the history of the Western philosophy, each of these words has served as a starting point for a new philosophy. Some of these points of departure are mentioned below.

---

* 'Schahoot' is associated with some other irreversible durations, like delayed or retarded actions.

** Its (superficial) translation reads: *"it came to pass"* or *"And it had been caused to be".*

*** היה

B.2 The Concepts of Time in Greek Antiquity****

The Greeks often hypostatized time into a single entity, cosmological principle or god; *Chronos*. For *Sophocles* (496–406 B.C.E.) time is all-mastering; it is both the supreme teacher and the supreme judge; it is also the consoler of man.

Anthropomorphic concepts of time were rejected by *Anaximander* (610–547 B.C.E.) who, together with *Thales* (640–547 B.C.E.) and *Anaximenes* (6th cent. B.C.E.), were regarded among the *"earliest scientists of the West."* According to Anaximander all coming-to-be, from all ceasing-to-be, takes place "according to necessity." In this connection, *Heraclitus* maintains that ". . . all things happen by strife and necessity . . ." (Appendix II).

According to some commentators,* even the gods, and hence time in Greek thought, originated from the *aether*. *Archytas of Tarentum* (400–350 B.C.E.), a close friend of *Plato* (428–348 B.C.E.) and a Pythagorean, associated time *with the movement of the entire universe: 'Time is the number of a certain movement, or also in general, the proper interval of the nature of the Universe."*

Zeno (490–430 B.C.E.), in comparison, devised his four arguments (or "paradoxes"), to show the impossibility of motion and time; that motion in continuous time and space had to be infinite, and, hence, impossible even though the distance covered was finite. He also claimed that motion was impossible when time and space were regarded as discontinuous and finitely divisible.

Archytas' time and Plato's notion of time are quite similar. Thus, for Plato ". . . *Time came into being together with the heaven . . .", ". . . in* order that Time might be brought into being, Sun and Moon and five other stars . . . were made to define and preserve the numbers of Time." We can, therefore, conclude that to both *Archytas and Plato, time and the universe are inseparable!*

According to *Whitrow**, time, unlike space, is not regarded by Plato as a pre-existing framework into which the universe is fitted, but is itself produced by the universe, being an essential feature of its rational structure. We may return to this topic in Lecture VII.

B.3 Aristototle, Plotinus, Crescas, Saadia, Maimonides and Albo

The Roman philosopher (of Egyptian birth) *Plotinus* (205–70), used the Platonic concept to create a mystic religion of union with the One through contemplation and ecstatic vision. Through *St. Augustine* (354–430), his theory entered into the traditional stream of Western philosophy. In his historical survey of all views *that make time dependent on motion,* he reproduces the view of *Aristotle* (384–322 B.C.E.) on time, which in his paraphrase reads; *"time is the number and measure of motion."* The original definitions of time by Aristotle are somewhat similar:

"... time is this, the number of motion according to prior and posterior"
"... time is not defined by time . . ."
"Not only do we measure the movement by the time, but also the time by the movement, because they define each other."

Time, according to Aristotle, is infinite and continuous, and it *cannot* have, as Plato thought, a "beginning" or an "end". For, he claimed, had there been a beginning to motion, a priori change or motion would have been required to start it. But this prior motion would itself have required a still prior motion, and so on *ad. inf.,* that is, *there must always have been motion, and consequently, time. Therefore, time cannot have a beginning, nor for similar reasons, can it have an end*

Rejecting the Aristotelian definition of time, Plotinus defines time as something *independent of motion;* in fact he identifies it with the life of the *"Intelligence"*, which is a kind of "extension" and "succession." It is varied in its nature and is a process of transition from one act of thought to another, the unity of which

---

**** Most of the historical concepts mentioned in this section were reviewed by P. E. Ariotti in *The Study of Time, II,* Springer-Verlag, Berlin, 1975 pp. 69–80. References to most of the citations mentioned here can be found in this paper.

* G. J. Whitrow, *The Natural Philosophy of Time,* Thomas Nelson & Sons, London 1961, pp. xi, 28, 28.

exists only by virtue of a certain kind of continuity. Finally, it is Plotinus' contention that the universe moves *within* this kind of "Intelligence," i.e., there are *two kinds of time*. One is indefinite time; the other definite. But both are independent of motion. The main differences between the Aristotelian and the Plotinian definitions of definite time are thus two fold:

1) According to Aristotle time is generated by motion; according to Plotinus, time is only made manifest by motion.

2) According to Aristotle, time is the measure of motion; according to Plotinus, time is measured by motion.

Plotinus' philosophy became the source of various philosophies in which the term *duration* appears as something *independent of motion!*

Traces of this influences may also be found in Jewish philosophy. The great philosopher *Saadia Ben Joseph* (882–942) maintains* that time is "that the world is within it;"—obviously an external time which is in opposition to the Aristotelian definition of time. Similarly, the Jewish philosopher *Hasdai Crescas* (1340–1412), author of *Or Adonai* ("Light of the Lord"), criticised the Aristotelian philosophy. He developed a new conception of "time-universe," influenced Pico delle Mirandola and possibly *Giordano Bruno*.

Crescas distinguishes between predestination, in the sense of fatalism, and causal determination, and admits that the human will, though free to choose, is determined by the causality of motives. *Spinoza* (see below, and in Lecture IX) mentions Crescas in a letter and seems to owe some of his ideas to him.

Crescas defines time as *"the measure of the duration of motion, or rest, between two instants"**,* thereby trying to free time from motion. As a consequence of his definition he arrives at the conclusion that there had existed time prior to the creation of the world.

On the other side we find the great Jewish philosopher *Maimonides* (Rabbi Moses ben Maimon, or, in short, 'Rambam') (1135–1204), who sought to reconcile Aristotelianism with biblical and Rabbinic teaching. His celebrated *Moreh Nevukhim ('Guide to the Perplexed')* contains the main tenets of his philosophical thinking and exercised considerable influence on both Jewish and Christian scholasticism.

Following Aristotle in viewing time as a product of motion, he concludes that it could not have existed prior to the creation of the world. Nevertheless, he states, we may have, in our mind, an idea of a certain duration, which existed prior to the creation of the world. He calls that duration "a supposition, or imagination of time, but not the reality of time." It is, therefore, in Maimonides' philosophy that the Plotinian time is (partially) combined with the Aristotelian time.

Joseph Albo (d. 1444), a pupil of Crescas, tried to harmonize Crescas with Maimonides; in his Ikkarim ("Principles"),*** he asserts that there are two kinds of time. One is *"unmeasured duration which is conceived only in thought, and which existed prior to the creation of the world and will continue to exist after its passing away."* This he calls *"absolute time".*

Albo identifies this time with the concept *"imagination of time"* that Maimonides has conceived.

Albo's second time is the Aristotelian one, which is numbered and measured by motion, and to which the concepts of past, present and future apply. It should be noted, however, that Aristotle stressed not only time vis-a-vis motion, but time vis-a-vis mind:

". . . when the state of our minds does not change at all, or we have not noticed its changing, we do not realize that time has elapsed . . ."****

I call this subjective time 'et', and then note, that when Aristotle states that ". . . change is always faster or slower, whereas time is not . . .", he is actually comparing 'et' with 'zmann.' Again, when he speaks about the magnitude of time, we may conceive it as 'schehoot', while the difference between 'meschech' and 'schehoot' is exemplified by his statement:

---

* Emunot we-De'ot, I, 4
** *Or Adonai,* I, ii, II
*** Ikkarim, II. 18
**** Physics, IV, 11

". . . it is not possible for a thing to undergo opposite changes at the same time, the change will not be continuous, but a period of time will intervene between the opposite processes . . ."*

Therefore, one may use the concept *'celestial meschech'* when contemplating on Aristotle's celestial time:

". . . the revolution of the heaven is the measure of all motions, because it alone is continuous and unvarying and eternal . . ."**

Within this kind of 'celestial meschech' there is no personal choice, no subjective 'et', nor even 'schehoot' to 'change our minds'. All is fully predetermined. Hence, it is only with respect to the concept of 'et' that human feeling, mind and thought come into the picture. What, then, is the origin of our 'et'? What past philosophers taught about this kind of time ordering? To the best of my knowledge *Anaxagoras* was probably the first name associated with such ideas. Not much is known about him and his work. But in connection with our subject some fragmentary pieces of records exist. Let us turn to them now.

B.4 Anaxagoras

At the beginning of time, according to *Anaxagoras* (about 500 to 428 B.C.),*** "all things were together." Moreover, he claims that "all things are together now" and "there is a portion or 'share' of every thing in everything." By an infinite number of "things" Anaxagoras explains the physical universe (on the assumption of an infinite number of *elementary 'stuffs'* as opposed to one in the *Ionian philosophy* and four in *Empodocles'*). For him these elementary 'stuffs' are *inseparable* from each other; and their inseparability leads to *the principle of infinite divisibility,* "because there cannot be a smallest, therefore all things are together."

Thus, according to Anaxagoras, the elementary 'stuffs' can undergo *unlimited division,* which, nevertheless, enables them always to remain associated together, i.e., there is an identical set of ingredients ("portions") in every lump of matter, though the portions may vary from one lump to another.

Modern science does not refute these ideas! In fact, today we know, that what we call elementary particles, are not elementary at all, for they have *internal structures* and can be *broken up* into smaller and smaller constituents, apparently with unlimited divisibility (e.g., molecules to atoms, atoms to protons, etc., protons to quarks, or to much smaller "lumps" of "matter" as modern *geometrodynamics* and *gravitational collapse* in massive stars imply (see Lectures I and III and §IV.13).

In fact, Anaxagoras opposes the notion of "atoms"—the indivisible particles. It is, therefore, quite *surprising* that in Western philosophy the name Anaxagoras is often linked to the earliest *'separation'* *of body and mind.* It is doubtless that this 'separation' was common in earlier thinking, though the records are not clear about that. Nevertheless, for Anaxagoras, everything, however small, *contains portions of all elements, except* as regards to mind, which, for him, is *the source of all motion.* In fact, emphasis on this 'separation' is clearly marked in *Plato's* writing, especially in the *Phaedo,* and in *Aristotle's* theory, especially in his *De Anima. It is in Aristotle that we find a theory that separates intellect from the physical body.* Both Aristotle and the Platonic Socrates complain that Anaxagoras, after introducing "mind", makes very little "use" of this concept. Since then, especially for *some religious motives,* this 'separation' has always been attractive as leaving open the possibility of "survival after death", etc. The body obviously perishes, but the mind, according to this, becomes immortal.

* Physics, VIII, 1.

** De Coelo, I. 2.

*** Only a few fragments of Anaxagoras' writings have been preserved, and quite divergent interpretations of his philosophy have been offered in succeeding ages, notably by Socrates, Plato, Aristotle and later thinkers. For the fragments see H. Diels and W. Kranz, *Die Fragmente der Vorsokratiker,* 6th ed. (1951–52). See also F. M. Cleve, *The Philosophy of Anaxagoras* (1949). Anaxagoras was celebrated for his discovery of the cause of eclipses. But he was prosecuted on the charge of impiety for maintaining that the sun was incandescent "stone" larger than the Peleponnesus). In science he had great merit. It was he who first explained that the moon shines by reflected light, though Parmenides may have known it before him.

B.5 Descartes

The problem became acute in the 17th century, which first saw the hope of presenting physical nature as a science of the whole. But an unsatisfactory compromise was adopted by *René Descartes* (1596–1650), who accepted orthodox Christian interpretations of "free will", indeterminism, etc. Therefore, he maintained three propositions, which taken together lead to logical contradictions. These are:

1. *The physical nature is a "closed system" in which all physical events are completely explicable in terms of physical laws;*
2. *The mind is a substance of a different nature and separable from anything physical.*
3. *Every mind is in intimate causal connection with some physical body.*

B.6 Spinoza and Leibnitz

These contradictions disappeared in the philosophy of Spinoza. Rejecting Descartes' theory, Spinoza used Descartes' method to demonstrate that free will is an illusion festered by ignorance. In his philosophy he explains in detail that it is the philosopher who can aspire to freedom of a sort by studying nature and accepting it; not as something opposite to a superior creator, but as the single whole of all that there is to conceive about. It is in this sense that Spinoza seeks to deliver individuals from "human bondage". "We are a part of universal nature," says Spinoza, "and we follow her order. If we have a clear and distinct understanding of this, that part of our nature which is defined by intelligence, in other words the better part of ourselves, will assuredly acquiesce in what befalls us, and in such acquiescence will endeavor to persist." Thus, insofar as man is an unwilling part and parcel of the whole of nature, he is in "bondage"; but insofar as, through understanding, he has grasped the nature of the whole, he is free. The implications of this doctrine are developed in the eloquent concluding sections of the *Ethics,* his most original work (Spinozism is further discussed in Lecture IX).

Leibnitz on the other hand, was anxious to avoid being thought to share Spinoza's concepts. Thus, his writings can be considered as a sustained effort to avoid falling into Spinozism. He owed much to Spinoza, concealed his debt, and even went so far as to lie about the extent of his personal acquaintance with the heretic Jew. Against Spinoza's all-embracing determinism he introduced *"monads"*, of which the human "soul" was an example. Each "monad" was "self-determining" and strict determinism was also avoided (§IX.2.17).

What makes these historical developments of special interest to this section is that in Spinozism, for the first time in modern philosophy, there is the beginning of a *scientific theory on the connection between mental and physical processes* (see §IX.2.18).

B.7 SUBJECTIVE CAUSALITY: FROM LOCKE TO HUME

It was Locke who accustomed the Western society to the false separation between *"secondary qualities"* (e.g., colors, sounds, smell, etc.) and *"primary qualities"* (those which are "inseparable" from the body and are enumerated as "solidity", "extension", figure, motion or rest, and number).

The "primary qualities", he maintains, are actually in bodies; the "secondary" in the percipient. *Thus in Locke's philosophy there would be no colors without the eye, no sounds without the ear, etc.* This philosophy has led many philosophers and scientists astray from a unified science of nature. It also served as an historical basis to later claims that have been voiced time and again in favor of the idea that *there is no physical world without the mind.*

Locke maintains that "since the mind, in all its thoughts and reasonings, hath no other immediate object but its own ideas, which it alone does or can contemplate, it is evident that our knowledge is only conversant about them".

Then came Hume who threw doubt on causality (see §IV.8 and §IV.22). He accepted Berkeley's abolition of the physical external world (Appendix III), but admitted no simple idea with an *antecedent "impression"*—probably his understanding of a state of mind directly *caused* by something *external* to the mind. But he could not admit such a definition of "impression", for he seriously questioned the very concept of "cause".

In Germany Leibnitz believed that everything in his experience would be *unchanged* if the rest of the world were annihilated. Kant, who was educated in the Wolfian version of Leibnitz's ideas, was also influenced by Hume. Kant takes Hume's claim that causality is a synthetic concept, but believed that it is known *a priori* (see also §IX.2.10).

B.8 SUBJECTIVE TIME: FROM KANT TO HEGEL

For Kant space and time are subjective concepts for they are part of our own apparatus of perception. Yet Kant believed that the *external* world causes *sensation* and that our own mental apparatus *orders this sensation in space and time.* For him *"things-in-themselves"*, (q.v. §IX.2.10), or *"noumena"*, the causes *of our sensations, are unknowable; they are not in space or time, they are not substances, nor can they be described by any of those other general concepts which Kant calls "categories".* (The problems associated with these concepts are re-examined in Lecture IX).

For Kant geometry is *a priori,* and space and time are not concepts; they are forms of *"Anschauung"* (which literally means "Looking at", or "view", but is sometimes translated as "intuition"). He distinguished among twelve *a priori* concepts; these are the twelve "categories" which he separates into four sets of three, namely;

1. of quantity: unity, plurality, totality.
2. of quality: reality, negation, limitation.
3. of relation: substance-and-accident, cause-and-effect, reciprocity.
4. of modality: possibility, existence, necessity.

According to Kant, these concepts appear to our mental constitution as *subjective* in the same sense in which space and time are—i.e., our mental apparatus is such that they are applicable to whatever we experience.

Thus Locke made the "secondary qualities" subjective. But Kant, like Berkeley and Hume, though in not quite the same way, makes the "primary qualities" also subjective.

Then came a man who tried to teach the universe his philosophy—*Hegel.* On one hand, he adopted Spinozism, for instance in maintaining that any portion of the universe is so profoundly affected by its relations to the other parts and to the whole, that no valid statement can be made about any part except to assign it its place in the whole.* On the other hand, his writings are based on Kant's metaphysics.

Hegel defines time as follows:

"It is the being which, in that it *is,* is *not,* and in that it is *not,* is. *It is intuited becoming.'"** What he means is that the "present" (which is) changes into the "past" (which is not) and how "its future" (which is not) "becomes present" (or is).

Some other accounts of Hegel's conception of time are as follows:

"Time, like space, is a pure form of sensibility or intuition: it is the insensible factor in sensibility . . . It is said that everything arises and passes away in time, and that if one abstracts from everything, that is to say from the content of time and space, then empty time and empty space will be left, i.e., time and space are posited as abstractions of externality, and represented as if they were for themselves. But everything does not appear and pass in time; time itself is this becoming, arising and passing away . . . Time does not resemble a container in which everything is as it were borne away and swallowed up in the flow of a stream. Time is merely this abstraction of destroying. Things are in time because they are finite; they do not pass away because they are in time, but are themselves that which is temporal . . ."

"The present is, only because the past is not; the being of the now has the determination of not-being, and not-being of its being is the future . . . The present makes a tremendous demand, yet as the individual present it is nothing, for even as I pronounce it, its all-excluding pretentiousness dwindles, dissolves and falls into dust."

* See §IX.2.9 and §IX.2.15.
** G. W. F. Hegel, Encyclopaedia *"Philosophy of Nature"*, Vol. I, pp. 229–240, ed., tr. by J. J. Petry (New York, 1970).

"The truth of time is that its goal is the past and not the future."

Hegel's accounts of time, while obscurely expressed, fragmentary, and frequently developed in relation to history, are nevertheless basic to his philosophy. For instance, his conception of "mind" is essentially an historical one and consequently time is a fundamental feature of it. At this point we leave Hegel. [But the reader may return to him in our discussion of the "Science of the Whole" (Lecture IX)].

B.9 HEIDEGGER, KIERKEGAARD, MARX AND SARTRE

Hegel's accounts of time have been the subject of various interpretations and modifications. Here I only touch upon their impact on the ideas of *"time-consciousness"* of such existentialists as *Martin Heidegger* (1889–1976) and *Soren Aaby Kierkegaard** (1813–1855).

Heidegger is generally regarded as the founder of existentialism. Working with Kierkegaard's 'destructive' analysis of traditional philosophies and Husserl's phenomenology he composed *'Being and Time'.*** In it Heidegger discusses Hegel's accounts of time within the context of his subjective *'ecstatico-ontological'* conception of time. Here 'past', 'present' and 'future' are described as 'ecstasies', and are defined in terms of human actual experience and not as points along a time coordinate. He considers the future in terms of human expectations, and the past by the subjective memories of the individual. He stresses that all philosophical interpretations of time are in terms of some systems, 'movements' etc., and concludes that Hegel's accounts cast little light on the problem of the origin of time.

Kierkegaard, on the other hand, attacks the Hegelian accounts of time by stressing that Hegel's 'present' is not the only reality, the 'past' not a mere 'no-longer-now', and the 'future' not a 'not-yet-now'. His account of existential time is based on subjective experience, 'present' experience, 'past' held in *"memory"* and a 'future' that is projected by our subjective decisions.

Hegel's philosophy of time and history influenced *Karl Marx* (1818–1883). But, unlike Hegel, Marx denies that socio-historical orders are rigidly predetermined ones. Thus, although he claims that the direction of socio-historical time is governed by class struggle and productive economic forces he asserts that the individual plays a role in directing the socio-historical temporal course. Nevertheless, Hegel's philosophy of temporal socio-historical courses and Marx's views are closely related and it is through Hegel's philosophy of history that time enters Marxism.

Hegel's philosophy of time has also influenced the French exponent of existentialism, *Jean-Paul Sartre* (1905–1980). Exploring the situation of 'anguish', which precedes 'commitment' to a course of action, Sartre emphasizes the role of time in consciousness. Employing Hegel's and Heidegger's accounts of time, he defines consciousness dialectically as "that which is what it is not, and is not what it is." In his book, *'Being and Nothingness'*,*** he brings out the changing character of consciousness through time, which is essentially Hegelian.****

B.10 SCHOPENHAUER, NIETZSCHE, BOLTSMANN, AND GÖDEL

The German philosopher, philologist and poet *Friedrich Nietzsche* (1884–1900) did not produce a systematic doctrine, but by virtue of his insight into the existential situation of modern man, his perception of the cultural flattening of the industrial era, and his ideas on the breeding of a new aristocracy has brought about a considerable impact on 20th century thought. He also developed a moral meaning attached to the concept of *'Eternal Return'*, or *circular time*.

* pronounced "ki∂rk∂gard" or "ki∂rk∂g∂r"

** English translation by J. Macquarie and E. Robinson (Oxford 1967). See pp. 377, 479–497.

*** English translation by H. E. Barnes, Washington Square Press, N.Y. 1966.

**** See also Lecture IX and a paper by W. Mays in the *Study of Time II*, J. T. Fraser and N. Lawrence, eds., Springer-Verlag, Berlin, 1975.

In dealing with Nietzsche's concept of time we must first distinguish among a number of concepts:

1) The linear time concept.
2) The circular time concept.
3) Continuous time concept.
4) Discrete time concept (see §IV.13 and §IV.12).
5) Beginning of time (see Lecture VII)
6) "The end of time" (eschatology).
7) "Eternity".

The later concept is essentially an antithesis of time, although it depends on the very concept of time. There have been many philosophical thoughts that are founded on a kind of time-less or unchanging basis. The Parmenides-Zeno tradition of the idea of the *One* that never changes, or moves, and can never be divided, provides a basis for eternity. Plato's world of Ideas floats in a kind of time-less eternity. The world of *nirvana* (Buddhism, Hinduism, Jainism) is a beatific spiritual condition attained by the extinction of desire in essentially a time-less eternity.

The circular time is another example of eternity. It represents *infinity* within *finiteness* and its roots can be found in India and in Greece. Pythagoras, Empedocles and Heraclitus, among others, believed in circular time. The Indian idea of circular time is associated with Vishnu's periodic reincarnation and with concepts about the chain of rebirth (Palingensis or Metempsychosis in Schopenhauer's terminology).

The idea of eschatology (end of time) may be traced back in time to the Zoroastrian who believed in a big conflagration that would end the world. The Judeo-Christian eschatology was probably influenced by Zoroastrianism, and with it the entire Western thought.

In his most famous work, *'Thus Spake Zarathustra'*, Nietzsche presents the irreversibility of time as the cause of anger for the *Will;* but Nietzsche wanted to prove that his Will is different from Schopenhauer's. Yet *Schopenhauer* (1788–1860) remained an influence from which he could not free himself. Schopenhauer's great anthology of woe, *'The World as Will and Idea'*, appeared in 1818.* Seldom had the problem of pessimism been described so vividly and insistently in philosophy. Almost without exception philosophers have placed the essence of the mind in thought and consciousness; not so Schopenhauer. For him Will is the essence of man, causality, and the world. And he declares:

"The act of will and the movement of the body are not two different things objectively known, which the bond of causality unites; they do not stand in the relation of cause and effect; they are one and the same . . ."

To Hume's question—What is causality—Schopenhauer answers: The Will! As Will is the universal cause in ourselves, so is it in things; *and unless we so understand cause as will, causality, according to Schopenhauer, will remain a magic word, and we shall be using meaningless and occult qualities like "force" or "gravity"!!*

For Nietzsche, in a somewhat similar way, the Will is a future-oriented causation. But his Eternal Return has the effect of transforming the past into the future, thus bringing the past within the reach of the Will. The idea of *Eternal Return* and *"Superman"* are then associated together, for the Superman is happy and wants *the repetition of himself*—naturally without heirs:

"Joy, however, wants no heirs, no children,
—Joy wants itself, wants eternity, wants recurrence,
wants everything eternally like itself."

"Did you ever want once to come twice,
did you ever say 'I like you, happiness! Hush! Moment!'
then you wanted *everything* to come back again."

The concept of *'recurrence'* has a particular interest here. As Dorothea Watanabe Dauer of the University of Hawaii has pointed out recently,** it is surprising that about the same time when Nietzsche

* English translation by R. B. Haldane and J. Jemp (Trubner, 1883, London).
** See her paper in *The Study of Time II,* Springer-Verlag, Berlin, 1975, pp. 81–97.

was considering the problem of irreversibility and the circularity of time, theoretical physicists, were struggling with a similar problem of time: irreversibility versus recurrence of physical phenomena as associated with theorems of the Austrian physicist *Ludwig Boltzmann* (1844–1906).

Now, Boltzmann's aim was to derive the *time–asymmetry* associated with the *Second Law of Thermodynamics* from the *time–symmetric laws of physics*. He never really succeeded. (For details see Lecture V.) But aside from his failure (which is not known today to many who still believe in Boltzmann's "proof"), his theorem, the so-called *H-Theorem,* encountered two objections which have been raised in the form of paradoxes. These paradoxes are due to Zermelo and Loschmidt. It may be instructive to state them breifly:*

B.10.1 ZERMELO'S RECURRENCE PARADOX

This paradox is based on Poincare's recurrence theorem which states: "For almost all initial states, an arbitrary function of phase space will infinitely often assume its initial value within arbitrary error, provided the system remains in a finite part of the phase space" (Poincare's cycle). **It means that if Boltzmann really succeeded in deriving the Second Law, that would amount to deriving one-way time from circular time, which is logically impossible.**** Now, according to Dauer, Nietzsche's argument for eternal return in time demonstates his knowledge of what was going on in the physics of his time—a fact on which there is now substantial historical evidence.

B.10.2 LOSCHMIDT'S REVERSIBILITY PARADOX

This is the second objective against Boltzmann's H-Theorem. It is based on the *Time Reversal Invariance* discussed in §IV.10 and it means that since the laws of physics are symmetrical with respect to the time inversion

$$t \rightarrow -t,$$

to each process there is a corresponding time-reversed process. But this results in contradiction with all evidence concerning actual (irreversible) processes. It means that if Boltzmann had proved that the entropy should increase in the future, then it must be possible to prove that the entropy must have been larger in the past, contrary to evidence (see Lecture V).

Summarizing, we point out that Nietzsche, starting from *irreversible* time obtains *circular time* (which implies *reversibility*), and ends up in a basic self-contradiction; while Boltzmann, starting from *reversible* time, wrongly claims to obtain an *irreversible* evolution, and also ends up in a basic self-contradiction.

Kurt Gödel (1906–) is also a proponent of circular time. Seeking mathematical solutions to Einstein's field equations (Lecture III) in the case of *"rotating universes"* (rotating relative to the totality of galactic systems), he obtained results which allow travel into any region of the past, present, and future, and back again, exactly as it is possible in other worlds to travel to distant parts of space. However, these solutions lead to absurdity. Moreover, the cosmological models considered by Gödel have since been found to be incompatible with evidence. Therefore, a philosophical view leading to such consequences can hardly be considered as satisfactory.***

B.11 BERGSON

Henri Bergson (1859–1941) was an influential philosopher among many *temporalistic-antimechanistic* thinkers. His irrationalism reached a very wide public, and his doctrine rests on the "liberation" of "men-

---

* See my CRT, Mono Book, Baltimore, 1970, pp. 3–5. (Paper by I. Prigogine).

** q.v. Lecture V.

*** Kurt Gödel, in *Albert Einstein: Philosopher-Scientist,* P. A. Schilpp, ed. Harper Torchbooks, N.Y., 1959, pp. 555.

tal intuitions" from the idea of space and the scientific notion of time (and on the relevance of an *"elan vital"*, a so-called "creative life force"). Bergson's irrationalism made a wide appeal, and effected a considerable influence on such scholars as *William James, Bernard Shaw, A. N. Whitehead and Georges Sorel.*

According to Bergson the 'separation' between "intellect" and "instinct" is fundamental. *It is the intellect*, according to Bergson, *which separates in space and fixes time, it represents "becoming" as a "series of states."* We are then told that the genesis of intellect and the genesis of matter are correlative, in other words, matter is something which mind has created on purpose to apply intellect to it. A few quotations will suffice to illustrate his irrationalism:

"An identical process must have cut out matter and the intellect, at the same time, from a stuff that contained both."

"Pure duration is the form which our conscious states assume when our ego lets itself *live*, when it refrains from separating its present state from its former stages."

"Within our ego, there is succession without mutual externality; outside the ego, in pure space, there is mutual externality without succession."

"Questions relating to subject and object, to their distinction and their union, should be put in terms of time rather than of space."

Thus Bergson regards time and space as profoundly dissimilar in the sense that *intellect is connected with space while instinct or intuition with time!*

According to Bergson time is the essential characteristic of life or mind. But here he makes an interesting distinction: What he calls *"mathematical time"* is a *form of space*, while the time which is of the essence of life is what he calls *duration*. The latter exhibits itself in *memory*. Then he distinguishes between memory via *"motor mechanism"* and memory via *"independent recollections."* (See also Lecture IX).

Bergson believes that apparent failures of memory are not actually failures of the mental part of the memory, but of the motor mechanism for delivering memory into action. Moreover, we are told that "true memory" is *not a function of the brain*. "Memory", he says, "must be, in principle, a power absolutely independent of matter." Against *"pure memory"* he puts *"pure perception"*.

"Pure perception", Bergson says, "which is the lowest degree of mind—mind without memory—is really part of matter, as we understand matter." Bergson could not accept strict determinism, the idea that the present moment is the product of the matter and motion of the moment before—and so on, until we arrive at the very beginning of the world, as the total cause of every later event, including life and intelligence. *That was a mystery in his eyes, an incredible process, a fatalistic myth! He simply could not accept a causal chain that leads from early cosmos to Shakespeare's plays! But on the same grounds, he could not even accept a causal chain that leads from apes to man!* In fact he rejects the causal evolution which Darwin and Spencer described, or which relativistic cosmology asserts.

To break the universal chain of deterministic evolution, Bergson argues, *one must relate time to life, to mind, to consciousness, to choice and to free-will!* Causality is, therefore, not the master law of nature. "Each moment", he claims, *"is not only something new, but something unforseeable . . .",* "for a conscious being, to exist is to change, to change is to mature, to mature is to go on creating one's self endlessly." What if this is true of all things? *Perhaps all reality is time and duration, becoming and change? How then shall we understand the essence of life and the universe if not by time and duration?*

Indeed, Bergsonism is mostly concerned with duration and movement. Bergson asked whether the being of which many past philosophers took cognizance by reflection, might not be one which endured, might not be time itself! He substituted Descartes' *"I am a thing which thinks"* with *"I am a thing which continues."* Thus, a major part of Bergson's work was to substitute duration for non-temporal values, and motion and change for static values. This was considered as a great revolution, which brought him international reputation and made Bergson one of the most influential philosophers. In 1927 he received the Nobel prize for literature.

Contrary to Spinoza and Hegel, Bergson convinced most Western thinkers that *philosophy and science should be studied in disregard to the whole of nature, and in disregard to universal systems*, i.e., they should address themselves to *separate, well-defined, detailed problems*, each of which should demand its *"separate point of view"*, though each should, somehow, be part of the "general philosophy of duration and change".

For him, the true nature of things, should be apprehended *only by intuition!* No wonder that a few keep regarding Bergsonism as nothing but a beautiful, yet mystical phraseology.

The intuitive Bergsonian concepts have, nevertheless, been adopted by many scientists, in particular those who founded *"Information Theory"*.*

Speaking about choice, "information", and "personal meaning", they formulated equations for the "amount of information" as identical in form with equations in statistical mechanics. They pretended that there may be deep-lying connections between the physics of time, thermodynamics, and Information Theory. As a result, many Western scientists believe that "In the beginning God created information, the rest of the world just followed . . .". Such anthrophomorphic foundations of physics represent today, more than anything else, a false view in the philosophy of science (§ IX.2).

Since Bergson claims that *everything living might be conscious, down to the lowest organism,* one is first driven to attack Bergsonism on the following grounds:

(1) Does 'natural' DNA possess a consciousness? Does it possess intuition?

(2) Do viruses and DNA possess consciousness?

(3) Are DNA and viruses capable of generating the time rate of their own physico-chemical reactions? Or, are these reactions controlled by *external* boundary conditions, timing, gravitation, "ordering," structures and a flux of energy?

Some of these fundamental problems were treated in Lecture IX. Nevertheless, most of them need a much larger space to be analyzed and, consequently, must be postponed to my next volume.

C. Ich-Zeit and the Principle of Covariance

Can we derive some relationships between "physical" time-concepts and the "time-concept of the mind?" Do other individuals experience the same "process" when all "observe" the "same" phenomenon? In particular, can the "ordering methods" of different observers become unified under the auspices of a single principle of covariance?

An individual experiences the moment "now," or, expressed more accurately, the "present sense-experience" *(Sinnen-Erlebnis)* combined with recollection of ("earlier") sense-experiences, thereby forming "series" of "irreversible", "time-ordered" experiences. The individuals' experience-series are usually *assumed* to be a "one-dimensional continuum" that is *common* to all other individuals. But is it?

Actually we are faced here with three questions, namely:

1) Does the value (extent of duration) of "my subjective time-period" (Ich Zeit) *duplicate* exactly the values of "subjective time-periods in other individuals when we *all* "observe *the same phenomenon,* or *experience the same sensation"*, but *independent* of our motion, acceleration, surroundings and past re-collections? i.e., independent of our (different) coordinate systems, states, histories, etc?

2) Does the transition from "my subjective time concept" (Ich Zeit) to the time-concept of general relativity give rise to the idea that there is an "objective ordering of time" that is independent of a "subject" ("Ich"), a "subjective observer" or of the phenomenon of life itself?

3) Are the directions and ordering principles of different "observers" in all possible coordinate systems, states, etc. subject to a universal principle of covariance?

The latter question has its roots in the formalism of the general relativistic *principle of general covariance,* (see Lecture III). According to this principle, which is also associated with Einstein's principle *of equivalence,* (Lecture III), *all ordering of events according to universal laws must preserve the same formalism or "form" under any general transformation of coordinate systems.* We learn from this principle that *if* an equation, law, or "ordering method," is valid in any *one coordinate system,* or for any "observer," *it is valid in all other coordinate* systems and for all other observers, provided certain transformations (or "translations") are *carried out,* i.e., it tells us that if a verified (tensor) equation holds in one group of systems, it holds in all other coordinate systems. However, *without* Einstein's theory of gravitation, this principle, by itself, is empty of content, for *any* equation can be *made* generally covariant by writing it

---

* See, for instance, N. Wiener, *Cybernetics,* MIT Press, 1948, pp. 30–43 and the criticism given in § IX.2.6.

in *any* one coordinate system and then working out what it looks like in another *arbitrary* coordinate system. Hence, it is *only* by introducing the modern concepts of curved geometry and *gravitation* (into our "subjective" time-concept and into a "generalized principle of covariance"), that we can introduce universal meaning and valid assertions into modern philosophical principles.

We must distinguish here between a *"strong"* and *"weak"* general covariance. By the former, I mean a statement about the effects of gravitation, mass, energy, pressure and momentum on local curved space-time, and about nothing else! By the latter, I mean a similar statement about *any* event, "observer" or phenomenon involving time, space, gravitation, matter, energy, pressure, momentum, etc. One example demonstrating the difference between these two views is given in Lecture VII with respect to our derivation of the "universal time" and Newton's laws of motion. Thus, according to the "weak" principle of co-variance, the answer to question (3) is affirmative, and, by the same token, it is negative for question (1), i.e., there is no objective-absolute time ordering; *it is only by performing proper transformations from one subjective time (Ich Zeit) to the subjective times of others, by taking into account their different motion, location, surroundings and past recollections, that all "subjective times" can be treated on "the same ground" and become sensible for valid conclusions.* In the main lectures we show that *gravitation* becomes *"the same ground"* for *all* subjective observers.

D. A FEW HISTORICAL REMARKS ON CAUSALITY*

Extensive literature is available on this subject and I do not intend to reproduce it here except for a few remarks. Thus, aside from the physico-philosophical concepts discussed in the main text, there are other notions of causality which may fall into three groups:

a) Random references in "scientific discussions," with little or no concern for what causality really means in physics.
b) Obsolete versions which, for most current theories, bear no physical consequences.
c) General philosophical versions.

I next outline briefly several notions from the last two groups:

Aristotle distinguishes four different senses of cause, namely:

1. The material cause, that which is constituent of that which is generated.
2. The formal cause, that is, the essence determining the form, pattern or creation of a thing.
3. The efficient cause, or the agent producing an "effect" (e.g. initiate change or motion or coming to rest).
4. The final cause or "purpose."

An example of the *material cause* may be found in protons, neutrons and electrons, which are material causes of the atom. On the other hand a "design" may be considered as a *formal cause* of the form, pattern, or creation of a machine or a building. The electric field may be considered as the *effective cause* of a current. The *final cause* is that for the sake of which the change occurs: we walk for the sake of health, Aristotle says, and so health is the final cause of walking. The use to which a machine is put is the final cause of building it. Being an engineer might be the final cause of students attending college. In physics we are normally not concerned with final causes, yet biology, engineering, and the social sciences are, to some degree.

Aristotle also points out that in a *pair of causes* each may be the cause of the other; for example, exercise is a cause (effective) of good physical condition, whereas good physical condition is a cause (final) of exercise.

---

* An interesting subject for meditation or class discussion is the informational role modern Maxwell demons (e.g. traffic lights in our cities, parking regulations, established working hours, effect of bell sound in bringing students to and from classes) play in our society, where, at the cost of a negligible amount of energy, they create "order," "organization," "reduced states of entropy," etc.—a state arrived at through prior expenditure of much larger amounts of energy in training the public to obey these signals.

An extension of Aristotle's causes may be what we call today *"feedback mechanisms,"* in which the "results" change the "causes" (see also Assertion 10 in the *Introduction*).

According to *Newton,* "to the same natural effects we must, as far as possible, assign the same causes." On the other hand, *J. S. Mill* maintained (in his theory of the plurality of causes) that an effect (e.g. "heat") may be generated by *various* causes; [today we know that "heat transfer" is subject both to conservation (causality), and to the second postulate (causation)—cf. Lectures III, IV and V].

The idea of effective cause as a firm basis for philosophy of nature was criticized by *David Hume.* All reasoning concerning "matters of fact" seem to be founded, Hume says, on the relation of cause and effect; for it is with that relation that we can go beyond the evidence of our senses or consciousness. Yet all causes and effects are *discovered* by experience. There is, however, Hume writes, some disposition to think it is otherwise with events which have become very familiar to us. While people argue that there is "necessity," or "secret powers," or "efficiency," or "agency," or *"connection,"* or *"productive quality"* which require that the effect follow the cause, Hume's argument is that no such entities are observed, or justifiably inferred.

Further, Hume writes: "Upon the whole, necessity is something that exists in the mind, not in objects, . . . necessity is nothing but that determination of the thought to pass from causes to effects and from effects to causes, according to their experienced union." And he adds: ". . . when, instead of meaning these unknown qualities, we make the terms of power and efficacy signify something, of which we have a clear idea, and which is incompatible with those objects, to which we apply it, obscurity and errors begin then to take place, and we are led astray by a false philosophy. This is the case, when we transfer the determination of the thought to external objects, and suppose any real intelligible connection between them; that being a quality, which can only belong to the mind that considers them."

While Hume himself fell victim to the kind of subjectivism he may have wished to avoid, his work helped remove some intuitive notions associated with the concept of causality in philosophy.

D.1 Materialists and Determinists

The terms causality and determinism are frequently used by the *"materialists",* who claim that only matter-energy is the *fundamental* constituent of the universe; that nothing supernatural exists *(naturalism) ;* that nothing which is purely mental exists; that the universe is not governed by intelligence, purpose, or final causes, but everything (including mental processes and emotion) is exclusively caused by non-mental, inanimate matter-energy causal processes; that the only objects science can investigate are non-mental physical or material-energy phenomena.

The *materialists* interpret thought and emotion as physico-chemical activities of the brain and body, and imagine that if all physical and chemical changes in the brain and body can be traced, it will be possible (at least in principle) to deduce all mental and emotional experiences. Consequently, if science shows that all matter-energy changes are linked by causal chains, mental and emotional events are also so linked, and there can be no room left for *"free will"* or *"caprice".*

The *determinists* conceive that all processes, including human acts, are causally predetermined by past events and acts (including such causal chains as those of heredity, environment, education, acquired habits, etc.).

By contrast, the *indeterminists* conceive that all processes, including human acts, are not entirely predetermined by the past, but that at every moment they may involve a certain amount of uncaused, capricious, and (in principle!) unpredictable events (cf. §IV.6–15 and Lecture V).

Spinoza, for instance, who was an out-and-out determinist, maintained that we think ourselves free in the same way that a stone in the air would think itself free if it could forget the hand that had thrown it. He drew the inference that if body states are strictly determined, then mental states are subject to an equally rigid determinism.

*Practically all first-rank philosophors and scientists—*Descartes, Spinoza, Leibnitz, Kant, Locke, Hume, Hegel, Mill, Alexander, Planck, Einstein, as well as many others—*were determinists!*

Many philosophers maintain that we are capable of doing what we wish (within limits) and so *"feel"* ourselves free. But they agree that this is *only* because we do *not* have sufficient information to understand that our wishes themselves are imposed on us by the causal laws of the past (which include *unconscious determinism!).* On the other hand, some advocates of quantum mechanics claim that it is not possible to know simultaneously the exact velocities and positions of a *single* elementary particle, and, even if we did it would still be impossible to *predict* what is going to happen next. (But see our reservations in Lecture V).

Others have attempted to reimpose strict causality on quantum theory by inventing *"hidden-variables theories"* (Lecture V). For that purpose they envisage a *"substratum"* (Lecture V), or "pregeometry" (Lecture IX), in which the causal links of events are—for the time being—concealed from us. Accordingly, it is possible that whatever the future may bring, lies uniquely and inevitably determined in this "substratum" of past configurations.

A subjectivism injected into microphysics by some quantum theorists is frequently based on the works of *Immanuel Kant*.* However a careful study of traditional interpretations of quantum mechanics and of Kant's philosophy shows that the superposition of microscopic states, which leads to *observer-dependence* in quantum terminology, is dissimilar in kinds to Kant's *a priori* role of ideas.

D.2 Russell's Functional Dependence

An extremely radical position was taken by *Bertrand Russell,* who rejects the word "cause" altogether. Instead, he claims that the law of cause and effect must be replaced by *functional dependence;* and that the constancy of scientific law consists not in *"sameness* of causes and effects" but in *"sameness of equations".* First he asserts that functional dependence is not *a priori,* or self-evident, or a "necessity of science," but an *empirical generalization* of a number of laws which are themselves empirical generalizations. Secondly, the *"principle" of functional dependence makes no distinction between past and future.* And ῾thridly, something like the "uniformity of Nature" (rather than "causality") is accepted on inductive grounds.

Russell's views have been popular among contemporary philosophers, although by no means uncritically accepted. But one must keep in mind that *"functional dependence"* is an abstract or purely mathematical notion. It does *not* require understanding of its physical or philosophical foundations, nor why it had been adopted in the first place. *It served Russell's purpose to remove any trace of physical mechanism from natural processes.* Furthermore, the idea that "cause" precedes effect in time, which is the basis of causation, is lost in the process.

Perhaps Russell's claims are consistent with the Action-at-a-Distance electrodynamics put forward by Wheeler and Feynmann. In that formulation (which is regarded as incomplete, having been formulated for a static (reversible) universe—see Part III), the radiative reaction of an accelerated charged particle is calculated on the basis of the assumption that an *advanced potential* has been emitted, so as to arrive at the particle just at the moment of acceleration. One may, however, find the level of mathematical proof of that formulation unsatisfactory and be dismayed by arguments that appear to be going in circles. One may also wish to argue against the rejection of order in the time sequence.

D.3 Whitehead's Philosophy of Organism

It is interesting to note also that Russell's collaborator, *A. N. Whitehead,* developed a *"philosophy of organism"* in which there is a "relation-in-being" precisely corresponding to cause-effect. In any case, *the very replacement of cause-effect relationships by any "functional dependence" is circulatory. Causality has been, and will remain, the root concern of all natural sciences.* Even in quantum mechanics, causality is an empirical and logical necessity, whether it is called functional dependence, causality, causation, macrocausality, or microcausality, or even if it is smuggled into the formulations without any due declaration (Lecture V).

We cannot accept "functional dependence" between, say, sun spots and economic cycles, since we do inquire for a "cause" and an "effect," and then conclude that there is one, but only between sun spots and terrestrial magnetic storms. Contrary to Russell's claims, work in empirical and theoretical science has the goal of providing cause-effect explanations.

---

* See Lecture IX and below for more details.

In fact, the notion of "non-temporal events", linked by a modified meaning of causality, is not a new concept in philosophy.

Thus, "causality" has also been associated with the following concepts:

1. A link between "events", "processes", or "entities" and "extra-experimental" but either temporal or *non-temporal* "entities" upon whose existence the former depend.
2. A relation between "a thing and itself" when it depends upon nothing else for its existence *("self-causality")*.
3. A relation between an "event", "process", or "entity" and the very *reason* or *explanation* for its being.
4. A relation between an *idea* and an experience whose expectations the idea arouses *because of customary association of the two in a sequence*.
5. A principle or category introducing one of the aforesaid types of order (i.e., it may be inherent in the mind, invented by the mind, or derived from experience; it may be an explanatory hypothesis, a postulate, or a "necessary" form of thought).

Any excessive use of circulatory terms is eventually subject to the "empirical law of diminishing returns." Unless we gain a more universal view of time itself, such words remain useless if not misleading. The same conclusion applies to other attempts to define "succession" of "cause-effects" and "events in time" (cf., e.g., *Leibnitz's view* that "succession" is the most important characteristic of time, defined by him as *"the order of succession"*, or *Whitehead's* definition of "duration" as "a slab of Nature" possessing "temporal thickness", a "cross-section" of the world, or "the immediate present condition of the world at some epoch").

*   *   *

THE PHILOSOPHY OF TIME & CHANGE: SOME HISTORICAL NOTIONS

A. Heraclitus and Plato

Heraclitus (c.540–c.475 b.c.e.) was the man who introduced the idea of "perpetual change" to Western philosophy. Disillusioned from the social system in which he lived (when the Greek tribal aristocracies were beginning to yield to the new force of democracy), he argued against the belief that the existing social order would remain forever. *"Everything is in flux and nothing is at rest"*, he declared; and added, *"you cannot step twice into the same river"*; even in the "stillest matter there is unseen flux and movement".

This emphasis on time and change, and especially on change in social life, is an important characteristic not only of Heraclitus' philosophy, and of Plato's and Aristotle's philosophies which followed it, but of modern historicism as well*. *That individuals, nations, cultures, and civilizations, change, is a fact which needs to be impressed especially upon those minds who take their social and physical environments for granted.***

Heraclitus believes also that strife or war is the dynamic as well as the creative principle of all change. "War", he says. *"is the father of all and the king of all things"*. *"One must know that war is universal, and that justice-the lawsuit-is strife, and that all things develop through strife and by necessity"*. Where there is no strife there is decay! But in this flux of change and struggle and selection, only one thing is constant, and that is the universal law of nature. "It always was, and is, and shall be". Moreover, it is the same for all things.

He had, however, another doctrine on which he set even more store than on the perpetual flux; this was *the doctrine of the identity of opposites*. It was linked to his theory of change. A changing thing must give up some property and acquire the opposite property.

Heraclitus belief in strife is also connected with this doctrine, for in strife opposites combine to produce a motion which is a *unification of the opposite states. There is a unity in the world, but it is a unity resulting from diversity:*

"Couples are things whole and things not whole, what is drawn together and what is drawn asunder, the harmonious and the discordant. The one is made up of all things, and all things issue from the one". "The opposites belong to each other, the best harmony results from discord, and everything develops by strife . . . "

This doctrine contains the core of *Hegel's philosophy*, which proceeds by a *synthesizing of opposites*.

* The meaning and limitations of historicism are discussed in Popper's classical book. *The Open Society and Its Enemies, I*: Plato; Routledge & Kogan Paul, London, 1945.
** It may be of interest to compare Heraclitus' and Plato's views on change and decline (see below), with Buddha's (ca. 563–483b.c.e.) last words: "All composite things decay. Strive diligently."

Heraclitus' thought was therefore passed on to modern Western philosophy by Hegel's philosophy and through it to all modern historicist movements, including the Marxist philosophy of change.

The doctrine of the perpetual flux, as taught by Heraclitus, has been painful to many minds. Many philosophers have therefore sought, with great persistence, for theories not subject to the empire of time. (e.g., Parmenides). However, as all evidence demonstrates, no ecosphere can be exempted (Lecture VI). Individuals, tribes, nations, cultures and civilizations can do no better. Heraclitus, indeed, was a thinker of considerable insight; many of his ideas have (through the medium of Plato) become part of the main body of Western philosophic tradition.

Plato (c.428–c.348 b.c.e.) adopted much of Heraclitus' philosophy of change. But he thought that *all social change is corruption or decay or degeneration.*

This "historical law" forms, in Plato's view, part of a law which holds for all created or generated things. Thus he believed that *all things in flux, all generated things, are destined to decay.*[+] He felt, like Heraclitus, that the forces which are at work in evolution are "cosmic" in origin. But whatever is the nature of these forces Plato believes that the law of historical destiny, the law of decay, *can* be broken by the *moral will of man,* supported by the *power of human reason.* He believed that *political degeneration* depends mainly upon *moral degeneration and lack of knowledge.* In turn moral degeneration is due mainly to *racial degeneration.* To prevent this corruption he strives for the perfect state, one which is free from the decaying processes, *because it does not change!*

These views differ from those of Heraclitus, who has shown a tendency to visualize the laws of development as *cyclic laws;* they are conceived after the law which determines the cyclic succession of the seasons. But can we speak today of cyclic *socio-historical change?* What did past philosophers and historians say about such a theory?

The first modern philosophy of change is usually attributed to Voltaire (1694–1778), who claimed: "Details that lead to nothing are to history what baggage is to an army, *impedimenta*; we must look at things in the large, for the very reason that the human mind is so small, and sinks under the weight of minutiae". For that reason he declared: "only philosophers should write history" and went on to seek a new method of arrangement of facts; *a unifying principle* by which the whole history of civilization could be understood *as a stream of natural causation* in the development of human culture. What is needed, according to Voltaire, is not the history of great lords, kings, and wars, but the understanding of the laws and global causation behind all human thoughts; behind culture, revolutions and movements.

"My object", he wrote, "is the history of the human mind, and not a mere detail of petty facts; nor am I concerned with the history of great lords . . . ; but I want to know what were the steps by which men passed from barbarism to civilization". In this grand perspective Voltaire portrayed a global picture of socio-historical causation; a philosophy of change which laid the basis of modern historical studies; a philosophy which has influenced many thinkers.[*]

[+] cf. *Parkinson's Third Law: 'Expansion means complexity, and complexity decay'.*

[*] Among these we find the English rationalist historian *Edward Gibbon* (1737–1794), author of *'The History of the Decline and Fall of the Roman Empire'*, a work of monumental erudition, accuracy of detail and shrewdness of judgment; the German historian *Barthold Georg Niebuhr* (1776–1831), author of *'Roman History'* (1811–32), which is considered an early example of modern "scientific" historical writing; and the English historian *Henry Thomas Buckle* (1821–1862), author of the *'History of Civilization in England'.*

Buckle asserted that *"the progress of every people is regulated by principles . . . as certain as those which govern the physical world".* He also equated the "progress" of civilization with the advance of human knowledge. However, his views were much criticized and his *History* was a source of considerable controversies. Yet, these controversies do not altogether impair the grandeur of Buckle's doctrine, and it remains an important witness to modern attempts to find principles and consistent historical causality in every social activity.

Today we already encounter in the literature such concepts as *"social causality"*, and *"social time structuring"*. These concepts generated some new interest in the history of these ideas,** demonstrating, time and again, that the story of socio-historical causation has not yet ended. In fact it cannot, in principle, be started before we resolve the old-new philosophical question of physical vis-a-vis mental "ordering of time". (Lecture IX).

Consequently, for the moment we leave this question open and proceed to demonstrate that *when causality and determinism are put out through the door they come in through the window*. The immediate problem is, therefore, to expose the nature of some false arguments which pretend to prove the rejection of causality and environment from "specific" natural events.

At one extreme view of history we find skepticism, which not only refuses to analyse patterns of historical processes, but also doubts the possibility of "ever knowing the causes of historical events". At the other extreme we find the impersonal school of historians which question the power of the individual to alter seriously the grand-scale course of events. This school claims that the exceptional person where he occurs, *is the projection or incarnation of a grand-scale evolution rather than an isolated phenomenon in itself.* Thus neither the individuals nor the nation-states can ultimately be regarded as "fully responsible" for grand historical changes, since the *collective* behavior of mankind is largely determined by factors of environment, culture, tradition, race, survival motives, social class, economics, technology, science, etc. (see below).

A more cautious view avoids such universals to hold that every event of human life, considered as an "open system", is an element in a complex causal chain. All human beings, according to this view, have complicated "built in" functions and subsystems but are also governed by the *external* world. *Thus, whether of universal direction in the environment, or as the result of innate forces, human action in prehistoric and in historic times must be thought in terms of complex* chains of feed-back type of *"causes and effects" between internal and external systems*. These concepts bring us to our next problem: *Is a single Law of universal evolution possible?*

For these reasons I use the concept "history" in its widest sense, namely *as all that has happened, not merely all the phenomena of human life, but those of the natural world as well*. It includes, in principle, everything that undergoes *"change"*, everything associated with the most difficult concept of all human concepts: *TIME*.

As modern science has shown, there is nothing absolutely static; *all nature is inherently rhythmic, non-static and evolving*; from the formation of galaxies to the gravitational evolution of stars and planets; from the birth and death of stars to the fine oscillations of molecular, atomic and subatomic objects; from the undulation of neutrinos, photons and sound waves to the tides of the seas; from biological clocks to social and industrial revolutions; from the metabolism of chemical elements by living organisms to the growth of knowledge into science, philosophy and industry.

Three "tendencies" characterize all these:

1) *Nature evolves toward a greater definiteness of aggregation, physical structure, complexity, asymmetry, memory, forms, and social structure.*[+]

2) *The whole universe and every part of it has its history* (lecture IX).

3) *The grand-scale idea of history has in a sense made the study of time in physics a branch of the general philosophy of change, structure and evolution. But it also transfers some general studies of history and evolution into the general philosophy of time.*

** H. Nowotny; *Time Structuring and Time Measurement: On the Interrelation Between Timekeepers and Social Time*, in *The Study of Time, II*; J. J. Fraser and N. Lawrence, eds. Springer-Verlag, Berlin, 1975.

[+] The concepts "aggregation", "structure", "asymmetry", and "memory" are discussed in Parts II to IV, mainly in association with the effects of *gravitation*. The concepts "social structure" and "social evolution" are briefly discussed below. Geological and astrophysical structures are discussed in Lecture I and in Assertion 9, *Introduction*.

The tendency to look at history vis-á-vis science has, however, encountered serious difficulties. It is, in the first place, the very notion that it is possible to discover patterns, historical causation, and grand regularity in the process of history, that is problematic.

B. Evolution: Uni-Directional or Cyclic?

The subject of evolution (see also Lectures VI to IX) raises many difficult questions. Some of these may be formulated as follows:

(i) Does a *"single law"* dominate *all* evolution; cosmic, galactic, geochemical, biochemical, social, etc.?

(ii) If we assume that a single, "unified field of force" dictates all evolution, then what about its *direction*? In other words, *is evolution unique, irreversible, a one-way street to the unknown, or is it essentially reversible, cyclic, and involving the growth and decay of all systems?*

(iii) How the direction of evolution change, if at all, when *multiple,* but *"independent forces"* dominate it?

We too often forget that such old and essentially classic questions are not only of interest to the student of history, sociology, biology, or philosophy, *but may also be the source of misconceptions in clashes between different ideologies. For there is a fundamental difference between a unidirectional evolution of a civilization and a cyclic one, which involves "rise and fall", "growth and decay", and the aggregation of parts into systems and wholes which must, eventually, dissolve and disintegrate.*

B.1 A FEW VIEWS ON MACRO-EVOLUTION AND HISTORICISM

Reflecting on the evolution of evolutionary theories, one cannot ignore the diverse assortments of *"doomsday theories"* that have been proposed by historians, philosophers and sociologists, notably by Heraclitus, Plato, Vico, Hegel, Spencer, Marx, Spengler, and Toynbee.

The Italian philosopher *Giovanni Battista Vico* (1668–1744) was the first to ask why there is no "science" of human history. He was also the first to write a systematic philosophy in terms of "the rise and fall" of human societies and to make use of the study of linguistics as a source of historical evidence. His philosophical work *The New Science* (1725) has had a wide influence, but only later, in the 19th century. There is, he thought, a sort of standing structure to history, one which repeats itself over and over in time. For Vico, history has no outcome. He also tried to teach the world the principle that the entire sweep of history can be described by man in a widely comprehensive scheme in which the life of actual societies is determined by causal relations.

B.1.1 Spencer's Cyclic Evolution

A considerably more sophisticated philosophy was advanced much later by the greatest English philosopher of the 19th century—*Herbert Spencer* (1820–1903). It is not intended to review here his broad philosophy (which has been frequently criticized—see below), except for a few remarks concerning the fundamental question of *direction* in evolution.[++]

[++] It should be noted that Spencer developed his theory of the evolution of biological species before the views of *Charles Robert Darwin* (1809–1882) and *Alfred Russel Wallace* (1823–1913) were made known. Wallace developed, independently, a theory of evolution similar to Darwin's (Both theories were published in 1858). It should also be noted that *Empedocles* (fl.445 B.C.E., in Sicily) developed a theory of evolution in which organs arise not by design but by selection. Nature "makes" many trials with organisms, combing organs variously; where the combination meets environmental conditions the organism survives and perpetuates its like; where the combination fails, the organism is weeded out; as time evolves, combinations are more and more intricately and adapted to their changing surroundings.

Spencer initially thought that biological evolution was caused by the inheritance of "acquired abilities" whereas Darwin and Wallace attributed it to *"natural selection"*. Later he accepted the theory that "natural selection" was one of the causes of biological evolution and even coined the phrase *"survival of the fittest"*.

Spencer holds that *alterations caused in a particular organism by forces external to it may be passed on by it to its descendants.* Furthermore, he did not think that the inheritance of "acquired abilities" or the "survival of the fittest" were the ultimate principles of evolution. In his view the so-called *'Principle of Change from Homogenity to Heterogenity' was more important than either and was itself a consequence of more fundamental principles.* For instance, his law of *"the persistance of force"* is considered a fundamental principle from which it follows that *nothing homogeneous can remain as such if it is acted on, because any external force must affect some part of it differently from how it affects other parts and, therefore, cause differences and variety to arise.*

Spencer's views mean that *any force that continues to act on what is homogeneous must bring about an increasing variety and structure; and that this new structure is passed on to ensuing organisms. In Spencer's view, this law of multiplication of effects (or "memory") is the clue to the understanding of all development, cosmic as well as biological. He maintained that an unknowable force continuously operates on the material world producing structures, variety, coherence, integration, specialization and individuation.*

Gases condense to form stars and planets; the earth becomes more structured and variegated; it then gives birth to simple biochemical structures and organisms; man evolves "from less complex species" and at first lives in undifferentiated hordes; then various social functions are developed, such as means of increasing the division of labor, and even knowlege is differentiated into the various sciences.*

Spencer rejected the notion of "special creation" and applied the idea, borrowed from the German embryologist *K. S. Von Baer* (1792–1876),** that biological development in the individual is *from the homogeneous to the heterogeneous*; from homogeneous universe to the heterogeneous solar system, to animal species, to human society, to industry, art, language, religion, and science.

Spencer may therefore be considered the most important pioneer of evolutionary theory. But in his emphasis on variety and differentiation he was unwittingly returning to the general philosophy which *Spinoza* had adumbrated two hundred years earlier.***

Spencer's famous formula of evolution reads****: *"Evolution is an integration of matter and a concomitant dissipation of motion; during which the matter passes from an indefinite, incoherent homogeneity to a definite, coherent heterogeneity; and during which the retained motion undergoes a parallel transformation".*

Using this generalization Spencer transformed a wilderness of facts into a single doctrine of universal evolution. Thus, the integration of biomolecules, plants and animal life; the development of the heart in the embryo and the fusion of bones after birth; the structuring of families into villages, cities, nations and civilizations; the division of labor and specialization in academic, industrial and political structures— all demonstrate his principle of integration in which the integrating parts form an entire new structure, in form, nature, function and operation.

Such integration results in a *diminishing degree of freedom for the individual parts, a lessening of motion in the parts, as, for instance, when the growing power of the state limits the freedom of the individual* or when an all-embracing science differentiates into fixed disciplines and an academic institution into "established" departments.

To conclude: *Whatever is transformed from homogeneous simplicity into a differentiated structure of complexity, and from parts into ever larger aggregations of forms and organisms is in the process of evolution* (cf. §IX.2.8.1).

* According to Spencer, there is nevertheless, an overarching unified knowledge, namely, philosophy. In his view philosophy is a *synthesis* of the fundamental principles of the *special sciences*, a sort of scientific *summa* to replace the theological systems of the middle ages.

** In his *History of Development* Baer suggests that all organisms and species arise from the same fundamental form: "Are not all animals in the beginning of their development essentially alike, and is there not a primary form to all?"

*** Spinoza is discussed in Lecture IX.

**** *Principles of Psychology* (1855), p. 367.

But at this stage Spencer makes a drastic addition; he endeavors to show that *"equilibration" and dissolution "must" follow evolution!!* Thus, he claims that societies must eventually disintegrate, cities fall into ruin. and life pass into the diffuse disorder which is death. *And so must the universe!*

In the "end" it too must die. As for states and civilizations, no government can be strong enough to hold the loosening entities together; social order will decline and with it individuation, structure, coordination, and functioning. Each civilization will decay and be resolved into the stage from which it evolved. Then another cycle may begin again, and the process may repeat itself endlessly. (Such an *"Eternal Return"* or *"recurrence"* probably influenced the philosophy of *Friedrich Nietzsche* (1844–1900); see also Appendix I).

Spencer's theory of social evolution has met with some opposition. For instance the French sociologist, *Gabriel Tarde* (1843–1904) maintains that *civilization results from "an increase of similarity" among the members of a group through generations of "mutual imitation".* Thus, in Trade's social philosophy evolution is conceived of as a process toward homogeneity. In addition, he held that about one person in a hundred is inventive, yet "invention is the source of all progress". Inventions are then imitated. But this, according to Tarde promotes opposition and the outcome of this opposition is ultimately adaptation, which, in turn, becomes "an invention".*

B.1.2 Doomsday Theories (Spengler and Toynbee)

The problem of cyclic evolution was re-examined later by two metahistorians, the German *Oswald Spengler* (1880–1936), and the British *Arnold Toynbee* (1889–1975). Their theories too have encountered severe criticism, especially amongst Western scholars. In the East, however, Spengler's famous *Decline of the West* (1918) is frequently cited in support of various Marxist views concerning *"the inevitable fate of the West"*.

According to Spengler's view, all civilizations are victims of the *same inevitable* laws of decay; their future is as predictable as the life cycle of an individual, i.e., *conception, birth, growth, and death.*

Both Spengler and Toynbee maintain that *the unit of historical study is not a nation but a civilization.* "Civilization" is considered as the crystallization of a preceeding *"culture"*. It aims at the gradual standardization of people within a special framework—masses of people who think alike, feel alike and thrive on conformism, people in whom the social instinct predominates at the expense of the creative individual.

"Culture", on the other hand, is different. It predominates in young societies and implies *"original creation"* and *new values, new intellectual and spiritual structures, new legislation, new moral codes, and most important, new science—a fresh look at nature!*

"Every Culture", according to Spengler, "is a four-act drama with an ascending movement of religion, aristocracy, and art, and a descending movement of irreligion, democracy, socialism, and the great city". The civilization of an epoch crystallizes always into the *"world city"* and "the province": megalopolis and its surrounding territory.

"World city and province, the two basic ideas of civilization, bring up a wholly new form-problem of history, the very problem that we are living through today with hardly any notion of its immensity. In place of a people true to a type, born of the soil and grown on the soil, there is a new sort of nomad, cohering unstably in fluid masses, the parasitical city dweller, traditionless, utterly matter-of-fact, without religion, clever, unfruitful, contemptuous of the countryman. This is a great stride toward the end". And Spengler adds: "The world city means cosmopolitanism in place of home, matter-of-fact coldness in place of reverence for tradition, scientific irreligion in place of the older religion of the heart, society in place of the state, 'natural' in place of hard-earned rights".

The megalopolis; the fluid megalopolitan populace; the places of amusement, of sport, and of football games; the 'educated' who makes a cult of intellectual mediocrity and a church of advertisement; the mass communications media and the multitude that blindly follow it, all these betoken the definite closing down of the culture and the setting in of a "parasitic" civilization, a civilization that characterizes the rejection of the previously matured culture, the end of an epoch.

"At this level", according to Spengler, "civilizations enter upon a stage of depopulation. The whole pyramid of culture vanishes. It crumbles from the summit, first the world cities, then the provinces, and finally the land itself whose best blood has incontinently poured into the towns".

Toynbee, like Spengler, turns to expose some of the characteristic phenomena of each civilization. Here we can only mention a few conclusions from his detailed studies comprising empirical data on many past civilizations. His conclusions are:

Civilization is essentially uncreative and "parasitic"—it feeds on the parent culture. Once a civilization has been launched, claims Toynbee, it must continually respond to challenges hurled at it by chance and circumstance. Some civilizations respond only to initial challenges but later fail to cope with new ones and become *"arrested civilizations"*. Others, falling victim to self-idolization, refuse to change with the changing times and become *"historical fossils"*.

A new civilization comes into being, according to Toynbee, through a burst of creative energy from *a small group of enthusiasts*—or "idea men"—called the *"creative minority"*. A new civilization is launched if the *"uncreative majority"* voluntarily adopts the views of the "creative minority".

Disintegration and decline of an entire civilization, says Toynbee, sets in when the "creative minority" *ceases to create. It then begins to rule by force, seeks power and entrenches itself by resorting to totalitarianism, absolutism, etc.*

Spengler calls the first stage in the life cycle of a civilization *a spring. It gives birth to a new world outlook and its people exist in a "precultural stage" characterized by mystical symbolism and primitive expressions. There is yet no developed philosophy, ideology or technology.*

Next, according to Spengler, comes *"summer"*, culminating in the *first philosophical frameworks* and new intellectual clubs. *Groups are formed with their own theories, clubs and banners, paving the way for the formation of "a new civilization".*

This period is followed by *"autumn"*, the zenith of intellectual creativeness, the era of big systems. *Power is shifted from elite to business, from property to money. It is the era of maturity, the final attempt to refine all forms of intellectual development.* There is only one way to go from here, and that is *down*.

The final phase, according to Spengler, is *"winter"*, the old age of civilization, characterized by the "World cities", *"materialism"*, a *cult of science, degradation of abstract thinking,* and a *dissolution of old norms. It is a dying culture, characterized by meaningless luxuries, outlets in sports, and rapidly changing fashions.* The West, said Spengler as early as 1914, was in its *"winter"* death phase, whereas the new Slavic and Sinic civilizations in Russia and China were in their *spring* phases. Thus, according to these theories, Western civilization is doomed to death, and the Slavic or Sinic civilization, having entered their proper phases, will evolve into their own "autumn-winter cycle". New civilizations will in time replace them in the eternal life and death dance of people and cultures.

B.1.3 On Dialectical Materialism

On the general level, some of the most important principles and views of dialectical materialism are as follows:

1) The world is material, and composed of what modern science would describe as matter-energy.

2) The material world constitutes "an interconnected whole".

3) Man's knowledge is derived from "objectively existing matter".

4) The world is constantly changing, and "there are no truly static entities in the world".

5) Changes in matter occur in accordance with certain universal regularities or laws.

6) The laws of the development of matter exist on different levels corresponding to the different disciplines of the sciences, and therefore one should not expect in every case to be able to explain such complex entities as biological organisms in terms of the most elementary physicochemical laws.

7) Matter is "infinite in its properties", and therefore man's knowledge will never be complete.

8) The motion found in the world can be explained by "internal forces", and therefore "no external mover is needed".

9) Man's knowledge grows with time, as is illustrated by his increasing success in applying it in practice, but this growth occurs through the accumulation of "relative"—and not "absolute"—"truths".

None of the above principles or views is original to dialectical materialism, although the sum of them is. Many of them date from the classical period of philosophy and have been maintained by various Western thinkers over the last few centuries.*

It should be noted that there are, in addition, two different schools in dialectic materialism: one is the *official ideology* as reflected in Soviet propaganda pamphlets. According to this view, not only does matter-energy obey laws of a very general type but these laws are identified in the *"three laws of the dialectic."***

The other school is an *unofficial* one. According to this school, matter-energy does obey general laws, but the three laws of the dialectic are *"provisional statements"* to be modified or replaced as science provides more evidence.

This philosophical view appears in the Soviet Union from time to time, and is held particularly among younger scientists and philosophers.

The history of dialectical materialism is to a large degree *"a story of exaggerations and amusing naivetes".* But the oversimplifications found in dialectical materialism, now evident, should not cause Westerners to forget that the "accepted science of today", from which we look back upon these episodes, does *not* contradict *all* the *initial* materialistic assumptions upon which these exaggerations were constructed.

Taking a broader view, which embraces a cluster of related doctrines, one notes that during this century the proponents of materialism, naturalism, new realism, pluralism and evolutionism have forced their detractors, by employing scientific discoveries in the physical and life sciences, *to revise* some of their arguments *in a more fundamental way than in the reverse case.* Such examples include various modified doctrines in *idealism, neo-idealism, vitalism, subjectivism, pragmatism, logical positivism, and neo-Hegelianism* (see below and in Lectures V and IX). In the domain of science some of the most fundamental issues involved in these arguments are related to the concepts of *causality, determinism, chance and probability, quantum mechanics, special and general relativity, cosmology and cosmogony, cybernetics, origin of life, physiology and perception.* (Most of these concepts are analyzed in the main text from a modern viewpoint.)

Dialectical materialism has usually been presented as though it were "a uniquely Soviet creation", "far from the traditions of Western philosophy". It is true that the technical term "dialectical materialism" may be credited to a Russian,[+] but it is also true that its roots extend from the beginning of the history

---

* cf., e.g., Lenin's dictum (The Impeding Catastrophe, 1917): "We do not invent, we take ready-made from capitalism; the best factories, experimental stations and academies. We need adopt only the best models furnished by the experience of the most advanced countries".

** These are: (1) The principle of the *Transformation of Quantity into Quality,* derived from *Hegel's* view that "quality is implicitly quantity, and conversely quantity is implicitly quality. In the process of measure, therefore, these two pass into each other; each of them becomes what it already was implicitly . . . . . " (2) *The Law of the Mutual Interpenetration of Opposites* (also called the Law of the Unity and Struggle of Opposites), derived from *Hegel's views* on "positive" and "negative", north and south poles of the magnet, "thesis", "antithesis" and "synthesis" (Cf. also *Kant, Fichte and Jacobi*). (3) *The Laws of the Negation of the Negation,* derived from *Hegel's views* on "negation", "positive concept", "affirmation", "old", "new" and "synthesis" (Cf. also, *Spinoza*). These concepts were somewhat modified and elaborated by Engels and have since remained virtually *unchanged* in the official dialectical doctrine of Marxist philosophy. An endless literature exists on these subjects, and most of it testifies that Soviet philosophers have borrowed heavily from *Aristotle* and *Kant*, adding various *Hegelian-Feuerbachian beliefs.* For additional remarks on Spinoza, Kant and Hegel see Lecture IX.

+ G. V. Plekhanov, frequently called the father of Russian Marxism. However, the term "dialectic" is an old classical Western concept (see Assertion 4, §3, *Introduction*). By "dialectics" *Engels* said that he meant the laws of all motion, in nature, history and thought. The *Hegelian dialectic* is concerned with logical subjective development in thought, from a thesis through an antithesis to a synthesis, or logical objective development in history by a continuous unification of opposites. In *Kantian philosophy dialectic* is a form of criticism concerned with metaphysical contradictions when scientific reasoning is applied to objects beyond experience. In general it is related to an art of examining the validity of a theory or opinion, especially (Socrates, Plato) by *"question and answer".* (But see Assertion 4, §3, *Introduction*)

of philosophy right up through the present-day currents of West-European philosophy. *We cannot, therefore, ignore the classical and Western origins of dialectical materialsim and completely surrender all its fundamental principles to Soviet propaganda!* Besides, we may also recall Schrödinger's words: "What is worth preserving preserves itself and requires no protection".

Both materialism and idealism, as such, represent two archaic schools of thought that originally were developed as attempts to answer the question: *What is the world made of? It is clear today that neither science nor the philosophy of science can ever give an absolute answer to this question.* (In Lecture V we see that the great controversy surrounding *quantum mechanics* touches these problems of idealism and materialism very closely as a philosophy of science. Quantum mechanics, according to many Western physicists, emphasizes *"the role played by the observer"*, or *"his mind"*, in introducing "irreversible effects" into the physical system, and thus they tend to favor philosophical idealism, positivism, subjectivism, etc.

Consequently, a few leading Western thinkers still maintain the view *which introduces the role of mind and "consciousness" into the very interpretation* of physical theories!)

B.1.4 The Historical Arrow of Time According to the Marxist Philosophy of Change

The problem facing any philosophy of change (or history) is the same, namely; *a test of an hypothesis in the light of a greater number of empirical events on which there is little doubt. But, for the first time in human history, a global constraint emerged in the form of energy-material crisis; a crisis that has drastically changed many previous convictions; a crisis that has demonstrated the most dominant role "the rest of nature" is and should be playing in any valid philosophy of change, evolution and time.* The philosophies mentioned before ignore this central theme, and should, therefore, be read with a critical eye. Yet, some of their kernels may become valuable in a broader view (cf. Parts III and IV). This applies, for example, to Spencer's philosophy.

Unlike any other civilization, our historical and scientific knowledge embraces a great portion of natural phenomena, the major activities of mankind in almost every region and condition on earth, as well as the history of all previous civilizations. *Knowledge is power,* for it represents, *in principle,* a concrete opportunity to have a better understanding of *what is required in order to prevent a decline and to master the course of events.* In practice, however, opinions differ substantially as to the causes of decline and the proper uses of remedial methods.

Obviously the problems raised by the crisis of our time are larger in scale and more complex than in previous crises of society. Previous crises did not involve as large a portion of the globe, as great a number of persons, as many nation-states, the threat of total world destruction, the global dwindling of essential raw materials, or the paramount role played by modern science and technology.

In spite of many past attempts no all-embracing, verified "philosophy of change" exists today. The most highly advertised attempts to develop such a philosophy is the so-called *historical materialism* or the *Marxist philosophy of history.* Marx (1818–1883) claimed to have found historical laws of motion on a theoretical par with nothing less than the Newtonian laws of motion in classical physics. He asserted that the discovery of these "laws" made prediction possible and his historical "laws" a strict science. The underlying causation of social evolution, according to Marx, is conditioned by the economic circumstances and arises from "the universality of *class struggle*". Accordingly, human actions are not really determined by "free choice" guided by personal ideals but are a "necessary consequence" of the underlying causes. *A primary cause is technology, for as it changes, economic difficulties are created over the division of labor, which, in turn, accelerate "the class struggle".* In addition, the established legal property relations become fetters on advancing technology, and these fetters are broken by the upsurge of the formerly oppressed class, which is given historical support by the advance of technology. But here I only stress the following points:

1) *Marxist determinism asserts that history is governed by causal laws which the human mind can recognize and which offer scholars a "science" to forsee the future of society in its most important and global characteristics.* Such *deterministic* ideas have *not,* however, been originated with Marx. They are part and parcel of the general issue of determinism, causality and the denial of "free will" *in classical Western philosophy* (Lectures IV, V, and IX)].

2) *What are the deterministic "laws" that govern history and can be made the basis of social prognosis?* The first of these laws, according to the Marxist philosophy of history, is supposed to determine *the direction* of historical process. In its general form, this "law" says that *economic developments are basic to social evolution in other areas.* Ideas and institutions, law and politics, even religious concepts and artistic expressions are parts of the social "superstructure", inevitably changing with the gradual transformation of the economic foundation. Thus even altruism, patriotism or other "idealistic beliefs" are themselves the products of "economic conditions" (and of their direct and indirect effects on the human mind).

3) Perhaps the most problematic points in this economic interpretation of history are the ambiguity of the concept *"economic change"* and the identification of the inherent *causes* of this change. If all the developments in the field of production, distribution and consumption are included, a maze of mutual causal relationships results and, with cause and effect undistinguishable in many instances, no social prognosis can be formulated. Marx himself nowhere gave an explicit and comprehensive presentation of his economic interpretation of history, although fragmentary expositions are dispersed over all his writings.

4) *What is therefore the cause for change?* To insert as much consistency into the writings of Marx it must be assumed that those changes which are regarded as basic to all other changes are for the most important part, *technological.*

Technological development, although its temporal rate of change is influenced by scientific, economic, cultural, political, legal, and other developments in the "superstructure", is very nearly irreversible-a one-way street toward greater human command over nature. Under all sorts of regimes and irrespective of prevailing philosophies, religions and creeds, technologists have improved their industry and control over nature. But this gradual advance depends on the physical experiment and its mathematico-philosophical interpretation! Hence the laters too become fundamental causes for social change! The most fundamental irreversible process in this "motion" is therefore the accumulated stock of applied knowledge—in memory, tradition, books, papers and computer files.* And most important, this stock of applied knowledge is bound to grow along the axis of time, and its very change is the backbone of human history.

Since technological development leads to greater production units, bringing together cooperation and interdependence between ever larger, multilevel, international companies, and therefore requiring ever larger masses of workers and capital, *only the community at large* (e.g., states and groups of states) *will finally be able to provide the organizational framework and the security of capital, production, distribution and marketing.*

Recent history testifies that to the extent that the socio-political consequences of this technological movement can be detected, claims to forecasts of social evolution may have little foundation. In this sense the techno-economical movement comes closely as a criterion of a social theory for it is testable, refutable, and falsifiable (cf. Introduction). Techno-economical evolution is, however, only a narrow slice in the global evolution; an evolution which depends, among other things, on limited, low-entropy, terrestrial resources that proceed toward less dowry. Thus, techno-economical activities may not only cease to grow, but may even decline. *Indeed, for the first time in modern history, the external world has introduced a global constraint; a central theme for any valid philosophy of history; a theme represented by the "rest of nature" and by interconnected thinking about global interdependence and the global ecosphere.*

C. Global Interdependence and Global Arrows of Time

In trying to comprehend the meaning of global interdependence we may first re-examine the previous assertions about the causal relationships of techno-economical evolution. These are:

---

* It is of interest to quote here the views of John Dewey (1859–1952), a U.S. philosopher and educator (Appendix III). On the transfer of our accumulated stock of knowledge through the process of education he says: "Speaking generally, education signifies the sum total of processes by which a community or social group, whether small or large, transmits its acquired power and aims with a view to securing its own continued existence and growth". (*"Education"*, in the *Cyclopedia of Education*, Monroe).

i) Technological progress, although its rate of change is affected by socio-economic, political and other developments, is very nearly irreversible;

ii) Under all sorts of institutional arrangements, and irrespective of prevailing religious creeds and philosophies, the human race has improved his control over nature, although with varying degrees of effectiveness;

iii) The gradual improvement of technology depends on physical experiment and its mathematico-philosophical interpretation;

iv) A maze of mutual causal relationships results and, with cause and effect indistinguishable in many instances, no clear cut definition of 'technology progress' can be made. Nevertheless, since the results of this 'progress' can be recorded—either in memory and tradition or, in historical times, in books, papers and laboratory files—*the stock of knowledge about nature is certain to grow, and thus, our technology moves along a one-way road; a road marked with a time arrow which points from less toward greater human command over nature, and from less toward more recorded knowledge about nature.* To the extent that other (irreversible) social consequences of this movement are distinguishable, the concept may be generalized and employed in other fields of science. The foundation of some irreversible evolutions can thus be given the impetus of global or even universal mathematico-philosophical interpretations. Indeed, attempts have been recently made to associate such global yardsticks with entropy growth and, consequently, to employ the mathematico-philosophical methods of modern irreversible thermodynamics in the development of new theories (Part III).

One such theory has been developed by N. Georgescu-Roegen,* who stresses the fact that *all species depend on the sun as their source of low entropy except the human race, which has learned to exploit the terrestrial stores of low entropy such as fossil fuels and minerals. He asserts that the entropy law (*Lectures IV and VI) *rules supreme over the economic process. The basic inputs into his economic model are drawn from the solar influx of low-entropy radiation, and from the terrestrial stocks. The output is high entropy in the form of dissipated heat, waste materials and pollution.*

According to Georgescu-Roegen, industry and mechanized agriculture allow a larger population to survive now *at the expense of a greater reduction in the amount of future life. Terrestrial dowry of low-entropy resources is thus decreasing along a one-way road; toward less dowry and greater amount of dissipated energy-matter. This irreversible global evolution has not been taken into account by "standard" economists who stress "reversible price mechanism" which offset scarcities.* Thus, Georgescu-Roegen claims that economic activity must not merely *cease to grow* (as the Club of Rome once suggested in its report *"Limits to Growth"*), but will eventually decline. A concept like "economic time arrow" or socio-economic progress *has, therefore, no universal basis. But time arrows associated with decreasing terrestrial dowry of low-entropy energy sources or with increasing dissipated energy-matter in our planet are not only sound and measurable but may also play an* increasing role in a broad philosophy of change.** They also demonstrate our main thesis: *No reliable theory can isolate evolution, economics, or techno-social processes from the global ecosphere, nor the global ecosphere from its vast surroundings* (Lecture VI).

* N. Georgescu-Roegen *The Entropy Law and the Economic Process*, 1971.

** Cf. the emerging conclusion about the connection between *environment and national survival.* The traditional equation of national security with military might is becoming increasingly incongruous as resource scarcities, overpopulation, overgrazing, overpumping, overfishing, deforestation, and the ravage of ecosystems are becoming more disruptive of economies and social structures around the world.

Despite abundant evidence for this assertion, national government continue to spend more on traditional military defence then on the development of new energy sources, ignoring the fact that a non-nuclear war cannot be fought today for long because it is too energy-dependent a phenomenon. [Which means that energy shortage increases the probability of fighting a nuclear war (and of polluting the terrestrial ecosphere)]. Indeed, in some cases the indirect threats to security may now arise less from the relationship of government to government and more from the relationship of government to resources and environment. This new trend requires reexamination of old equations of national security—an act that involves a broad and entirely new interdisciplinary approach to the subject.

STRUCTURALISM AND THE DIVIDED AMERICAN THOUGHT: A SHORT GLOSSARY OF TERMS

We are to be idealists only north-northwest, or transcendentally; when the wind is southerly we are to remain realists . . .

America is not simply a young country with an old mentality; it is a country with two mentalities, one a survival of the beliefs and standards of the fathers, the other an expression of the instincts, practices and discoveries of the younger generations.

. . . no modern writer is altogether a philosopher in my eyes, except Spinoza . . .

George Santayana

* * *

Most Americans associate philosophy with eggheaded professors and boring books that are practically powerless and time-wasting. Their common misuse of the word "philosophy" papers over the broad range of contradictory philosophical views which actually govern their society but are nevertheless subterranean and hidden from superficial examination. The pervasive domination of their hidden philosophy cannot be easily detected. The easiest to detect is a strong and pervasive *antiphilosophical attitude*, a hostile attitude which has prevented the cultivation and recognition of a *"formal philosophy"*. Beneath the surface their philosophy does not advance much, but it does not disappear either. Above the surface it is often associated with such diverse "schools" and "slogans" as *pragmatism, idealism, subjective idealism, positivism, neo-Kantianism, logical positivism, logical empiricism, scientific empiricism, structuralism, physicalism, existentialism* and more recently, especially among young people, with *nihilism, neo-anarchism, hedonism, syndicalism, the "New Left", neo-nationalism* (including militant black nationalism) and *cryptofacism* (including the Ku Klux Klan revivals).

A dominant "American pragmatism" is an "idealism" with regard to the reality of universal concepts, a common materialism with regard to the reality of particular "things", and "pragmatic" with regard to the contradiction between the two (see below).

Classical idealism asserts the "reality of mind" *over* the "reality of matter". Classical materialism asserts the opposite. This contradiction produces instability and the danger that the two halves will split apart. And often they do. Then we witness the use of double standards, such as described previously, or the emergence of crude pragmatism, nihilism, neo-anarchism and the like. But, most important, these contradictions have been accumulated with time and have increased the gap between thought and practice; between philosophy and science; between education and society at large. How? The problems associated with these contradictory processes are briefly discussed below together with a few remarks on their origin and on their over-all effect on Western thought.

A. On the Origin of the American Dichotomy in Thought:
Chauncey Wright→the Harvard "Boxing Master"

What is the seminal source of the pragmatic current of thought in America? What are its main maxims? Who contributed most to the early development of the American dichotomy in thought and practice? These questions are briefly discussed below. (In later sections we briefly discuss logical positivism and structuralism.)

During the speculative doldrums of the decade 1850–60 America was on the threshold of a philosophical upheaval which would strongly affect the course of thought and practice for the next century. During this crucial decade, a group of Harvard persons, including *Wright, Green, Peirce, James, Fiske and Holmes*, met informally "to discuss philosophical problems". For our purpose, the first key individual in this group was Chauncey Wright who was called the "boxing master" of the group, since they all had to

468

face his bold claims in assessing modern science and various European philosophies (mainly those of Spencer, Darwin, and Mill). It is therefore imperative to examine briefly, with a critical eye, some of the most decisive claims of Chauncey Wright (1830–75).*

One of his main claims is to keep modern science and philosophy "distinct" and separably apart!!! For him modern science is primarily a methodological discipline. For instance, he insisted that *no consequences whatsoever can be drawn for cosmology from modern science.* Issues which *"belong properly"* to cosmology *cannot* be treated by appeal to *any scientific theory*, no matter how far it is generalized!

In discussing Spencer's philosophy and Darwinism, he claims that the scientific concept of evolution *must be restricted to biology and to some physiological studies of psychology!!* In his view the Spencerian claims for universal explanatory significance of evolutionism** have no basis, nor will they ever be supported by modern biological research!!! *Moreover, he convinced his followers, that the leading concepts in physics are too closely specified by strict mathematical formulas and specific experimental data in that disciplinary area to ever furnish a sound basis from which a universal law of evolution or cosmology might ever be derived!!! Hence scientific thinking must be restricted to its own territory!!! No efforts should be made to use its results in the evolution of philosophical problems!!! Modern science, according to Wright, should therefore turn a nihilistic face toward anybody who dares to base philosophical laws on modern science!* Amazing as it is, his (negative) philosophy of science has since ruled the roost of American academe (see below).

Wright prefers to reformulate philosophical problems in terms of practical reason and moral belief.*** He was, however, unable to create a constructive philosophical work of any lasting significance. Nevertheless, he had greatly influenced the development of American pragmatism, mainly through the works of C. S. Peirce (1839–1914) and W. James (1842–1910) (see below).

B. A Few Maxims of Pragmatism

Pragmatism is a loose doctrine of meaning which owes its early development as a movement to Peirce and James.

Other prominent pragmatists include *J. Dewey* (1859–1952), *C. I. Lewis* (1883–1964) and the Italian *G. Papini* (1881–1956).

Pragmatism as a method does not aim at any *final philosophical conclusions;* it mainly implies a *general direction of thought*, in a given time, place and culture.****

The *pragmatic maxim* was first stated by Peirce in 1878: *"Consider what effects that might conceivably have practical bearings you conceive the objects of your conception to have. Then, your conception of those effects is the whole of your conception of the object".*

Instead of asking whence an idea is derived, or what are its premises, pragmatism examines its results;

---

* cf. C. Wright, *Philosophical Discussions*; and E. H. Madden, ed., *The Philosophical Writings of Chauncey Wright.*

** q.v. Appendix II.

*** cf., J. Collins, *Crossroads in Philosophy*, Henry Regnery Co. Chicago, 1962, p. 156.

**** Cf. Burks, A., *Peirce's Theory of Abduction*, Philosophy of Science XIII, 301–306 (1946); Boler, J., *Charles Peirce and Scholastic Realism*, (Seattle, 1963); Gallie, W., *Peirce and Pragmatism*, (Edinburgh, 1952); Buchler, J., ed., *Philosophical Writings of Peirce*, (New York, 1955); Thompson, M., *The Pragmatic Philosophy of C. S. Peirce*, (Chicago, 1953). P. P. Wiener, ed., *Values in a Universe of Chance: Selected Writings of Charles S. Peirce*, (1839–1914). C. Hartshorne, P. Weiss, and A. Burks, eds. *Collected Papers of Charles Sanders Peirce*; J. Dewey, *The Influence of Darwin on Philosophy and Other Essays in Contemporary Thought*; L. S. Feuer, *"H. A. P. Torrey and John Dewey: Teacher and Pupil" American Quarterly, 10*, 34 (1958) and *Journal of the History of Ideas, 19*, 415 (1958); John Dewey: *"The Superstition of Necessity"*, The Monist, *3*, 362 (1892/3); *"The Ego as Cause"*, The Philosophical Review, *3*, 337 (1894).

it shifts the emphasis from principles, first things and first causes toward last things, effects, consequences and verified experimental data. Early scholasticism asked, "What *is* a thing? What are its inherent causes?" and lost itself in endless disputes; modern pragmatism asked, "What are its consequences?" and turns contemporary science into experimentalism.

"The true", said James, *"is only the expedient in the way of our thinking, just as 'the right' is only the expedient in the way of our behaving"*. Thus, according to James, *truth is a time-dependent relation, an evolutionary process which "happens to an idea"*, *a subjective relation which depends on "human judgment" and "human needs!"* It is the name of whatever proves itself to be satisfactorily consistent in the way of *thinking at a given time in the history of mankind*. Accordingly, *human history and human needs affect the foundations of a scientific theory*. We shall return to his crucial problems later.

Pragmatists (and many contemporary scientists) insist upon the feasibility and desirability of resolving physical and metaphysical problems on the basis of the "practical distinctions" which prevail at the time of a given dispute. Accordingly, if no "practical difference" can be found between two conceptual alternatives, at a given time in a given culture, then both mean practically the same thing, and all dispute is idle! *"To find the meaning of an idea"*, said Peirce, *"we must examine the consequences to which it leads in action; otherwise dispute about it may be without end, and will surely be impractical"*. But while Peirce thought of pragmatism as akin to the *mathematical-physical method*, James's motivation and interest were largely moral and religious. (I demonstrate in Lecture V that James's work in psychology has influenced many pragmatic physicists, including the famous quantum theorist-positivist *Niels H. David Bohr*.)

A number of difficulties are associated with classical American pragmatism. "An idea", we are told by James, "is 'true' so long as to believe it is profitable to our lives". And he adds that "our obligation to seek truth is part of our general obligation to do what pays".

James's doctrine is, therefore, an attempt to ignore all extra-human events. "We cannot reject any hypothesis if consequences useful to life flow from it", he declares. The result is an incredible subjectivistic contradiction which is characteristic of most fallacies in modern American thought. In fact, James's doctrine assumes that there is no *fact, experience* or "truth" except where there is life, in particular, human life. Obviously, taking this doctrine too seriously may eliminate important doctrines of science, especially the ones associated with geophysics, astrophysics and cosmology (see also Parts III and IV). Universal laws and cosmic processes are forgotten; all that is required in such a pragmatism is a belief in last *humanistic effects* and the final effects of the universe upon the human creatures inhabiting this tiny planet. No wonder that even the Pope rejected the pragmatic defenses of religion. Indeed, with James's definitions even Santa Claus exists, for this hypothesis works satisfactorily in the widest sense of the words "useful", "profitable", and "good".

This is not the end of the trouble. Another fundamental problem is the imposition of *"a cultural now"* on the pragmatic maxim! This imposition represents a refusal, *in principle*, to discuss possible *future* empirical evidence or prepare for it. *Therefore, this unofficial but dominant doctrine is responsible, at least in part, for the degradation of abstract thinking amongst the Western scientists who adhere to it; it is the major force behind the present aversion to philosophical inquiry in Western science!!!* But it is not the only one, as I demonstrate in Appendices IV and VI and in the Introduction to Volume I.

A number of subtle relationships exist between pragmatism and "scientific empiricism"; relationships which raise a number of problems, a few of which are mentioned below:

1) Although Peirce's maxim has been an inspiration not only to later pragmatists but to operationalists* as well, Peirce himself felt that it might easily be misapplied so as to *eliminate important doctrines of science*.

2) Not all pragmatists are in complete agreement. Neither Peirce nor Dewey, for example, would accept James's views. James, in turn, does not believe that some of their methods entail his specific phi-

---

* See below.

losophical doctrines: *"individualism"*, *"neutralism"*, *"indeterminism"*, *"meliorism"*, *"pragmatic theism"*, *"supernaturalism"*, etc. In fact, he states that pragmatism is independent of his new philosophy of *"radical empiricism"* and inclines toward the *anti-intellectualist* bent of the Italian pragmatist *Papini*.

3) Although pragmatists (Peirce, James, Dewey) frequently attack *"crude empiricism"* they nevertheless describe themselves as empiricists, so that *today pragmatism is often regarded as synonymous with empiricism*. Note, however, that the term "empiricism" has been used very loosely and confused with numerous related beliefs, assumptions, practices, attitudes and definitions, e.g., *sensationalism* (as, for example, in Hume), *positivism, idealism, transcendentalism* (as, for example, in Kant), *pragmatism, scientific empiricism, Lockean empiricism, phenomenalism, intuitionalism*, etc. It is frequently employed as an ambiguous combination of propositions, practices, methodologies, beliefs, postulates, and "approximations" about *"observation"* (see also Lecture V), *experience, origin of ideas,* and the *limitations of knowledge"* (Lecture IX).

4) Since colonial times the doctrines of *Francis Bacon* (1561–1626), *Thomas Hobbes* (1588–1679) and *John Locke* (1632–1714), the three famous *British empiricists,* have exerted much influence on American thought. In addition, the "theory of knowledge" which holds that "ideas are reducible to sensations", as in Hume (1711–1776), and that experience is the "final criterion of the reality of knowledge" has gained much popularity in the U. S. and is frequently called *sensationalism* or *radical empiricism* (James).

C. Experimentalism and Operationalism

According to Dewey *"experimentation enters into the determination of every warranted proposition".* Hence, in his view the very process of inquiry is experimentation. Thus, *causal propositions* (see Lecture IV) become heuristic, teleological, not retrospective. Moreover, laws are *predictions* of *future* occurrences provided *"certain operations"* are carried out. (Dewey's *"instrumentalism"* is scarcely distinguishable from *"experimentalism".*)* In addition, Dewey holds that if "certain operations" are performed, then "certain phenomena" having "determinate properties" will be observed (cf. Assertion 10 in the Introduction).

In this connection the U. S. physicist-thermodynamicist-operationalist *P. W. Bridgman* (1882–1961) holds that *"we mean by any concept nothing more than a set of operations: the concept is synonymous with the corresponding set of operations".* Thus if the operation is *(or can be)* carried out, the proposition has meaning; if the consequences which it forecasts *occur*, it is *"true"*, or has *"probability"*. But such an *operationalism endangers important scientific and philosophical propositions, for they might be excluded as "meaningless" on "operational" grounds. Moreover it ignores the fact that measurements presuppose theories* (see also Assertion 10, Introduction and §D below).

In the face of these and other limitations a large number of U. S. philosophers have attempted to formulate modifications and refinements. Many of them joined the movement called logical positivism (§D below). But before we turn to this movement I want to introduce a few historical remarks about earlier American thinkers. The title "first American philosopher" is usually given to *William Ames,* who never reached America (he died en route). He was one of the formative influences on early New England thought. In 1622 he gave his inaugural address entitled *"Urim and Thummin"*, arguing for curriculum reform and decrying the low state of student morals. He called for the ethical expression of religious belief and wrote a textbook entitled *The Demonstration of True Logic.* Following Ames' death his wife brought his books to America. Since then many American philosophers have devoted much effort to "Logic" and its endless derivatives—hardly in any logical proportion to the actual needs of their society, science or ethics.

American idealism, which began witn the early ecclesiastic thinkers, had its faint beginnings in the works of *Jonathan Edwards* (1703–58). Previously, in Europe, *Bishop George Berkeley* (1685–1753) had insisted that "matter" is nothing but the complex of "man's sensations". As a pluralistic idealist

---

* "Instrumentalism" is closely related to "operationalism" (see below).

reflecting on the spatial attributes of distance, size, and situation possessed, according to *Johan Locke* (1632–1714), by *"external objects"* in themselves apart from man's perception of them, Berkeley concluded that the discrepancy between the visual and the tactual aspects of these attributes deprived them of all "objective" validity and reduced them to the status of *"secondary qualities" existing only in and for "consciousness"*. Indeed, he claimed that "existence", except as "presence to consciousness", is meaningless. Hence, nothing can be said to exist except minds ("spirits") and "mental content" ("ideas"). But he also posited the "existence" of "universal mind", of which the content is the so-called objective world.

Berkeley gathered much support in London for his project of a college in Bermuda *"for the education of young colonists and Indians of the American mainland"*. His general idea succeeded beyond his wildest expectations, albeit with some modifications. Instead of a college in Bermuda, Bishop Berkeley spent three years (1729–31) in Newport, R. I., establishing what has since become a mainstream of American thought. In fact the mainstream of American thought has not run too far astray from the original mental banks raised for it by Berkeley (see Lecture V and below). There were, of course, certain "practical" objections to Berkeley's philosophy, most succinctly expressed when Samuel Johnson kicked a stone ("I refute it thus!") to show that sensations *are caused* by something material. In a sense, the tradition of American philosophy has ever since been based on a double standard—one for thinking and another for kicking.

American materialism started in the same period with *Cadwallader Colden* (1688–1776). Matter itself, he declared, was unknowable, but we could know it through its *"actions"*. It is *"through action"* that "things impinge" on our *"senses"*.

Later, just as idealism influenced *Ralph Waldo Emerson*(1803–1882), so did materialism influence *Thomas Cooper* (1759–1839). It was left to the great American philosopher *Charles Sanders Peirce* (1839–1914) to establish an entirely new doctrine based on Colden's concept of action. Peirce was able to bring both idealism and materialism under the same roof. He attempted to combine them in a "philosophy of experience" called *pragmatism*. He suggested that *it is possible to accept both the transcendentalism of principles and the naturalism of material events if one supposes that both are developing from a world of "chance" to one of "order", and if one insists on seeking principles through the method of experiment practiced by the physical sciences*.

However, in the hands of his followers, the two traditions diverged again, and no one has since tried to combine them again on a theoretical basis. Today, one can therefore detect this split philosophy at work among Americans in such famous adages as *"Trust in God and keep your powder dry"* or *"Praise the Lord and pass the ammunition"*. It is detectable also in American philanthropy vs. capitalism; idealistic righteousness vs. the acquisition of material goods; perfumed toilets and deodorized culture vs. neglected slums, polluted rivers and unaesthetic advertisements on main streets; idealistic support of the United Nations vs. preparations for conventional and/or atomic wars; millions of cars, big homes and big science vs. religious affiliation and spiritual exaltation; material gains, hard work, money in the bank, and anti-intellectualism vs. a strict moral code, a belief in world peace, college education, and a kind of hidden scientism.

Americans are aware of this duality; they have a deep-seated feeling of insecurity which is a direct result of their split philosophy. Pragmatism and its various by-products are therefore dangerous tools that work well in the short run but may backfire under sustained conditions of external stress.

But my main point is as follows. World War II ended with a *shift* in the center of Western academia: from intellectual Western Europe to pragmatic America—a shift which resulted in the global spread of the American language, "Americanism", "big science", "science funding", the export of the idealistic-materialistic-pragmatic-subjectivistic-positivistic approach, and drastic modifications in the methodology of Western science and philosophy. *It is this split philosophy of Americans which, to my mind, is causing the crisis in Western academia* (Appendix VI).

D. Positivism, Logical Positivism, Structuralism, and the Preoccupation with the Analysis of Language

Positivism was developed in Europe and was first associated with the doctrine of the French mathematician-philosopher *Auguste Comte* (1798–1857). His Cours de Philosophie (1830–42) was one of the seminal works of the 19th century. Comte was most indebted to Saint-Simon (1760–1825), who had emphasized three stages of intellectual development, the need for a new and secularized spiritual order to replace supernaturalism, the belief that social phenomena can be explained by a set of unifying laws, the conviction that the aim of social studies should be ameliorative, and that the outcome of this systematization should be the guidance of sound social planning.

Developing further the ideas of *Saint-Simon, Turgot, Burdin*, and others, Comte asserted that the evolution of human thought and knowledge passes through the three main stages:

First — the *teological*, or *fictitious* in which supernatural beings are found to explain all natural events.

Then, *metaphysical*, or *abstract*, in which these supernatural beings are replaced by "abstract" forces and veritable entities inherent in all beings and capable of producing all phenomena.

Finally, the (pure) *scientific*, or the (pure) *positive*, in which phenomena are described in terms of mathematical formalism and scientific laws. *"Reasoning and observation, duly combined, are the means of this knowledge"*. (These are Comte's own words; and they have often been forgotten by positivists discussing rationalism — see below and also §IX.2.14.)

Comte's philosophical writings were probably the most powerful inspiration behind the development of scientific orientation in philosophical thought in contemporary times. In these undertakings he contributed most to the basic social science of sociology and established it in a systematic fashion. In this he was second in modern times only to Herbert Spencer and his philosophical undertakings that stress the role of evolution in all of nature (Appendix II).

Comte derived many of his ideas from Plato, Aristotle, Hume, Spinoza, Kant, Gall, Turgot, Condorcet, Burdin, Saint-Simon and Montesquieu. From some of them he took over the idea of *historical determinism*, from others the idea of progress in the intellectual development of mankind.

Comte classified the sciences *in the order of decreasing generality and increasing complexity*. Hence they appeared to him in the following genetic series: *mathematics, astronomy, physics, chemistry, biology*, and *sociology*. He claimed that each of these sciences depends upon all the sciences before it; *sociology was therefore the apex of the sciences, and the others had their cause for existence mainly because they could provide explanations for social phenomena*. Thus positivistic sociology "reduces" social facts to "laws" and "synthesizes" the whole of human knowledge. It is in this way, according to Comte, that sociology emerges as well equipped to guide "the reconstruction of society". *Philosophy, according to Comte, is not something different from science and its purpose is to coordinate all sciences with a view to the improvement of human life. He also held the idea that the religious impulse would survive the decay of revealed religion and ought to have an object.*

Comte gave sociology its name and divided it into "social statics" and "social dynamics". He anticipated the organismic school of sociologists (by regarding society as a *"collective organism"*) and established the theory of *Social Causation* (by showing how individual acts and motives are determined by their causes). His influence on European and American thought was profound yet felt often only indirectly.*

In his comprehensive efforts to establish the priority of the positive or scientific method he left his mark on the intellectual development of such thinkers as *John Mills, J. A. Hobson, Werner Sombart, Max Weber, Thorstein Veblen, L. T. Hobhouse, Robert Michels, Graham Wallas, Ludwig Gumplowicz, Gustav Ratzenhofer, A. W. Small, A. F. Bentley, Lester F. Ward, Ludwig Stein, C. A. Ellwood, R. Congreve, and Fredrik Harrison* (see also the Table below).

---

* Positivism in legal philosophy confines itself to *positive law*, i.e., *the law that actually is valid in a certain country at a certain time*. It excludes any higher law. The *Algemeine Rechstslehre* (general theory of law) in Germany, analytical jurisprudence in England, and the American *"legal realism"* are varieties of *"legal positivism"*.

Thus positivism in its narrow philosophical sense describes the philosophy of Comte. *In a broader sense* the term is commonly applied to the *empirical philosophers*. (J. S. Mills "experience philosophy" is positivistic in this sense.) *In this broader sense positivism is therefore intimately connected with American pragmatism and experimentalism.* All three emphasize the achievements of science, yet all three acknowledge the fact that philosophical and metaphysical questions arise within the sciences;—questions which are unanswerable by strict experimental methods.

This was part of the basis for the development of a later movement called *Logical Positivism*. It differs from Comte's positivism in holding that all *rational* doctrines are *"meaningless"*. In fact it claims that the great unanswerable questions about causality and determinism are unanswerable "just because they are not genuine questions at all".

What then is left for philosophy? Facing this difficult question at the end of World War I, Ludwig Wittgenstein found a simple answer: THE SOLE REMAINING TASK OF PHILOSOPHY IS THE CRITIQUE OF LANGUAGE! How and where did he arrive at this idea?

Isolated from the external world in a prisoner-of-war camp in Italy (he served in the Austrian army during World War I), Wittgenstein tried to capture the external world and imprison it with him in the camp. How? Simply by convincing himself that by concentrating on the internal world of the mind he could replace the inaccessible external world with a rich logical analysis of what he and his fellow-prisoners had to say about the world. Very practical indeed. And he recorded these ideas in a little book that was so forcefully written that it has since revolutionized much of Western philosophy. He called this book *TRACTATUS LOGICO-PHILOSOPHICUS*. But released from his physical captivity, he rediscovered the external world, and then released himself again; this time from his mental captivity. He became a worker and accepted jobs as a gardener, builder, amateur architect, teacher, etc. During these years he abandoned philosophy and even came to doubt the one he had written. He was in fact quite surprised when a large part of Western academe accepted his definition of philosophy.* He had no choice but to accept a chair of philosophy at the University of Cambridge [becoming first a Ph. D. in 1929, then a fellow of Trinity College (1930–36), lecturer (1930–35) and professor of philosophy (1939–47)]. *And never again did he publish in his lifetime.*

Indeed the *Tractatus* profoundly affected logical positivism. Yet Wittgenstein claimed that his ideas were usually misunderstood and distorted even by those who professed to be his disciples. For instance, the idea that all "meaningful propositions" are *"truth-functions"* of "elementary propositions" which stand in *"a picturing relation"* to reality was interpreted by logical positivists as requiring the "dependence of meaning" on verifiability *"through sense experience"*—an interpretation which has no ground in Wittgenstein's book.

Whatever is our opinion of the *Tractatus*, the fact remains that Western philosophy cannot rise now to what it was before its publication. Many Western philosophers have since considered the *Tractatus* their "Bible" and much of British and American "activities in philosophy" in our times has been affected in one way or another by Wittgenstein's theory. The Einsteinian methodology,** which became to be known at about the same time with the *Tractatus*, was therefore rejected as "meaningless",

There were many reasons for the wide acceptance of the *Tractatus*; the "inward" trend in Western philosophy did not start with Wittgenstein's book. It originated with various socio-political doctrines which affirmed idealism, pragmatism, transcendentalism, subjectivism, and the "role of man's mind

---

* This was not without a *political* reason. Considering the central role philosophy plays in the growing camp of Marxism and world communism Wittgenstein's dwarfish philosophy was just "the proper answer", according to certain Western political minds, to curb the influence of dialectical materialism. It allowed them to declare the latter simply as "meaningless"—in science, in society, in politics, in philosophy.

** Cf., e.g., Lectures IX, V, Appendix VI and the Introduction to Volume I.

and consciousness" in the universe. Worst of all, the denial of a full-fledged philosophy in the service of Western science, society, politics and ideology has become a hidden weapon in Western political manipulations against the growing influence of dialectical materialism and Marxism.

Thus, leaving the study of the vast external world to the care of the empiricists, and the study of unifying principles of human knowledge to the care of dialectical materialists, most Western philosophers decided to isolate themselves by turning to the internal worlds of the mind and the *"critique of language"!* This shut philosophy out of the greater part of the natural sciences and *a priori* eliminated the potential for significant advances in original physico-philosophical thought.

This trend was accelerated by events in the world *external* to the Western academe which led this new movement—first from Vienna and later from the U. K. and the U. S. These external events included the collapse of the Austrian Empire and consequently Vienna's loss of political and intellectual leadership. It is therefore no wonder that many members of the *Vienna Circle* (see below) had sufficient reason to direct their minds *"upon themselves"* rather than face external events. The rise of Soviet communism and dialectical materialism only contributed a political impetus to that fateful trend. *The outcome was inevitable. Western thought declined, philosophy was cornered, scientism flourished, and ideology ended up in a dead end street.*

But perhaps the greatest damage was done to physico-philosophical endeavors of Western academe. It was Einstein who led the intellectual battle against these subjectivistic trends. But he lost the battle (at least in his adopted homeland, which not only isolated him in the linguistic sense but also in all his attempts to crush the "psychological ghosts" that were introduced by idealists, logical positivists, subjectivists, and theologians into the "scientific interpretations" of the quantum theory. q. v. Lecture V).

* * *

Many other names are associated with logical positivism and I do not intend to mention them all here except for a few passing remarks on my way to the more interesting subject of structuralism.

1. Logical positivism is the label commonly used in reference to Wittgenstein's *Tractatus* as well as to his later, modified works (which were published posthumously by his followers: *Remarks on the Foundation of Mathematics*, 1956; and *Philosophical Investigations—German text with English* translation, Oxford, 1953). It therefore differed somewhat from the philosophy which the *"Vienna Circle"* (see below) based, in part, upon the *Tractatus*. According to the "Vienna Circle" *all genuine philosophy is a "critique of language"*. Yet, according to some of its members, *its result is to show that "all genuine knowledge about nature can be expressed in a single language common to all sciences"*. They also made the common claim that *"philosophy is not a theory but an activity"*. Emphasis was also given to the study of the limitations of *"true sentences about nature"!* Thus, any *"necessary truth"* was concieved as derivable from a rule of language or mathematics. *This was most easy to show in mathematics but quite impossible in the physical, the social, and the life sciences.*

2. Wittgenstein's *Tractatus* introduced a *"theory of meaning"* derived in part from the logical studies of *Guiseppe Peano, Gottlob Frege, Bertrand Russell* and *A. N. Whitehead*. It gave the "Vienna Circle" its logical foundation and attracted such thinkers as *Moritz Schlick, Rudolf Carnap, F. Waismann, Otto Neurath, Herbert Feigl, H. Reichenbach* and *W. Dubislav*. Most of the members of the Circle moved to the U. S. where they also collaborated with *Philipp Frank* and *Charles Morris*, as well as with *A. J. Ayer, G. E. Moore* and *Wittgenstein* in England (cf. the Table below).

3. *Logical Positivism* (sometimes called "Logical Empiricism", "Neo-Positivism", "Analytic Philosophy", etc.) as a movement is sometimes associated with the movements called *Scientific Empiricism* and *"the Unity of Science Movement"* (W. Dubislav, K. Grelling, O. Helmer, C. G. Hempel, A. Herzberg, K. Korsch, H. Reichenbach, M. Strauss, C. Morris, E. Kaila, J. Jorgensen, A. Ness, A. J. Ayer, J. H. Woodger, M. Boll, E. Brunswik, H. Gomperz, F. Kaufmann, R. V. Mises, L. Rougier, E. Zilsel, E. Nagel, W. V. O. Quine and many others).

But the general attitude and views of *Scientific Empiricism* are only in partial agreement with those of logical positivism. *While logical positivism dogmatically dethrones all philosophies except the philosophy of language, this movement analyzes the biological and sociological aspects of language as was earlier the case in pragmatism. (For falsifications deriving from certain uses of logical positivism in quantum mechanics, continuum physics and relativity theory* see Lectures V, and IX).

Logical Positivism and Structuralism

Recent activities in the "critique of language" have given rise to an *entirely modified methodology* that is commonly called *Structuralism.* Developed mainly in France and in the U.S. it variously stresses a number of methodological viewpoints:

a) *Social and cultural behavior can be explained in terms of "languages" which contain "codes", "lexicons", and "syntax". Thus, it assumes that there is an "underlying connection" between social and linguistic phenomena, and that social as well as cultural systems can be investigated by means of a communication system.*

b) *"The study of the whole" is logically preferable to "the study of parts". Here structure is essentially the manner in which parts are organized and interrelated in the whole.* (For instance, the position of symbols in a whole system of symbols is emphasized.) This assumption represents a partial return to an old classical tradition in philosophy; *a rational tradition which was borrowed from Spinoza and Hegel but applied only in its narrower sense* (cf. Lecture IX).

c) *All approaches in structuralism assume that "structure" is a kind of "internal" attribute that needs no external causes to explain its properties.* (This is a clear departure from Spinozism, cf. Lecture IX.) People are endowed with *"innate attributes"* for "the perception of structures" (cf. Kantianism and "innate ideas" in Lecture IX). It is therefore frequently assumed that linguistics should also be investigated in terms of "innate attributes".

d) *One must distinguish between superficial "appearances" and "deep", "hidden" and "underlying structures" which actually determine unconscious processes, "internal intelligence" and the "human spirit". This is a clear "antipositivistic" assumption, which means that one may develop useful general theories by employing abstract, or even semimetaphysical concepts that, at the present time, are not accessible to direct observation and verification.*

e) *Priority is usually given to the study of synchronic (simultaneous, static) structures, i.e., structuralism seldom deals with time-evolving (diachronic) structures.* (This limitation has been greatly criticized in recent years.)

f) *Structuralism borrows methods from the exact sciences and applies them to the study of social and human phenomena, e.g., linguistics, psychology, anthropology, history, sociology, and the arts. Thus, it attempts to transform research methods from one field to another when the resulting benefits and the emerging illuminations, rather than the immediate experimental verifications, recieve the highest priority. But structuralism stresses that in this process linguistics should be considered as the most developed" and "the most scientific" field among the humanities, i.e., it should be used as a sort of a guiding "master science"!*

While none of these claims is entirely new, the sum of them is. Nevertheless, structuralism does not represent a new philosophy nor does it use a single unified methodology.

Two thinkers lead this movement: *Claude Lévi-Strauss** in France, and *Avram Noam Chomsky*** in the U. S.

---

* Cf. his *Anthropologie Structurale;* Plon. Paris, 1958; *La Pensée Sauvage;* Plon. Paris and also *Structuralism,* ed. by J. Ehrmann, Anchor Eooks, N.Y., 1970. In most of these books Levi-Strauss advocates structuralism in ethnological interpretation and in the analysis of myths.

** Cf. his *Language and Mind;* Harcourt, Brace & World, 1968; *Aspects of the Theory of Syntax* (1965); MIT Press, Mass. 1969; *Syntactical Structures* (1957); Mouton, The Hague, 1968, *Current Issues in Linguistic Theory*; Mouton, The Hague, (1964); see also Michael Lane (ed.), *Structuralism, a Reader*; Jonathan Cape, London, 1970.

The comprehensive works of Lévi-Strauss and Noam Chomsky have been guided by most of the methodological viewpoints mentioned above. Yet they vary in scope, applications, and details.

The revolutionary work of Lévi-Strauss deals mainly with such subjects as mathematical logic, linguistics, anthropology and music, and is partly derived from the works of *Ferdinand de Saussure*,*** the neo-Kantian *Ernst Cassirer*,[+] *N. Troubetzkoy*,[++] *R. Jakobson*[+++] and others.

Generally speaking it combines the methodological assertions mentioned above and is generated from symbols by virtue of their position in a system of symbols. Lévi-Strauss borrowed from Cassirer the idea of analogies in research methodologies in different fields and the idea that man's attributes to the generation of symbolic systems may be employed as a master model for a general theory of anthropology. Saussure was the first to point out (about 50 years ago) *that language is structure!* His work gave rise to the development of research on the *structure of sounds—phonology* (the science or doctrine of elementary sounds uttered by the human voice in speech, including its various distinctions of subdivisions of tones. In a broader sense it includes the study of the development of sound changes within a given language and the study of phonetics or of phonemics).

Much of this development was carried out in Prague (in the Twenties and the Thirties) by *Troubetzkoy* and *Jakobson*. Early structuralism then moved from phonology to syntax (Chomsky and his followers). Saussure's doctrine has therefore opened up a new era in modern linguistics—the era of structuralism. As a result a few schools of linguistic structuralism were developed in Prague, Copenhagen, France, and the U. S.

It is the last school which concerns us here, mainly because it has been advanced somewhat against the mainstreams of traditional American thought based on logical positivism and dogmatic empiricism.

The *Chomskian revolution in transformational linguistics* has been somewhat influenced by European structuralism. But it has also evolved as a genuine American response to the excessive empiricism which had dominated American linguistic studies under the influence of *behavioral* psychology*, and as a respone to some fundamental difficulties that *"phase-structure grammer"*, syntax, and various linguistic interpreations encountered.

From our point of view the significance of the Chomskian revolution is in the reintroduction of some classical rational philosophical increments into the "territorial domains" of logical positivism. These amendations are diametrically opposed to the dogmatic attitudes which have dominated philosophical studies in the U.S. till recently.

To my mind the unorthodox assumptions of Chomsky and his followers are precisely the most important causes of the success of their revolutionary studies. *And they give us hope that American philosophy will, after all, combine traditional modes of philosophical inquiries with current modes of research in the "hard sciences".*

E. Pragmatic Scientists and Physics

A frequent claim made by "applied scientists" is that new interpretations of the foundations of the exact sciences have no practical applications and are, therefore, of "purely philosophical interest". This attitude is thoroughly misconceived. First, the present separation of causes of natural phenomena into numerous "independent" variables, postulates, theorems, laws and theories is unjustified from the practical point

---

*** *Course in General Linguistics,* Philosophical Library, N.Y., 1959.

[+] *The Logic of the Humanities,* Yale University Press, 1966, *Language and Myth,* Dover Publications, N.Y., 1953: *The Philosophy of Symbolic Forms* (1923–1929, 3 Vols., R. Manheim, New Haven, 1957).

[++] *La Phonologie Actuelle,* Psychologie de language, 1933.

[+++] Cf. Jakobson R. and M. Halle, *Fundamentals of Language,* Mouton, The Hague, 1965.

* *Behaviorism* is closely related to *operationalism*. It maintains that since all test-statements describe behavior, our theories too must be stated in terms of possible behavior.

of view if one is seeking to eliminate the *applicative difficulties created through this separation.* Secondly, new conclusions at the fundamental level revolutionize the methods of measurement of terrestrial time-keeping,[+] temperature, entropy, Doppler shifts, velocities, mass, energy, "elementary particles" and many other important quantities. For instance, in practice there are as many *different* temperatures as there are *methods for measuring them.*[*] But none of the available temperature definitions is universally valid. Available definitions depend on the *practical purposes of the measurement!* Insofar as normal laboratory experiments are concerned, the difference creates no problem and, consequently, there is no dispute. But in applications such as the mechanics of the upper atmosphere, shock waves, hypersonic flight, molecular beams, relativistic thermodynamics and astrophysics, the disagreements in the observed values of temperatures, their various physical meanings and the different domains of application create *unsolved* practical problems! No one has solved these riddles yet, partly because any such attempt leads to philosophical inquiry. Similar problems are encountered with *entropy* (a term which seems superfluous to many and which, according to Truesdell, "suggests nothing to ordinary persons and only intense headaches to those who have studied thermodynamics but have not given in and joined the professionals") and with most of our fundamental concepts in the exact sciences.[**]

The importance of purely abstract thinking and philosophy has been stressed by many great scientists of the past and by historians of science. Defending the role of philosophy in science, Einstein wrote:

> The hypotheses with which [a new development in physics] starts become steadily more abstract and remote from experience. On the other hand it gets nearer to the grand aim of all science, which is to cover the greatest possible number of empirical facts by logical deduction from the smallest number of hypotheses or axioms. Meanwhile the train of thought leading from the axioms to the empirical facts of verifiable consequences gets steadily longer and more subtle.

My wonderful American colleagues praise Einstenian contentions but consult the first available economist in charge: Very few engage in serious research without first applying for a proper "grant" from public or governmental institutions.

Einstein himself was a philosopher-scientist. Admirably, he actually used philosophy to create what is now a significant portion of modern science, making contributions which are today utilized in almost every "practical" domain of our society.

F. Chronological Index in American Thought

The following is a non-inclusive chronological index of European scholars who visited America or influenced the development of American thought from colonial times to the present. It also includes (native-born and long-term resident) American scholars. The Europeans are marked with an asterisk. Note the decline in the frequency of asterisks in recent times

1521–1626	Francis Bacon*	1617–1688	Ralph Cudworth*
1564–1642	Galilei Galileo*	1623–1662	Blaise Pascal*
1571–1630	Johannes Kepler*	1632–1677	Baruch Spinoza*
1576–1663	William Ames	1632–1714	John Locke*
1588–1679	Thomas Hobbes*	1638–1715	Nicolas Malebranche*
1596–1650	Rene Descartes*	1642–1727	Isaac Newton*
1599–1662	Samuel Hartlib*	1644–1718	William Penn

[+] See Section §III.4 in Lecture III.

[*] E.g., absolute (positive) temperature; absolute (negative) temperature; electron temperature; kinetic temperature; inverse temperature; global temperature; hidden temperature; spin temperature; Ott's and Planck's temperatures in relativistic thermodynamics, etc. (see my *MDT* and *CRT* for details).

[**] The non-scientific role of entropy in the exact sciences is discussed in Lecture IV.

1828–1906	Henry Conrad Brokmeyer	1857–1911	Alfred Binet*
1828–1920	Roberto Ardigo*	1858–1947	Max K. E. Planck*
1829–1900	Charles Caroll Everett	1858–1932	G. Wallas*
1830–1875	Chauncey Wright	1859–1938	Edmund G. A. Husserl*
1831–1879	James Clerk Maxwell*	1859–1938	Samuel Alexander*
1831–1918	J. Lachelier*	1859–1947	Pierre Janet*
1832–1920	W. M. Wundt*	1859–1941	Henri Bergson*
1832–1918	Andrew Dickson White	1859–1952	John Dewey
1833–1911	Wilhelm Dilthey*	1860–1944	G. F. Stout
1833–1885	Elisha Mulford	1861–1934	J. M. Baldwin
1833–1899	Robert Green Ingersoll	1861–1924	James Edwin Creighton
1833–1901	E. K. Dühring*	1861–1947	Alfred North Whitehead
1834–1916	George Holmes Howison	1861–1925	Rudolph Steiner*
1834–1926	Charles William Eliot	1862–1943	David Hilbert*
1835–1909	William Torrey Harris	1863–1916	Hugo Münsterberg
1835–1910	Mark Twain	1863–1930	Mary Whiton Calkins
1836–1882	Thomas Hill Green*	1863–1931	George Herbert Mead
1836–1903	Francis Ellingwood Abbot	1863–1952	George Santayana
1838–1916	Ernst Mach*	1864–1937	Ferdinand C. S. Schiller*
1839–1897	Henry George	1864–1937	Paul Elmer More
1839–1914	Charles Sanders Peirce	1866–1952	Benedetto Croce*
1840–1900	Thomas Davidson	1867–1927	E. B. Titchener
1842–1906	Eduard von Hartmann*	1867–1940	F. J. E. Woodbridge
1842–1901	John Fiske	1868–1936	Arthur K. Rogers
1842–1918	Hermann Cohen*	1869–1949	James R. Angell
1842–1910	William James	1869–1950	John Elof Boodin
1842–1914	Ambrose Bierce	1870–1937	Alfred Adler*
1843–1925	James Ward*	1871–1938	William McDougall
1843–1931	Harald Höffding*	1871–1960	Ralph Tyler Flewelling
1844–1924	G. Stanley Hall	1871–1938	William Stern*
1844–1900	Friedrich Nietzsche*	1872–1944	Walter T. Marvin
1844–1906	Ludwig Boltzmann*	1872–1970	Bertrand Russell
1846–1924	Francis Herbert Bradley*	1873–1939	Edward G. Spaulding
1846–1926	Rudolph Eucken*	1873–1946	Edwin B. Holt
1847–1945	Ernst Cassirer	1873–1958	G. E. Moore*
1847–1922	Georges Sorel*	1873–1962	Arthur O. Lovejoy
1847–1910	Borden Parker Bowne	1873–1966	William Ernest Hocking
1848–1894	G. J. Romanes*	1873–1953	William Pepperell Montague
1848–1915	Wilhelm Windelband*	1874–1928	Max Scheler*
1848–1925	Friedrich L. Gottlob Frege*	1875–1944	James Bissett Pratt
1848–1923	Bernard Bosanquet*	1875–1961	Carl Gustav Jung*
1849–1936	Ivan Petrovich Pavlov*	1876–1902	Ernst Schröder*
1851–1933	Felix Adler	1876–1957	Ralph Barton Perry
1852–1919	Paul Carus	1877–1940	David Ross*
1852–1913	Shadworth Hodgson*	1877–1946	James H. Jeans*
1853–1921	Alexius Meinong*	1878–1933	Durant Drake
1853–1928	Hendrik Antoon Lorentz*	1878–1958	John Broadus Watson
1853–1932	W. Ostwald*	1878–1953	Walter B. Pitkin
1854–1926	Albion Woodbury Small	1878–1965	Martin Buber*
1854–1924	Paul Natop*	1879–1955	Albert Einstein
1854–1912	Henri Poincaré	1880–1943	Max Wertheimer
1854–1927	Benjamin Ide Wheeler	1880–1947	Morris R. Cohen
1855–1916	Josiah Royce	1881–1956	Giovanni Papini*
1856–1906	William Rainey Harper	1882–1936	Moritz Schlick
1856–1926	Emil Kraepelin*	1882–1944	Arthur S. Eddington*
1856–1939	Sigmund Freud*	1882–1961	Percy Williams Bridgman
1857–1939	Lucien Levy-Bruhl*	1883–1955	Jose Y. Gasset Ortega*
1857–1929	Thorstein Veblen	1883–1931	Kahlik Gibran

POLICY AND PUBLICITY:
A CRITIQUE

Let us come now to references to authors, which other books contain and yours lacks. The remedy for this is very simple; for you have nothing else to do but look for a book which quotes them all from A to Z, as you say. Then you put this same alphabet into yours . . . And if it serves no other purpose, at least that long catalogue of authors will be useful to lend authority to your book at the outset.

Miguel de Cervantes

We haven't the money, so we've got to think.

Lord Ernest Rutherford

You damn sadist', said Mr. Cummings, 'you try to make people think'.

Ezra Pound

Work, Finish, Publish.

Michael Faraday

A. Erosion of Academic Quality and Abstract Thinking

Any civilization depends upon innovation and knowledge a creative minority has been able to generate together with the incalculable effects these have had upon everyday thinking and conduct (Appendix II). Since innovation and accumulated increases of knowledge are the basis on which we progress, it is pertinent to analyze how, where and when they occur or are subdued. *Here we come to highly significant questions concerning the nature and course of scientific development, the obstacles which have lain in the way of philosophic revolutions, the sources of success and frustration in innovation and the roots of illusions and misapprehensions that many intelligent people still harbor with respect to "freedom of speech, press, research and inquiry".*

Most of us are well aware of mankind's hostility to change. This dogged obstinacy in clinging to habits, this suspicion of the unfamiliar, are exactly what might be expected given our slow evolutionary origin. It may therefore be regarded as a natural trait, a trait which has slowed down the pace of change in civilizations while, at the same time, increasing the security and permanence of each achievement. Such a trait is behind the common lack of inventiveness, the reluctance to adopt new ideas and the tenacious hold to old ones. It may also be linked to the roles sacredness and tradition have played in the histories of science and philosophy. *It is certainly behind the solid walls of cultivated prejudice and inherent stupidity which even the most trivial of innovations is faced with in the sciences and the arts.* Yet, as we have stressed in p. 433, it plays a major role in the structure of science and the sharpening of new ideas. As knowledge and ingenuity increased, people departed further and further from their wild life; but their fundamental traits remained essentially unchanged. All such fundamental traits and natural characteristics are hereditarily transmitted no matter how much or how little people may be educated and "civilized".

Civilization, on the other hand, is precarious; its knowledge must be assimilated anew by each generation. This knowledge can increase, but it may also decline as history testifies. It is a property that can be lost immediately and completely should humanity be subjected to an *overwhelming brainwashing. And this is precisely the danger inherent in the overwhelming use of certain standards of mass communications.* To make the matter clear, let me examine a few fundamental questions.

Mankind's curiousity and search for knowledge is not a cultivated habit of modern society. The

484

dangers of attack made preliminary scouting a valuable asset in the survival of early man. Men were by nature wont to pry and try and fumble, long before they scientifically analyzed and experimented. Obviously, discoverers and explorers must generally be exceptionally curious and active.

A man watching TV, for instance, is neither active nor free. He becomes captivated and passive. *Overwhelming exposure to mass communication degrades his capacity for abstract thinking* (an important symptom of a declining culture?), *it dissolves his old norms* (another such symptom?) and *it stimulates his attachment to meaningless luxuries* (an additional symptom?). Here children, Hottentots and university professors react essentially in a similar manner. They become susceptible to any change and receptive to rapidly changing fashions.

We are faced here with an *apparent paradox*. On the one hand, individuals are reluctant to endorse or accept new ideas when innovators present them on a person-to-person basis; on the other, they tend to endorse or even adopt those same ideas when they are presented to them through the media. How do TV, radio and the newspapers succeed where *bona fide* science so often fails? *In what sense does the Western system of mass communications become responsible for the declining standards of academia?*

B. Scientists as Laymen

Laymen tend to regard commentaries, news and other information delivered to them through the mass communications media as something reliable and unquestionable as the "accepted" rituals and solemn words a tribe has had throughout history. What I mean is that the mass communications media represent the accepted *"tribal culture"* which people have sought to identify with from the early dawn of humanity. With the overwhelming mass communications that we are witnessing today the whole world becomes like the playground of a single tribal culture. What excessive observance of the religious ritual was in the past the *"television moloch"* has become in the present.

Most contemporary scientists are fairly narrow in their professional occupation; They are quite uncritical and provincial in their overall philosophical outlooks. Outside the domain of their discipline they tend to respond to the media as laymen do. On matters lying outside their profession they seldom question the *reliability* and *integrity* of contemporary journalists, reporters, and the like. Moreover, they are subject to a good deal of confusion concerning the basic meaning of "freedom of speech", "freedom of the press", "freedom of research" and "academic freedom".

The pursuit of freedom is admirable as a human goal. But the notion of freedom lacks philosophical exactitude. In contrast to current beliefs, one must first distinguish between "freedom of speech" and "freedom of research".

For instance, in the West no one would suggest today that a law be passed prohibiting people from criticizing ethnic groups. But most people might object to giving scientists public funds to go around making such criticisms. One must also distinguish between "freedom of the press" and freedom of the media to conduct their business *without assuming responsibility for the consequence of their decisions*. Surely there is a major ethical issue here which *supersedes* any question of information and the freedom to publish it without fear of government sanction or censorship. It is a question of *self-control* and *self-government*.

Journalists, reporters, commentators and the like have been quick to argue that an ever-growing literature on these subjects exists, a literature which deals with the history, problems and modern techniques of mass communications and with the increasing sense of social responsibility advocated for journalists and reporters. But the fact cannot be eluded; as matters now stand, there is no abiding code of ethics in this profession, nor is there any *independent* administration of justice, courts of justice, or jurisprudence which, *without* government control, can democratically *impose* internal responsibility. This, and not just reporting and Nielsen ratings, should be the chief criterion of journalism and broadcasting. Governments in democratic countries can *and should* demand such self-governing, independent, internal jurisdiction.

C. Destruction of Selectivity in Academe

Journalists and commentators are also quick to criticize "scientific *elitism*". They employ this concept in a pejorative and disparaging sense, as implying values and academic procedures *"in conflict with the fundamental democratic aspirations of Western society"*. This charge is misleading the public and is far more dangerous when repeated in the mass communications media.

Indeed, the negative aspects of elitism, those associated with self-appointed and self-serving privileged groups, run counter to certain basic tenets of our civilization. But in this respect journalists, editors and commentators themselves are not essentially less culpable in their actual conduct than certain clans of scientists. In fact, they help build up scientists as a guild of masters and apprentices and a hierarchy of leaders who rule through conformity to fashions and the media.

The scholarly community must place a very high value on *selectivity* in order to encourage work of the *highest quality*; it must organize itself so as to stimulate the *unusual* and the *original in contrast to the popular!* Selectivity is necessary for the advance of science, and, *therefore, some* people *are* bound to be disappointed. Their complaints must be treated seriously but without fear and terror. The assessment and maintenance of quality is of fundamental importance for scholarly activities. Any alternative is not likely to work better.

The function of *quality control* is to determine which individuals deserve special recognition through the award of a prize, which work is qualified for publication in the best professional and scholarly journals, which faculty members to promote and which to pass over or dismiss. Its meaning is simple: *polarization of quality in science, technology and philosophy.*

But does such quality control also require the polarization of financial support and the use of exclusive funds? Not necessarily, as we shall try to show later.

D. Participation in Decision-Making

In the 18th century Thomas Jefferson (1743–1826), the third president (1801–9) of the United States, in arguing that local government, or the government closest to the people, was best, could claim that citizens knew most about their local governments and least about the national one. Obviously this claim is not valid today for the media has disrupted this precious organism by placing its irresistible forces at the service of centralization. Similarly, efforts to decentralize national administrations have received little practical support from the media. Thus, almost everywhere we now observe the decline of genuine subnational systems and the growth of national and international bureaucratic systems. (But see also a brief discussion of the Hegelian philosophy in Lecture IX.)

The Western media is also responsible for the decline of a number of important Western doctrines associated with survival.

During the last decades we have been witnessing endless debates in favor of the reduction of (Western) defense spending, a position that has been strongly supported by the media. But why cannot the Western media maintain a continual debate over its own national and social responsibilities?

Today there is no longer much doubt among top-level decision-makers that almost any decision on the national or international level is bound to be nullified or even reversed without the full cooperation of the media.

We do not suggest any formal conspiracy of the media to destroy the West. Rather, the fault can be laid to a combination of technological advances in mass communications and the arrogance of certain journalists and commentators who have spent their energy to pandering to low-level gratification, vices and weaknesses.

The incredible way the media has campaigned against recent defense budgets and the development of essential means to defend the West may best be illustrated by the U. S.-U. S. S. R. race to develop certain advanced technologies. Both the Western media and the scientists captivated by such propaganda are the ones who have contributed the most to the present situation in which Soviet technology is ahead of the U.S. in a number of critical areas. A pity. For the journalists and commentators it is not only a question of self-control and self-government, it is, above all, a matter of *active participation and "real sharing" in national decision making. Surely there is room here for a major change; for these people must be given equal responsibility in making national decisions.* Any alternative leads only to total disaster.

E. Words, Deeds and Critics

Word, as we know, underlies intelligent thinking. Yet the new Moloch's vocabulary is being corrupted in part by a curious style intended to invest even the most banal ideas with importance. This tendency has, gradually, affected academe. Indeed, one can easily detect its influence in lectures and even in matters of university finances. All professions have their jargon, but the language of academics seems nowadays to lead further and further into the woods of meaninglessness. It is symptomatic, perhaps, of the present condition of academe when university presidents resort to such words as: "We will divert the force of this fiscal stress into leverage energy and pry important budgetary considerations and control out of our fiscal and administrative procedures". Indeed, the modern bureaucrats of academe earnestly adopt such a *pseudoprecise* jargon (and probably hope that it will give their words some authoritative heft). *But the facts cannot be eluded. Such dictions only confirm the Confucian maxim: "If language is incorrect, then what is said is not meant. If what is said is not meant, then what ought to be done remains undone".*

A most dangerous route is taken when words are viewed as deeds! Indeed a real crisis in communication results when such word-deeds are *immediately* followed by superficial commentaries *(taken by the public as well-proven evidence without even questioning the subtle motivations of their making or of their selection at a given time in a given context).* Today, almost anyone, even in the lowest cultures, can learn "the propaganda trick" and air it with a force beyond what he can naturally give account for. Frequently the person behind this action does not even consider himself responsible for the results of *his own* influence. He is *trained* to believe that he is "beyond feedback", "beyond judgement" and "beyond trial"; for he "only comments on, or selects information, facts and deeds which *other* persons have made or have been linked to".

Indeed, in a few minutes, a TV editor can rear more glorious fashions or dig deeper intellectual traps than any made with dozens of classical books over centuries. Yet, today there is essentially no one in our society to stop him except himself. Hence his education, his intellectual capability, his ethics, and his social responsibility must be well balanced and much above the average. So much so that even a most skeptical public can trust his judgement though normally it doesn't see or hear him. But perhaps it should. *Perhaps media policy-makers and editors should report to the public, on a regular basis, the pros and cons of their (and other) policies. Perhaps they should also share responsibility with other decision makers; in academe, in defense, in industry, in government; for they are part and parcel of their society; not above it as priests are, nor separated from it as they sometimes pretend.*

Indeed, editors, journalists and commentators are the priests of the publicity Moloch and, hence, of our society. They should, therefore, strive for the highest possible standards; in culture, in education, in theory, in practice. They should be given the best in philosophy; the best in science, technology and the humanities; the best in ethics. They, no less than academe, must be the guardians of selectivity and excellence. They, no less than government and judiciary, must be the wise protectors of society and order. But can they do all that? We may never know if we don't give it a try. Of them, the least we hope for, is to stop creating a world of ideas which transcends that of the facts; because *in the mass media what pass nowadays for facts, are indeed so subtle that we are less and less confident in our ability to separate the subjective functionings of individual opinions and cultural biases of editors and commentators from facts and objective science. Anyway, there is so much room for changing the present practices and eliminating some of the quandaries in which these publicity priests have found themselves involved, that no simple answer can be given to the aforementioned question.*

F. Unity and Structure

The French political philosopher Baron de la Brede et de Charles Louis de Secondat *Montesquieu* (1689–1755) wrote that governments are likely to be tyrannical if they are responsible for administering large territories, for they must develop the organizational capacity characteristic of despotic states. Indeed, it is modern mass communications that make it possible today for large territories to be governed democratically. *But modern mass communications harbor the danger of an even greater tyranny than Montesquieu foresaw: the tyranny of the Mass Communications Moloch* (MCM).

The need to devolve and transfer the authority and control of political power is reduced when a central government can communicate directly with citizens in all parts of national or regional territories. But this development lessens the importance as well as the reliability of *subnational institutions*. At the same time it gives rise to large, impersonal, national bureaucratics. But most important, the rise of exceptional leadership is seriously impeded by the rise of the media for its current methodology prevents, *a priori*, the potential rise of another Roosevelt, Churchill, or even an Einstein. Indeed, with the current methodology the West can never again raise exceptional leadership other than that of the MCM.

But at this point a skeptic may ask: "How important is the preservation of law, order, and academic selectivity? And how important is the 'sharing responsibility' of these priests in 'the preservation of state' and in the uplifting of Western thought?"

If effective harmony of the whole is preferred to disharmony; between man and nature, or man and men, or man and himself, then life in society requires the concession of some part of the individual's sovereignty to the common order of the whole. *Despite the philosophical adventures of Sophists and Nietzscheans, all human conceptions revolve about the effective coordination, optimal structure and orderly conduct of the whole.* A group *survives* in competition or conflict with nature or with another group, or with itself, according to its *unity and power*, according to the ability of its members *to function together* in accordance with *a common* ideology, law, structure, and order. Every individual is a microcosm in the large structure of the whole; let these microcosms fall into harmony with the whole (of a much subtler nature than that of the interactions whereby the planets are held together in their orderly movement), and the individual survives and evolves; let them lose their interconnectedness with their surroundings, and instability of the whole structure begins. This, and not the worship of the publicity Moloch, is the first law of "free press" which every modern society must seek if it would have life.

But there is another fact that must be taken into account.

The *"accumulation of knowledge"* increases our understanding and control of nature. But at the same time it is likely to generate trouble. Take the national level for instance. Here history testifies that *a nation cannot be strong unless it has some ideology or faith*. A complete explanation of socio-historical, physico-chemical, or socio-economic forces could hardly inspire hope, devotion, or sacrifice; it could not offer much to the distressed, nor courage to the embattled. *A living ideology (or faith) can do all this.* Moreover, it can moderate greed and passion, especially when coupled with an ethical philosophy. *Indeed, a nation is thrice armed if it fights with faith and ideology.* The question is 'What ideology?'

G. Science and the Media

Most scientific findings are *initially published* in professional journals *as having (approximately) equal weight. Therefore (initially), it is all to easy to miss the significance of particular new departures.* Hence, an early selection of news items reflects an opinion. Here a broad perspective is more important than a news rush.

The public should be in a position to question both scientists and journalists. Here one task of the scientific communicator is to persuade the public to regard science as part of his cultural environment and, at the same time, to *demythologize* science, for only then will it be regarded with *objective skepticism*, as indeed it should. Such communications should have a didactic purpose, point out both the glories and the dan-

---

* Peter Farago, *Science and the Media*, Oxford University Press, Oxford, 1976.

gers of science and also dwell on the cultural, historical and other aspects of scientific developments. But, most important, they should not vulgarize science or philosophy by drawing oversimplified parallels between a few isolated facts and everyday experience. Science as well as technology deserve more respect, the respect of a scholarly skeptic.

Both science and the arts represent attempts to order man's perception of the universe; they deal with different aspects of nature, but the commonality of their principles is greater then is usually appreciated.

But whereas the interpretative arts are considered worthy of *national and local support*, the communication of advanced scientific and technical ideas to those who are not scientists is usually left to the care of individual media, haphazard presentation, sporadic efforts, narrow interests, one-time crusades, and fashion.

Ideally the communicator should be a scholar with a broad perspective. He should not only know the relevant facts which constitute the news item, but also grasp and master their significance in other spheres. However, with few exceptions, Western scientists refuse to take on such tasks. The participation of Western scientists at the level of public communications, as distinct from their presence on legislative, economic, or political committees, is regarded not so much as wrong but as simply not done.

But the refusal of scientists to participate in public communications is contrasted by their willingness to communicate among themselves by means of "scientific papers". This too encounters a number of problems.

Books rule the world, or at least those nations in it which have a written language; the others do not count.

Voltaire*

H. A Vested Interest in Publications?

According to Spinoza, "humility is very rare" (*Ethics* III, App., def. 29) and as Cicero said, even the philosophers who write books in its praise take care to put their names on the title page.

In contemporary scientific communities, increased "production" of papers—the so-called scientific papers—is considered a basic *"measure of achievement"*. This is partly the result of the continuity of ideas, which links the present with a past in which a scientific publication indeed meant novelty and scholarship. Partly it is a product of the elaborate obscurantism of "modern scholarship" in which a primitive *"publish-or-perish"* syndrome influences academic promotions. Partly it reflects an erroneous view of the problem of achieving scientific progress. And partly, we have seen, the preoccupation with the number of publications is forced quite genuinely by the *nexus* between quantity and the subtle process of winning apparent status.** Thus, somehow along this "modern" road, quantity is gradually replacing quality.

The pursuit of discovery is admirable as a scientific goal. But the notion of discovery lacks philosophical exactitude; there is agreement neither on its substance nor its source. *We know that the great majority of present-day publications contain nothing but mathematical or verbal funambulisms—printed matter devoid of any original or valuable idea, pages which very few "professionals" ever read.*

Today there is a vested interest in scientific publication; since many scientists derive non-scientific benefits from their publications, they are apt to be unwilling to forego these benefits even if these publications conflict with their own standards of morality.

Since modern science began, about 15,000,000 scientific papers have been published, and we are now adding about a million new papers every year, which will double the number in 15 years. These are published in some 40,000 journals, each carrying an average of 25 per year. About 15 million papers imply the

---

* A modern skeptic may put this another way: Oil rules the world, or at least those nations in it which have a profitable industry; the others do not count. I prefer to reverse it: TV rules a declining world, or at least those nations in it which worship it; the others *do* count for they *still* read books.

** Some skeptics put this another way: *"If you cannot be rich, be famous"*.

existence of about 4 million authors, most of whom, because of exponential growth, are alive today. Consequently, there is approximately one journal for every 100 authors.

If the average scientist reads 5 journals, each journal will reach about 500 potential readers. But this figure is illusory. A now classic paper by Urquhart[+] shows that more than 3000 of the total of 9120 different scientific periodicals available in a library (in this study the Central Science Library in London) were not called for at all during the period of investigation (the year 1956); 2274 were used only once. The most popular periodical had 382 requests, 60 periodicals were requested more than 100 times each, and half the requests could be met from the top 40 journals.

Such studies indicate that although about 40,000 journals exist, half the reading that is done is confined to only about 200 of them.

The besetting sin of all journal creators has been to imagine that theirs is the journal to end all journals in their field. But one even doubts whether any group, like the founders of such a journal, remains entirely faithful to it beyond the publication of the first issues. Some supporting members of the group may, however, become so active in order to gain attention as to justify the cynical claim that they publish more papers than they actually read.

Statements about "good researchers" vis-a-vis "poor researchers" may apply if we are able to produce a *"globally reliable yardstick of quality"* rather than the admittedly crude count of *quantity*. Since the *"amount of usage"* provides a reasonable measure of the *"scientific importance"* of a journal or a man's work, one can consider the use of a paper in terms of the *references* made to it in other papers. The fact that authors preferentially cite their own papers, those of their special friends, or those of authoritative figures which confer status on their work, should not be confused with the tendency of the researcher of average conscientiousness to give credit to papers which *have* provided the foundation for his work.

In contrast to current belief, a false paper does not attract many citations, at least not in the exact sciences. It is interesting to note that the data on references made in a single volume of *Chemical Literature* show a halving of the number for every 15 years of increased age.[*]

In this connection Price[**] shows that the actual amount of literature in *each* field of science is growing exponentially, doubling every 15 years. Consequently, to a first approximation, the number of references of a given date is *in proportion* to the total literature available at that date. Thus, although half the literature cited will in general be less than a decade old, Price concludes that *any paper, once it is published, has a probability of being used at all subsequent dates!* He also demonstrates that *highly distinguished scientiests have most frequently been highly prolific authors.* These somewhat surprising results appear to give more credit to high productivity rates in publications than one is willing to admit. Perhaps each paper, once reviewed and accepted for publication, does represent at least a quantum of useful scientific information. Perhaps scientific papers are not produced merely to be counted by chairmen, Deans and promotion committees. The inherent driving force to publish may be motivated by *bona fide* intentions; and the prime object of the scientist may not, after all, be publication itself but *communication* with his fellow scientists. Maybe some overlapping and duplication results from the vastness of science and from its diversification. Indeed, it may even be desirable that important contributions be published a few times over, each time in a different and slightly independent mode. Nevertheless, it is reasonable to assume that *additional "socio-economic factors"* are involved in the process of scientific publication.

But before going on I would like to stress a few additional points about our preoccupation with the publication of scientific papers.

Any *direct* onslaught on the identification of quality with scientific publication has a *drawback*. Scholar-

---

[+] D. J. Urquhart, "Use of Scientific Periodicals", *International Conference on Scientific Information*, NAS—National Res. Council, Washington, D. C., 1958, pp. 277.

[*] P. L. K. Gross and E. M. Gross, *Science*, 66, 385 (1927).

[**] D. J. de Solla Price, *Little Science, Big Science*, Columbia University Press, N. Y., 1963, p. 79.

ly discourse, like politics, has its deeper rules and they must be respected. Thus, the *definitions* of quality, creativity and novelty in science (and *what* and *who* add to it) have to be sidestepped and bypassed. Instead, the present argument has been directed to seeing how extensively our present preoccupations, most of all with the production of papers, are molded by tradition and inertia. Released from these preoccupations, we become free to survey some of the ideas which serve and/or drive science and society *at large*. These at least have a plausible relation to the welfare of society and the state.

I. Time and Policy

Can the Western states survive for long if industry, science and education are not oriented to reach ideological, social or national goals? Should government authorities and planning commisions employ new philosophical ground-rules in seeking the welfare of society? Such questions are so perplexing that no educated person can deny having given them some thought, though as a student he had been trained not to ask such questions. Thus, no global framework, nor any workable doctrine has been developed in the West to guide the perplexed.

Harnessing industry, research and education to attain planned ideological, social or national goals is one of the most intriguing and complex problems that confront present and future societies. *Most of us are aware of this problem.* Yet, somehow, we entertain the hopeful notion that some sort of *"National Policy"* or *"Master Plan"* underlies government and public intervention in science and industry, and that there is a clear determination of *priorities* on the part of government and other public agencies. It is, however, a naive belief.

There is no "National Policy". There is no "Master Plan". There is no Ideology. Each nation and each administration has habits; there are "established traditions" of moving things, personnel and funds, but no abiding overall doctrine which makes it all the more urgent to seek to understand our commitments, to select priorities and generate master programs, and equally as urgent, to abide by a set of guiding rules that will serve to resolve major questions in the conduct of *government, academe, media,* and *industry*.

But whether or not there is a doctrine or a "Master Plan", decisions have to be made, which raises the question: *How* are they being made *now* and how *should* they be made *tomorrow*?

Indeed, the decisions governing growth or decline in various social, scientific, and industrially-based areas are the outcome of a complex series of interactions. Competing forces in this decision-making process may be traditionally, politically, economically, humanistically, or environmentally driven. Frequently final decisions are made on grounds far removed from genuine national, international, and humanistic goals, or from relevant scientific-technological facts.

How, one may ask, are individuals or governing boards supposed to overcome these problems?

In trying to comprehend these problems one may first attempt to gain a grand-scale perspective on the history of civilizations, general philosophy, and the philosophy of science. Politicians too need that, at least if they wish to preserve the independence or perhaps even the very existence of their states.

Secondly, one must begin by examining *what is possible in the Western states today and what is not.* Totalitarian planning, in the sense of *Soviet "five-year-plans"*, lies far behind and beyond most Western states. "No Doctrine" and "No Planning" is another extremity which only leads to anarchy, decline and collapse. Leaving the problem entirely in the hands of governmental bureaucrats is another impossibility (which can only lead us back to the present crisis in the Western system). But does a fourth possibility exist? This question was taken up in the Introduction. Here we examine it in the world of academe where young researchers often claim that the present funding and planning of R & D includes a form of hidden corruption in which *peer group assessement* of grant applications makes sure that certain group participants and their close associates get their share out of the general but limited budget. Recent attempts by, say, the (U.S.) N.S.F. to refute these claims have not been objective nor convincing (see also below). Hence one cannot ignore nor belittle these dangers. Another claim is concerned with the support of *large scale* fundings that lead to increasing fragmentation by the formation of interested groups. These groups may sometimes tend to become mere federations of individuals interested only in the rewards associated with large fundings.

Large funding presents a dilemma to the individual; should he refuse managerial position, and thus remain only an individual researcher without the power of a big team at his command; or should he accept this position and become so loaded with bureaucratic problems that he has no time for promising intellectual-scientific research and may, therefore, loose his self-esteem as a scholar. These dangers have been felt mostly in U. S. universities; but they are now common in other Western countries.

What can be done to prevent the possibility of suppression of young or unorthodox applicants whose research might invalidate the work of members of the peer groups?

The following guidelines may, perhaps, be useful in certain situations:

i) *Multiply independent sources of large research funds.*

ii) *Channel, whenever possible, large national development funds to academic centers for applied research through industry and the public, i.e., through public and industrial assessement of the value of the proposed R &D.* It is also instructive to keep in mind the fact that in national planning the most important factor is not the ratio of R &D expenditures over GNP per capita, the number of published papers etc., *but their proper or selective distribution in the various branches of (quality) research applied to the available resources of the country and to its needs. Moreover, the most significant factor for the development and for the successful adaption of revolutionary theories, discoveries or technologies is the long-range, daring judgement of a few people from various backgrounds who are willing to work together without much regard for protocol and administrative orthodoxy, who can respond quickly to the practical experience of the men of action, and who can see both the ideological and practical necessity and urgency of the implementation of a new idea.*

We may reexamine the *"promotion-qua-publication policy"* in contemporary academia. As matters now stand, a young talent who wants to help his country is bound to sacrifice his academic career. The Western university has not yet invented the method to reward him for his contributions. *Rewarding a scientist who has made contributions to national projects may, almost paradoxically, have a positive effect on the quality of scientific-technological work for no longer would young R & D personnel be compelled to publish in order to be rewarded.* This would affect scientific publications in one way: towards higher standards.

iii) Scientists tend to overlook the fact that defense-oriented research and development saves them from the partial *technological stagnation* that is *inherent* in *"a pure consumer goods economy"*. But when pressed they all agree that such research carries with it a hideously subsidizing support for "healthy" scientific and technical progress. The history of science is crowded with cases in which great and exciting advances have originated from intimate cooperation with the military. The improvement of *military* equipment almost always paves the way for the arrival of better *civilian technology*. Indeed, *defense problems, advanced technology, food, and the depletion of the raw materials which nature has stocked in the earth's crust will continue to occupy our minds for a long time to come.* Thus, if the *effectiveness* of our investment in research and education persists at its present low level, and if we fail to assign priorities in academe to meet the aforementioned demands, the outcome will indeed be tragic (see also my AIG).

J. Distorted Reviewing and the Success of Unfunded Scientific Discoveries

The problem of *"quality versus quantity"* associated with *"scientific publications"* touches upon *a deeprooted dilemma:* Does peer review promote the best research? Can it detect potential discoveries and breakthroughs? Should all progress in contemporary science be subjected to peer review?

There are many different versions of review, but each has the same goal: to help decide how the limited funds available can best be spent to advance both knowledge about nature and the various, interconnected goals to which this new knowledge contributes. The underlying assumption in selecting the reviewers is that individuals who have themselves succeeded in research and development (and often in raising funds) can make *"a better guess"* in "the distribution of funds". The same methodology applies to publication in scientific journals. Here the editors pick a referee to judge the merit of a submitted manuscript. This method of choosing reviewers and referees has played an important role in shaping

:ontemporary science and technology. It has probably succeeded in technology more than in the exact
sciences, but it has totally failed in regard to the physico-philosophical inquiry (Lectures III and V).

*Reviewers, panels and committees may not be the best friends of original thought. They may tend to be
biased against really innovative proposals, those which go against the dominant fashion in the field and
which could lead to major conceptual changes. Often the reviewers are themselves the biggest spenders in
the field, not necessarily the best researchers. They belong to an inner clan of "jobholders", or are members
of one "invisible college" or another. They must live and let live. They have no special interest in recognizing
potential breakthroughs, nor in attaining social and national goals. But perhaps they cannot.*

The greatest successes in science and philosophy (excluding engineering) *have been accomplished with
little or no financial support.* During the first half of this century a small number of physicists engendered
most of our progress in fundamental physics — and this *without* the present system of financial support
and grant-supported research (cf. the biographies of Einstein, Planck, Schrödinger, Bohr, Debroglie,
Boltzmann, etc.). *The four greatest discoveries of the century — the structure of the atom, quantum mechanics, the theory of relativity and the structure of DNA — were accomplished without reviewers, panels,
committees and grant-giving agencies.* Unlike the first discovery, the scientific history of the last three is
quite well known. So let us examine the mode of discovery associated with the first one.

By the turn of the century, the concept of the atom had taken on a deeper meaning than it had possessed
in Dalton's day, a century earlier. For instance, J. J. Thomson suggested an atomic model in which each
atom was pictured as a globule of positively charged fluid with the electrons embedded in it, rather like
seeds in a watermelon.

A new basis was provided in 1901–1911 by Ernest Rutherford who noted that on passing through a
thin film of mica or metal, a stream of α-particles was somewhat broadened, or scattered, by its collisions
with the atoms of the film. This scattering was to supply information on the disposition of an atom's
mass, believed to be fairly evenly distributed, as in Thomson's globule of positive fluid. Later Rutherford
conceded:

"... I had observed the scattering of α-particles, and Dr. Geiger in my laboratory had examined
it in detail. He found, in thin pieces of metal, that the scattering was usually small, of the order of
one degree. One day Geiger came to me and said, 'Don't you think that young Marsden, whom I
am training in radioactive methods, ought to begin a small research?' Now I had thought that, too,
so I said, 'Why not let him see if any α-particles can be scattered through a large angle?' I may
tell you in confidence that I did not believe that they would be, since we knew that the α-particle
was a very fast, massive particle, with a great deal of [kinetic] energy, and you could show that if
the scattering was due to the accumulated effect of a number of small scattering, the chance of an
α-particle's being scattered backward was very small. Then I remember two or three days later
Geiger coming to me in great excitement and saying, 'we have been able to get some of the α-
particles coming backward ...'. It was quite the most incredible event that has ever happened
to me in my life. It was almost as incredible as if you fired a 15-inch shell at a piece of tissue paper
and it came back and hit you. On consideration, I realized that this scattering backward must be
the result of a single collision, and when I made calculations I saw that it was impossible to get
anything of that order of magnitude unless you took a system in which the greater part of the
mass of the atom was concentrated in a minute nucleus. It was then that I had the idea of an atom
with a minute massive center carrying a charge."*

Today, in *Big Engineering*, where the choice is frequently between several large projects, panels may
be necessary because of the size and complexity of the financial decision to be made. However, in other
fields, panels just may not work well. They merely generate vicious cycles of spending and respending,
missing the *bona fide* breakthroughs. In part this is due to the many different and intricate interests of the
referees. In part it is due to the absence of personal commitment to success, even with a risk. For that
matter, it is by no means clear whether our present most expensive method of fundamental research

* (From Needham, J. and W. Pryel, eds., *Background to Modern Science*, Macmillan, 1938, pp. 61–74).

merits the expectations of both the public and the scientists. The system must therefore be revised. Shrinking budgets, elitism, and dedicated Small Science and Philosophy can only help us recover from the present crisis. But they cannot, by themselves, bring about a recovery, for they must be combined with the tenets of an entirely new methodology. The central question is therefore: *What methodology?*

K. A Concluding Letter*

Mr. Taxpayer
Inflation City
The Planet Earth

Dear Mr. Taxpayer:

The news media informed me recently about a profound transformation in your willingness to support my research. In view of the fundamental "uncertainties" inherent to the minds of all newsmen (as prognosticated by Heisenberg's "uncertainty principle" and Bohr's "Complementarity Principle"), I hereby inform you that these news can never be rigorously confirmed.

Yet, to be on the safe side, I hereby review the situation in my scientific subfield.

During the first half of this century my (poor) colleagues have made most of our scientific progress, supporting themselves with "Wisdom and Philosophy" (cf. the biographies of Einstein, Planck, Schrödinger, Bohr, Debroglie, Boltzmann, etc.). An indeterminable quantum transition occurred after my generation had decided to replace old-fashioned research by the modern "Grant-Supported Research".

But according to the new regulations you are not entitled to "know" that you paid my "Grants" [since by Quantum-God-given truths a "rigorous distinction between the 'knowing subject' and the 'passive object' (the latter is my work) is impossible"]. To be more specific, according to Heisenberg, "The partition of the world into observing and observed systems prevents a sharp formulation of the law of cause and effect". Since cause-and-effect relations exist between your support and my research, I cannot let you observe me spending your money. Got it? Most of my fellow scientists did.

Thus, instead of letting you break God-given quantum laws, I compensated you by letting you "observe" plenty of our unobservable ghost particles, including the "worm-holes", "tachyons", and geons (supplemented with a few strange particles). I also delivered to you endless publications, which, if not significant, at least demonstrate our mathematical funambulism. So why should you now become unsatisfied with my prolific work?

Of course, according to these "interaction-observation with consciousness" (please ask E. Wigner to explain them to you. He would love to), you are entitled to make an irreversible cut in your support. In that case, I would have no choice but to revert to old-fashioned research of "Wisdom and Philosophy"—an act that would indeed endanger the activities of my contemporary colleagues. Anyway, do you know about any dusty old Patent Office in Bern for rent? Or, perhaps a lenses-polishing cellar in Amsterdam?

Respectfully yours,

Mr. Quantum Theorist
Head
The Uncertainty-Memorial Laboratory
Ghost-Particle City
The Contemporary University at the Intellectual Desert
Vacuum State, 17,000,000,000 years.

Postmaster: This is our zip-code in (cosmic) time—(Not in space!)

Copies: To the Journal of Irreproducible-Results, Section D: "On the Irreversible Transformation of Physics into Psychology".

* Reprinted, with permission, from the author's note in the *Journal of Irreproducible Results, 23,* No. 4, 4, 1978.

THOUGHT-PROVOKING AND
THOUGHT-DEPRESSING QUOTATIONS

A. Skepticism and Skeptics

And I gave my heart to seek and search out by wisdom concerniig all things that are done under Heaven . . .
I have seen all the works that are done under the sun; and, behold, all is vanity and vexation of spirit . . . .
For in much wisdom is much grief; and he that increaseth knowledge increases sorrow.

Attributed to King Solomon
Ecclesiastes, I

One should keep the need for a sound mathematical basis dominating one's search for a new theory. Any physical or philosophical ideas that one has must be adjusted to fit the mathematics. Not the other way round. Too many physicists are inclined to start from preconceived ideas and then try to develop them and find a mathematical scheme that incorporates them. Such a line of attack is unlikely to lead to success.

A. M. Dirac

The right to search for truth implies also a duty. One must not conceal any part of what one has recognized to be true.

Albert Einstein

It is impossible to demonstrate the non-contradictoriness of a logical mathematic system using only the means offered by the system itself.

Kurt Goedel

He who can properly define and divide is to be considered a god.

Plato

A man said to the universe:
'Sir, I exist.'
'However,' replied the universe,
'The fact has not created in me
A sense of obligation.'

Stephen Crane

B. Determinism

"We ought then to regard the present state of the universe as the effect of its anterior state and the cause of the one to follow . . . an intelligence which could comprehend all the forces by which nature is animated

495

and the respective situation of the beings who compose it. . . . would embrace in the same formula the movements of the greatest bodies of the universe and those of the lightest atom: for it, nothing would be uncertain and the future, as the past, would be present to its eyes. The human mind offers, in the perfection which it has been able to give to astronomy, a feeble idea of this intelligence."

Pierre Simon de Laplace

"It is only in the quantum theory that Newton's differential method becomes inadequate, and indeed strict causality fails us. But the last word has not yet been said. May the spirit of Newton's method give the power to restore unison between physical reality and the profoundest characteristic of Newton's teaching—strict causality."

Albert Einstein[14]

"The average plain man who is no philosopher will probably consider that the springs of human action are too varied, too intricate and too complex to be summed up in any single formula. His own philosophy is not very clear-cut, but may perhaps be described as one of determinism for others and freedom for himself.."

James Jeans

"Mr. Average Man thinks over his past, and proclaims that if he were young again, he would choose a different profession. He may insist that he would be free to make his own choice, but all he means is that if, at the age of eighteen, he had had the knowledge and experience of life which he now has at fifty, he have acted differently. Of course he would, and so would we all, but this is no evidence of freedom. If Mr. Man now had to make his choice again, with precisely the same knowledge and experience as he had at eighteen, he would review the situation in the same way as he did before, the same considerations would be thrown into the scales, and the balance would again swing in the same direction as before. He will not claim a freedom to act from pure caprice, but only a freedom to yield to the strongest motive—the freedom of Newton's apple to fall towards the earth rather than towards the moon, because the earth attracted it more forcibly than the moon. And this is not freedom of any kind; it is pure determinism.

As Hume said, to have made a different decision, he would have had to be a different man.

Or perhaps he may claim he is free to choose in trivial matters, as for instance whether he will ask for black or white coffee. Perhaps he usually asks for black, and if on some rare occasion he asks for white, he may imagine that in so trivial a matter his choice was wholly undetermined. But a psychologist will tell him that, even here, he can only yield to the strongest motive, no matter how weak these motives may be."

James Jeans

"The principle of causality must be held to extend even to the highest achievements of the human soul. We must admit that the mind of each one of our great geniuses—Aristotle, Kant or Leonardo, Goethe or Beethoven, Dante or Shakespeare—even at the moments of its highest flights of thought or in the most profound inner workings of his soul—was subject to the causal fiat and was an instrument in the hands of an almighty law which governs the world."

Max Planck

"There is in the mind no absolute or free will; but the mind is determined in willing this or that by a cause which is determined in its turn by another cause, and this by another, and so on to infinity".

"Men think themselves free because they are conscious of their volitions and desires, but are ignorant of the causes by which they are led to wish and desire".

Baruch Spinoza

C. Space, Time and Knowledge

"Space and body are not really different".

Baruch Spinoza

Ubi materia, ibi geometria.

Johannes Kepler

Science is partially-unified knowledge;
Philosophy is completely-unified knowledge.

Herbert Spencer

Those who can think learnt for themselves and not from the Sages.

Kuan Yin Tze (8th cent.)

It is man's social being that determines his thinking.

Mao Tse-tung

To non-European civilizations, the idea that numbers are the key to both wisdom and power, seems never to have occurred.

Arthur Koestler[18]

But by measure and number and weight, thou didst order all things.

Solomon[15]

The experiences of an individual appear to us arranged in a series of events; in this series the single events which we remember appear to be ordered according to the criterion of "earlier" and "later", which cannot be analysed further. There exists, therefore, for the individual, an I-time, or subjective time. This in itself is not measurable. I can, indeed, associate numbers with the events, in such a way that a greater number is associated with the later event than with an earlier one; but the nature of this association may be quite arbitrary.

Albert Einstein[19]

Epochs are labeled by temperature, which is roughly the same as red shifts, $T = (3°K)(1 + Z)$. In theoretical studies of expanding universe models, including the closed mixmaster model . . . a convenient epoch label in Einstein's equations has turned out to be equivalent, namely $\Omega = -\ln(V^{1/3})$ where V is the volume of space at the given spoch. I find this Ω-time to be very attractive as a primary standard; the enduring measure of evolution throughout the history of the Universe is its own expansion.

C. W. Misner[10]

D. Laws and Outlaws.

. . . the law that entropy always increases—the second law of thermodynamics—holds, I think, the supreme position among the laws of Nature.

A. S. Eddington[3]

I hesitate to use the terms "first law" and "second law", because there are almost as many "first laws" as there are thermodynamicists, and I have been told by these people for so many years that I disobey their laws that now I prefer to exult in my criminal status and non-condemning names to the concrete mathematical axioms I wish to use in my outlaw studies of heat and temperature. The term "entropy" seems superfluous, also, since it suggests nothing to ordinary persons and only intense headaches to those who have studied thermodynamics but have not given in and joined the professionals.

C. Truesdell[8a]

. . .**the H-theorem and the Clausius-Duhem inequality are not the same** . . . This does not show that either principle is right or wrong.

C. Truesdell[8];

I think that this Conference, and the general tenor of what we read in various books, would support the following view: Whenever several of us are asked to solve a routine problem, we write down the same equations and give the same final answer. However, if we are asked to justify or to motivate the equations (such as the First Law, the two parts of the Second Law, etc.), we seem to disagree and to produce different basic systems. Is this a psychological phenomenon? Is this just a question of semantics? Where does it come from?

J. Kestin[11a]

A careful study of the thermodynamics of electrical networks has given considerable insight into these problems and also produced a very interesting result: the non-existence of a unique entropy value in a state which is obtained during an irreversible process . . . ,I would say, I have done away with entropy. The next step might be to let us also do away with temperature.

J. Meixner[3]

I think Professor Kestin's question clearly shows that, at present, one can no longer consider thermodynamics as a closed field. In other words, it actually has never been so, ever since its foundation. I suppose it has always been viewed in a much wider framework, and I think this point of view is still very actual.

R. C. Balescu[11b]

Will the two principles of Meyer and Clausius provide a sufficiently solid foundation for some time to come? No one has any doubt about this; but what is the origin of this confidence?

H. Poincare[12]

With classical thermodynamics, one can calculate almost everything crudely; with kinetic theory, one can calculate fewer things, but more accurately; and with statistical mechanics, one can calculate almost nothing exactly.

E. Wigner[11c]

E. How To Win a Grant?

To: Universal Creation Foundation*
REQUEST FOR SUPPLEMENT TO U.C.F. GRANT No. 000–00–00000–001
"CREATION OF THE UNIVERSE"

This report is intended only for the internal uses of the contractor

Period: Present to Last Judgment
Principal Creator: Creator
Proposal Writers and Contract Monitors:
    Jay M. Pasachoff and Spencer R. Weart
    Hale Observatories
    Pasadena, California 91109

BACKGROUND

Under a previous grant (U.C.F. Grant No. 000–00–00000–001), the Universe was created. It was expected that this project would have lasting benefits and considerable spinoffs, and this has indeed been the case. Darkness and light, good and evil, and Swiss Army knives were only a few of the useful concepts developed in the course of the Creation. It was estimated that the project would be completed within four days (not including a mandated Day of Rest, with full pay), and the 50% overrun on this estimate is entirely reasonable, given the unusual difficulties encountered. Infinite funding for this project was requested from the Foundation and granted. Unfortunately, this has not proved sufficient. Certain faults in the original creation have become apparent, which it will be necessary to correct by means of miracles. Let it not be said, however, that we are merely correcting past errors; the final state of the Universe, if this supplemental request is granted, will have many useful features not included in the original proposal.

*Reproduced with permission of *The Journal of Irreproducible Results* (June 1973).

PROGRESS TO DATE

Interim progress reports have already been submitted (*"The Bible," "The Koran," "The Handbook of Chemistry and Physics,"* etc.). The millennial report is currently in preparation and a variety of publishers for the text (tentatively entitled "Oh, Genesis!") will be created. The Gideon Society has applied for the distribution rights. Full credit will be given to the Foundation.

Materials for the Universe and for the Creation of Man were created out of the Void at no charge to the grant. A substantial savings was generated when it was found that materials for Woman could be created out of Man, since the establishment of Anti-Vivisection Societies was held until Phase Three. Given the limitations of current eschatological technology, it can scarcely be denied that the Contractor has done his work at a most reasonable price.

SUPPLEMENT

We cannot overlook a certain tone of dissatisfaction with the Creation which has been expressed by the Foundation, not to mention by certain of the Created*. Let us state outright that this was to be expected, in view of the completely unprecedented nature of the project. The need for a supplement is to be ascribed solely to inflation (not to be confused with expansion of the Universe, which was anticipated). Union requests for the accrual of Days of Rest at the rate of one additional Day per week per millennium ($Dw^{-1}m^{-1}$) must also be met. Concerning the problem of Sin, we can assure you that extensive experimentation is under way. Considerable experience is being accumulated and we expect a breakthrough before long. When we are satiated with Sin, we shall go on to consider Universal Peace.

We cannot deny—in view of the cleverness of the Foundation's auditors—that the bulk of the supplemental funds will go to pay off old bills. Nevertheless we do not anticipate the need for future budget requests, barring unforeseen circumstances. If this project is continued successfully, additional Universe—Anti-Universe pairs (Universes—Universes †) can be created without increasing the baryon number, and we would keep them out of the light cone of the Original. By the simple grant of an additional Infinity of funds (and note that this proposal is merely for Aleph Null), the officers of the Foundation will be able to present their Board of Directors with the accomplished Creation of one or more successful Universes, instead of the current incomplete one.

PROSPECTIVE BUDGET

Remedial miracles on fish of the sea	
Remedial miracles on fowl of the air	
Creeping things that creep upon the earth, etc.	∞
Hydrogen	n/c (created)
Heavier elements	n/c (nucleosynthesis)
(Note: The Carbon will be reclaimed and ecologically recycled).	
Mountains (Sinai, Ararat, Palomar)	∞
Extra quasars, neutron stars	∞
Black holes (no-return containers)	∞
Miscellaneous, secretarial, office supplies, etc.	∞
Telephone installation (Princess model, white, one-time charge, tax included)	$16.50
Axiom of choice	optional

* "I cannot but conclude the bulk of your natives to be the most pernicious race of little odious vermin that nature ever suffered to crawl upon the surface of the earth"—*Jonathan Swift*

SALARIES

Creator (1/4 time)   at His own expense
Archangels
   Gabriel   1 trumpet (Phase 5)
   Beelzebub   (Low sulfur)
   Others   Assist. Halos

Prophets
   Moses   stone tablets (to
                                                                               replace breakage)
                                                                                          finite

Geniuses
   n.b. Due to the Foundation's regulations and changes in
   Exchange Rates, we have not been able to reimburse Euclid
   (drachmae), Leonardo (lire), Newton (pounds sterling), Descartes
   (francs), or Joe Namath (dollars). Future geniuses will be
   remunerated indirectly via the Alfred Nobel Foundation.

Graduate students (2 at 2/5 time)   reflected glory.

MONITORING EQUIPMENT AND MISC:

1 200-inch telescope (maintenance)   finite
Misc. other instruments, particle accelerators, etc.   Large but final
Travel to meetings ∞
Pollution control equipment   +40%

Total $\infty + 40\% = \infty$
Overhead (114%) ∞

Total funds requested ∞

Starting date requested:
   Immediate. Pending receipt of supplemental funds, layoffs are
   anticipated to reach 19.5% level in the ranks of angels this quarter.

I hate quotations. Tell me what you know.

Ralph Waldo Emerson

References

1. Whitrow, G. J., *The Natural Philosophy of Time,* Thomas Nelson & Sons, London 1961. a) p. 27; b) 28.
2. *The Nature of Time* (T Gold and D. L. Schumacher, eds), Cornell University Press. a) p.4; b) p.233; c) p.235; d) p.185; e) p.184; f) pp.25,26; g) p.189; h) p.189; i) p.-93; j) p.277; k) p.145.
3. Meixner, J., in *A Critical Review of Thermodynamics* (E. B. Stuart, B. Gal-Or and A. J. Brainard, eds.) Mono Corp., Baltimore 1970 pp.37,47.
4. Landau, L. D. and Lifshitz, E. *Statistical Physics,* Addison-Wesley, London 1969, pp,29–32.
5. Beauregard, O. Costa De "On World Expansion and the Time Arrow", *Modern Developments in Thermodynamics,* Gal-Or, B. ed., Wiley, N. Y. 1974, pp.73,77.

6. Kampen, N. G. Van, in *Fundamental Problems in Statistical Mechanics* (E.G.D. Cohen, ed.) North Holland, Amsterdam 1962. (Quoted from Ref. 12a).

7. Penrose, O., *Foundations of Statistical Mechanics*, Pergamon Press, Oxford 1970, a) p.42.

8. Truesdell, C., *Rational Thermodynamics*, McGraw Hill, N. Y. 1969, a) p.11; b) p.164; c) p.12; d) p.21.

9. Landsberg, P. T., *Thermodynamics with Quantum Statistical Illustrations*, Interscience, New York 1963; *Pure and Applied Chem. 22* (3–4), 1970 a) p. 550.

10. Misner, C. W., *Phys. Rev. 186* (no. 5), 1328–1333 (1969). (A similar time has been advocated by E. A. Milne, *Kinetic Relativity*, Oxford University Press, New York 1948) (q.v. also ref. 3.).

11. *A Critical Review of Thermodynamics* (E. B. Stuart, B. Gal-Or and A. J. Brainard, eds. Mono Book Corp., Baltimore 1970. a) p.509; b) p.510; c) p.205;, quoted by L. A. Schmid; d) p.441.

12. Poincare, H., *Thermodynamics*, Paris 1892. (Preface)

13. Eddington, A. S., *The Nature of the Physical World*, Cambridge 1928. p.133.

14. Einstein, A., in *Nature*, p.467, March 26, 1927.

15. Wisdom of Solomon (Apocrypha) 38 : 15.

16. Jaynes, E. T., *Am. J. Phys. 33*, 392 (1965).

17. R. J. Boscovich. *Theoria Philosophiae Naturalis*, Vienna 1758. Par. 552. English transl. J. M. Child, 1922 (La Salle, III : Open Court).

18. Koestler, A., *The Sleepwalkers*, Hutchinson, London, 1959.

19. Albert Einstein, *The Meaning of Relativity*, pp.1, Princeton Univ. Press, N. J., 1956.

CRITIQUE OF WESTERN METHODOLOGY

We have mastered the physical mechanism sufficiently to turn out possible goods; we have not gained a knowledge of the conditions through which possible values become actual in life, and so are still at the mercy of habit, of haphazard, and hence of force . . . With tremendous increase in our control of nature, in our ability to utilize nature for human use and satisfaction, we find the actual realization of ends, the enjoyment of values, growing unassured and precarious. At times it seems as though we were caught in a contradiction: the more we multiply means the less certain and general is the use we are able to make of them.

John Dewey

> Man is stranger to his own research;
> He knows not whence he comes, nor whither goes
> Tormented atoms in a bed of mud,
> Devoured by death, a mockery of fate;
> But thinking atoms, whose far-seeing eyes,
> Guided by thoughts, have measured the faint stars,
> Our being minggles with the infinite;
> Ourselves we never see, or come to know.

Voltaire

Knowledge is one. Its division into subjects is a concession to human weakness.

Sir H. J. Mackinder

503

1. The Decline of Intelligentsia

Young people look around in vain for social and intellectual leadership and for the beginning of a new era when a new workable theory about the important questions of our time has effected the whole society. What they find, however, is nothing but provincial opinions of the same kind as any one else's. They look for the grand perspective of a satellite; they are ready to settle for that of an eagle; but what they find is that of a gopher. They look for a doctrine that affects the life of one's culture, of one's science, of one's state. Instead, they find lecture-room jargon, and pale restatements of the views of predecessors.

They search for new leadership in thought, methodology and practice; they hope to develop a better world, with new grand visions; *but they find inertia, nihilism, careerism, scientism and academicism* (see below). They ask: Has the time for the grand goals passed? Has the West played out, exhausted, and used up all its non-materialistic potentials? Has we been born in the winter of our civilization, or in the spring of a modern new era?

Some search in vain for an "original thinker" who has made a name by advancing a workable theory about an important question of our times. They ask: Has any contemporary intellectual intervened effectively, with a comprehensive doctrine, or a single compelling ideology, in education, science, industry, socio-economic structure, defense, mass-communications, or in any other important actuality? Why is it that the mere idea of calling upon one of them to prove his intellectual eminence in the media, in government, or in social planning, is enough to evoke our pity? *Indeed, what are the origins of the decline of intellectual leadership, of the triumph of mediocrity, of the emerging double standards, of the corrosion of national will, of the frantic hedonism-nihilism, and of the preference of immediate comfort and prosperity above long-term survival? But most important, can the ruling parties and the so-called intellectual elite of the West spearhead the task of stemming the tide of decline in our vital domains of thought and practice? And, if they can, will the prevailing methodologies help them deal with the totality of our problems? Do they have the proper philosophy to do so?*

2. The Search For a Methodology

The gradual disappearance of exceptional leadership (Appnd. IV), the declining standards of academe (Introd.), the rise of the mass-communications tyrrany; have they led to the emergence of an (often undeclared) "Western doctrine" that "there is something inherently wrong with (a democracy) having a *philosophy-doctrine at all"?* The answer to this question may be a valuable clue as to *why the West may not be able to win or even maintain a status quo with its opponents: the Soviets and their allies have a working philosophy-doctrine; we have none.* Moreover, we possess a deep-rooted aversion to develop or employ one.

The roots of this aversion as well as its devastating effects were examined in Appendix III and in the

Introduction. But suppose this aversion may be overcome through some proper methods of education. Then we are immediately faced with the need to start a search for proper candidates. This is by no means a simple task, especially in an age characterised by the dominant influence of science on almost any field of thought and practice.

But a search for candidate theories cannot begin without a proper *methodology*. We need it because without a proper methodology we may not be able to adopt a proper philosophy nor orient much of our activities in academe, government and industry. What I mean is explained below.

3. The Einsteinian Methodology Vs. Skepticism.

In searching for a proper methodology we may begin by re-examining what is possible today in modern science, *and what is not*. Systematic methaphysics lies far behind and beyond most of us, and a workable Western philosophy has been dried up long ago. But a third possibility, corresponding to modern scientific skepticism, still remains. And it can be approached by the hitherto unused methodology associated with the physico-philosophical and ethical methods of Einstein (cf. the *Introduction* and Lecture IX).

Classical skepticism is usually the negation of philosophy, declaring it to be useless. Not so with Einstein's skepticism. He regards philosophy as the removal of borders between science and science-based philosophical inquiry. Where classical skepticism is led to renounce classical philosophies, Einstein's leads to a critical examination and to a cautious selection of some of their kernels in the light of new advances in human knowledge.

The present procedures of scientific inquiry is to resolve the observed phenomenon, or the debatable problems, into *isolable* elements; these are, then, *added* together, practically or conceptually, *"to represent"* the actual phenomena, or the real problems.

Experience has shown that this separation-isolation of parts and causal chains, and their summation and superposition, works well in the early stages of many sciences. However, for the long run it simply does not work. In all the sciences we are now confronted with *"wholes"*, with organizations, with mutual interactions of many elements and processes, with "large systems", etc. They are *non-additive*, and, therefore, cannot adequately be dealt with by present analytical methods. For one cannot split them anymore into isolable elements and linear causal chains. Thus, compared with the approach of classical science, the methodology of interconnectedness requires new concepts, models, theories—whether the problem is that of an atomic nucleus, a living system or a socio-ecological organization. *Mutual interaction, instead of linear causality; structured complexity, instead of summation of undirected and statistical events; asymmetric buildup of interconnected evolution, instead of summation of isolated historical events—these, somewhat loosely, characterize the new methodology that we need develop* (see below).

But perhaps the most important need is for a *unified study* in which the participants combine such a methodology with an old-new kind of knowledge. They should not "just know something about every discipline", nor should they be a collection of "experts" from different disciplines.

Despite the reasonableness of these assertions, an effort along these lines confronts major difficulties in virtually every area of traditional Western thought. They involve tradition, philosophy, empirical evidence, faculty, students, curriculum, leadership, government, policy, and a code of ethics.

My first task was to sketch this methodology and to try to put it into practice in resolving (apparently) "different disciplinary problems" that are associated *with a single central concept. That concept is the most complicated of all human concepts. It is time and its direction:* atomic and cosmic, thermodynamic and electromagnetic, evolutionary and social, classical and relativistic, objective and subjective, micro-scale and macro-scale, quantal or continuous, physical and biological.

Time and its direction are therefore the most fundamental and the most universal expressions of any science-based world picture.

With such ideas any claim of absolute or separable philosophies falls to the ground while the fundamental principles of science and philosophy become a single indivisible whole (Lecture IX and below). Thus, any difference can only be of a terminological-methodological character, whether related to the essence of the subject, or to the methods of theory and practice. Moreover, such interconnected ideas make the close union between philosophy and the physical, biological, social, and historical sciences very fruitful. Indeed, the emerging results from this union are frequently surprising as well as rewarding.

3.1. Everything Is Connected To Everything Else.

It should be stated at once, in formulating the methodology, that I subscribe to the adage: *individuum est ineffabile. Thus the philosophy presented in these lectures does not undertake to argue away the creativeness of the individual mind. It takes only one single fact for granted, namely that no mind is an island; that it is an open system, in itself part and parcel of a greater system.*

In reality we find no (singular) abstract man, but only (plural) interrelated men—men who have been shaped by long and diverse evolutionary influences and educated into definite and specific beings, differing from generation to generation, from climate to climate, and from land to land. It is therefore impossible to divorce man *qua* part and parcel of the physical world from man *qua* member of society; nor can one separate man *qua* perceiver of the external world from man *qua* linkage in a long geo-chemical, biosocial evolution (Lecture IX). What do I mean? In trying to explain these ideas I introduce the following assertions:

1) *Each discipline of modern science, each view of a disciplinary philosophy, is only a narrow slice through the complex structure of nature. While each can illuminate some features of the whole evolutionary system, that picture is distorted, false, and misleading to a degree. For in looking at a selected set of events, chosen a priori on the basis of a traditional field of study we ignore a good deal of the rest, and, therefore, may arrive at the wrong conclusions. In the actual world everything is linked to everything else. So must fundamental studies be.*

2) No physical or social system can, in principle, be analyzed as "entirely isolated". Therefore, we cannot reach *complete* knowledge about a given object in itself, for it cannot be isolated from the "rest of nature". A specific example is the *"first law of ecology"* which reads:

"Everything Is Connected to Everything Else".

That is fine for ecology, a skeptic may say, but are cosmology, philosophy and ecology linked? Are social evolutions and the vast physical environment related to each others? Is there any evidence to support such claims? The answer to these questions is quite subtle. In trying to formulate it one may stress the following facts: *Each living species is suited to its particular environmental subsystem, and each, through its life processes, affects the physical and chemical properties of its immediate environment.* These links are bewildering in their variety and very subtle in their cause-and-effect intricates. Thus, as the links between one living system and another, and between all of them and their global surroundings, change, *the resulting dynamic interactions that sustain the whole, affect all individual lives.* The *origin* of these changes must, therefore, be sought and examined within a proper scheme of global interdependence. Human understanding of global links comes hard because the modern person has been *trained* to examine only *isolated, separate phenomena, each dependent upon a "unique" or a "singular situation" and each studied within the "framework" of a specific "discipline"* (e.g., geology, archeology, biological evolution, anthropology, micro history, social evolution, etc.). *Interconnected thinking* (see below) *is very difficult, for human beings participate in the environmental system as subsidiary parts; and this very participation distorts the perception of the whole.* Moreover, modern society has, with the aid of technology, made *"its own"* environment. Therefore, some planners *believe* that no longer they depend on the global environment provided by nature. This illusion led astray many planners: they neglected the sun's heat, the air, the seas, the soil, the varying terrain, the climate, the forest, the intricate grid of roads, rivers, lakes, springs, villages and cities. They warm, and cool themselves with "man-made" machines; they travel on "man-made" roads, through "man-made" villages and cities, using "man-made" vehicles; they reach their "man-made" homes; eat "man-made" food, and read about "man-made" evolution, laws, industry, transportation, etc.

But these illusions are easily shattered, for like the people they house or carry, the buildings and vehicles, themselves, are products of the earth's environment. So are the machines, the roads, the villages, the cities. Without the earth's environment, "man's own environment", like himself, could not exist (§ IX.2.3).

The existence of the present energy-environmental crisis warns *against* the illusory hope that by *reducing history* to a simple set of individual histories, *the sum will somehow add to the actual whole.* We must, therefore, stop training young people to think about "separate", "uncaused" historical events, for we are confronted by a situation as complex as the whole physical environment of the earth and its vast array of living structures; as complicated as the temporal and spatial complexity of the external physical world and as intricate as the functioning and structure of living individuals, tribes, nations and civilizations.

INDEX